D0225845

• Henry Norris Russell •

• DAVID H. DeVORKIN •

Henry Norris Russell

DEAN OF AMERICAN ASTRONOMERS

PRINCETON UNIVERSITY PRESS
PRINCETON AND OXFORD

PUBLISHED BY PRINCETON UNIVERSITY PRESS, 41 WILLIAM STREET,
PRINCETON, NEW JERSEY 08540
IN THE UNITED KINGDOM: PRINCETON UNIVERSITY PRESS,
3 MARKET PLACE, WOODSTOCK, OXFORDSHIRE OX20 1SY

LIBRARY OF CONGRESS CATALOGING-IN-PUBLICATION DATA

DEVORKIN, DAVID H., 1944–
HENRY NORRIS RUSSELL : DEAN OF AMERICAN ASTRONOMERS /
DAVID H. DEVORKIN.
P. CM.
INCLUDES BIBLIOGRAPHICAL REFERENCES AND INDEX.
ISBN 0-691-04918-1 (CLOTH : ALK. PAPER)
1. RUSSELL, HENRY NORRIS, 1877–1957. 2. ASTRONOMERS—
UNITED STATES—BIOGRAPHY. I. TITLE
QB36.R78 D49 2000
520′.92—dc21
[B] 99-059472

THIS BOOK HAS BEEN COMPOSED IN ELECTRA

DESIGNED BY CARMINA ALVAREZ
COMPOSED BY EILEEN REILLY

THE PAPER USED IN THIS PUBLICATION MEETS
THE MINIMUM REQUIREMENTS OF
ANSI/NISO Z39.48-1992 (R1997) (*PERMANENCE OF PAPER*)

HTTP://PUP.PRINCETON.EDU

PRINTED IN THE UNITED STATES OF AMERICA
1 3 5 7 9 10 8 6 4 2

TITLE PAGE PHOTO: HENRY NORRIS RUSSELL AT THE PEAK OF HIS CAREER
CA. EARLY 1930s. PHOTOGRAPH COURTESY E. SEGRÈ VISUAL ARCHIVES,
AMERICAN INSTITUTE OF PHYSICS.

• TO THE MEMORY OF •

LYMAN SPITZER,

MARTIN SCHWARZSCHILD,

THEODORE DUNHAM,

ADRIAAN WESSELINK,

SUBRAHMANYAN CHANDRASEKHAR,

AND MARGARET RUSSELL EDMONDSON.

CONTENTS

• PREFACE •

HENRY NORRIS RUSSELL helped make the physics of the atom central to astrophysical practice during the first half of the twentieth century. He was, in effect, America's first modern astrophysical theorist. Through circumstance and training, Russell always believed that observational astronomy had to become more problem-oriented and theory-driven and less an open-ended empirical exploration: a natural philosophy more than a natural history. For fifty years he held to this view, and as he gained power and authority in the profession, largely through practicing what he preached, he fostered a culture that slowly adopted this attitude. He demonstrated that to be competitive in astronomy, one needed the tools, training, and outlook of the physicist. Here I explore the "fine structure" of Russell's life in science to reveal how he influenced change in American astrophysics. Fine structure reveals how and why scientists make different choices at different times, and what those choices dosclose about the changing nature of science.[1]

Russell learned his trade at a time when mathematical physics was the dominant explanatory framework in astronomy (as manifest in mechanics, gravitation, and the theory of fluid bodies), but it was also a time when astronomical research was still largely devoted to data gathering. The image of an astronomer was that of a life of constant vigilance spent in an observatory chamber, in Caroline Herschel's immortal words, "minding the heavens."[2] Surveying the work of a modern observatory in 1891, David Gill, director at the Cape Observatory, felt that it took "special natural gifts—patience and devotion," to be an observing astronomer: "Nothing apparently more monotonous can be well imagined if a man is not 'to the manner born.' " He praised the disciplined meridian circle observer, whose nightly duty "presents no dreamy contemplation, no watching for new stars, no unexpected or startling phenomena."[3] When Russell was young, the legacy left by Kirchhoff's revelations about the chemical identity of the solar spectrum and his elucidation of a thermodynamics of radiation brought the spectroscope into the astronomer's chamber. A few astronomers tried to adapt the laboratory techniques of the chemist and physicist to obtain empirical knowledge of the physical makeup of planets, stars, and nebulae. Beyond verifying that true gaseous nebulae exist, that stars are great glowing globes surrounded by tenuous gaseous atmospheres, and that the elements found on earth exist in the heavens, little additional headway was made. Theories of the atom were hardly settled, and the astronomical application of experimentally derived radiation laws was fraught with uncertainty. Astronomers typically classified how stars differed from one another in spectroscopic appearance, and speculated over what these differences revealed about their constitution, composition, history, and fate. Most astronomers were excited by the capability of observing motions in the line of sight spectroscopically, or the "radial velocities" of celestial objects, because this activity furthered traditional goals, required little in the way of theory, and informed the "sidereal problem," uncovering the structure and dynamical history of the sidereal universe.

When Russell arrived on the scene, there had been a few prominent workers, such as G. Johnstone Stoney, a physicist at Queen's University in Dublin, and the spectroscopist J. Norman Lockyer of South Kensington, who suggested models for the atom based upon astronomical evidence, but both models were idiosyncratic and failed to attract a following.[4] Unlike chemists, therefore, who enjoyed a robust experimental tradition and had tried to parlay it into a theory of atomic structure, astronomers had no indigenous atomic model to protect. Thus the challenge Russell and his cohorts encountered was of a very different order from that experienced by Wilhelm Ostwald and his followers.[5] Spectroscopic astronomy, like chemistry, had indeed been largely taxonomic, concerned primarily with appearance and standarization. But unlike chemists, spectroscopic astronomers generally avoided deductive analysis and some questioned adopting a theoretical framework at all.[6]

In America in 1900 a robust community of observational astronomers was slowly forming around the efforts of a few particularly energetic and ambitious individuals who knew how to tap the fortunes of American philanthropy. By the end of the century America boasted the most powerful telescopes in the world and, largely through apprenticeship training, was building one of the largest organized workforces as observatory directors adopted managerial techniques that maximized the mass production of data.[7]

Flush with the excitement of having the biggest telescopes and the finest skies, American astrophysics was then almost completely observational and highly empirical, and some people, like Russell, sensed its limitations. The venerable Austrian physicist Ludwig Boltzmann described one major American factory observatory as an ocean liner rather than a place of science: "[T]he officers and crew are always doing the same routine jobs, over and over again . . . Gargantuan masses, but not one new thought!" Boltzmann scoffed at the high moral ground taken by the devoted spectroscopic observer, yet it was clear at the turn of the century that a theorist, as one astronomer later observed, would have found himself "out of place at an observatory."[8]

Russell managed to run against this tide while taking great advantage of it. Sharing with his contemporaries a penchant for understanding the sidereal universe in evolutionary terms, in an age energized by Darwin, Russell paid more than lip service to searching out evidence for how the Sun and stars were born, lived, and died. It was his plan of action. He began as a mathematical theorist, testing, improving, and applying orbit theory against observations to improve knowledge of the masses and densities of the stars. He soon found ways to manipulate spectra, magnitudes, and motions into new means of understanding the physical nature of the Sun and stars, ultimately looking to modern physics for an explanatory framework for a star's behavior.During his career he enlarged his set of conceptual tools in response to intellectual pressures created by the types of questions he and his contemporaries were asking. As he made these changes, he became a symbol for how the tenets of his science had to change as well, as a human endeavor, as a practice, and as a culture.[9]

These chapters explore how Russell made linkages between widely disparate observations and was tireless in his attention to detail, how he marshaled enormous amounts of evidence to buttress his theories and in so doing convinced others that his propositions and his techniques warranted attention. The Hertzsprung-Russell Diagram, the Vogt-Russell theorem, Russell-Saunders coupling, the Russell Mixture, and the Russell method all are powerful conceptual devices that provided a good portion of the foundation for the quantitative study of the physical structures and the "lives" of the Sun and stars. And all bear witness to Russell's scientific style and his output—not only product, but the means he employed to produce and display that product in order to achieve consensus and trust among his colleagues. What is fascinating about Russell's means of establishing consensus, because it reveals as much about Russell's self-assertiveness as about the nature of the community of workers he was part of, is that, as he claimed in 1937, his work rested upon "a tissue of approximations." We will pay close attention to what Russell meant by this, and how he made it his trademark in a community that had always prided itself upon precision and exactitude.[10]

Twice proposed for the Nobel Prize, Russell garnered every major award available to astronomers of his generation. He was the most cited American astronomer in the 1920s; at his retirement in 1947, the American Astronomical Society created its premier lectureship in his honor.[11] The enormous scientific authority Russell enjoyed at the peak of his career was based largely upon his abilities and accomplishments, but involved as well politics, patronage, and personal ambition. As a theorist well aware of the limitations of observational evidence, Russell was a gifted and creative opportunist always seeking the most efficient and effective ways to study the stars. Never at rest and never satisfied, he probed and prodded others to action, always ready to give advice and not shy to request aid from others. As a result, his abilities and promise attracted the attention of two highly exceptional and powerful people, Edward Charles Pickering and George Ellery Hale. Each helped Russell realize his ambitions, as they were, in fact, two of the three astronomers most responsible for creating and shaping the American astronomical community. With Simon Newcomb they founded the Astronomical and Astrophysical Society of America (now the American Astronomical Society). Hale also founded the *Astrophysical Journal* and both the Yerkes and Mount Wilson Observatories. Pickering was Harvard College Observatory director, second president of the AAS (for fourteen years), and leading producer of astrophysical data in the world at the turn of the century.[12] In 1908 Pickering viewed Russell as his key to interpreting the masses of data his staff had been collecting for decades. And after World War I, Hale wanted Russell to help his Mount Wilson staff appreciate the interpretive power of modern physics. Both opened their observatories to Russell and brought their considerable influence to bear to make him a national figure in science.

Russell was an elite, East Coast establishment intellectual who was a traditionalist in social and organizational matters. He always seemed to believe that, or at least act as if, his talents warranted him a special role in American astronomy, where the hard-won data being generated at America's observatories would flow to him for analysis. Just as Charles Darwin knew "how to make people help him" and botanist Asa Gray was a clearinghouse for observations made by others, Russell came to view himself as a "headquarters scientist" who knew best how to exploit the data being collected at observatories.[13] His early successes in this role emboldened him to advise observatory directors about the right data to gather—those best suited to answer the deepest questions posed by astronomers of his generation.

Although he built no institutions and edited no journals, Russell's awesome command of the subject made him compelling to those, like Pickering and Hale, who were the builders. He was a constant adviser and referee for editors and observatory directors alike, wrote an influential astronomy column for *Scientific American* for more than forty years, and created ad hoc networks to solve specific problems. He coauthored a seminal textbook that served as an entry point to professional astronomy for several decades. His aggressive nature and enthusiasm for novelty kept him at the forefront of stellar astronomy for much of his life and provided him with enormous intellectual authority.

There are many clues in Russell's personality and the peculiar circumstances of his training at Princeton and Cambridge that will account for how he established his identity, fought for acceptance, gained authority, and became to his contemporaries "Dean of American Astronomers." Not so much an honorific as an acknowledgment of his intellectual role in the community, the epithet "Dean" best describes his status at the height of his career. Yet he remained a maverick well into his mature years, unsettled in what he viewed was a "land of settled habits," constantly trying to move American astrophysics away, slowly and painfully, from its naturalist, empirical roots.[14] But Russell also became a traditionalist in his later years, comfortable in the world he created, a physical universe dominated by a mechanistic Bohr atom. As such he was a transition figure whose life embodies critical clues to the nature of disciplinary change.

Much about Russell's personal life remains accessible, though we cannot retrieve truly intimate thoughts and passions because the bulk of his personal correspondence was destroyed by his family after his death in 1957.[15] Unavailable are letters to his parents, and, most disappointing, to his wife, someone who no doubt managed a large part of his life, making it appear seamless to his professional colleagues. Allusions in his professional correspondence hint at her considerable role as protector, manager, and fan, and to the trials of being a husband and father. This evidence also suggests that Russell was an exceedingly anxious and driven individual, and that his professional and personal lives were closely entwined. He consistently overworked himself, and those near him, beyond the point of exhaustion, and he suffered many periods of collapse. Internally and to only a very small circle of confidants and family, he was constantly tormented by self-doubt and indecision. Externally, he was the surest of individuals.

• ☆ •

Biographies of lives in science are artifacts of motivation and documentary evidence. My motivation has been to seek out the process through which astrophysics became physical. Russell was a prime mover in this process, and this fact forms the primary reason for recounting his life as a scientist, recovering the factors that made Russell do what he did, and rationalizing the way he chose to do it. The biographical framework constructed here is designed to convey Russell's enormous influence within astronomy, but it is also appropriate to show how he practiced a particular form of astronomy that heretofore has not been sufficiently examined by historians. Social factors that rationalize his influence in the community of American astronomers, a community he did not create but did help shape, will be important as explanatory agents, and will help to illuminate Russell's impact on American astronomy and provide context to his life. But the organization of his biography must reflect his science, since it was around his science, the specific problems he chose to attack, that he organized his life.

Thus the structure of the book departs from a strict chronological narrative at the point where Russell's career became multithreaded. Links between his many lines of research will provide guidance, but they are not meant to hide the fact that his intellectual life became complex and quite messy as he accreted new interests. Nor is the parallel construction of the latter chapters of the book meant to promote the idea that Russell maintained a compartmentalized life. His was the life of an opportunist, making connections wherever and whenever possible. The links established here are intended to remind us of that fact.

• ACKNOWLEDGMENTS •

LONG AGO my father gave me his copy of volume 2 of Russell, Dugan, and Stewart's *Astronomy*, and I soon found the subject to be far different from my childhood fantasy. Why was so much physics involved? Didn't astronomers simply look through telescopes and record what they saw? Most, but not all, astronomy professors I encountered focused on physics, using astrophysical problems as chew toys. Astrophysicists-turned-historians like Jack Meadows introduced me to the period when astrophysics did not yet possess a rational basis. Spencer Weart opened the world of oral history to me, and through the American Institute of Physics I obtained a National Science Foundation grant in 1979 to organize Russell's papers at Princeton, a daunting task that benefited from the guidance of Joan Warnow and Jean Preston, and was carried out with enthusiasm and great skill by Peggy Kidwell, who along the way provided insights that enrich this biography. In 1977 A. G. Davis Philip invited me to co-organize a historical session to commemorate Russell's centenary, and through this we brought together Russell's family, students, and colleagues to explore the many facets of his life in science.[1]

Ralph Kenat proved to be a wonderful collaborator in the mid-1980s. He helped unravel the often complex paths Russell took to make his contributions to stellar spectroscopy and atomic physics. Karl Hufbauer has always been my best critic as the most perceptive observer of the intellectual and social landscape of twentieth-century astrophysics. Encouragement and criticism by Paul Forman, Michael Dennis, and John Lankford at different times helped me grow. In 1990, John Bahcall invited me to the Institute for Advanced Study to finish working through Russell's papers, and to start writing. During my stay in Princeton, Martin Schwarzschild, Lyman Spitzer, Louis C. Green, William O. Harris, Noel Swerdlow, and Sarah Brett-Smith all helped give my work a positive direction, and Margaret Best at the Institute helped make my stay a pleasant and productive experience. Archivists at the Seeley Mudd and Firestone Libraries showed me every courtesy: Jean Preston, Nanci Young, Ben Primer, Ann van Arsdale, Robert Chavez, and Don Skemer all were terrific, as was Jane Holmquist at the Princeton University Observatory Library.

Russell's professional papers housed in the Firestone Library at Princeton form the primary material for this work. More than fify linear feet of records were removed from his home and office by his family, and deposited between 1957 and 1969, with additions by Charlotte Moore Sitterly in 1974. It is clear that the collection, now amounting to some 23,000 letters, was culled at some point, since very little exists in the collection prior to 1908 (no outgoing and only a smattering of incoming letters), and there are significant gaps at critical points throughout (Russell wrote constantly during his extensive travels), some of which have been filled by material found elsewhere. There are copious working notes, manuscripts, and holographs of lectures. But no diaries or other records of a personal or intimate nature have been found, though "bushels" of such materials did exist at some point in the past.[2] A single exception is an extensive trip diary written by his father in 1892, intended to be read by his congregation and kindly made available by his granddaughter, Margaret Edmondson Olson.

There is no evidence that Russell ever kept a diary himself, although the simple fact that so much was preserved during his lifetime indicates that he had every desire to be remembered. There is little question that Russell wanted to be the subject of biography, even though no direct evidence has been found to substantiate this. It is also impossible to speculate on how he would have reacted to the biography now before you. We do not know if the destruction of his personal correspondence was an act he would have approved, or lamented. His professional correspondence appears to be candid about various concerns for family and friends, his hopes and desires, his uncertainties and frailties. But his letters also have a guarded quality; the issues he raised and his feelings about them were often obscured by literary or religious aphorisms. Given what is known about his overall nature, however, his commentary reveals his view of life, or at least the view he wished to project. Naturally, he said different things to different people at different times, which hints at the extent to which he could dissemble as he constructed his persona.

Records surviving in the Princeton Astronomy Department, as well as in Astronomy Department Records archived at the Seeley Mudd Library at Princeton, do help to fill in many gaps. Another useful but admittedly qualified source comprises oral histories taken with family members, students, and colleagues. They lend wonderful flavor, if not precision, and do help to craft a three-dimensional figure. Further afield, among the many archives I consulted to gain perspective on Russell from third-party correspondence, most valuable were the collections housed at Princeton as well as the Yerkes Observatory (thanks to Judy Bausch), the Mary Lea Shane Archives of the Lick Observatory (Mary Shane and Dorothy Schaumberg), the Harvard Archives (Clark Elliott), the Lowell Observatory Archives (Antoinette Beiser, Henry Giclas, Marty Hecht, Mary Lou Evans), the Mount Wilson Observatory Papers at the Huntington Library (Ron Brashear), numerous collections at the American Institute of Physics' Niels Bohr Library (Carolyn Moseley), Friends Academy archives (Mary Anne Rearden), the Presbyterian Historical Society Archives, the Princeton Theological Seminary Archives (William Harris), the Oyster Bay Presbyterian Church (Margee Styles, Richard Duvall), the Trenton, New Jersey, Hall of Records, the Nassau County Office of Administrative Services Record Services Division, the Carnegie Institution of Washington Archives (Susan Vasquez and Shaun Hardy), and the Freedom of Information Act section of the Office of Public and Congressional

Affairs, U.S. Department of Justice. The many other archives I consulted here and abroad are listed in the bibliography, as are the names of those who kindly submitted to oral history interviews. Libraries and librarians were equally helpful, but none more so than Brenda Corbin and Gregory Shelton of the U.S. Naval Observatory Library and Mary Pavlovich of the National Air and Space Museum library.

I also want to thank Joanne Palmieri for alerting me to important materials in the Shapley Papers, Cathy Cooke for her grasp of E. G. Conklin's world, David Cassidy and Marjorie Graham for their knowledge of F. A. Saunders's life, John Hammond for his advice about the history of Oyster Bay, and, most especially, to Henry Norris Russell jr., Margaret Russell Edmondson, Frank Edmondson, and their daughter, Margaret Edmondson Olson, for their advice, support, and encouragement. The Edmondson and Olson families kindly read drafts of the manuscript, as did Karl Hufbauer, Peggy Kidwell, Mike Neufeld, and Marc Rothenberg, whose insights, guidance, and collegiality are much appreciated. Portions were also read by Robert Multhauf, Dorrit Hoffleit, Alan Batten, Bert Moyer, Von Hardesty, Martin Schwarzschild, Lyman Spitzer, and John Bahcall, all of whom provided valuable suggestions. I am indebted to Carola Jeschke and Henke Boute for their translation services courtesy of the Smithsonian's Behind-the-Scenes Volunteer Program. Tracy Goldsmith, Alex Magoun, and James David provided wonderful assistance during the course of my research. Finally, I want to thank Paul Routly for his collaborative aid over many years, helping to inventory and abstract Russell's letters, and searching for useful commentary in the collections of other astronomers.

Trevor Lipscombe's advocacy of this book within Princeton University Press is deeply appreciated, along with his guidance and good spirit during the review and contracting processes. I benefited greatly, as will my readers, from Lauren Lepow's thoughtful and conscientious copyediting and Carmina Alvarez's discerning eye and visual style. David Campbell made the process seamless.

Some of the episodes, notions, and conclusions in this book first appeared in articles in the *Journal for the History of Astronomy*, *Isis*, *Physics Today*, *Scientific American*, and the *Quarterly Journal of the Royal Astronomical Society*, as well as in essays written for various conferences and occasions, especially Philip and DeVorkin (1977). All have been substantially altered and many have been considerably fleshed out; a few have been summarized. I thank the editors, reviewers, and readers of these books and journals for their advice and counsel in the long process of bringing this biography to life.

Beyond the support provided by the NSF and the Institute for Advanced Study, I want to acknowledge the American Institute of Physics Small Grants Program, the Dudley Observatory Pollack Award program, and the Women's Committee, the Scholarly Studies Program, and the Research Opportunities Fund of the Smithsonian Institution.

• Henry Norris Russell •

• CHAPTER 1 •

Religious Heritage

Russell's Mail

Working at home on a November day in 1934, Henry Norris Russell, chair of Princeton University's Department of Astronomy and director of its Halsted Observatory, wrote to an inquirer about heaven and the immortality of the soul. The lean, white-haired, fifty-seven-year-old theoretical astrophysicist admitted that "Heaven is not a place but a state." The ardent presbyter added, "What else it may be, I do not know, but this universe of ours looks to me like a good enough job to give me a thorough-going confidence that the Designer of it will look after our interests elsewhere quite adequately."[1] A few years later, scrawling on his ever-present pad of legal-size paper, he answered yet another inquiry from his growing circle of lay readers seeking divine guidance from science: "[T]he hypothesis that the order of nature and the evolutionary process have arisen 'unaided by any creative or universal intelligence' appears to me to be more improbable than the alternative."[2]

Russell usually answered his voluminous mail at his maternal ancestral home, 79 Alexander Street, then at the southwestern edge of the undergraduate campus near the railway station. It was, moreover, in sight of the Princeton Theological Seminary, where his father had trained. Seclusion was necessary for the angular and restive professor. He did not avoid others; but in the company of others he could not sit still. In the manner of many professors of his day he preferred to write in his book-lined study at home, a home he had known since the age of twelve.

By this time in his life, Russell was well known beyond the small circle of astronomers and physicists in which he moved. Associate editor and regular columnist for *Scientific American*, essayist for a growing audience that read American religious periodicals, coauthor of a well-received and wide-ranging astronomical textbook, and frequent speaker on college campuses, in community auditoriums and church sanctuaries, Russell was a deeply religious person, eager to speak on religion from the perspective of a scientist. Son, grandson, and brother of Scottish Presbyterian ministers, Russell held "Sunday Seminars" at Princeton, lectured for Princeton's undergraduate Philadelphian Society, counseled Princeton undergraduates in Dodge Hall on praying to a personal god, spoke frequently before congregations in Princeton, Philadelphia, and New York, and wrote passionately on how scientific inquiry supported a moral and revealed universe. Russell was a religious modernist striving to keep religion alive on an increasingly secular campus.[3]

By the last decade of the nineteenth century and the first of the twentieth, the rise of the sciences fostered the reductionist belief that a unity of truth was at hand. Science, or the scientific method, became the model for the addressing of social issues and was celebrated as the most effective way to train the mind and build character in students, leading them on to productive and effective moral lives. The environment Russell was born and raised in, and would eventually take for his own, promoted a "passion to reveal the secrets of nature by probing deeper and deeper into the physical world." Such striving had deep religious and moral underpinnings, connected to a natural theology that looked for traces of the divine in nature.[4] As "a man brought up in religion from infancy," Russell entered the world of science and prospered, "still a convinced adherent of the faith in which I was reared: I believe in a God who is the Ultimate Reality."[5]

Russell's outlook was typical for his day: the physicist Robert Millikan, eldest of three sons of a Congregationalist minister, held deeply religious convictions and, like Russell, wrote on science and religion. Forest Ray Moulton, a celestial mechanician, was also well known for his ardent attempts at reconciliation.[6] Among prominent American scientists contemporary to Russell who were raised in professional families, 21 percent had cleric fathers and 39 percent had cleric grandfathers, typically Congregationalist, Methodist, or Presbyterian. Presbyterians accounted for 20 percent of the prominent scientists listed in *Who's Who* for 1931.[7]

Russell promoted a "Scientific Approach to Christianity" that reflected modernist views instilled in him both by his parents and by his Princeton intellectual heritage. The latter was defined most clearly by Joseph Henry, the highly influential Princeton natural philosopher who later became secretary of the Smithsonian Institution, and by Princeton's eleventh president, James McCosh. McCosh, one of the founders of the Free Church of Scotland who was later much absorbed in reconciling Darwinism with design, infused Princeton (then called the College of New Jersey), with his view "that science was immensely useful to the defense of faith." His Scottish commonsense philosophy led him to argue, as had Joseph Henry before him, that disciplined intuition was necessary to those who would interpret the sensible world in order to comprehend God's design for the universe. Both Henry and McCosh argued that pure empiricism was an unlikely path to new knowledge; a more direct route began with framing a hypothesis.

McCosh acted to strengthen mathematics instruction at Princeton as a counter to what he believed was a growing tendency to rely only upon experience. He was convinced that the "very nature of things" could be "perceived at once by intuitive reason."[8]

Russell's career as the first astrophysical theorist in America was, in part, a reflection of McCosh's liberalization of Princeton and his elevation of science as a practice based upon a balance of intuition and observation. So we begin our reconnaissance of his life by looking at the intellectual and religious worlds in which he was born and raised.

James McCosh's Princeton

McCosh believed that the physical and natural sciences promoted a "religious understanding of the world." He built a "theory of the universe conditioned by Christian revelation," and carried it to New Jersey and the village of Princeton, where in 1868 he became the eleventh president of the college. McCosh took up his new post confident that he could rebuild the College of New Jersey around a synthesis of science and theology by establishing a course of secular training.[9]

When McCosh arrived at the college, the highly respected intellectual Charles Hodge, Joseph Henry's contemporary and friend, led the Princeton Theological Seminary, the seat of "Old School" Presbyterianism in America. Yet the two had much in common, mainly an enthusiasm for natural theology. Along with many American intellectuals, including Henry's cousin and former seminary student Stephen Alexander, the college astronomer, Hodge had little difficulty with the Nebular Hypothesis as long as it was "used merely as a scientific explanation of the origin of the solar system illustrating God's design and providence." Clearly then, how a scientific theory fared depended upon the "spirit in which [it] was taught."[10]

Hodge differed from McCosh by trying to limit and control the scientific practices that McCosh wished to broaden. In the 1870s, as McCosh and Hodge debated the meaning of Darwinism, McCosh hired professors of science who would foster reconciliation.[11] Revelation existed through the power of human intuition to perceive it, McCosh believed: evidence in the rocks, and in organic life, revealed "a proof of design and of a designing mind."[12] But intuition was acquired only through discipline. Thus both McCosh and Henry played down the simple accumulation of facts and promoted hypotheses as "the great instruments of discovery."[13] This was McCosh's plan for Princeton, heartily endorsed and pursued by those he brought to the campus, and later by those who trained there. He rejected the "classifying, theory-dodging tendency" that still characterized much of American science in the latter half of the nineteenth century. And much of his faculty, especially the astronomer Charles A. Young, whom he hired in 1877, felt the same.[14]

When McCosh retired in 1888, he left Princeton with a growing campus, a noted scientific and engineering faculty, laboratories, a large observatory, and the second largest college library in the United States.[15] Research flourished in pockets at the newly revitalized college, although its primary mission remained undergraduate training for the church and professions.

The Family and Ministry of Alexander Gatherer Russell

Henry Norris Russell described himself as "a man half of Puritan and half of Lowland Scots stock." His father, Alexander Gatherer Russell, was born in Musquodoboit, Halifax County, Nova Scotia, on 6 October 1845, the son of a minister and a descendant of Elgin weavers. His mother, Eliza Hoxie Norris Russell, was born in 1848 and was well-educated, adept in mathematics, and interested in the world. Both father and mother instilled a strong sense of moral responsibility in their children.[16]

Alexander Russell moved to Princeton to enter the seminary in the fall of 1872, after taking a degree from Dalhousie College, teaching for a year, and undergoing religious training at the Presbyterian Seminary in Truro. In Princeton, he boarded in the Norris family home at 79 Canal Street.[17] The head of the household was Henry Lee Norris, descended from Salem sea captains dating back to Old Colony Puritans. Norris had been a rubber merchant in New York when, in 1847, he married Maria Schaeffer Hoxie, a schoolteacher who had distinguished herself in 1840 as the first recipient of a mathematics medal from the Rutgers Female Institute of New York. They soon moved to Para (now Belem) in the empire of Brazil, where he established a rubber manufacturing and trading business.[18]

Norris eventually became American consul in Para, but after a decade he moved the family to Edinburgh, where he became manager of the North British Rubber Company. Maria Norris saw to it that all her children received a proper education; their two daughters, Eliza Hoxie and Ada Louise, were respectively awarded a first and second in mathematics for their performance in a "ladies class" given by Edinburgh professors in 1868, following in the footsteps of their mother.[19] After retirement, Norris moved his family to Princeton in 1874, and by the time Russell took a room with them in their stately Gothic home at 79 Canal, the Norris family enjoyed a comfortable life from substantial investment income. Fiscally conservative and religiously progressive, the Norris family attended the Second Presbyterian Church in Princeton, which was less traditional, more evangelical, and more open to townspeople who found the First Presbyterian Church too formal and austere. There was then a growing schism over the doctrine of predestination, the denial of the role of human will in the salvation of souls. Both sides of Henry Norris Russell's family were reluctant heirs to the theology of John Calvin.[20]

Alexander Russell's three seminary years were defined by Charles Hodge, but the landscape he entered was being reformed by McCosh. Liberal influences continued to play upon Russell after graduation in 1875 as he preached in Brooklyn and searched for a permanent post. He delivered his first sermon at the First Presbyterian Church and Congregation of Oyster Bay, Long Island, on Easter Sunday, 16 April 1876, and he continued for several weeks to minister to the congregation before he was finally installed as pastor in July 1876. Thus established, Alexander married Eliza Norris on 1 August 1876, removing her from the Norris home and hearth, seeing promptly to her admission into his church and attending to the birth of their first child, Henry Norris, just over a year later on 25 October 1877.[21]

Oyster Bay, its harbor and environs, was a country hamlet founded by Dutch colonists in the mid–seventeenth century. It was known for its asparagus and cultivated oysters. A protected yacht harbor made Oyster Bay a desirable summer colony on the north shore of Long Island. Prior to the arrival of the Long Island Railroad branch from Glen Cove in 1889, access had been by ferry, stage, and train, or by side-wheel steamer, whose arrival in Oyster Bay harbor after three hours on the Sound from New York City was the event of the day. Alexander Russell's congregation, newly fitted out in its Romanesque church sanctuary on East Main Street, near the harbor, included laborer and fisherman families, but was heavily underwritten by wealthy, prominent, and powerful summer residents who sought out the same liberal ministrations they had come to favor at the Fifth Avenue Presbyterian Church in New York. Russell's most prominent parishioners were Theodore Roosevelt, Sr., and his family, including the future president of the United States, who knew Oyster Bay from the time his grandfather built a summer home there in the early 1870s.[22]

Alexander Russell was an active and tireless promoter, holding multiday marathon services and fund-raising campaigns to pay for the construction of the new sanctuary. He delivered two different sermons every Sunday and led a midweek prayer service. More in keeping with the style of an urban congregation, Russell adopted a liturgical rather than a revivalist form of worship. He made music both a frequent and a fervently joyous part of his ministry.[23] Despite his efforts at raising membership and revenue, the church remained poor in the early 1880s during Henry's most formative years. The home where Henry was born and spent his first nine years of life was on flat land in the poorer district close to the harbor. But as the local economy picked up later in the decade, the church revived, owing largely to Alexander Russell's unflagging efforts. A new and gracious manse was built on church grounds on East Main above and behind the church, overlooking the town and harbor. It was both home and office for the father, and a new world of status and privilege for his three sons, Henry, Gordon MacGregor (b. 1880), and Alexander (b. 1883). Even so, reflecting his father's lifestyle, Henry would always maintain severe economies, most clearly manifest in his directorship of the Princeton University Observatory.[24]

The pastor set a challenging example for his three sons, most of all for Henry, who as the eldest adopted his father's trait of self-absorbed devotion to duty. Local stories recount how Alexander worked himself to exhaustion and neglected his own health in his concern for others.[25] Life was a constant campaign for good causes, including public libraries and schools. Life also had to be met head-on, vigorously, courageously, with a morally uplifting spirit. William L. Swan, church organist and ardent yachtsman, knew the elder Russell from his first years in Oyster Bay. "He was very diffident, and never 'played to the gallery,' " Swan recalled, and was "most methodical and orderly."[26] Other local clergy remembered Russell as "open, fair, true, genuine, aboveboard; nothing kept back . . . In him there was no guile."[27]

Alexander Russell, in the spirit of James McCosh, was also a reformer. In 1882, Russell was elected Stated Clerk of the Nassau Presbytery. He soon won the respect of the wider community and used it as a platform for doctrinal change. In 1888 and again in 1889 he was a major campaigner in the Nassau Presbytery's bid to the General Assembly to revise the severe Westminster Confession of Faith.[28] Russell preached that "the Confession was the result of a time when men's minds were turned from the beneficent side of God to the austere side, and when life itself was full of lashes, gibbets, scaffolds, and dungeons." Sensitive to preserving its mandate as a missionary movement, and mindful of the increasing influence of Darwinism in America and of its many influential Presbyterian defenders and interpreters, not the least being James McCosh, liberal pastors like Russell called for change: "[S]aid revision to set forth on the one hand the sovereignty of God, and on the other most fully the love of God to man as it shines so brightly in the Gospel."[29]

Russell's passion helped to secure the endorsement of fifteen presbyteries, but the Nassau petition was not carried at the General Assembly in 1889, nor in 1900 when Russell and his presbytery tried again, allied with the Union Theological Seminary.[30] They partially succeeded only in 1903, faced by continued opposition from the Princeton Theological Seminary. By then, Russell's congregation had profoundly changed as the demographics of Oyster Bay shifted and as some of the church's most prominent members, like Theodore Roosevelt, migrated across the street to Christ Episcopal Church. Roosevelt's second wife, Edith Carow, was a devout Episcopalian, and their move stimulated some "church shuffling" among the elite. This shift, along with the transformation of Oyster Bay from a village polarized by vast disparities of wealth to one where a rising local professional and merchant class was emerging by the turn of the century, meant that more working-class members of the community took leadership roles in the church. Still, Reverend Russell enjoyed the continued support of the elite; at his twenty-fifth anniversary of service in 1901, Roosevelt attended his reception as a friend of the family and ardent supporter of the faith.[31]

Home Life

Life in the manse orbited Alexander Russell's twin foci of worship and pastoral care. The home of the minister was "dominated by a strong-willed father who valued education along with Christian faith." And as the eldest, Henry likely aspired to the power and authority Alexander Russell projected.[32] But Henry was also a product of his mother Eliza, who, as a minister's wife, had responsibilities well beyond the family but tried to preserve an island of personal privacy and respite for her sons. Thus while she assisted her husband's calling by serving as secretary of the regional Missionary Society, corresponding secretary of the Christian Endeavor, and treasurer of the Ladies Aid Society, and served as well as a teacher in the church Sunday School and as an active member of the Woman's Club, among a whirlwind of other expected duties,[33] she spared her sons the onerous duty of memorizing the *Shorter Catechism*: "[M]y mother had a remarkable gift for connecting her rules for everyday behavior with the Ten Commandments . . . The emphasis was even more on moral principle than on religious obligation."[34] Even so, as a proper family of a Presbyterian minister, the Russells often went door-to-door, holding prayer meetings in the street, "saving souls."[35]

What little can be reconstructed of Russell's daily life in the 1880s and early 1890s is based on impressions left by family members. They paint a picture not unlike that of the young Theodore Roosevelt during his Oyster Bay summers over a decade earlier: "of a slight, tousled boy browned by the sun, clothes in disarray, who could barely keep still."[36] All three sons were attended to by a succession of live-in nannies when they were young, then were trained at a local Dames' school. Henry early on was tagged as a sensitive, precocious boy. He was two and a half years old when he first saw his brother Gordon, and remarked that his brother "moves his head like a pendulum." By the age of three, as family lore attests, he had learned to read through self-study with a picture dictionary. Playing at Bay Head on the New Jersey shore in his fifth summer, when the family was visiting his maternal grandparents, Henry and Gordon were each given a penny. But when one penny slipped through the slats of the boardwalk, Henry claimed it was Gordon's, because "[t]he dates were different." There was sibling friction between Gordon and Henry, since Gordon was a prankster and tease, and Henry "was a serious child and he didn't know how to handle it." In September 1890 Henry would be sent to Princeton Preparatory School, whereas Gordon and Alex studied at a Friends' Academy in nearby Locust Valley, subsidized by funds from wealthy parishioners.[37]

In Oyster Bay homes of that era, most sounds were those that people made: food being prepared, children running about, and the occasional noises of an active fishing harbor. Any music heard was live—although Edison's phonograph was shown on the streets of Oyster Bay in 1889 as a curiosity, drawing a large crowd—and Alexander Russell loved to sing. There was the ticking of a grandfather clock, the creaking and slamming of doors, and the calming sounds of a small brook close by.[38] The Russell boys had their games and their friends. One game Henry later recalled was a team-based variant of hide-and-seek. Responding to a childhood friend who harbored the memory decades later, Russell thought back:

> I remember very well the paper chase, which you spoke of and how I enjoyed hiding in the bushes and hearing you all go rushing past down the hill on the false trail, which we had carefully prepared. I appreciate very

much also the things which you say about my father. One may well be proud to be his son.[39]

Henry did recall his father years later as a "successful country pastor," a robust, healthy man. He would note his fondness for music and that he participated in choirs around New York for years, serving as one of the directors of the New York Oratorio Society. Alexander's musical enthusiasms were not passed on to his eldest son, however, so Henry often liked to recall that his father was drawn to mathematics and languages, much as Woodrow Wilson recalled his own pastor-father. Both his father and mother read voraciously. He described her as "bookish," but for Henry, her most prominent legacy was a "strong trait for mathematics."[40]

Henry also claimed to have acquired his interest in astronomy in Oyster Bay: "I recall my parents showing me the transit of Venus in 1882, when I was five years old."[41] The apparition was, indeed, quite visible to properly shaded eyes, and its transit across the Sun's disk in 1882 was a widely publicized affair in the New York area.[42] Young Henry learned of the coming event from his father, who sermonized on the meaning of its prediction on 3 December, three days before the transit.[43]

Alexander Russell discoursed more than once from the pulpit on the wonders of heaven and nature, and of the human desire to know. From the titles recorded by William Swan among the 2,500 sermons his minister preached, one can find astronomical themes. Favorites were seasonal changes and upcoming eclipses.[44] At times, Swan found Russell's sermons uninteresting, "more like an essay than a sermon." But Russell's sermons could be moving, particularly when delivered without text or devoted to the questions of temperance and church reform. In these times, Alexander Russell's sermons were a "triumph." He was a "Priest to be proud of."[45]

"Reformation principles were postulated rather than argued" at home, Henry recalled late in life, owing to "the diplomacy necessary for a small town minister." But his father was certainly not reticent to discuss the "Revision" with casual acquaintances in his son's presence when they were aboard a steamer bound for Europe in 1892, and of course he was often quoted in the newspapers.[46] His father's failure to prevail stayed with Henry throughout his life. He revealed his youthful confusion in 1925 as he recalled how his father spoke fondly of the Unitarian William Ellery Channing, who led "thousands of good men [to] cast those unwelcome tenants, Predestination and Original Sin, out of the back doors of their minds, bade them with execrations never to show their detested faces within their souls again, and settled down in a house swept and garnished."[47]

"I 'joined' my father's church at fifteen," Henry recalled years later, "after some tribulation, (*vide Pilgrim's Progress*), which might have been spared except for Victorian reticence on my part." By then Henry had been living with his maiden aunt in Princeton for over three years. He was growing into a broodingly serious fellow, confused about his identity and his readiness to profess his faith. After all, Bunyan's Christian chose to forsake his family for his faith and, like Henry, had left his family early in life. Bunyan's allegorical tale, required reading in a Presbyterian family, is filled with "rejection of family and of past limitations" and the rejection of "old loyalties" to gain the "ideal reward."[48] Upon entering the college, Henry identified himself as Presbyterian and, like his father, intended to teach, but little else survives from that time to inform us of his motives, fears, and desires.[49]

A Summer Abroad

Starting in the fall of 1890, Henry spent his school years in Princeton and returned each summer to Oyster Bay. After his second year, Henry accompanied his father and mother on a "dream of a life time" trip to England and Europe, a gift from the congregation. His two younger brothers stayed home with Aunt Ada Louise Norris in Princeton, which gave this summer special meaning for Henry. The pastor had never been to Europe, so this was a voyage of discovery for father and son. Accordingly, Alexander Russell prepared a three-hundred-page diary of their journey. Meant to be read by his parishioners, it still provides glimpses into the pastor's character, his perception of the world, and his relationship with his son.[50]

They left in late June 1892, toured Edinburgh, London, and much of northern Europe, and returned in early September. Every minute of their seventy-some-day odyssey was filled with some form of purposeful activity. During the ocean voyage as passengers on the first-class deck of the SS *Aurania*, Alexander and Henry investigated the workings of the ship, from its control room and machine room right down to the "stoke hole" where the stokers fed the steam boilers. The pastor became a navigator, noting the influence of eastward motion on the length of the day, the variation of the compass as they changed in longitude, and the drop in temperature of the seawater indicating that they had left the Gulf Stream. He was a manager, approving of the "discipline and routine" of the ship's complement, clearly relishing an orderly world where good people had good jobs to do and did them right.[51]

The pastor soon became a naturalist, watching dolphins and whales and recording the rare bird skimming the waves. He was always watching, observing, adventuring, and recording. His detailed observations continued as they passed the lighthouse at Fastnet Rock in the Irish Sea on the last leg to Liverpool, cleared customs, and searched out the nearest cathedral for Sunday worship. The same pattern continued as they traveled north to Edinburgh, Eliza's home, where they toured the sacred and historical sites of that fabled city, including the ancestral rubber manufacturing mill where Norris had labored. Alexander Russell wrote his most emotional passage after visiting a cathedral in Liverpool where, overcome by the clothing of the children and the familiar songs and prayers, he looked out the window and "saw the ivy—the English ivy all around it, my eyes filled with tears as I said to myself, 'At last, at last I am in dear old England.' " The pastor was certainly not a Jacobite.[52]

Their journey had been carefully planned around two goals: to experience both the natural and civilized wonders of their cultural and religious homeland, and to give Mrs. Russell a chance to see her family. Even more than Edinburgh, sites of religious significance in London were high on the pastor's list. But he also filled the pages of his diary with descriptions of the contents of the British Museum, the architecture of the city, and the technology of transportation. As he had aboard the *Aurania,* he described in knowledgeable detail the workings of the railroads and clearly loved to experience them with his son; he swooned when a prize-winning locomotive steamed into St. Pancras, and was delighted with the working models of the steam-driven world displayed at South Kensington.

In the crypt at St. Paul's, they found the great names of British military history but admired the artists and ecclesiastics. In Westminster Abbey the pastor became a pilgrim. It "came to me like a revelation as of something often before imagined but never before realized."[53] Reeling off the names of the interred, he was most passionate about the Poet's Corner and the great masters of music. But his goal was to reach the inner sanctum of the abbey, which, after knocking on the right doors he managed to enter with Henry: the very chamber where the seventeenth-century divines crafted the Confession and where the Old Testament was revised in the nineteenth century. England was Alexander Russell's spiritual homeland; on their last day in London, bound for the Antwerp ferry, he strained his neck trying to catch a glimpse of Bunhill Field Cemetery, where John Bunyan was buried. He had looked for Bunyan's church in Sheffield from their London-bound train.

Henry accompanied his father through most of his travels, but only once in his father's diary did Henry gain a voice. During a visit to the Museum of Antiquities in Antwerp, which sat atop dungeons dating from the Inquisition, the pastor led Henry and a few other boys down into the dark and dank chambers. At the deepest point, the boys blew out their candles to shiver in the closeness of the place. Russell's father left them in total darkness for a few moments but then returned to the sound of Henry's voice: "Here comes the officer to lead us out to execution."[54] Hardly an executioner, Alexander Russell, in his zest for life, posed a challenge for a fourteen-year-old whose physical energies were limited; later in their travels Henry would retreat more and more with his mother to their hotel room while his father continued to explore.

Eliza Russell had spent some months in Heidelberg in her youth and found some of her old friends and acquaintances still healthy and hardy, though her husband did not find militarized Germany a comfortable place. Students paraded their dueling scars from their adventures, which disgusted him: "There is evident room for reform in the peculiar customs of German universities." He found the military fortifications in Strassburg ominous and oppressive.[55]

In Switzerland, they turned to vigorous hiking. The mechanics of the cog tramway up the Rigi were as fascinating as the rocks and strata. On the descent, with Lake Lucerne spread out before them, the pastor became annoyed by cigar smokers in the car, "smoking as though the whole world was theirs to pollute . . ."[56] Here again he taxed the energies of Henry and his mother, though Henry did his best to keep up. Arriving at the Matterhorn, while the bulk of their tour group looked to the comforts of the hotel, Alexander marched off to survey the territory, recording it in stream-of-consciousness detail. It is just this type of focused energy and exhaustive attention to detail that his eldest son would adopt in his research. The pastor loved to be up to meet the sun, and to witness its setting, just as Henry would do decades later when he visited places of natural beauty, like the Lowell and Lick Observatories.

There were also fortuitous contacts made on the journey. When Pastor Russell led a Scottish service in Lucerne in late July, they found Francis Landy Patton in the congregation. Patton, who was then president of Princeton, lingered with the Russells and others, discussing matters of the day. From their exchange the Russells seem to have been on familiar terms with Patton.[57]

Letters from home started arriving in early August, mainly from Aunt Ada, reporting that all was well with Gordon and Alex.[58] Reassured, the family moved on to Paris, another whirlwind of detail. Once back in London, Alexander led his family over familiar ground, adding a trip to Windsor, where he was so taken by the countryside that he put Eliza and Henry on the return train and set out on foot for their hotel. They had missed the drill inspection at the Tower of London on their first visit and now were in luck: "Henry who had never seen anything of the kind before, was so much interested that he insisted upon staying till the inspection was completed." Obviously captivated by the pomp and circumstance, Henry watched as each man was inspected, and was amazed at the tiny infractions that could render a man unfit to guard "Her Majesty's Tower."[59]

Oxford overwhelmed Mr. Russell. Just as his son would later write and publish reams of detailed argument supporting his theories and conclusions, exhausting his detractors with specifics, the father never seemed to tire of describing meticulously his impressions of the world around him. His industry peaked as he encountered the Church of St. Mary Magdalene, the Martyr's Memorial, and the old entrance to the City.

Eliza and Henry were by now totally exhausted. Henry developed a sore throat in Oxford, and his health withered in Chester, where his father had planned to walk some six miles to Sunday services to hear William Ewart Gladstone, the obsessively tenacious Scotsman who had been Queen Victoria's prime minister four times and was freshly triumphant as the newly designated first lord of the treasury. "But I must wait till the morning & see how Henry may be."[60] Alexander missed Gladstone, summoning a doctor and searching for the prescribed medicinal. After the crisis had passed, the doctor predicted that the ocean voyage would act like a tonic, doing Henry a world of good.

They had to rush to board the Cunard Steamship *Gallia* in Liverpool, which seemed not as nice as the *Aurania*; it was smaller and "there is a large list of steerage passengers, some of whom at least have come from a land where soap is not plentiful! I understand that many are Russian Jews."[61] Their passage home on the salon deck was a rough one but delighted the pastor no end. He stayed on deck to watch as "the great vessel rolled & plunged among the yeasty . . .

monstrous . . . waves." One wave reached over the bow and sent spray to the top of the smokestack. He recounted every step the crew took to batten down the ship, taking considerable pride in his own knowledge of proper procedure.

The pastor's stream-of-consciousness narrative, created from memory, reveals his mental capacity for detail, a quality Henry would later exhibit. His narrative continued to 1 September and their arrival in New York, recounting concerts, athletic events, and conversations with the captain and crew about the state of sea travel. Nearing the docking area, aboard a tender taking the salon passengers to shore, leaving those in steerage because of a cholera scare, the Russells reunited with the spires of the "twin cities" of Manhattan and Brooklyn and bade farewell to their trusted crew. Thus ended "our most delightful and never-to-be-forgotten Summer Trip."[62]

The months they spent together were a very special time for Henry; he would repeat the experience with his own family thirty-seven years later. His parents' decision to leave his younger brothers at home indicated his place; Gordon and Alexander may have been too young for such a vigorous trip, or limited funds might have excluded them. But at the very least, for a sensitive adolescent who had spent much of the previous two years apart from his parents, the trip must have reassured him that he was still part of the family.[63]

This brief window on family life provides a chance to speculate about the father and his relation to Henry. The pastor was wholly engrossed with the event the day, be it ocean travel, inspecting the many historic churches and cathedrals of his ancestral homeland, or, we might surmise, attending to the daily needs of his congregation in Oyster Bay. Both his love of music (chiefly liturgical singing) and his devotion to his calling find expression in his record of their trip. As with Woodrow Wilson's father, who led his eldest son on tours of local mills and factories to expose him to the workings of the world, and would test him to determine how well his son had absorbed the experience, Alexander Russell constantly guided Henry through his world, from the mechanics of the drive system of the steamer, the art of navigation, and ocean sounding, to the details of rail transport and the geology of the Alps. In all of this the pastor's enthusiasms were endless; education and exposure to life were the means to develop mind and character, clear thinking and accurate expression.[64]

But the diary also reveals a father centered on his own passions. Save for the visit to Eliza's father's workplace and her family and friends in Edinburgh and Heidelberg, it was the pastor's show. Since this was a public document, little about the state of his family emerges, beyond Henry's or Eliza's health. Even though Henry's name appears explicitly many times as his father's companion in discovery, nothing is revealed of his inner thoughts about his son, nor of his son's temper or disposition. The one exception was Henry's utterance from the dungeons, enacting the role of resolute character in the face of certain death. Delighted by such manliness, the pastor recorded his son's words for the world to read, as if in them he found affirmation that his eldest son had come of age. This exception, combined with what is known about Henry's student life at

Princeton, suggests a reason for the trip beyond the father's personal desires for adventure.

Pastor Russell was, by all indications, a typical Victorian, eager to experience real life and to establish moral regeneration through vigorous and disciplined physical and mental exercise.[65] The Edinburgh natural philosopher James David Forbes "prescribed travel to shape the raw material of a young man's moral nature." Vigorous exertion strengthened both manly character and scientific perception. Proper Victorian males knew both "nervous exhaustion" and "vigorous, resolute immersion in direct experience of nature."[66] The pastor was no exception, and the father wanted the same for his son. "The roast beef of hard industry gives blood for climbing the hills of life," Joseph Ruggles Wilson wrote his son Woodrow in 1878; the pastor could have said much the same to Henry.[67]

Of the Russells' three sons, however, Henry was the least involved in active sport and seemed to avoid physical exertion as much as possible. He was evidently a nervous and rather anxious child, and so if his father shared Forbes's vision that adventurous, vigorous travel would bring his eldest son around, the European tour was his instrument to help his son mature.[68] The diary clearly reveals how Henry was challenged by his father's capacity for exertion. Strongly implied too is that Henry lived in a world defined by his father: that when one was in the pastor's presence, one was focused by his personality and the force of his enthusiasms. Just as Woodrow Wilson's father's "principal avocation, as well as vocation, was talking," Pastor Russell was rarely at a loss for words. Henry would inherit this trait once he found his own position in life.[69] But Henry probably felt closer to his mathematically inclined mother than to his clerical father. It was primarily the mother's role to manage the home and create its overall atmosphere, though she did so in deference to the wishes of the vigorous and ever-busy father.[70]

Indeed, Henry too would lead a vigorous life, but one of the mind, not the body. All the energy, intensity, and self-absorption manifest in his father's diary reemerged in his son. Some of the outdoorsmanship survived too, though in the form of less strenuous pleasures like nature study. Henry did eventually assume the discipline and character that Forbes's vision of a Victorian male demanded, but he would do it in his own way.

If the European trip had some influence in drawing Henry closer to the family, he still took over a year to decide to join his father's Oyster Bay congregation as a full member of the church; he did so only on 1 October 1893, ten days after he had entered Princeton, and three weeks before his sixteenth birthday.[71]

The Family Legacy

Typically cryptic about the influences upon him, Russell still took pride that his father had fought to remove "so much polemic sixteenth century phraseology" from the Presbyterian Confession of Faith, just as his family took pride in his mother's moral responsibility and mathematical prowess.[72] As the eldest son, Russell assumed his father's theological liberalism, and the intervening years to adulthood did little to change his outlook. His 1925 "Fate and Freedom" Terry Lectures on Religion, delivered at Yale, rejected Old School Calvinism's use of the deterministic aspects of the exact sciences to resurrect predestination, "under which the whole course of Nature and of human events has been decisively settled in detail before the beginning of the world."[73] Russell warned that

> [b]elief in the determination of events by influences antecedent to the human will has indeed come back, fortified by a mass of evidence derived from the field of science and possessing its redoubtable authority. We can no longer escape determinism by changing from one school of religious belief to another, nor even by rejecting religion altogether. The fight must be fought out squarely on scientific ground.[74]

The 1920s was a particularly turbulent time when Presbyterian theological orthodoxy was counterattacking the newly established liberalism, in no small part exacerbated by the weakening of organized religion on the American college campus. In 1925, typically for the time, Russell claimed "full acceptance of the mechanistic theory of nature . . . not as a demonstrated natural law, but as a working hypothesis." His challenge was to show that mechanistic determinism in the inorganic world could do no harm to religion, morality, government, or art, and in fact did not promote any sense of "inexorable predestination." Science, Russell argued, was bound to explain the properties of systems and was not free to attribute properties ad hoc; nor could it assume that the ultimate explanation of these properties could be known from their superficial characteristics. "The old hope of finding final and perfect statements of the truth about a simple universe has fled . . ." In its place was a striving for closer and closer approximations to the truth, "a steadily increasing accuracy of approximation in the description and interpretation of an incredibly and magnificently complex universe." In its constant striving, science had freed itself from absolutes, whereas theology had not. Thus Russell felt that there was much theology could learn from science in an age of relativism.[75]

Russell did, however, rely on his theology for his basic conception of nature, which persisted even as modern physics revolutionized the physical universe. In light of the new mechanics of Schrödinger, Heisenberg, and Dirac, Eddington and others proclaimed in the 1930s that determinism in the physical universe had "really been eliminated."[76] Not for Russell. Strict mechanistic determinism still described the macro universe, though he was willing to reinterpret it as a world governed by statistical mechanical laws, and one that appeared wholly mechanistic in the aggregate but retained free will for the individual, whether that individual was a particle or a person. To Russell, statistical determinism was "more impressive than any dogmatic assertion of certainty . . ."[77] But just as clearly, he remained deeply wedded to the old beliefs. Trying to clarify his conception of the statistical properties of nature for a colleague in 1941, during a set of conferences convened in New York under the rubric "Science, Philosophy and Religion," Russell explained privately that his fondness for determinism was not derived from his professional life: "Personally an ultimately deterministic hypothesis satisfies my feelings, but my reasons are based on theology and not at all on physics."[78]

Publicly, Russell delighted in the recitation of Scripture, finding it both uplifting and useful for illuminating points made in argument. Publicly he also eschewed its formalisms just as he rejected the rhetoric and rigid teleological demands of the organized church. Although he would become an elder in the First Presbyterian Church of Princeton, his rejection of compulsory daily chapel on campus in the 1920s made him too liberal for even his closest friends at the seminary, one of whom approved of his status as church elder but felt that he could never be a seminary faculty member.[79] Inwardly, however, Russell could be a Calvinist. Pondering the causes of war in 1943, Russell could find no rational analysis. War, he felt, was attributable not to "human unreason, but to human depravity; but if I said this in public, so many people would regard me as a variously-qualified Calvinist that I forbear!"[80]

Russell outwardly reflected his parent's liberalism in both his religion and his science. Just as he loved Scripture and literature, or neat mathematical tricks, he rejected religious formalism as well as the formalisms of mathematical rigor; for him the process of science was an intuitive one. But theory had to be tempered by observation. For Russell, the universe was so complex that one could not hope to determine "detailed behavior by exact mathematical analysis."[81] He believed that the scientific spirit was the best guide to understanding. It fostered right ways of thinking: respect for observations and the ability to evaluate their efficacy, or the ability to scrupulously distinguish between nature itself and the theoretical models created to describe nature. "Perfect certainty may not be attainable, but a degree of assurance can be reached which justifies our venturing our fortunes, or lives, or even our souls in deliberate and reasoned commitment to a trustworthy faith."[82] In Russell, faith and science came together, very much his parents' legacy but also seamlessly strengthened by his institutional roots.

Religion and Science at Princeton in the 1890s

When Russell entered the College of New Jersey, term opened "in the Chapel with the reading of the Scriptures, and address and Prayer by the President."[83] Daily chapel was compulsory, and Francis Landy Patton was now in the pulpit, succeeding McCosh in an 1888 backlash to McCosh's secularizing reforms. Deeply committed to restoring the traditional virtues of an ecclesiastical campus, the zealous heretic-seeking cleric and polemicist offended some alumni and rekindled divisiveness between science and religion on campus.[84] Although he fought secularization, he assured critics that Princeton would continue as a center for research, because "the professor who has ceased to learn is unfit to teach." Fine words, but Patton was in deep opposition to McCosh's vision, disputing evolution in any form, certain that it was "a device for banishing God." Patton, preoccupied with moral and right thinking, let academic standards slip, which, purposely or not, made Princeton more popular than ever in the 1890s.[85]

Patton did not prevent the teaching of evolution. Although some, like geologist Arnold Guyot and the mathematician John T. Duffield, were skeptical about strict Darwinian evolution, others—like George Macloskie, who studied under McCosh in Belfast and came to Princeton in 1875 to take up the professorship of natural science—"vigorously promoted science among orthodox Presbyterians and consistently urged the value of a chastened Darwinian perspective . . ."[86] Macloskie and McCosh were strong defenders of evolution, as were their followers, chiefly Charles A. Young, the astronomer, and the physicist Cyrus Fogg Brackett. Young lived in the director's residence next to the students' observatory on Prospect Street, and Brackett was his nearest neighbor and "most intimate friend here." Together they occupied what Princetonians chidingly called "atheist's corner."[87]

Young keenly knew that powerful campus forces were not comfortable with science. He may have lived at "atheist's corner," but he had studied theology and preached that astronomy revealed the handiwork of God. At a time when the majority of Princeton faculty were clerics and when "[a] deeply religious feeling permeated" official campus life, Young was recalled by students older than Russell as "[d]eeply religious himself, of pure and simple faith." When many were fearful that science would overrun religion, Young was well-known as a defender of the view that "there was nothing inconsistent in the revelations of scientific research and true religion."[88] In lay sermons he celebrated the belief that as astronomers have come to know "more and more of the material universe," they have also revealed the "Glory and Majesty of the Creator."[89] Astronomy, Young hastened to point out, could say nothing about God's "moral attributes," nor could it demonstrate "his Providence and Holiness, his Justice or his Mercy." For evidence of these, Young added, one must look to moral law "written upon the human heart" and to the course of history.[90]

Young believed, as Russell would all his life, that there were limits to knowledge. Faithful to a scientific tradition imbued with moral and religious virtue, Young presented a "self-effacing style of 'humility.' "[91] For Young, science was both incomplete and mutable, but it was preferable to a literal Genesis because it "seems to me far less honorable to the Divine intelligence and power than that which supposes him to have contrived the matter out of which the worlds are made, that from a chaotic nebula should have resulted the present stately cosmos by the simple operation of the laws He first imposed."[92]

McCosh hired Young from Dartmouth in 1877 to complete the long-delayed Halsted Observatory, which would soon boast a 23-inch refractor, three inches shy of the largest in the world. Young also complemented McCosh, sharing his sympathy for responsible speculation and theorizing. Truth was attainable only through the systematic combination of observation and intuition. Pure empiricism was not viable, and a priori reasoning was barren of reality. McCosh's Scottish commonsense philosophy demanded the flexible combination of intuition, to arrive at working hypotheses, with deduction from observation, to test those hypotheses. Just as McCosh influenced those he trained at Princeton, such as Woodrow Wilson in the 1870s, those he drew to the college as faculty, like Young, were of kindred spirit and would, in turn, influence others, like Russell.[93]

Astronomical instruction, Young proclaimed in his 1890 *Elements of Astronomy*, a book Russell would soon read, not only trained the mind but helped one to appreciate "the dignity of the human intellect as the offspring, and measurably the counterpart, of the Divine; able in a sense to 'comprehend' the universe, and know its plan and meaning."[94] Russell likely heard Young say more than once in class, when the spirit moved him, "God hides that we may find, and we exult in finding like little children with whom their father plays hide and seek."[95] Such evangelical campaigning was typical among senior American scientists of Young's day, clearly capitalizing on a growing enthusiasm for science and the potential of its products to benefit society.[96]

Russell as Product

Like Woodrow Wilson, Russell was a product of the intellectual and religious environment fostered by McCosh. He also reflected his father's enthusiasms for the world and his theological liberalism, his mother's sense of moral responsibility, and Young's belief in the compatibility of science with religion. Russell ardently believed in the need for balance between observation and intuition in scientific practice, as McCosh and Young taught. He also sympathized with the need for secularization and pluralism on a campus that remained one of the most traditional in America.[97]

Russell, fostered by the Princeton that McCosh had built, was a secular scientist living in a revealed universe. In his earlier writings, Russell was comfortable expressing a deep-rooted belief in the presence of design in nature, not continually acting, but evident from first principles, an attitude reminiscent of McCosh's age and a vestige of the enthusiasms of natural theology.[98] He could also indulge in cosmogonical speculation, up to the limit established by God's design. To retain that freedom, he knew, as did his mentor Young, that it was his responsibility, as well as his legacy, to campaign for science in the service of a revealed moral universe, and that the revelations of science would, in the long run, serve the religious life. Russell also knew, and advocated more than once publicly, that theorizing in science was never enough, nor was it an end unto itself. Nor was open-ended observation, uninformed by theory, an effective means to conduct research. The balance he sought in science stood as a guide for a better world, a world in which religion remained a force.

In times of crisis, Russell invoked the lessons of scientific practice to shore up his moral and religious universe. In 1916, with the world plunging into war, he called for a "Scientific Approach to Christianity." In the mid-1920s, in the face of fundamentalist attacks, he showed how the scientific imagination—disciplined, expanded, and liberated in the pursuit of nature—could come to the aid of our limited human conceptions of the world in the understanding of God. In the

1930s as he worked to shelter European refugee astronomers on American soil, and near the end of his life, recovering from heart attacks and the horrors of a second great war, Russell remained convinced that science could make "valuable contributions toward the resolution of some old theological difficulties."[99]

Throughout his life Russell would campaign for many causes, ranging from the abolition of compulsory prayer on campus to the defense of modern physics even when it espoused indeterminacy. He also argued fervently for "practical human self-determination" and remained uncomfortable with metaphysical explanations.[100] Believing that "[t]he only credible God is one who is responsible—to use our absurdly inadequate human word—for the entire universe," Russell claimed that science was the salvation of religion not only for what it revealed about God's handiwork, as Young claimed at the turn of the century, but for the tools it provided to aid human beings in the struggle to perceive that handiwork.[101] Scientific training was essential if one were to gain a rational perspective on the doctrinal tensions that were part of everyday religious life. As he preached in 1925, a religious person trained in science, "especially in physical science," would not be alarmed at differences in theological attitude. The scientist, Russell argued, could cope with diversity and in fact welcomed it. A time of controversy was, to the scientist, not a time of doubt but "an exhilarating one of rapid advance." Only scientific training could provide the tools necessary to enable one to be "very suspicious of the proposition that certain statements of theological theory—or even of religious truth—possess absolute, plenary, inspired accuracy."[102]

Russell expressed the tentativeness typical of liberal Protestantism: that ultimate truth was not attainable.[103] In the 1920s, when the Victorian ideal of a unity of truth had finally been laid to rest, Russell campaigned for greater tolerance toward religion as a set of ethical principles. Late in life, though inwardly a determinist, he identified his public position as "relativist rather than absolutist," feeling that "a physicist can take no other." No theory was permanent, and views constantly changed, improved, iterated closer to the truth, but never reached the truth. Just as he believed that "man was created 'to glorify God and to enjoy Him forever,' " he did not believe that humans would ever be "capable of attaining an absolute knowledge of God—or of anything else."[104] It was not only human to strive for perfection, it was mankind's duty, even though no measurement could ever be perfect, "owing to human imperfection."[105]

As we examine Russell's life as an astronomer, looking for clues for what astronomy meant to him as a calling and how he planned to answer that call, we must appreciate that from the first he saw the universe as deeply moral and religious, subject to the natural laws created by God. Though both he and the world around him changed in many ways, he never abandoned that vision. Nor did he abandon his religious background and upbringing, which though supportive rather than determinative of his life course, still shaped the choices he made about his career and his scientific practice.

• CHAPTER 2 •

Russell at Princeton

Princeton Prep

Pastor Russell's support for public education did not extend to his sons. After attending a local Dames' school, Henry was sent away at age twelve to attend the Princeton Preparatory School in 1890. His brothers completed their schooling at the Friends' Academy in nearby Locust Valley, whose rigorous curriculum and reputation were more than adequate for entrance to a good college.[1] No doubt Henry was sent to Princeton Prep to follow in his father's path, but the Friends' Academy's policy of compulsory worship in 1890, when Henry reached prep school age, might have been a factor as well. This rule was suspended in August 1891, which opened the gate for Gordon and Alex.[2] A family recollection also suggests that Henry's dour intellect became a target for Gordon's teasing, and that the parents thought the brothers would develop better if separated. Money was not an issue: wealthy parishioners helped to pay the Friends' Academy tuition, and Henry's housing in Princeton was free. "I did not earn my own way," Russell recalled, "even in part. Legacies to my mother from her parents had made our economic position fairly comfortable."[3]

How Henry reacted to his move to Princeton is not known. He would live with Aunt Ada Louise Norris, his mother's younger spinster sister, who was hardly a stranger. And Princeton itself, a small town of segregated shanties and comfortable Victorian structures standing amid farmland and marshes in central New Jersey, was not all that different from Oyster Bay, except that it lacked his family, friends, and the sea. The town itself was centered on a wide dirt road called Nassau Street, which divided its commercial center from the growing college to which, in many ways, the town was attached. Though in the latter part of the century Princeton was starting to attract wealth spilling over from Philadelphia,

Newark, and New York City, it was still a town where Princeton faculty defined community society. It was a comfortable academic culture: neighbors among the faculty socialized, cared for others of their class, and were cared for by a large population of housekeepers, nannies, and cooks.[4]

The Norris home, a substantial three-story Gothic structure that had been the residence of John Thomas Duffield, a Princeton mathematics professor and ally of Charles Hodge, became Henry's home for the rest of his life. Located on one of the main routes into Princeton Township from the Boston Post Road, the home at 79 Canal (renamed Alexander Street by 1896) was reported to have the largest spruce tree in Princeton. It did boast an apple orchard, fig trees, and a gracious porch around three sides. The high-ceilinged home was filled with "eight beautiful antique clocks . . . and the walls [were] adorned by family portraits dating generations back."[5]

Not much would change in the house for years, except the people who lived in it. Aunt Ada owned the property jointly with Henry's mother. She immersed herself in Henry's upbringing, tutoring him during his preparatory school years and later in his liberal arts studies in the college. Neither her facility with languages nor her piano playing seems to have influenced Henry. On the other hand, she, like her sister and now Henry, had a definite flair for mathematics. And Aunt Ada more than maintained an active intellectual curiosity. She was one of the founding members of "The Present Day Club" in Princeton, inviting some twenty-nine women to meet in her parlor in late January 1898 to create an "organized center of thought and action among the women of Princeton and to stimulate an interest in science, literature, art and social and ethical culture." Charter members were among notable Princeton faculty families including Duffield, Fine, Hodge, Magie, McCosh, and Paxton. Ada Norris became club president in 1899 and helped to organize their series of lectures, workshops, seminars, and tutorials. She even engaged her friend C. A. Young to speak about the state of astronomy. After his mother's death in 1903, Aunt Ada would become Henry's closest companion until his marriage in 1908. As a member of the Daughters of the American Revolution and of the Society of Mayflower Descendants, she never let her nephew forget his heritage.[6]

Princeton Prep, about a mile up Nassau from the town center, was founded in 1873 to supply the college. It was purchased in 1888 by John B. Fine and became a private school under his management.[7] When Russell entered in 1890 as a nonresident student, he was among some three dozen other boys, thirty of whom lived on the small bucolic campus. Henry spent three academic years at Princeton Prep in a course of study likely indistinguishable from what he could have secured in Locust Valley. In Princeton he developed an interest in chemistry and, out of that, photography. Aunt Ada tolerated his experiments and the stains.[8] Henry returned to Oyster Bay in the summers, but what his life was like, whether he worked or studied or merely rested, is unknown, save for the record his father kept of their European tour. At the end of that summer, Henry readied himself for the College of New Jersey's battery of entrance examinations. These provide a glimpse of Russell's training.

Russell was a common applicant, for there is no indication of special treatment. He took oral and written exercises in English, Latin, and Greek, as well as grammar, composition, literature, history and geography, arithmetic, algebra, and plane geometry.[9] Apparently Russell was well prepared by Princeton Prep and by Aunt Ada, for he passed preliminary examinations in Greek, algebra, arithmetic, and geometry in September 1892. He was reexamined in June 1893 only for grammar, composition, Latin, and geography, and again passed, entering the Academic Division, declaring for mathematics and stating that he planned to become a teacher.[10]

Living with his aunt at 79 Alexander Street, Russell avoided most of the typical costs of student life. Because he was a minister's son, tuition was waived. What fees remained were easily managed by Aunt Ada. From the standpoint of living expenses alone, Princeton was the Russell family's wisest and most economical choice. There is no evidence that Henry inclined to any other.[11]

Matriculation in the College

On the day college opened and Russell entered, 20 September 1893, the faculty gathered at 3:00 P.M. in the chapel. Russell, one month shy of his sixteenth birthday and one of the youngest members of his entering class, encountered President Patton once again, but the faculty "swelled with McCosh students, fully twenty-four of fifty-three." Most of these shared McCosh's vision of the modern university, "and several had a resolve to carry it out." In concert with younger faculty, influential alumni maintained McCosh's hold on the college.[12] Yet many on Princeton's faculty regarded their institution more as a "big boarding-school" than as a competitive college.[13]

Princeton in the 1890s was not Pastor Russell's Princeton. Reflecting national trends, dress was more casual, once-dominant literary societies now competed with athletics, and a generally active and boisterous campus life was filled with glee, banjo, and mandolin clubs and sports. Princeton's thespians, the famed Triangle Club led by Booth Tarkington, traveled widely. The *Princetonian* was now a daily student newspaper and the nerve center for campus life. At Princeton, the "acquisition of status and friends" was as important as formal training. Princetonians were generally not middle-class strivers but the sons of wealthy families seeking out a diverting college experience. Class spirit was all-important. Hazing was rampant.[14]

In addition to compulsory daily prayer in the Marquand Chapel, Princeton required religious instruction on Saturday mornings, divine services superintended by Patton at 11:00 A.M. Sunday mornings followed by Sunday school in local churches, and class prayer meetings later that day followed by more church services that evening. Russell's teachers never "attempted any proselytizing in connection with their teaching," but the heavy-handed theological indoctrination spurred rebellion among his classmates, intensifying the violence of their pranks.[15]

Russell entered the Academic College, rather than the narrower, practically oriented, and decidedly weaker School of Science, and declared his major area to be mathematics.[16] His first two years of study were determined: Roman history, Latin composition, extended tutorials and recitations. Bible was compulsory, as were algebra, geometry, trigonometry, and English. Russell's first electives included German and a second course in geometry. He was taught by junior mathematics faculty during his first year and scored firsts. He faltered a bit in English and Roman history and slipped badly to a third in Latin composition. Among his 136 freshman classmates, Russell was rated in class 2 (ratings went from 1 to 6) in the first session, 1893–1894, although he recovered in the second session, placing at the top of the first class. He stayed at the top for the remainder of his undergraduate years. Serious to a fault, Russell typically read all the assigned texts well before term opened.[17]

In June 1894 the dean of the faculty expressed deep alarm over the number of absences from morning prayer and repeated cases of intoxication among undergraduates, but Russell's name never appeared on the lengthening lists of delinquents.[18] Under Aunt Ada's watchful eye, he was fully occupied devouring newly required courses in zoology, botany, chemistry, and mechanics in his sophomore year, and faced advanced Greek and "Sophomore Essays and Disputation." Elementary mechanics was taught by William Francis Magie, then second to Cyrus Fogg Brackett in the four-man physics department and a founder of the American Physical Society. Magie used Selby's standard text *Elementary Mechanics*, which Russell had already absorbed by the beginning of term. Years later Magie still liked to tell a story of how Russell took pleasure intimidating his classmates with his complete knowledge of the course. A Princeton legend recounts that Magie took Russell aside one day after class "and told him that he would prefer it if he did not come to class . . . it would be upsetting to him and to the other students, because he knew too much." If this advice was given, it was not heeded, for Russell rarely missed class.[19]

After completing his last required course in mathematics, "Conic Sections," under Henry Burchard Fine in the fall term of his sophomore year, Russell took electives in differential and integral calculus that spring, again under Fine. These had been required courses in past years, but, reflecting a national trend, the trustees had reduced the mathematics needed for graduation. In response, Fine redoubled his efforts to make mathematics accessible to his students. Russell recalled him as "a most inspiring teacher" who, though he led the department, still paid attention to sophomores. Fine always remained after class, "discussing matters, elementary to him, but new and fascinating to his students." Russell was always among the few who habitually lingered.[20] He kept impeccable notes; firm and neat, carefully penned theorems, proofs, and worked examples reflected a methodical mind encountering revealed truth. Devoid of personal commentary or doodling, these notes were verbatim recitations from text and were subject to Fine's inspection.[21]

The son of a Princeton Theological Seminary graduate, brother of Princeton Prep headmaster John Fine, and a disciple of McCosh, Fine trained in classics and mathematics at Princeton, subsequently studying algebraic geometry in Leipzig under Felix Klein and then Leopold Kronecker in Berlin. He brought the German style of education back to Princeton in 1885 as instructor in mathematics, rising quickly to full professor in 1889. Although he maintained enthusiasms for algebra, numerical analysis, and non-Euclidian geometries, he was mainly a teacher, writer of textbooks, and campus politician. When Russell entered the college, Fine was a rising force as one of Woodrow Wilson's closest allies on the Princeton faculty.[22]

Fine was always on the lookout for talent and discovered Russell when he administered the preliminary entrance examinations in algebra in September 1892. Russell did so well that he was dismissed from the regular examinations in mathematics given to all freshman in June 1893. Henry soon became "one of Dean Fine's enthusiasms." With Fine's attentions, Aunt Ada's tutoring, and the mathematical legacy of the Norris family ever-present, Russell's first passion had to be mathematics.[23]

Russell dispatched all the standard courses that rounded out the true Princetonian. For his sophomore disputation, Russell chose (or was given) the task of arguing for women's suffrage: the question was much in the news, as two western states now allowed women to vote and three others were facing referenda. He stated that women were "physically, intellectually and morally fit" to share the franchise with men. Deriding detractors as "un-American," Russell offered a rather pedestrian argument that did nothing to dispel contemporary feminine stereotypes. His effort garnered no prizes that year but met the requirement.[24] His heart was not in the task; when asked the same question at graduation, for a statistical summary in the *Nassau Herald* report of the class of '97, Russell wrote dispassionately, "Yes in abstract, no in concrete." Even so, this was a liberal view compared to those of his fellow Princetonians at a time when only four states, all west of the Mississippi, had granted women the full franchise.[25]

Russell's junior year was dominated by more specialized courses in mathematics, physics, psychology, and chemistry. In addition to required courses in logic and political economy, he took at least two mathematics courses per term, and he read Maxwell's *Theory of Heat* and Preston's *Theory of Light* under Magie, entering the world of thermodynamics. He also took his first astronomy course under Charles Augustus Young.

Encountering C. A. Young

The low-pitched nocturnal rumblings and occasional high-pitched scraping whine of the Halsted Observatory dome were audible from Russell's second-floor bedroom on Alexander Street. Now, entering Young's "Elementary Astronomy" as a junior elective, he was about to meet the master of that fabled place, and to learn what astronomy was all about.

One can well imagine Russell, clutching his small brown notebook and Young's *Elements*, heading across campus to the Observatory of Instruction at the foot of Prospect Street, where astronomy classes were held. Entering the low, rambling wooden complex to take Young's popular course, Russell found the white-bearded patriarch behind a broad wooden desk adorned by rudimentary armillaries and globes. Behind him a pointer and chalk tray framed a worn slate board where the old professor, slowed and weakened by age and illness, made astronomy come alive. The rabbit warren of small rooms known also as the "Students' Observatory" was a quaint place filled with clocks, transit telescopes, a classroom, library, and offices, all capped by an old dome sheltering a 9.5-inch Clark refractor. Here Young was the genial master of a two-person department.

Russell wrote furiously his first day in Young's class. Though his notes maintain a spontaneous flavor, he most likely worked them over, checking against the *Elements* and a growing pile of texts his classmates watched him lug across campus, to the library and home. Thus he recorded that Young spoke of astronomy as a science of accuracy and simplicity. The earth's motions were known to 1/20th second of arc and required "astronomical instruments accurate to 1/500,000 of an inch." And it was a science where everything was governed by the laws of gravitation, laws that had been tested time and again to reveal a universe of exactness, beauty, and mathematical elegance.[26] "No other science" Russell knew from reading the *Elements* the previous summer in Oyster Bay, "so operates to give us on the one hand just views of our real insignificance in the universe of space, matter and time, or to teach us on the other hand the dignity of the human intellect as the offspring, and measurably the counterpart, of the Divine; able in a sense to 'comprehend' the universe, and know its plan and meaning."[27] For Young, mathematics imbued astronomy with its power and certainty: "Astronomy more than any other science puts a man in his proper place with reference to the universe."[28] Astronomy in Young's day was "one of the last bastions of natural theology."[29]

Young's popularity was not, however, due to piety: "Short, with a well-rounded paunch and eyes that twinkled like the stars he looked upon," one admirer recalled, Young was "a profound inspiration and a lovable instructor."[30] Russell remembered him as "Twinkle Young" mainly for his "active and effective sense of humor."[31] When President Patton slipped up during one of his more tedious sermons, blurting out, "No, my brethren, we can no longer hope for such a sudden change, after a clearly misspent life, that the leopard can change his skin or that the snake can give up his spots," Young couldn't control himself. Though Patton quickly subdued the sanctuary, Russell watched while Young's "head was bowed lowly and reverently and his shoulders shook, and shook and shook."[32]

In his lectures, Young defined the various types of astronomies pursued in that day. First, there was descriptive astronomy: the facts—time, the seasons, the constellations and star lore, what the planets look like through a telescope, all interesting, he claimed, but lacking meaning. Then came spherical astronomy, triangulating the orientation of Earth in space. From there sprang practical astronomy, predicting the places of the Sun, Moon, and stars for the navigator, and theoretical astronomy, the fundamental laws revealed by the motions of the heavenly bodies. Here also one found the theory of instruments, observations and their corrections, and the mathematical bases for what Young called "astro-mechanics," which taught how to "apply the law of gravitation . . . to explain motions in space . . ." And then, finally, came the newest astronomy of all, "astro-physics," which "[i]nquires into composition and properties of the heavenly bodies."[33] Here was where Young's heart resided.

Although a leading practitioner of spectroscopic astronomy, Young lived very much in the traditional Newtonian universe. He was not a mathematical astronomer like his contemporary Simon Newcomb. He rarely computed orbits and never contributed to the theory of orbital mechanics. But he reveled in the power of mathematical astronomy to create predictive models, mindful that they required constant testing against observation. For Young, astrophysics meant the precise measurement of the positions of spectral lines, for chemical identification and for physical motion. In 1884, as retiring president of the American Association for the Advancement of Science, Young celebrated the fact that "[r]esiduals and minute discrepancies are the seeds of future knowledge, and the very foundations of new laws."[34]

Finding those discrepancies, not in positions of stars or planets, but in the positions of spectral lines, had made Young's reputation. Graduating in 1853 from Dartmouth, where he trained under his father, Young then taught classics and studied theology before he was appointed professor of natural philosophy and astronomy at Western Reserve College, a post he held until 1866, interrupted only by his service as a captain in the Ohio Volunteer Infantry in 1862.[35] Young succeeded his father at Dartmouth in 1866 and during his decade there established the basic spectroscopic nature of the solar chromosphere by observing the solar atmospheric profile during eclipses. At the instant of totality during an eclipse in Spain in 1870, he recorded that the normally dark Fraunhofer lines of the solar photospheric spectrum suddenly "reversed" from dark to bright, just as Kirchhoff's laws had predicted. Well into the twentieth century, solar physicists regarded the "reversing layer" as a finite region in the solar atmosphere where line absorption took place. Young also gained wide respect for his careful chemical identifications of lines in the solar spectrum, sometimes correcting identifications made by others.[36]

In 1876, sensing that Princeton's astronomer, Stephen Alexander, was close to retirement, McCosh's trustees began to search for a graduate of the college to teach astronomy, one who was "at the same time a recognized scientist and a true Christian." Finding none who filled the bill, they spread their net wider and soon settled on Young, who they claimed not only was "one of the two or three greatest astronomers in America" but enjoyed a worldwide reputation and was a proven teacher. They liked his "earnest childlike piety" and accepted him even though he was not Presbyterian but an "orthodox Congregationalist."[37] Young, however, would not come unless the college would provide a great telescope to fill an observatory chamber that had been built by Alexander with funds provided by General Nathanael Norris Halsted. The board of trustees quickly accepted Young's terms, and so Young left Dartmouth in June 1877.[38]

By the time Russell took his courses in the mid-1890s, Young was internationally renowned as a pioneer solar spectroscopist.[39] His courses, therefore, were rich in what was then called the "new astronomy." But beyond the measurement of spectral line positions, astrophysics, as some called the practice, had little of the elegance, exactness, or rational foundation enjoyed by Newtonian mechanics. The physics and chemistry of light, heat, and matter, sciences of the terrestrial laboratory and of the spectroscope, were only just beginning to be applied to the Sun and stars. "Here indeed lies the fundamental peril," Agnes Clerke observed in 1903, "and at the same time the essential prerogative of astrophysics." It relied on the "risky expedient" of extrapolation, pushing into the unknown from the narrowly confined known, assuming the continuity of physical laws. It was still a qualitative science that kept traditional astronomers at a distance, since it could provide only a " 'first approximation,' to be subsequently corrected and controlled." Clerke, echoing James Keeler's remarks at the founding of the Yerkes Observatory in 1897, argued that the discovery of new laws, or the explanation of phenomena in terms of known laws, is one of the most important objects of astrophysics.[40]

Even though his own reputation was made largely on the basis of careful line measurements and chemical identifications, not speculation about the physical conditions of the Sun and stars, Young found disciplined speculation wholly appealing. He worked and taught in a manner that would deeply influence Russell, allowing him to ultimately reject the need for relentless rigor and numbing precision in favor of approximate and iterative methods.

After the installation of the 23-inch Clark refractor in 1882, complete with a large Hilger spectroscope from London, Young used it eclectically. He observed Venus's atmospheric spectrum and resolved the dense spectra of sunspots into a myriad of fine dark lines, thereby casting off speculation that spots were composed of "soot." He was always faithful to observe passing comets, and in the mid-1880s he carefully observed markings on the planets with his new telescope, including the recently discovered "red spot" on Jupiter.[41]

What Young did not do was institute a systematic program of observation with the Halsted refractor. Even though it was the fourth largest telescope in the world, it had no great encompassing project nor systematic observing program. He was an opportunist during the discovery era of astronomical spectroscopy.[42] As discovery yielded to extension and refinement, practitioners like Samuel Langley and Henry Rowland developed systematic programs to map the solar spectrum, which took years of effort and occupied the lives of their students and assistants.[43] Much the same happened in stellar spectroscopy when the largest American observatories concentrated staff effort on the orderly production of spectral classifications, magnitudes, and radial velocities.

Young did none of this. His researches remained exploratory and, to an extent, serendipitous. He did not establish his observatory as a factory and never created programs that channeled his students into narrowly focused lines of work. Even though he respected contemporary views on the authority of the observatory director, he never imposed that authority to forward his own research.[44] Indeed,

Young approved most of the young George Ellery Hale, whose mission was to close the gap between the laboratory and the stars. He not only relished Hale's independence in taking such a path but was pleased when Hale let his "observers . . . use their instruments and funds in any line they please."[45] Even after he had secured a new and more powerful Brashear universal spectroscope in 1891 that could perform high-dispersion observations of sunspot or chromospheric spectra, Young limited his work, with the help of an assistant, to reobserving and improving his lists of known chromospheric lines. A former student, Edwin Frost, published the list as a "partial revision" in an appendix to Frost's translation of Scheiner's *Astronomical Spectroscopy* in 1894.[46]

Under Young, Princeton was highly regarded as a training ground for the new methods of spectroscopic astronomy. "The best equipped observatory for teaching purposes that I know of is at Princeton," claimed Asaph Hall, Jr., in a 1900 survey.[47] At the time, most major American observatories taught through apprenticeships rather than through formal instruction, which naturally created narrowly defined careers. This situation was a concern to Hall, who perceived "a tendency sometimes towards the factory system, which is to be regretted" because, even though it may be necessary for efficiency, it limited students' vision and could even contribute to the "extinction rather than the extension of research." College observatories, on the other hand, fostered a broader curriculum, Hall believed, because "it is just as important to train men as to carry on investigations . . ."[48]

Princeton under Young was the only fully equipped spectroscopic observatory where the "factory system" was not in place. Careers were therefore defined neither by the research goals of the director nor by the instrumentation. Russell would benefit greatly from this environment. He received personal attention from a personable mentor who believed in the authority of the observatory director but did not demand it for himself.[49]

During the years Russell was a student, Young devoted himself more to teaching and textbook writing than to research. He had already published well over a hundred scientific papers and scores of popular magazine articles and was a prolific writer of a wide range of texts, starting in 1881 with *The Sun* and then *General Astronomy*, which first appeared in 1888 as a upper-level text that required familiarity with algebra, geometry, and trigonometry. *General Astronomy*, which devoted much space to "the Sun's Light and Heat," defined the standard in 1901 as "the best semi-mathematical descriptive astronomy extant."[50] Each new edition of Young's texts was fresh with frontier science. He frequently added paragraphs in fine print to reveal deeper issues; such a vehicle, Young once told a correspondent, "breaks the continuity of lesson-getting [but] on the other hand it 'opens windows in heaven' letting the pupil get glimpses here and there of wider horizons."[51] Russell would be a prime beneficiary of this open-ended view of astronomical practice. Training under Young, Russell came to know his heroes.

George Darwin was one, whose mathematical powers had "opened a new field of research" and shed light on the dynamical history of the Earth-Moon system.[52]

Astronomers, Young argued, would do well to follow in Darwin's footsteps, as they would if they took to heart the gravitational contraction theory for solar energy linked with Hermann Helmholtz's name. This path led to a theory of how stars like the Sun changed as they aged, which, for Young, was the most provocative of all questions dealing with the Sun: the origin and maintenance of its heat. Young acknowledged that gravitational contraction provided for only a short solar lifetime, "and many of our geological friends protest against so scanty an allowance."[53] But it was the best theory available and could lead anywhere. Young had faith that spectroscopy would someday inform the deepest problems of cosmogony and stellar evolution by revealing the physical nature of the Sun and stars.[54]

Young was always on the lookout for prudent ways to close the gap between the laboratory and the stars. He was intrigued by Norman Lockyer's idea that the chemical elements could be broken down under the extreme heat of the stars—his theory of "dissociation"—which, though not popular, was for Young "ingenious speculation" that opened pathways worth following. Lockyer's scheme "ought to be true," Young felt, predicting that "there is more than a possibility that its essential truth will be established some time in the future."[55]

Russell's class notes reflect these passions. Russell in fact would bring together two of the most fruitful and provocative lines of inquiry Young spoke of: the dynamical theory of the origin of the Earth-Moon system, by George Darwin, and Lockyer's speculations about the lives of stars interpreted through his theory of dissociation.

Of all the factors that link teacher and student, most important was Young's eclecticism. He never focused on one specialty, nor was he burdened with campaigns. Although he is remembered as a solar spectroscopist, in fact he published equally along traditional lines—micrometric and transit observations, planetary and cometary detail, chronometry, the motions of planetary satellites, eclipses, and solar transits. The few hours he spent at the telescope were not part of a routine but were prompted by specific problems he wished to solve.[56] Nowhere could a blueprint for Russell's scientific career be stronger than in that of C. A. Young.

Upper-Division Courses under Young, Reed, and Fine

Russell scored very highly in all his junior year courses, but highest of all in astronomy, where his gained a perfect score in the largely descriptive, elementary course. Only his friend and classmate Nick Stahl, whom Russell always thought to be the "brightest man in class," came close in Young's course, scoring in the ninety-first percentile.[57] Russell, no doubt, became the twinkle in Young's eye.

Russell's astronomy notebooks differ significantly from those surviving from his mathematics courses. They were filled with diversions, provocative statements, contemporary issues, and new ideas. Soon he was taking notes from *General*

Astronomy and was recording Young's taste for current events, such as William Ramsay's 1895 discovery of helium on Earth. This was exciting news, for Young had long puzzled over Lockyer's detection of the strong emission feature in the yellow region of the solar chromospheric spectrum, which Lockyer had called helium after discovering it during the Indian eclipse of 1868.[58]

Russell also gained his sense of the moral utility of science under Young. Echoing the "scientism" of his times, Young preached that astronomy trained the mind. As a science promoting accuracy in measurement, as well as precision in mathematical technique, astronomy stimulated "exactness of thought and expression, and corrects that vague indefiniteness which is apt to be the result of pure literary training."[59] Russell's notes also reveal how Young could reconcile solar radiation with a Christian moral economy: "Waste of solar energy not real, but from the point of view of human ignorance. About 1/100,000,000[th] is intercepted by the solar system." The rest merely escaped, to a purpose known only to God.[60]

The notebooks also record Russell's personal enthusiasms. He took little notice of physical demonstrations in optics but underlined, capitalized, and emphasized those observations, concepts, and subjects which raised his adolescent excitement level: the detection of a prominence equal in height to the solar radius drew two exclamation points ("!!"), and he boldly printed out "C O M E T S" to highlight Young's vivid descriptions. Russell devoted the most attention to Young's discussion of the power of celestial mechanics, and to the "Age and Future ('life') of the Sun." Helmholtz's contraction theory took center stage.[61]

Russell's notes record only cursory attention to classical topics like time, the seasons, or the calendar, and labored over topics in solar physics, stellar constitution, and evolution; this fact marks Young's introductory course as highly unusual for that day, if Russell's notes are representative. Young went beyond mere recitation: speaking of the equatorial acceleration of the Sun, he wanted his students to realize that it was unexplained, and that the explanation when found would doubtless reveal how the Sun was put together. Young, in both his lectures and his textbooks, was also not afraid to address "what-ifs," unknowns, and curiosities requiring fuller study. As he did in lecture, in print Young gave a balanced description of Lockyer's theory of the dissociation of the elements, whereas other texts of the day typically left out such conjectural discussions.[62]

In his senior year Russell took more astronomy and physical science, more mathematics; he also attended President Patton's lectures on ethics, the required capstone to a Princeton education and an experience common to American college campuses, in which all knowledge was drawn into a "Christian framework."[63] Patton's lectures left an impression on Russell, for they informed many lectures he gave in later years. In 1940, as Russell spoke before members of the philosophy faculty and their students in an upper room in Murray Dodge Hall, holding forth on the "Philosophical Viewpoint of an Astronomer," an old classmate in the audience thought back to their senior year and to many cold nights when they debated Patton's message: warmed by a cylinder stove and dimly lighted by a kerosene lamp, "[w]e sat on the floor and held bull sessions far into the night, discussing the problems of the universe and of mankind. Henry was sometimes there."[64]

Russell never dropped below Henry Fine's horizon. He breezed through the theory of functions, where he was exposed to number theory and number systems, the theory of the complex plane, matrix manipulation, determinants, and the theory of groups. By the end of the course, Russell had filled several fifty-page, four-by-seven-inch, cloth-covered notebooks with techniques for assessing series convergence, making linear transformations, or exploring Riemann geometry. No erasures, cross-outs, or doodling here, even though such normal excursions dotted his other course books; he clearly enjoyed geometry in his musings, since at times one finds in his other notebooks doodles of right circular cones, paths in potential wells, and sophisticated figures of rotation mixed in among computations of solar eclipses and fragmentary notes taken while he was wading through Calderwood's ponderings over Kant's metaphysics. Passages like "Is there a supreme principle of reality [morality] superior to experience?" were festooned with rotating figures in equilibrium reminiscent of Darwin's models. Evidently Patton did not review his students' notebooks, but Fine did. Russell's notebooks, filled in on the recto pages and then backfilled on the verso pages, demonstrate that as he progressed from elementary to general astronomy in his senior year, his involvement increased exponentially. This was due mainly to Young, his favorite professor.[65]

All students learned to make observations of standard stars transiting their meridian to determine time and longitude.[66] Russell spent a good portion of the fall semester of his senior year learning to use the sextant and transit. He also learned how to determine clock error and fundamental star positions. Russell's notes here are sporadic and often perfunctory, and even have some chemistry laboratory notes mixed in, along with sketches of the Moon. Clearly these were the tasks of an apprentice and not the tools Russell wished to collect.[67] Russell always enjoyed ideas and geometrical manipulation more than using physical equipment; observing with a meridian circle and recording chronograph was a right of passage, not a privilege.

Taylor Reed taught the practical courses. Reed, ten years Russell's senior, came from a pioneering family in central Pennsylvania, and like most Princetonians was Scots-Irish, a Presbyterian, and a Republican. He graduated from Princeton in 1886 and became Young's assistant in 1888 when he obtained his master's degree, after two years as a fellow in experimental science and an instructorship in mathematics. He dabbled a bit in micrometric observations but was primarily a tutor. There is no evidence that Young ever directed Reed, or earlier assistants, to any pursuit other than meeting their classes.[68]

In the spring, Russell continued with Reed's lectures and exercises. For the first time he had to use a preciously delicate Rutherfurd grating for solar observations. As far as technique was concerned, Russell noted only Reed's warning: "*DON'T DROP IT.*" He managed not to and carefully recorded the appearance of the solar spectrum, noting the Fraunhofer lines and confirming the character of Young's recently improved solar prominence observations. By the end of

March Russell was making double star observations with a small filar micrometer and also recorded the appearances of stars of different classes of spectra. In April Russell sketched a solar prominence.

Russell was less involved with his Students' Observatory exercises than with his formal coursework. When Young lectured on the Nebular Hypothesis, Russell recorded it as "an explanation of astronomical evolution." Unlike some astronomers, who accepted Laplace's theory of the origin of the solar system but balked at the theory of natural selection, Young held a wholly naturalistic view of the universe. His textbooks, and Russell's notes, comment frequently on continuous development, presently acting and not confined to original cause. Young told Russell's class that there were two views as to how the solar system came to be:

(1) That it was made so on purpose.
(2) That it came about by continuous development and evolution from the past.[69]

To which Young appended the comment that "[t]here is no atheism about the latter view." Young felt obliged to say this, of course, because even though Princeton had been inoculated with McCosh's enthusiasms, under Patton it was again a place where science had to remain vigilant.[70]

Young suggested that his class look to his textbook for reasons why the Nebular Hypothesis was the best model extant to describe the origin of the solar system. He noted only that "the main idea is almost certainly true," but little advance had been made since the days of Laplace in working out the details. He did not mention the efforts of his predecessor, Stephen Alexander, who had captivated popular audiences at midcentury with his evangelical exposition of the Laplace theory, and who used it to support various idiosyncratic ideas about the behavior of nebulae. Young preferred to concentrate on George Darwin, who "has shown mathematically that a cloud of meteors if big enough [would] in the long run behave just like a gas made of molecules."[71]

To Young, the major issues challenging the Nebular Hypothesis were mechanical in nature: how to get the various rings to form out of a contracting proto-Sun, then how to get planets out of those rings, and finally, and most problematic, how to properly distribute the angular momentum in the system. The problem was (and would continue to be for many decades) that the condensation process into planets seemed excruciatingly slow and the transfer of momentum too small to account for the distribution of angular momentum in the present solar system, where the Sun contains the vast proportion of mass, but the planets have the momentum. The Sun's age of some eighteen million years based upon the Helmholtz-Kelvin contraction theory, which provided heat through the expenditure of the Sun's gravitational potential, was not adequate for the timescales required: "[T]he solar system may be much older . . ." Young offered no solutions, nor did he suggest that these inconsistencies negated the theories of Helmholtz or Laplace: "These results not anything like certain."[72] But, as Russell recorded, he was optimistic that somehow answers lay in the direction George Darwin had taken, developing mathematical theories describing the dynamics of rotating liquid figures in equilibrium.

George Darwin would soon attract Russell's attention, but for the moment his world was confined to classwork, exams, and senior projects. He filled his notebooks with predictions for solar and lunar eclipses, which he calculated to a hundredth of a second to test the limits of the lunar theory.[73] Later in the spring, with warmer weather making night duties more comfortable, Russell examined visually the spectroscopic appearances of thirty-five of the brightest stars and several nebulae using a two-prism spectroscope attached to the 9.5-inch refractor. He described in rather pedestrian fashion the appearance of prominent absorption features for the white, yellow, and red star classes, noting the major Fraunhofer lines that defined their classification into the traditional sequences created by Secchi and Vogel. Following what was by then decades-old practice, Russell carefully noted the appearance of lines and bands and the way the bands were fluted, cascading to the red or toward the blue end of the spectrum. As both Secchi and Vogel believed, these differences marked the primary characteristics of the two major classes of red stars, but the physical causes for these differences remained unknown.

Russell did not concern himself with the causes of the different spectra. His job was to describe what he saw, and to turn his observations into his senior thesis in practical astronomy, "Visual Observations of Star-Spectra," which he completed on the last day of the semester. Other than being spectroscopic, Russell's treatment was indistinguishable in character from student reports in the 1880s under Young, which were quantitative calculations of the position of the Students' Observatory at Princeton. Presumably, Russell performed these rituals too, though they have not been preserved in his papers.[74]

Graduation "Insigni cum Laude"

On 9 June 1897 the faculty met to prepare the final honors lists. Russell stood first, followed by John Henry Nichols and Nicholas Stahl, whom Russell had counted among his few college friends. The faculty awarded Russell the rare appellation "insigni cum laude," standing above summa and magna; this, according to some on the faculty, was an honor given last to Aaron Burr, though at least one other was granted in 1891.[75] Russell also garnered the highest honors that the mathematics and physics departments bestowed on students. He was selected to deliver the Latin Salutatory at commencement, which at Princeton then stood above the Valedictory Salutation, the latter given that year by the very popular philosophy major and class secretary John Henry "Pop" Keener.[76]

Russell's Latin Salutatory on the 16th was brief and to the point, though not barren of emotion. He rejoiced that this was the 150th anniversary of the college's founding, and that the occasion had attracted alumni and "illustrious men" from everywhere, including George Darwin and J. J. Thomson, who were invited for the sesquicentennial and to witness the college's transformation into the "new

Princeton University." He properly mourned the passing of his class, reverently recognized professors for sharing their wisdom and friendship, and saluted the Princeton Tiger's honor on the athletic field and in the debating chambers.[77]

In the pomp attending graduation, Henry was splashed over the newspapers from New York to Philadelphia. A special "Illustrated Supplement" to the *New York Tribune* celebrated five top seniors but reserved special attention for Russell:

> the first honor man of the class . . . one of the most remarkable scholars educated at Princeton in many years. He has been the undisputed intellectual leader of his class during its entire undergraduate course, no other member having the temerity to compete with him for that honor after his ability became known during the first few weeks of his college course.[78]

Henry Fine was quoted as saying that Russell was "by far the best mathematician who has ever come under his instruction."[79] The *Tribune* confirmed that Russell was not athletic, nor did he favor any of the competitive aspects of the literary and oratorical life of the college, "his tastes inclining him rather to quiet study and investigation."[80] Russell's hometown Oyster Bay newspaper picked up from the *New York Times's Illustrated Magazine* and the *Tribune*, celebrating Russell's heritage: "That he should have proven himself a remarkably bright boy, comes logically from the fact that he has a remarkably bright father and mother. We believe in an aristocracy of brain and blood."[81] Russell was also featured in the *Philadelphia Public Ledger*, and his name echoed through smaller dailies around the region. This kind of attention may have been manufactured by the sesquicentennial and Princeton's campaign to mark its emergence as a national center of academic excellence.

If Russell represented an "aristocracy of brain and blood" to elitist Long Island north shore editors, he was also the rather awkward, mildly dyslexic stay-at-home boy cherished by the class of '97 as "Our Bright and Shining Star," pictorially celebrated by Keener in the 1898 edition of the *Record*. "For him we predicted a distinguished career . . ."[82]

Russell enjoyed a special place among his 191 classmates. Among them, 41 men declared for law, 23 for business, and 15 each for medicine, civil engineering, and the ministry. There were 117 professed Presbyterians, 34 Episcopalians, 10 Methodists, 2 Catholics, and 1 Jew.[83] Russell claimed allegiance to Clio Hall but was not an active debater. He belonged to none of Princeton's fraternity-like eating clubs and played no organized sports, though he professed to like mountain climbing (as his father did) and Remington bicycles. But when asked his favorite sport, he snapped back, "I'm no sporting reporter." At graduation he identified himself as six feet tall, spread over a lean 140-pound frame, with gray eyes, dark brown hair parted in the middle (the popular style), and ever-present wire-rim glasses. His favorite beauty was brunette, and favorite woman's name Elizabeth—his mother's, of course. He had no nicknames that he admitted to, although his classmates variously called him "poler," slang for a grind, or just plain "Henry."[84] He was devoted to the campus Philadelphian Society, the undergraduate religious organization, read *Harper's Monthly* and the *New York Trib-*

Fig. 2.1. After the publication of his first paper, Russell was celebrated as "Our Star" in *Record of the Class of 1897 of Princeton University* Number 1, opposite p. 46. Courtesy Princeton University Archives, Seeley Mudd Library.

une, and liked Sir Walter Scott as an author and Robert Lowell as a poet. His favorite Shakespeare play was *Julius Caesar*. He played neither cards nor billiards, did not smoke, chew, or dance, was never summoned before the faculty, sent home, or "conditioned," and professed no knowledge of beer. He occupied one room, had no idea what college expenses were "since I live at home here," and corresponded with one girl. And he was generally opposed to bloomers.[85]

The unidentified girl could well have been Lucy May Cole, daughter of a prosperous New York lawyer, who summered with friends in Oyster Bay. Henry had apparently known Lucy for several years; they dabbled clumsily at tennis, according to family lore, playing on a court his father had built behind the manse.[86] Henry would not marry Lucy until 1908. There was, as yet, a lot of ground to cover.

Of course, the "Class Prophecy" was merciless. Henry would soon accept a chair in mathematics at Bryn Mawr College and would succumb "in a few weeks to the wiles of one of the many sirens who had beset him immediately upon his arrival."[87] His classmates, "being proud of him . . . loved to tease him." They would "gather round him as he hurried, overladen with books, across the snow-covered campus, . . . trying on various pretexts to make him late for class."[88]

> Can you not see his slender, boyish figure, topped with a cabman's ear-flapped cap, bottomed with heavy "artics"? And always, in our jollifications at his expense, he entered good-humoredly, himself providing much of the fun by his apt retorts,—a friendly banter right up to the end of our life together.[89]

Henry could handle mild teasing as long as it did not go too far. His nickname of "poler" was apt, for he did study intensively, and this worked to his advantage as a defense. Usually finding himself way ahead of the game as an exam approached, Henry would relax just before the day of reckoning by conspicuously biking or walking around campus, calling to his classmates, "Come for a ride with me in the country."[90]

Although many of Henry's classmates thought he was the most awkward and anxious man in class, they voted him the brightest, and they warmly and generously honored and handsomely supported him in years to come. He had few close friends and hardly socialized beyond the occasional intellectual rump session, though he was an ever-loyal fan cheering on the Princeton Tiger. During summers at Oyster Bay, however, which were, indeed, Henry's time for recreation and recuperation, he would sometimes invite classmates home. Like his father, he favored walking, cycling, sailing, or paddling around in a dinghy, if one was available. One friend who visited was Ed Axson, younger brother to Ellen Axson Wilson and brother-in-law to Woodrow Wilson, who was then a popular professor of political science and jurisprudence at Princeton and one of the most powerful men on campus.[91] August heat waves in Princeton made the Russell home on Long Island Sound a nice place to spend some time. To the Wilson family, Russell was a terribly serious, determined drudge. Axson, a Magie student in physics, had once gained access to the campus telescope, and after finding Jupiter for his older sister, she remarked that he was "as intelligent about it all as Russell."[92]

Russell's recollections of his undergraduate life were less than objective. In 1937, the "dignified, genial, fast talking, white-haired Chairman" told the *Daily Princetonian* that the golden nineties were "truly golden," and life was "simple and autonomous." Princeton was "the best place on earth," and there was no "weekending, little chance to spend money, no automobiles, no Sunday trains, rich and poor living much alike and carriages sinking hub-deep in the mud of Nassau and Washington streets." Russell most likely stayed home on weekends, though not his classmates, since Manhattan was a frequent weekend playground. He even believed that "the club problem which entrenches Princeton in moon-madness was not serious . . . it was Tarkington's Princeton, a self-sufficient Princeton. Most important of all, everyone knew everyone else and we were happy." Russell denied that Princeton was ever a place where loafing was an elective for juniors and a "required course in Senior year." It was, he felt, "just as sound intellectually then as now." But Russell always set his own standards and led his own life; he was hardly representative of his class.[93]

For professors of mathematics and astronomy on the prowl for talent to fill Princeton's new graduate school, Henry was a treasure worth grabbing and holding. On graduation day, therefore, Russell knew he was not leaving Princeton. He apparently had no intention of doing so, for he had elected to compete in a special exam in geometry, calculus, and the theory of functions that won him the J.S.K. Mathematics Fellowship for his first year of graduate study.[94]

• CHAPTER 3 •

Graduate Years: Entering the Profession

The American Astronomical Community in 1900

Russell moved seamlessly from undergraduate to graduate life at Princeton at a time when the astronomical profession was still small and, like other physical sciences in America, was just beginning to organize. The first issue of the *Astrophysical Journal* appeared only during Russell's senior year. A national astronomical society did not exist until Russell was a second-year graduate student. There were then fewer than three hundred practitioners, mostly teachers; the Ph.D. was becoming more common but was still not the only way to enter the profession; and only a few score workers actively published their research in observatory reports, or in a handful of journals. Six observatories possessing large refractors were considered well-equipped "for the highest astronomical research." One dozen observatories possessed refracting telescopes of larger than sixteen-inch aperture, then considered a very respectable size for competitive research.[1] American astronomy was, in large part, parochial and hierarchical, with elementary instruction in the science centered on college campuses that were slowly assuming the status of research universities. The bulk of research and advanced training took place at observatories, many loosely connected to university campuses. Professional training was therefore akin to apprenticeship. An observing astronomer's life was defined by a single instrument, technique, and specialist activity. Specialization was taking hold, creating boundaries. American astronomy still looked to Europe for direction, but its burgeoning observational facilities and mass production of astrophysical data were making it a new world power, just as America found itself testing the temptation of empire.

During his graduate student years, Russell must have been well aware of these events and trends: of the inauguration of the *ApJ*, the dedication of the Yerkes Observatory, and the founding of the Astronomical and Astro-Physical Society of America. Young played a symbolic role in these ventures, endorsing younger rising stars like George Ellery Hale who created all three institutions. Hale built much of the modern infrastructure for observational astrophysics and would eventually offer a new direction to Russell's career.[2] But Russell's initial direction came from Young and Princeton.

Publishing for Recognition

The college's sesquicentennial and the creation of Princeton were accompanied by calls for a graduate college and a formal program.[3] Graduate instruction was sporadic, and few departments provided it consistently. Princeton was a place where "[t]here were very few avid graduate and postgraduate students from whom to draw inspiration and fervor."[4] Most students came for advanced training, not a degree: Young invited Edwin Frost to spend a year with him in 1887, and Frost repaid him by proofing Young's *General Astronomy*.[5] Dayton C. Miller had worked under Young on the orbit of Comet 1889V and may have earned the first doctoral degree in "mathematical astronomy" granted at Princeton, in 1890.[6] Russell's would be the third.

Russell formally entered the nascent graduate college in September 1897. There were just over one hundred students, including Stanley Chester Reese, who as a second-year graduate was the first holder of the new Thaw Fellowship in astronomy.[7] Reese was the only advanced graduate student in astronomy and, like Russell, had won the J.S.K. Fellowship the previous year as a Princeton senior. As J.S.K. Fellow, Russell was officially a student in the graduate Department of Mathematics and Science, which included astronomy.[8] Only when he was advanced to candidacy in the spring of 1899 did Russell declare for astronomy, with mathematics and logic as secondary areas.

Russell plowed through courses on the theory of functions, the theory of numbers and higher algebra, linear and partial differential equations, solid geometry, and higher metrical geometries. Graduate instruction in astronomy consisted largely of directed individual study and research. The only formal course was "Theoretical Astronomy and the Calculation of Orbits," a full-year tutorial based upon Watson's *Theoretical Astronomy*, Klinkerfues's *Theoretische Astronomie*, and Oppolzer's *Lehrbuch zur Bahnbestimmung der Kometen und Planeten*. It was the prelude to a thesis.[9] Young based his tutorial upon Watson's standard text, which he would have his student read in order to construct "ephemerides from given elements and the calculation of elements from observed positions." He would meet with his student "once or twice a week," and mostly examine the accuracy of computations the student had carried out the previous week.[10] Celestial mechanics, or the computation of orbits, was standard fare. The first four Princeton theses in astronomy, starting with Miller's and ending with Charles Lane Poor's, were all studies in the practical application of orbit theory. Russell was no exception; Russell's and Reese's theses both applied basic planetary perturbation theory to study the motions of asteroids.

At the end of Russell's senior year and during his first graduate semester, Taylor Reed left for a sabbatical and was replaced by Edgar Odell Lovett, an instructor in mathematics in the School of Science, who had developed an interest in celestial mechanics. Lovett was a highly charismatic and popular faculty member who attracted admirers like Patton and Wilson. His enthusiasms for teaching and planning equaled Fine's. Lovett's entrance into Princeton astronomy further strengthened its connections to mathematics.[11]

Young and Fine encouraged Russell to explore special problems and carry them to publication, which Russell was eager to do. In his first year Russell examined mechanisms of reducing mathematical functions to graphical form for rapid solution, following techniques he had learned during his senior year. He tried to publish one of them in his first semester; the paper was accepted but the journal failed before his paper was published. By the spring of 1898 he had refined his methods on the basis of Salmon's *Higher Plane Curves*.[12] His first successful publication was in fact a graphical method to solve the orbits of visual binary stars. For Russell, mathematics had become a means to explore astronomical problems.

At the end of the century more than 3,500 visual double star systems were known, and dozens of techniques, both analytical and graphical, had been developed to solve their orbits. Visual doubles were very lethargic, however; periods of revolution could take centuries. Few of them had accurately measured trigonometric parallaxes, so very few systems could be studied adequately to derive masses. Young's 1895 *General Astronomy* noted that only about fifty systems had been analyzed, and only six had reliably known masses. In Russell's time, graphical methods by Klinkerfues, Thiele, and Glasenapp were popular, as was an analytical solution by Kowalsky, which required considerable calculation and was sensitive to observational errors.[13]

Russell looked for a way both to simplify the reduction procedure and to make it less sensitive to observational error. He knew that Klinkerfues's method was the fastest, but Thiele's method required less computation, so Russell combined the virtues of the two. He circumscribed the true elliptical orbit by a circle and used it to convert the observed orbit (a projection on the sky) into the actual orbit. The beauty of this idea was that as soon as the projection of the circumscribed circle was drawn, the major axis of the true orbit was known. Russell made one observation of the well-studied system Eta Cassiopeia with the 9.5-inch equatorial, combined it with older observations, and then used his new method to demonstrate that it equaled Thiele's and Klinkerfues's in accuracy but was faster, "as the graphical work is very simple."[14]

Russell thought he had made a real contribution, and wanted to announce it quickly, so in the spring of 1898 he sent versions to both the *Astronomical Journal* and *Popular Astronomy*. His class recorder noted that "[t]he discovery has been favorably commented upon by Professor Young and other astronomical experts."[15] But within weeks of submitting his papers, Russell learned that he had been scooped by H. J. Zwiers, an assistant at Leiden University Observatory, who had published essentially the same method two years earlier as the lead article of the February 1896 issue of the *Astronomische Nachrichten*, a prominent publication central to the discipline. Even though they knew of Zwiers's work, the editors of the *AJ* did not hesitate to publish the young American's effort.[16]

Although Russell later claimed that the issue of the *Astronomische Nachrichten* containing Zweirs's paper was at the bindery when he created his own method, the central library cataloged the thick bound volume (containing issues as late as 15 June 1896) in 1902, and it is unlikely that the tome would have been at the binder's during the entire interval. Of course, Russell had barely completed his junior-level course in descriptive astronomy when Zweirs's paper appeared, and there is no evidence that he ever systematically scanned back issues of journals. At the time, moreover, astronomers routinely ignored or failed to notice recently appearing work, and the American editors definitely wanted to support a promising student of C. A. Young. Even so, as Russell soon realized, Zwiers's method provided closer comparison of theory and observation. Although Zwiers maintained priority, later reviewers usually mentioned Russell's effort too, but this was gratuitous.[17]

Despite the scoop, Russell's classmates celebrated his publication in the *Astronomical Journal*: this is why class secretary Keener drew a star around his graduation portrait for the 1898 *Record*. It speaks not only to the affection his class had for him but to their confidence in him. Young remained wholly confident in Russell as well and urged his precocious student to apply for the Thaw Fellowship in Astronomy for his second year, which was tantamount to declaring himself for astronomy. Young had, however, failed to keep track of the official deadline date for application and barely got Russell's papers in on time. Russell received the Thaw Fellowship when he passed his master's degree examination in June 1898.[18]

The extra work Russell put into research beyond his regular tutorials took its toll. But so did a tug back to mathematics from Henry Fine. The issue at hand during his 1898 summer was whether Russell should teach a section of freshman algebra. Fine, summering in Great Barrington, Massachusetts, wrote Russell in Oyster Bay sometime in July proposing that he do some teaching to help broaden his résumé in anticipation of a future job search: "If you are successful, the experience will be immensely valuable to you and give you a new confidence in yourself and us a new confidence in you." Fine also assured Russell that he would be protected; if Russell could not handle the "incipient disorder," Fine would take his class over personally, and it would never be recorded that Russell had failed. "You will know something about yourself by that time. On the other hand, if you do no teaching this year before you go abroad for study we shall have to say: 'His scholarship is admirable, but we cannot say how good a teacher he is likely to make.' "[19]

Russell wanted Young's opinion, because the Thaw Fellowship was for research. Young agreed that Russell needed teaching experience to land a good job but was more concerned about Russell's health. Under normal circumstances, teaching freshman algebra was a trivial exercise. But Russell's state of mind, in Young's opinion, was anything but normal:

> I have no doubt that *if* you can do it without drawing too much on your strength and nervous energy, the experience will be extremely useful to you. But you will be likely to find the work pretty trying until you can learn to adapt your rate of talking to the rate of comprehension of an average boy, and can acquire a more quiet and "tutorial" manner than you are likely to possess at first. I won't preach though.[20]

Young did not preach, but he was crafty: "On the other hand," he added, "note my *if*." Young worried that Russell's state of health was too fragile "to warrant you in undertaking any extra work, and that of a new kind likely to be *worrisome*, if not really very laborious. You had better not undertake anything that might unfit you for your next year's work abroad. But as regards this you must take the best medical advice."[21]

Both Young and Fine clearly worried about their bright fragile boy. Ever the deadly serious drudge, rushing across campus, being teased by his classmates, Russell was no match for the typical obstreperous freshman. He could take only so much teasing before he would break. At home, his boisterous brother Gordon had often tested his limits; in adult life, Russell never allowed teasing in his own home: "he had no tolerance" for pranks and practical jokes.[22] Also in later life, when faced with career choices, Russell drove himself to deep depression and self-doubt. He literally tortured himself when choices had to be made. Young, more than Fine, must have seen this coming and tried to avoid a collapse, especially since both fully expected Russell to travel abroad in the following year.[23] Fine was very strong on foreign study; it was considered the elite path into a professional life in science.[24]

Russell, however, stayed at Princeton for the full three-year term. He did not take up Fine's offer to teach and spent the next academic year engaged in research and preparation for the thesis. Others in his class were tapped by Fine as instructors in the next two years, including O. D. Kellogg, who was then just a senior. Kellogg went on to complete his Ph.D. in Europe in 1901 and then returned to Princeton as instructor for the next three years. Russell's path would not be so smooth.[25]

According to his publication record, Russell's attention ranged from descriptive studies of planetary detail to the elements of minor planet orbits to the determination of the densities of stars in eclipsing binary systems. They all reveal Russell's emerging style.

The Venusian Atmosphere

Through the winter and spring of 1898–1899, Russell sporadically observed Venus, Mars, and Jupiter with Princeton's telescopes. He knew that Venus possessed a deep atmosphere because when it was observed at inferior conjunction, its crescent extended farther than would be expected if it were a bare sphere like the Moon.[26] Russell wanted to determine how deep the Venusian atmosphere was, and where the solid surface lay by performing an analysis of the extension of the cusps of the crescent. Chastened by his recent experience of being scooped, Russell took great pains to search out historical records; he soon satisfied himself that he had something new to offer. First, he found that earlier analysts had neglected the fact that the Sun's rays were not parallel but were a function of the apparent solar semidiameter. Those who did apply a correction otherwise made some type of numerical blunder. And second, no one till Russell had taken the trouble to collect all known observations into one consistent analysis, or to consider what happened to sunlight physically as it passed through the Venusian atmosphere.[27]

Measuring the extension of Venus's crescent requires viewing the planet as close to the Sun as possible. Young helped him mask the 23-inch Halsted lens to cut glare when Venus was within a degree of the Sun, and Russell found that the cusps projected some thirty-seven degrees beyond the theoretical limit for a solid sphere. He even glimpsed "the complete circle of the planet."[28] His analysis could distinguish among the separate influences of refraction, absorption, and scattering. Using a model based on Earth's twilight characteristics, he concluded that scattering and absorption dominated over refraction. Venus's atmosphere was therefore "hazy."[29] Russell made everything fit quite nicely, and this time he was not scooped.

Russell had managed to synthesize diverse sources of data and generate a theoretical model. Given what little was known of scattering theory in that day, Russell did about as much as anyone could; at that point he was in no position to add to theory but competently applied simple geometrical arguments and analogies based upon observations of Earth's atmosphere to come to reasoned conclusions about Venus. Although the parameters predicted by his model were found years later to be way off the mark, he was the first to argue correctly that scattering, not refraction, was largely responsible for the extension of the cusps.[30]

With Young's endorsement, Russell's Venus paper appeared without revision in the *ApJ* in May 1899. He later looked back on this paper as his "first independent contribution."[31] Closer to the time of publication, Russell reported to John Henry Keener, class secretary, "I confess that I took up work relating to Venus of my own free will." The "romantic titles" in his vita, Russell feared, which by then included notes on predicted places for the recently discovered minor planet Eros, should not be misconstrued as desires for the comforts of flesh: "I am really not to blame for that."[32]

Romance was probably far from Russell's mind in the spring of 1899 as he prepared for examinations to advance to candidacy, worked steadily on Eros, and, after his exams were over in May, studied the influence of irradiation on the apparent size of the Jovian disk.[33] In June, since there were apparently no second-year graduate students in astronomy following Russell, he received the Thaw Fellowship again, and Stanley Chester Reese was awarded the Ph.D. for his thesis on Jupiter's gravitational influence on a minor planet.[34] Reese's thesis was in some respects a preamble to Russell's, the following year, on Eros, but it was not carried to such great lengths. Russell's 1899 summer, however, was devoted to neither the wiles of Eros nor to Venus, as far as the record shows, but to the densities of a fascinating type of variable star patterned after a star the Arabs called the Demon.

The Densities of Algol Stars

Once every sixty-eight hours, Algol "winks," losing some 80 percent of its brightness for several hours; otherwise it behaves normally. For over a century, speculation was that Algol was eclipsed by a dark body in orbit around it, and by the time Russell took notice, Algol was a well-studied object and was, by consensus, an eclipsing system.[35] Techniques for orbital analysis of eclipsing systems by E. C. Pickering, François Tisserand, Hugo von Seeliger, and others required many hours of painstaking, tedious calculation. Russell was not interested at the moment in improving their methods; he wanted to find a shortcut to determine the densities of Algol-type stars, a commodity that had drawn much attention because density was an indicator of the relative ages of stars.

Stars were believed to form out of clouds of matter and condense through self-gravitation, so the denser stars were believed to be older. If one found a way to determine densities, this could help decide the temperature history of stars: did stars, as they aged, heat upon contraction, or cool? At the time, how density varied with age was not a debated point, but how temperature varied was unsettled.[36] Russell must have been alerted to the problem during Young's lectures, but he definitely knew of it in 1899 when Young publicly charged that a new law T.J.J. See claimed for his own was in fact the formulation of J. Homer Lane and had been known for decades.[37] "Lane's law," as refined by Ritter and Lord Kelvin, suggested that temperature in a gas sphere was inversely proportional to radius; hence as a star contracted under its own gravitation, its surface temperature would rise. Thus knowing the relative densities of stars and equating these to their relative temperatures was tantamount to determining their temperature histories and served as well as an empirical check on Lane's law.[38]

At home in Oyster Bay in the summer of 1899, Russell developed the idealized case for spherical bodies and circular orbits, where the gravitational analysis depended only upon density and time (the dimensions of the bodies dropped out neatly). He compared the orbital period and length of eclipse in seventeen Algol-type systems to derive an upper limit for mean density, the limit being when the two stars eclipsed each other through their full diameters. Russell sent his manuscript to Young, who told him that the basic idea had been worked out by James Clerk Maxwell. The idea was not new, Young advised, but "your application of the principle is new and well worth printing."[39] Accordingly, Russell was careful to note how his results compared to those of others who had preceded him in the effort, such as Vogel and Dunér, and discussed possible errors in his idealized model. His conclusion was also not new: "[I]t is evident" he argued, "that the Algol-variables as a class are much less dense than the Sun, probably less than one fourth as dense."[40]

Appearing immediately before his paper in the *ApJ* was a similar effort by a South African. Alexander Roberts had submitted his paper six months before Russell's and provided a somewhat less elegant treatment. His Algols, three of which were on Russell's list as well, were also less dense than the Sun. His result was "striking, but it is in no way new," because it supported the prevailing view that stars cooled as they aged.[41]

Both Roberts and Russell expressed doubt that the densities derived for any one system were meaningful. Russell's selection of candidates, however, indicates that the systems he chose had spectral classes generally earlier (bluer and younger) than solar-type stars. He found no correlation between temperature (qualitatively based upon Harvard spectral class) and density for his individual systems, but the two systems with the latest (or oldest) spectral class also had the highest densities relative to the Sun. Russell could have noted that his Algol densities did show a weak correlation with advancing age, but he chose not to speculate.

Once again, however, Russell had been scooped. It was a hot field, and by now one can see that such things attracted Russell. He was struggling to become a player so was doubly unsettled when he was not mentioned by name by Hugo von Seeliger, who wrote a short review criticizing Roberts for not properly citing earlier work by himself and others.[42] By then, however, Russell was already advanced to candidacy, declaring for astronomy with subsidiary topics in mathematics and logic. He had completed all requirements, including preliminary exams in his three major subjects, as well as in French and German. Only his thesis remained.

Eros

His choice of thesis—an analysis of the orbit of the minor planet Eros—was not remarkable, but his execution was. As he later claimed to Keener, he was working on Venus's atmosphere when he heard about "a new and interesting asteroid," whereupon "I rashly started my thesis; and then the discoverer of the thing named it Eros."[43] Eros was, indeed, "[t]he astronomical sensation of the year." Soon after its discovery on 13 August 1898 and its provisional designation DQ, as the 433d minor planet discovered, astronomers realized that Eros's mean distance from the Sun was less than that of Mars, and its highly eccentric orbit brought it very close to Earth. This meant that Eros could be used to triangulate the distance between Earth and the Sun, the solar parallax.[44]

Using published sources, Russell joined a growing crowd of computers who provided provisional ephemerides and predicted positions.[45] By early 1899, DQ had been named Eros, and international cooperative plans were afoot to observe it to refine the solar parallax. Russell soon followed his first paper with more predictions good through June 1899.[46]

Russell could not obtain his Ph.D. for calculating ephemerides. But just as Reese had studied the perturbations of a minor planet by Jupiter, Russell developed a refined model for Eros's orbit, including the gravitational effects of the nearer (and therefore more influential) planets. As he noted in the manuscript draft of his thesis, perturbations by Mars provided the most interesting case study because their orbits interlocked.[47] Calculating such perturbative effects not only would lead to improved predictions for positions but would provide refined knowledge of the masses of the perturbing planets themselves. Russell at first planned to complete a full study of all the effects of Mars's perturbations on the motion of Eros, specifically how it caused all the orbital elements of the minor planet to deviate from ideal Keplerian motion in a central force field. But at some point he realized, or was told, that this was far too ambitious for a thesis, and so he settled on how Mars influenced the major axis of Eros's orbit, and corresponding terms in the mean longitude.[48]

Russell addressed what is now called a restricted "problem of three bodies." He derived the perturbative influence by Mars using Fourier series expansions, but he soon found that they would not converge to provide a neat solution, which threatened to increase the amount of computing "fully fivefold." He therefore decided to "use some method of analytical expansion whenever practicable," such as mechanical quadratures, a standard device for the approximate evaluation of mathematical functions using either interpolation or least squares; Russell chose Urbain Le Verrier's method of interpolation, a standard procedure.[49]

His analysis exhibited an economy of computation that also characterized his Algol and visual binary work. He paid close attention to sources of error, seeing no need to create a theory of motion that went beyond the accuracy of the observations. He therefore neglected small long-period terms "whose determination did not seem worth the labor until the elements of Eros are known with greater certainty than at present," and discussed how his numerical approximations still yielded a theory capable of handling the best observational data "without a prohibitive amount of labor."[50]

Russell did not decide upon his course of action all at once. He tried various methods and frequently updated his observational data. His fragmentary computational notebooks reveal many weeks spent calculating long series expansions, much of it crossed out, some lightly, some heavily as if in frustration. His final calculations ran some thirty pages.[51] And as he labored away in the winter and spring of 1900, once or twice he went to the telescope just to look at the sky for relaxation. In March he observed the Moon, searching for features visible by reflected earthshine. Russell also took up another venture, which was to last for the next forty-three years: he started writing a monthly column for *Scientific American*. His first signed effort was the "Heavens in April," which appeared as he completed his thesis.

Finally, by late April, the deed was done. Russell presented his thesis, and fifty dollars, to his readers, Young and Lovett, who declared it "acceptable." On 9 May, Russell defended his work in the observatory classroom. He faced Magie, who presided in the absence of President Patton, and a panel consisting of Young,

Lovett, Fine, and Hibben. Questions ranged beyond his thesis to astronomy and physics in general and, with Fine and Hibben, to mathematics and logic. Russell prevailed summa cum laude, a new distinction the faculty conjured up for the occasion.[52]

Russell never lived down the fact that he had worked on an object called Eros. His anxieties and denials to his classmates only inflamed their fun; a 1933 account called his thesis "[t]he greatest embarrassment of Dr. Russell's career," after which he found himself a world-renowned authority "on that intriguing subject . . . He still blushes when he thinks of it." For someone who professed to harboring a "Puritan conscience," who in adult life did not drink coffee or tea, smoke, or indulge in any way beyond a bit of sherry before dinner on special occasions, any association with Eros, no matter how innocent, was discomfiting.[53]

Graduation

At graduation Russell had nine technical publications under his belt and was a monthly contributor to *Scientific American*. He had completed what was a traditional course of training in astronomy, heavy on the mathematics, and including a smattering of physics and practical work at the telescope. He had felt the rush and the challenge of competition and found his research anticipated by others. And he had completed a thesis that was very much in line with those Young had directed in the past, even though it was far and away the most ambitious.

It was Young's policy, and the nature of the profession, that astronomers must demonstrate mathematical proficiency.[54] Russell's work on Eros was the best and most visible application of the traditional art and took nothing away from his marketability. Although spectroscopic astronomy had certainly gained prominence by 1900, it remained largely descriptive or required instrumentation that was rare at teaching facilities. By 1900, Young's spectroscopic equipment was not competitive; he was well behind the spectroscopic curve defined at the Harvard, Yerkes, and Lick Observatories. And the "new astronomy" still had its vocal detractors, including Seth Carlo Chandler, the editor of the *Astronomical Journal*, J. R. Eastman, and Simon Newcomb. Unless one had access to radial velocity data, there was little one could do in spectroscopy to demonstrate facility in mathematical computation, which was the core of the profession.[55] Even the most ardent supporters of astrophysics, such as the chronicler Agnes Clerke, admitted at the turn of the century that "[w]hile its conclusions are of vital importance and of intense interest, they result from deductions in which the premises are insufficient, and are proportionally uncertain."[56] Finally, the Eros work best utilized the training Russell had received under Fine, who remained very interested in his best student.

Three days after his exam, Russell sent a note on the perturbative influence of Earth upon Eros's orbit to the *AJ*, a discussion left out of his thesis. He found that it would be very sensible in time, amounting to a shift of its mean longitude by a full degree in a century. He showed how continued observations of Eros would lead to an independent determination of Earth's mass, which would then provide a nongeometrical method for deriving the solar parallax. From Oyster Bay, Russell wrote up a second paper on his general method.[57] It took him some time to wind down, and when he did, he crashed.

Russell quickly left Princeton for Oyster Bay after graduation in June 1900, without, apparently, any plans for postdoctoral study in Europe. In fact, his future was very unclear. Despite Fine's advice, Russell had not yet taught a course, gave no indication of wanting to, and had no experience at some prestigious foreign university. In May, he had accompanied Young as part of a nine-person Princeton solar eclipse expedition to North Carolina and acquitted himself nicely, providing timing predictions. What else he did, or what plans he made, remain unknown, but evidently he was not in the best shape.[58] We can only speculate without documentation, but it is unlikely that Russell or his family ever considered alternative careers for him, such as the ministry or even the application of his skills in business. Military service was probably not an option. In those heady expansionist days since Teddy Roosevelt led the Rough Riders up San Juan Hill in July 1898, Oyster Bay folk like the Russells must have been more interested than most in the aftermath: subduing the Philippines, which took several years. Although Russell's father clearly despised militarism and the trappings of war, as his 1892 trip diary makes clear, Henry was always partial to Kipling, especially "The White Man's Burden," which appeared in the United States in *McClure's Magazine* in February 1899 just as the Philippine-American War began. Kipling did not hide the great human cost of empire, a factor not lost on Russell's father, but the message brought home to Henry was more likely one of high moral purpose and righteous action. This was also a time, after all, when his home life was less than tranquil, for the pastor and the presbytery of Nassau once again had taken up the issue of the revision of the Westminster Confession of Faith.[59]

In his report to the class secretary in January 1901, anticipating his class reunion in June, Russell was cryptic as usual about his hopes and desires. He likened his graduate years to those experienced by "the hero of Booth Tarkington's first masterpiece—the Senior on the cover of the 'Tiger'—in his struggle for the elusive diploma." This trial had led him to his present state, "at home on Long Island, resting by doctor's orders, and trying to recover from the fatigue incident to the first three . . ." Russell admitted that "while it is not altogether delightful to be laid on the shelf, still, Oyster Bay is a pretty good sort of a shelf, and I am not so flat on my back that I cannot enjoy life." The self-image Russell projected, evoking the masthead Tarkington had created for the third year of the *Princeton Tiger* in October 1892, was hardly what the Board of Editors had in mind. Though the freshman had a bit of a struggle to catch hold of the rope, once he had it in hand, he enjoyed a rather smooth progression to manhood and emergence as a dapper, sophisticated member of society.[60] Russell evidently saw a constant struggle, a herculean effort that made heroes out of boys. The effort

Fig. 3.1. Cover page from the *Princeton Tiger* for 28 April 1898. Russell viewed the depiction of the senior as struggling for the "elusive diploma." He expressed no outward concern for the Spanish-American War, however. Courtesy Princeton University Archives, Seeley Mudd Library.

had put him out of commission for more than six months, seeing doctors in New York City. He was then about to leave on a three-month recuperative trip to Italy with his mother, starting in Naples and Capri, and then moving on to Rome. "I hope that when we return I shall be on my feet again, and able once more to keep up with the procession." Russell, pulled in two directions by mathematics and astronomy, trying to satisfy both muses, had suffered beyond exhaustion and broke down. He had a lot of thinking to do about his future.[61]

• CHAPTER 4 •

Postdoctoral Years at Cambridge

Resting, Writing, and Thinking about Binary Stars

When Taylor Reed left Princeton for a job at General Electric in 1901, Young asked around about a replacement. Naturally he looked for a Princeton man. He asked about Stanley Reese, then at Yerkes, but Edwin Frost gave him bad marks. "I suppose," Frost mused in March, "that young Russell is too fresh for the present."[1] Frost was right. Russell was spending much of the year "resting at home and traveling to recover health broken by overwork." He was then recuperating in Capri, tended to by his faithful Aunt Ada, who had replaced his mother at the last moment to be guide, companion, and nurse. They would soon move on to Rome and were not planning to return until late April. He managed to attend his class reunion in June, where he stood sullenly alone and to the side for the group photograph, though he did speak on "Ninety-Seven Everywhere under the Sun." What else he did that that summer and fall in Oyster Bay is unknown. He continued to write and publish, however, and that record survives.[2]

Russell's Princetonian ties were responsible for his association with *Scientific American*. The Munn family, prominent patent attorneys in New York City, had owned the magazine since the mid–nineteenth century and were staunch Princeton loyalists. Charles A. Munn hired Russell in the spring of 1900 to replace Garrett P. Serviss, giving the young graduate student an outlet he evidently relished. The column started as a simple description of the heavens, adorned by a map, but soon Russell started to report on progress in the field. He wrote about Eros, noted favorable oppositions of Mars, and mused about the Martian canals and the life that might exist there. Russell's columns reveal not only considerable sensitivity to sky lore and popular observing technique but a strong mission to present science as a consensual process. He wrote many of his columns from Princeton, which was still very much part of his life. Russell's association with *Scientific American* grew with time. In 1926 Orson D. Munn called him "an associate editor" who "handled, supervised or prepared all the astronomical and astrophysical material that has appeared in the magazine." The relationship lasted for forty-three years, through three generations of Munns.[3]

Scientific American became Russell's therapeutic outlet, but he did not abandon deeper explorations. During the summer and fall he examined the theoretical behavior of a self-gravitating sphere of gas, developed an improved technique for determining the orbits of spectroscopic binary stars, and analyzed several systems. These works demonstrate Russell's interest in using his analytical and intuitive talents in astrophysics and celestial mechanics to study the evolution of stars and stellar systems, an interest that ultimately led him to Cambridge University and George Darwin.

Lane's Law Revisited

In November 1901 Russell filed away a set of calculations he had been working on that dealt with the cooling of a gaseous sphere, noting only that he had stopped the work "on account of the uncertainty regarding anticipation by others."[4] Likely sitting at home in Oyster Bay watching the leaves turn brown and the landscape decay into the uniform gray of a Long Island winter, he found himself in the right frame of mind to consider the ultimate heat death of the Sun and stars. He had been toying with one of the great questions of astronomy in the last half of the nineteenth century: how the Sun and stars are born, how they live their lives, and how they die. There is no question that he knew he was retracing steps taken by J. Homer Lane, August Ritter, and Lord Kelvin, who explored how density, pressure, and temperature varied with radius in a gas sphere in convective equilibrium. Lane showed in 1870 that temperature varied inversely with radius, which Russell appreciated when he explored the densities of Algol-type stars.[5] Now Russell tried to extend the theory of convective gas spheres and the finite energy store of the Sun and stars.

He explored the behavior of a spherical mass of homogeneous gas undergoing contraction where no heat is gained or lost to the system, and developed equations to describe how the temperature and pressure in the gas varied with distance from the center of the star. The equations he derived were at first complex and cumbersome, and led him to seek approximations that would simplify the expressions and provide him with a physically interesting result, expressed in terms of the ratio of specific heats and the mean molecular weight of the gas. He carried his derivation through to the point at which he began to consider conditions where heat might be gained or lost to the system, but did not complete his argument.

Russell's little packet of calculations lacks a single citation. He could have written them out while in Italy and well away from an astronomical library, but he was back in Princeton for his yearly reunion with his class in June and had plenty of time to read the literature before he decided, in November 1901, to close down this line of inquiry. He might have been influenced by Young, who lacked any conviction that Lane's law actually did apply to real stars like the Sun. He knew that a student of Kelvin's had concluded that the total lack of knowledge of the internal character of the Sun and stars, especially knowledge of the ratio of specific heats or of the applicability of the perfect gas laws, made it impossible to say anything quantitative about the internal structure and lives of stars. Russell may well have been trying to address such questions, but, realizing that it was really not going anywhere, dropped the exercise for one more promising.[6]

Binary Orbits and Evolution

In late 1901 and through early 1902, Russell became interested in spectroscopic binaries. These systems, known to exist since the late 1880s, could not be seen visually as double but exhibited either one or two sets of spectral lines that shifted periodically back and forth in a stellar minuet. Spectroscopic binaries became popular objects with the application of photography to spectroscopic observation and because they were short-period systems—from days to weeks. Moreover, because they had very short periods, "these are precisely the systems of highest cosmogonic interest," Agnes Clerke proclaimed in 1903, since they were believed to be "at the outset of their evolutionary careers."[7] More and more spectroscopic binaries were being discovered each year, and it was hoped that they would form an evolutionary link to visual binaries, thought to be older because of their longer periods.

Russell knew that existing reduction methods were sensitive to the quality of the radial velocity measures of orbital motion. The popular graphical method of Rudolf Lehmann-Filhès in Berlin required precise knowledge of the points of maximum and minimum orbital velocity, which were not always observable. Russell therefore devised a fast and efficient technique that used data equally from all portions of the velocity curve. As he had in his previous work, Russell applied Le Verrier's methods for handling series formulae to represent the entire velocity curve in terms of a simple trigonometric series. Russell was not the first to try this but was the first to carry the analysis out using higher-order terms to handle highly eccentric orbits, systems that he believed would show the greatest evolutionary changes and were among those most commonly observed. He concentrated on ways to evaluate the higher terms as simply and efficiently as possible. Admitting that his analytical procedure took a bit more effort than the graphical method of Lehmann-Filhès, he argued that it was more accurate and could handle a wider range of orbits.[8]

Russell tried out his method on a few well-observed systems and immediately ran into trouble because observers of spectroscopic binaries rarely published exact times of observations. Russell needed to know the hour, not simply the day. He interpolated, assigning errors for the uncertainties, trying to fit the observations to a smooth curve, but never satisfied himself that he had overcome the problem. Nevertheless, in mid-April he sent a short manuscript to the *ApJ* reporting on his results and lambasting observers for failing to provide accurate timings.[9] The observers he attacked were powerful gatekeepers: H. Deslandres at Meudon and W. W. Campbell, the Lick Observatory director.

George Ellery Hale wisely rejected Russell's manuscript, suggesting that he write directly to Campbell or Deslandres rather than criticize them publicly.[10] Russell accepted Hale's wisdom and wrote Campbell. He was now most interested in the peculiar Zeta Geminorum system, which was on the Lick observing program. In May 1902 Campbell agreed to take observations for Russell when the system was again in the skies. He was aware that "the orbits must undergo considerable change owing to the great perturbations that evidently exist." He asked Russell about tidal effects, evidently knowing little about how to handle them himself.[11]

This was Russell's entrée. Tidal effects were not important, he told Campbell. What was really going on, he suspected, was "an actual example of the periodic solutions of the problem of three bodies which have been discussed by Darwin and Poincaré." The odd light variations could be due to the pulsation of one of the components, he explained, following a suggestion by Johnstone Stoney, who had argued that the light curves seen in some globular cluster variables indicated "free vibration nearly equal to that of the forced tidal oscillations."[12] Russell's intuitive explanation was on the mark, since the brighter component of Zeta Geminorum is a Cepheid variable. At the time, however, the pulsating model was hardly popular and would not be so until Russell's student Harlow Shapley revived it.

What Russell really needed, beyond Campbell's data, was a better understanding of theory, which he hoped to gain soon: "I hope to spend next winter in Cambridge under Professor Darwin," he told Campbell, "and, if the problem is too much for me with my present methods, I may attack it again then." Campbell knew that there were too few American astronomers capable of handling a full dynamical analysis of these puzzling systems, so he was happy to encourage Russell to study with Darwin.[13] By the spring of 1902, Russell was ready to resume his career. In the two-year hiatus between the expected date of his foreign study and when he actually went, Russell had gained an agenda.

Move to Cambridge

Cambridge was a clear choice for Russell. Wilson and Fine were both Anglophiles, as was Russell's father. Darwin and J. J. Thomson had been leading lights among scientists attending Princeton's sesquicentennial. Most attractive was Darwin. Fifth child of Charles Darwin, Cambridge wrangler, fellow of Trinity, Plumian Professor of Astronomy and Experimental Philosophy at Cambridge since 1883, and immediate past president of the Royal Astronomical Society, Darwin was the leading student of the application of the theory of the forms of equilibrium for rotating, self-gravitating fluid masses, which was just what Russell was looking for. Cambridge opened up a new world for Russell, and indeed he fully intended to "settle in." After his first year as an advanced student, during which he took in as much as he could of Cambridge life, his attentions centered more and more on the observatory out Maddingly Road, where he loved to walk on the weekends. There he remained on and off for two more years as an "honorary research assistant" with a strong desire to continue as long as possible.[14]

It appears that Russell entered Cambridge formally, but at the very last moment. His record of application to the registry of the university postdates his arrival and was written on registry stationery. He embarked for England in August 1902 armed with letters of introduction from Patton and Wilson as insurance in light of his late entrance. Patton, always cautious, suggested that Fine write directly to J. J. Thomson, and that Young write to Darwin. Wilson and Patton expressed great fondness for and faith in their Henry and assured the Cambridge dons of his high standing. Russell's brief application stated that he had come "with the intention of pursuing research in Gravitational Astronomy and kindred studies and having in mind at present especially the problems presented by the system of Zeta Geminorum." He presented his credentials and copies of five original articles, offering them "as evidence of fitness to pursue the subject of astronomy." Wilson sent Russell on his way with his "warmest good wishes" and identified Russell to the Cambridge registrar as a personal friend and "as one of the most capable men of recent years at Princeton."[15]

Thaw Fellowship funds had been saved in 1899 for Russell's travel. He also had support from his mother and from Aunt Ada, who accompanied him to Cambridge and helped him get set up in his rooms at 69a Trumpington Street. He entered King's College, where Robert S. Ball, director of Cambridge University Observatory, was a fellow. As always, Aunt Ada eased his way and allowed him to concentrate on the job at hand. Aside from travel and lodging, there were no heavy fees. He was formally admitted into King's as an "Advanced Student" in mathematics, a position that allowed him access to courses of advanced study and research.[16] By then he had already attended Darwin's first lecture on the calculation of orbits, and Ball's second lecture on planetary theory, deeply regretting that he had missed the first lecture. During his first year, Russell attended lectures by Ernest W. Hobson on spherical harmonics, by E. T. Whittaker on general dynamics, by A. R. Hinks on photographic reduction techniques, and by Darwin on the calculation of orbits, dynamical astronomy, and the figures of equilibrium of a rotating body.

Russell encountered Cambridge when, as he recollected years later, "Cambridge was no tiny place": "J. J. Thomson was in his glory. Rutherford had just come, and James Jeans was the youngest member of the trinity. Arthur Eddington was an undergraduate . . . I found that though I had come from a small place to a big one I was perfectly able to carry on."[17]

Russell came to know Eddington and Jeans a bit then, and coursed through Cambridge's colleges making friends and spending evenings singing songs and learning limericks that he would gleefully recite ad infinitum years later in his acquired Anglophilic accent.[18] He was invited to meetings of the newly established $\nabla^2 V$ Club and attended its fifteenth meeting on Friday, 2 December 1902, held in Owen W. Richardson's rooms at Trinity, where the group discussed E. T. Whittaker's solution (presented at the previous meeting) for $\nabla^2 V = 0$, the mathematical expression for an inverse-square force field in the absence of a source. Russell started attending as a visitor but was elected to membership in May 1903, having been nominated by H. A. Wilson, who was interested in the theory of the convection of heat. In November 1903, Russell hosted a meeting and spoke on "stellar statistics." This event garnered only a cursory note in the club's minute book, in contrast to the reaction after the twenty-fifth meeting in Jeans's rooms in the cloisters at Trinity on 12 May 1904, where he spoke on the atom "as a system of electrons in statistical equilibrium." Jeans's remarks stimulated a "long and interesting discussion," the highest accolade of the recording secretary. The club was a social exercise in mathematical physics and later became one of Eddington's outlets; it was clearly a place where Jeans's elegance was appreciated more than Russell's approximate methods.[19]

Cambridge, Russell also reminisced in his "Spiritual Autobiography," "was my introduction to the greater world": "Since I first settled there, England has never been a foreign land to me. The unvarying courtesy and kindness which I received from everyone, in my college (King's), from my teachers, and from a growing circle of friends, made this automatic."[20] He regarded Cambridge as "an expansion of previous experience." For the first time, he came to know people "of a wide variety of beliefs with which I had previously had little acquaintance." Anglo-Catholics, agnostics, Tories, all were new to him, and there were north-country Quakers like Eddington, who joined the $\nabla^2 V$ Club on 1 November 1904. He encountered Jeans, who did not return to Cambridge from an enforced medical absence until the Michaelmas term 1903. Russell soon found others who shared his regard for Darwin and his convictions, which were "clarified and strengthened by later developments in physical science."[21]

Darwin applied the power of mathematical physics to problems in geology, producing "a new strand of applied mathematics and physics of the earth . . ."[22] Starting in the late 1870s, Darwin examined how the tides could influence the evolution of the Earth-Moon system. His major conclusion was that in past time Earth and the Moon must have been closer together and, originally, were one body which, through rapid rotation, had split apart.[23] Darwin, like the French mathematician Henri Poincaré, studied equilibrium configurations of rotating fluid bodies, focusing on how their shapes were changed by rotation, searching for conditions promoting stability or rupture. His "fission theory" of the origin of the Earth-Moon system dominated for over a half century. It had a physical timescale built in, so Darwin was able to use it to date the age of the earth, eventually convincing Kelvin that his severely short timescales had to be revised upward.[24]

Russell recognized in Darwin's methodology the essence of a truly new astronomy—not only a spectroscopically based set of new observational tools, which constituted the new astronomy for people like Campbell and Young, but a wholly new conceptual framework upon which the physical universe might be described, tested, and rationalized. Thus one sees among Darwin's followers and in his brand of mathematical physics one of the main threads that would someday constitute theoretical astrophysics. Above all, Darwin's approach was physical, as opposed to the more mathematically pure styles of Poincaré, G. H. Hardy, or Ball. As Darwin's sometime associate and student S. S. Hough observed in 1914, "Darwin's orbits give clear and tangible illustrations." Darwin's mathematics modeled the real world as simply as possible, by reducing the number of physical parameters involved, usually dropping higher-order terms to "facilitate the analysis." Next, he would test his simplified model by using real data and add back rejected terms until the fit was reasonable, at least to within limits of observational accuracy. If all this did not work, he would then resort to mathematical quadratures to gain insight into the physical meaning of his model.[25] Russell had found both a kindred spirit and a powerful guide.

Darwin had already established a direction for Russell. Binary systems, Darwin had suggested in his 1897 Lowell Lectures in Boston, could be studied to good profit if the researcher considered the tidal interactions between the stars. Tidal friction increased the eccentricity of the stars' orbits, and tidal action tended to be most important when the masses of the stars were similar. Evidence of the generality of his theory, Darwin argued, came from "[t]he fact that the orbits of the majority of the known pairs are very eccentric . . ." He described the history of a double star system starting with the "rupture of a nebula in the form of an hour-glass, into two detached masses." The state of evolution of binary star systems was somehow related to both the eccentricity and the size of the orbit. Darwin saw much heuristic value in his theory: "[W]e must await its confirmation or refutation from the results of future researches with the photographic plate, the spectroscope, and the telescope."[26]

Darwin defined many of Russell's career interests over the next decade. During an early lecture in January 1903, Russell wrote furiously as Darwin opened with a review of the classical areas investigated by dynamical astronomy, and seized upon his mission:

> [T]hen comes the theory of the tides and that of the figures of equilibrium of rotating liquid and gaseous bodies. The last subject has hardly been tackled yet. This leads to theories of cosmogony—the origins of planetary systems and double stars. The recent progress of double-star and spectroscopic astronomy affords a new field for Dynamical Astronomy. The average binary star consists of two bodies of nearly equal mass and must be discussed in quite a different way from the ordinary planets.[27]

More than the lectures of Ball or Whittaker, Darwin's dealt with just what Russell had been searching for. Darwin posed practical problems and then worked them through, identifying the mathematical techniques he was drawing upon, where they came from, and how they could be applied with greatest efficiency. No mentor could have fit Russell's goals better than Darwin, especially as his lectures quickly turned to rotational disruption and instability and criteria for fission. The previous November, as Darwin traced out Jeans's analysis of "an infinite cylinder rotating about its long axis," Russell copied down the resulting figures in his notebook, quipping to himself: "looks like a 'ping pong racket,' " adding almost breathlessly, "Though not proved it is very probable that it actually splits into two as the process goes on. But we don't yet know how big the satellite is."[28]

In close binary systems one must be able to describe the combined potentials of two fluid ellipsoids almost in contact at the formative point of their evolutionary careers. Darwin discussed this requirement in his lectures, often resorting to spherical harmonic solutions. But his treatment assumed knowledge of spherical harmonics and did not dwell on the internal logic or basic theory of the technique, which was why Russell read under Hobson as well. Ball's lectures, usually given in the Plumian Professor's Room in the museum at King's, provided a heavy dose of mathematical physics applied to a broad range of physical problems. Ball, an inveterate geometer trained at Trinity College, Dublin, was very partial to mechanics and loved to study rigid body motion. His lectures were devoted to applications of perturbation theory to the problem of the stability of the planetary system, the chief question driving celestial mechanics at the end of the century.

Russell's notes from Ball's and Hobson's lectures, like those he had taken under Fine, were orderly and precise. For Russell, pure mathematics held its own fascination, but it was one divorced from physical reality, and hence from the passion he felt when applying mathematics to the real world.[29] Darwin's lectures were more discursive. He would digress to comment on recent work, and his lectures were often bibliographical essays. Darwin paid homage to Poincaré, for instance, but advised his students to read Karl Schwarzschild's far more accessible treatment of Poincaré's work rather than suffer through the originals.[30] He took great pains to give intuitive pointers to predict when his rapidly rotating, highly flattened ellipsoids would fission. First the ellipsoid would become pear shaped, but it would remain stable until it developed three unequal axes. Using Jeans's unpublished ideas to describe the situation, Darwin developed a determinant whose solution was so nasty, Russell wrote, that "it would be 'cruelty to children' to work this out." Accordingly, as so many students had been directed before him and since, Russell added, "We can do it for ourselves."[31]

Russell found Darwin a delight. He long cherished a memory of watching Darwin and his friend Sir David Gill verbally jousting during a luncheon hosted by H. F. Newall. Darwin chided Gill, "I cannot conceive how any one can derive pleasure from making a meridian observation." Gill shot back, "with some vigor," "Then you're not a born astronomer." "Oh no," Darwin replied very sweetly, as Russell retold the story, "I'm a born fool." "And with that," Russell recalled, "the matter rested!"[32] The lesson of the luncheon was clear. Darwin clearly enjoyed poking fun at the sanctimonious morality of a life devoted to repetitive, mind-numbing observing, knowing full well that his old friend's career had flourished from such labors, and that Gill was the very astronomer who had proclaimed in 1891 that "[n]othing apparently more monotonous can be well imagined if a man is not 'to the manner born.' "[33] Years later, when George Forbes's biography of Gill appeared, Russell half-jokingly told Frank Schlesinger that he was anxious to read it, if only to find out "whether I am a born astronomer or not."[34] Russell clearly identified with Darwin.

During his first year, as he promised Campbell, Russell dove into the intricacies of mathematical modeling that would help him plot the evolutionary courses of close binary systems like Zeta Geminorum. With the year barely half over, Russell mastered Darwin's fission theory and applied it to binary star systems. But he also started to think about his future, looking for a way to prolong his stay in Cambridge, and found that his interests coincided with those of Arthur R. Hinks, one of Ball's assistants at the observatory. During the winter term, Russell took in Hinks's lectures on practical astronomy and learned that Hinks needed mathematical help.

Hinks, four years Russell's senior, trained at Trinity College and was appointed second assistant at the observatory by Ball in 1895, after he had gained the rank of senior optime in part 1 of the mathematical tripos. Ranking below some thirty others in his year who had achieved wrangler status, Hinks was competent, though hardly a remarkable mathematician at Cambridge in those years.[35] Accordingly, under Ball, Hinks's role as second assistant was to take care of the practical details of observational research. He was assigned to develop photographic techniques for determining the positions of celestial objects, and so when Eros came along, Ball redirected him to apply his techniques to Eros to refine the astronomical unit, what came to be known as the "Eros Campaign."[36] But Hinks's preoccupation with Eros meant that another long-standing project Ball had initiated was languishing, even though the telescope designed for it had been completed and all preparations made. Both Hinks and Ball were looking for someone willing and able to explore how photography could facilitate the determination of the distances to the stars.

Remaining at Cambridge as a Carnegie Assistant

Russell would continue to think about the evolution of binary star systems in his spare time, if he had any, but his livelihood starting in the fall of 1903 was as Hinks's assistant, though officially he was supported as a "research assistant" of the Carnegie Institution of Washington. How Russell obtained this support, and the conditions under which he hoped to work in this capacity, reveal that he hoped to extend his association with Cambridge as long as possible.

Darwin had already applied to the new Carnegie Institution of Washington to establish a program of geophysical research "intermediate between observation and experiment." He felt that the program had to be led by a "grinding mind, forming theories to be proved or disproved by observation."[37] In February 1903, Darwin and Ball asked the Carnegie Institution again for a two-year grant "to provide a Research Assistant in the Observatory" to revive A. R. Hinks's dormant photographic parallax project. "At the present moment," Darwin reported, "there is an excellent opportunity of securing the assistance of an exceptionally able man," who was "anxious to undertake" the work, but unable to do so "as a volunteer assistant for the two years at least which the work would take." Darwin took "pleasure in adding, that during the few months he has spent in Cambridge he has shown himself full of zeal to justify his College reputations, and to take full advantage of his opportunities for work here."[38]

Hinks asked Hale and Campbell for endorsements, claiming that Russell was the key to their plans: "And as you know he is particularly keen on binaries. Hence these proposed additions to our observing lists."[39] At the least, support for a "research assistantship" for Russell meant supporting the observations of additional stars Campbell might want:

> He is an uncommonly good man, and we have an uncommonly good telescope here which can't be worked to the full because my hands are deep in the Solar Parallax reductions. So if by the aid of the Institution we can secure Russell's help for a couple of years we shall be in a fine position to get ahead with the plans formed long ago, but which seemed to be postponed to the distant future.[40]

Hinks said much the same to Hale, knowing that Hale had his own experimental photographic parallax program at Yerkes, headed by Frank Schlesinger.[41] In such experimental programs, Hinks added, one had to choose survey fields that would provide useful data whether or not a measurable parallax was found: "[W]e are earnestly praying that the Carnegie Trustees may think we are poor but deserving."[42]

As it turned out, "poor but deserving" had little to do with anything. What made a difference was that the Carnegie Trustee Executive Committee decided in November 1902 to support annually some twenty-five "young investigators who have shown exceptional ability." Each Carnegie "Research Assistant" would receive not more than one thousand dollars for a year's work, the appointment renewable for a second year.[43] Thus Darwin and Ball's proposal was thrown in with 126 other applications, and Simon Newcomb was asked to comment on Russell. Newcomb found Russell fully qualified for an assistantship based upon his published record: "These papers are highly original and show him to be a man of great ability." Newcomb called Russell's Venus work a breakthrough, "the first explanation of one of the enigmatical phenomena observed during transits of Venus over the sun's disk."[44]

In March, Russell was awarded the assistantship. There was some haggling over what Russell was to be called: the Carnegie Institution appointment was for an independent researcher, whereas Russell wanted to be an official member of the Cambridge staff, and there was also a question of where the results would be published. But in the end, Russell accepted the Carnegie appointment and the Carnegie, advised strongly by Dudley Observatory's Lewis Boss, agreed that Russell would retain freedom of action as to where he and Hinks published their results.[45] Although his connection to Cambridge was not as solid as he had hoped, it bought him two more years. Accordingly, Russell remained an independent investigator, dependent wholly upon Carnegie funds, and so would have to return to the United States when his money ran out.

Photographic Parallaxes, Solar and Stellar

Starting in January 1903, Hinks's lectures focused on photographic reduction techniques for solar and stellar parallaxes. He proved to be a fervent English partisan, at odds with the way the French wanted to manage photographic reductions: "M[aurice] Loewy is the leader of the French school of photographic methods; I am a follower of the English school; and between the two there is a radical difference of view, not unlike a difference of political faith, which is hardly a subject for argument."[46]

At the turn of the century the technique of photographic astrometry still had to prove itself against traditional visual methods, and it did not help matters that there was little consensus among the advocates of photography on how to reduce photographic positions to star positions and how to show that these were an improvement on visual methods. Much of the controversy centered on the reduction of photographic plates for the monumental *Carte du Ciel*, inaugurated in 1887 to record millions of stars to the fourteenth magnitude, one hundred times fainter than existing visual catalogs. An International Congress convened to create a Permanent International Committee to coordinate the program and to decide upon reduction procedures.[47] When Eros was discovered, the Permanent International Committee assumed authority for managing its reconnaissance, even though in the thirteen years of its life, the cooperating observatories of the *Carte du Ciel* had made little progress toward a consensus on reduction methods.

The *Carte du Ciel* and its attendant *Astrographic Catalogue* form the social framework for an understanding of the problems and the practice of precise photographic astrometry at the point Russell entered the field in 1902. Although he was engaged to produce stellar parallaxes, the problems were the same. Russell eventually aided Hinks in all three activities: stellar parallaxes, the solar parallax, and reduction procedures for the *Astrographic Catalogue*. Every aspect of the photographic technique was being contested: How many comparison stars were needed on a photographic plate to determine the position of the program star, or of Eros? Could the stability of photographic emulsions be trusted, or did one need to photographically impress a standardized grid pattern (a reseau) on each plate to calibrate them? Did photographic positions have to be reduced to the fundamental coordinate system defined by meridian circle observations? What was the best way to compare measurements between adjacent fields, or at different epochs, or from different instruments? How should atmospheric dispersion, refraction, Earth's precession and stellar aberration be handled, among a universe of nagging systematic effects? By 1900 most of the fifteen observatories participating in the *Carte du Ciel* had completed their observations, accumulating some 13,000 photographic plates out of a projected 17,000; but only a very few actual charts with useful positions had been created, and even these were controversial.[48]

Hinks championed the reseau method created by H. H. Turner at Oxford, which required a comparatively small set of comparison stars to produce plate constants that would then yield the position of the program object. Turner's method was well suited for stellar parallaxes, Hinks argued, as well as for the reduction of Eros photographs and for any measurements of position of isolated objects that did not require reference to the fundamental frame. The Permanent International Committee, based in Paris, disagreed, arguing that the Eros plates all had to be tied into the fundamental frame defined by meridian observations. Russell's job was to develop Hinks's photographic parallax program, but in effect he became responsible for demonstrating the superiority of Turner's reduction procedures and then executing them using the 12-inch Sheepshanks polar Coudé telescope that had been installed at Cambridge in 1898. Manufactured by Grubb, the Sheepshanks had a sophisticated three-element lens, the largest in existence, and its mounting, designed by Turner and A. A. Common, was a simplified version of the two-mirror Coudé that Maurice Loewy championed at Paris. In theory it avoided the flexure problems inherent in long-focus refractors and produced a brighter image than did the classical Coudé.[49] Hinks's charge was not only to develop the Eros observing program for the Sheepshanks but to use it as a standard against which he could assess the observational quality of photographs from the other participating observatories.

Russell learned quickly that photographic astrometry was a hotbed of dissension. Hinks was not relutant to rail against the "French school of reducers" in his practical astronomy course, where Russell learned the basics of plate reduction using the reseau and how to account for sources of systematic error. He became proficient in operating Hinks's precision measuring engine, which had a divided glass scale in the eyepiece field driven by a short micrometer screw. This combination had just enough travel to cover a plate area defined by adjacent reseau marks.[50]

Hinks wanted to show how Turner's reduction formulae could account for all known errors, but had failed to provide credible theoretical arguments to support his conviction. He had been unable to convince his critics of the efficacy of his techniques, since they all were working at levels of accuracy that tested the precision of their telescopes, measuring machines, and reduction procedures. There were just too many variables and too many unknowns. In his capacity as coordinator, he also faced the practical problem of developing a version of Turner's method that could allow for the comparison of plates taken by such dissimilar devices as small refractors, wide-field astrographs, visual micrometers, and reflectors.

As Russell took in Hinks's lectures, he and Hinks mapped out their strategy. By September 1903, before his Carnegie money arrived, Russell had developed a new highly efficient method of reducing plate positions on the reseau system. He also designed a magnitude reducing mask, or "color-screen," to equalize the brightness of the parallax star and its field stars.[51] Hinks was delighted, telling Hale and Campbell how "Russell with his superabundant energy" was establishing the "theoretical basis of the linear reduction formulae" to simplify the procedure. Hinks was emboldened by Russell's analysis and felt he could now stand up to his critics who had argued that Hinks's methods were, at best, only equivalent, and not superior to their own, which required fewer plate constants. Until they were won over, it was all still "a subject of controversy."[52] In their first summer of work, Russell and Hinks had completed what Hinks called their "manifesto" on reduction techniques. Russell had done this work without Carnegie support, since the funding was delayed. Ball therefore directed the Observatory Syndicate to make Russell an "honorary assistant in the Observatory," just the connection Russell always wanted.[53]

Designing the Stellar Parallax Program

After Russell developed the theory supporting Hinks's reduction procedures, they set observing priorities. At the top of the list were stars with well-determined parallaxes, to test the technique. Next came visual binaries with well-determined orbits whose masses could be calculated if their parallaxes were known, "which is of value from an astrophysical standpoint."[54] Third, he added variable stars, common proper motion stars, star clusters, and planetary nebulae, the last surviving from Hinks's original priorities. They also included stars that were likely to have large parallaxes, such as bright stars and those exhibiting large proper motions. Among the last two categories Russell and Hinks selected stars chiefly of Vogel spectral types I and III, white and red stars, choosing them over stars of the intermediate type II, or solar type yellow stars.[55]

J. C. Kapteyn and the Irish astronomer W.H.S. Monck had shown that as a class, stars of type II seemed to be closest to the Sun because they exhibited the largest mean proper motions. Monck suggested, however, that this result was illusory, and that the dim red type III stars were closest, but the overall sample was biased by an admixture of intrinsically bright red stars at very great distance. Hinks and Russell mentioned none of this in their paper, but, consciously or not, they selected just the types of stars that would provide a test for the existence of an admixture of stars among those in type III, showing that it contained both bright and faint stars. In effect they designed a test of Lockyer's theory of stellar evolution, which envisioned young red stars as huge bloated swarms of colliding meteors, and old red stars as dim burnt-out cinders that had spent their gravitational budgets.[56]

Norman Lockyer had long championed the idea that the evolutionary lives of stars included a heating and a cooling phase, and that stars in both phases could be found in the sky. In 1902 he published a major catalog of the stars illustrating his theory, which he called his "Meteoritic Hypothesis."[57] Of the 55 stars on Hinks and Russell's observing lists, 21 were in Lockyer's catalog, and of those, 19 were just those which would test his theory. Parallaxes would yield absolute brightnesses, which could then be converted to an estimate of diameter and density for the few systems that were spectroscopic or eclipsing binaries in the sample. Hinks and Russell made no mention of testing Lockyer's theory in this fashion, and it is doubtful that they would have admitted to such a goal, given that they were trying to prove the photographic process. However, Russell was well aware of Lockyer's catalog and his theory. In 1924 he recalled, "[I]t occurred to me that one of the most conspicuous results of [Lockyer's theory] would be a change in the total brightness of the star."[58] This was exactly what Russell and Hinks could derive from their sampling of Lockyer's stars. From 1903 on, in his reports to the Carnegie on the results of their parallax work, Russell hinted that he was indeed thinking about Lockyer's ideas, teasing out the stages of evolution from his parallax work.

To obtain maximum parallax advantage, stars had to be observed when they were orthogonal to the Sun. So the best times to observe were just after dusk and just before sunrise. This required two widely spaced observing sessions per night. Russell also had to be on call continuously, since there were not many clear nights in the dank Cambridge climate. It is for these reasons, Russell recalled, that he moved into the Hinks household sometime in late 1903. Since no lodgings were available within easy walking distance of the observatory for people with night duties such as his, the Hinks family "promptly invited me to live with them for a year—nominally as a paying guest, actually as one the family." By that time, Hinks had been promoted to chief assistant and had gained a house on the observatory grounds.[59]

But there was a far more personal reason for Russell's move. On 26 September 1903 Russell's mother, at age fifty-eight, died of cirrhosis, a chronic liver condition that may have been caused by a childhood bout with yellow fever when the family lived in Brazil. It was a slowly degenerative illness, but it turned critical and she died quickly.[60] There was no time to bring Henry home to be by her side at the end, since he was somewhere on the Continent, probably vacationing in Italy. Her two younger sons were at hand since Gordon was then a teacher in the Brooklyn Heights School, not far from home, and Alexander was a summer tutor to President Roosevelt's children at Sagamore Hill. A local paper reported that the Reverend J. H. Hobbs, who presided at the funeral, made "reference to the son in Europe who knew nothing of his mother's death [, which] brought tears to many eyes."[61]

The day after his wife's funeral in Oyster Bay, which included President Roosevelt among the mourners, Alexander Russell left on a steamer from New York to bring the news personally to his eldest son. The pastor was already on the high seas when Eliza was buried in Princeton. A telegram would not do; evidently the family feared that the shock might push Henry, stricken with the belief that he had forsaken his mother, over the edge once again. And well it might have, for as Russell recalled years later, this was "the severest personal blow which has yet befallen me." It was just at this time that Hinks invited the distraught twenty-six-year-old into his household. It is no wonder that years later Russell keenly remembered the kindness: "As I review a longish life, I am overwhelmed with the number of deeds of pure goodness which have been done to me, and this was one of the chief."[62]

Getting back to work slowly that fall, Russell still enjoyed the many intellectual diversions Cambridge offered, such as the $\nabla^2 V$ Club, but even these took second place to his primary duties when he started to observe. The long focus of the Sheepshanks lens and the inefficient emulsions then available required exposures of at least three to five minutes to record enough comparison stars for the reduction procedure. Guiding during the exposures was a difficult and tedious process, but the color screen he had developed seemed to work quite well. Hinks and Russell proudly displayed their filter at a Royal Society soirée.

Observing and Reducing the Data: Striving for Efficiency

When he wasn't observing, Russell continued to refine the reduction procedures, finding more shortcuts for calculating the positions of the stars on the plates, using Turner's standard coordinates.[63] The Carnegie Institution granted a second year of support after Lewis Boss confidentially advised that Russell was "approaching his task in a workmanlike manner" and had many obstacles to overcome with an instrument like the Sheepshanks equatorial. Boss regarded Russell as a "young man of unusual ability and promise" who should be given a "fair chance during two years to show what he can do."[64]

As plates were taken and the reduction procedures were put in place, Russell started measuring, searching for systematic errors everywhere. By the summer of 1904, Russell had covered 47 of the 55 parallax fields with 118 measurable plates, each containing four multiple exposures. By then, over 80 plates had been measured, and 24 fields had been completely reduced to form the equations of condition. In August, Boss continued to describe Russell's reduction techniques as "ingenious" and "especially efficient in securing economy of labor in the measurements and computations."[65] Efficiency was second only to accuracy. Russell keenly appreciated the need to reduce measurement and reduction labor without sacrificing accuracy, to make his techniques useful to others, and to keep from exhausting himself. Since the plates were all taken when the stars were on the meridian and near quadrature, he had to measure only one coordinate of parallactic motion, "thus halving the labor of measurement without material sacrifice of accuracy."[66] He was also able to demonstrate that as long as his parallax star was at the geometrical center of his comparison stars, he could avoid the labor of reducing positions by least squares. Russell further found that his systematic setting errors were so small that he needed to measure the one coordinate in only a single direction. Thus "the labor of measurement (which is the severest part of the whole work) has been again halved," making his method twice as fast as Schlesinger's competing method.[67] His economies of measurement and reduction were so great, he claimed, "that a single observer can measure and reduce his photographs about as fast as he can take them—at least in this climate."[68]

This was no idle statement, for Russell keenly knew, as did Schlesinger, that they were competing not so much between themselves as with those who still advocated visual parallax observations with meridian circles or heliometers. Efficiency was an important indicator of success, but so were reliability and accuracy. Thus Russell took pains to point out that the heliometer method, which depended heavily upon the relative positions of stars seen within the split telescopic field, could not detect the spurious motions of the comparison stars.[69]

By now Russell had become a strong moderating influence on Hinks, who was prone to overly heated debate on the floor of the Royal Astronomical Society. Russell had been elected a member in 1902, endorsed by Ball, Darwin, and Hinks, and did whatever he could to convince Hinks of the need to work collaboratively.[70] At a meeting in the spring of 1904 he helped to settle what had become a rancorous debate over measuring machine calibrations and sources of systematic error. As Turner observed ruefully, "We do not seem to be as yet at the end of the small discordances which are occurring in the measurement of photographic plates." But, he added, it would be worthwhile to further "enquire into the source of error that Mr. Russell has spoken of." Russell suggested a means of self-referential calibration that actually nullified many distortions when they were small.[71] But more to the point, his analysis raised hope that a technical solution was possible, which ultimately helped Hinks in his dealings with his competitors.

In his second report to the Carnegie, Russell announced that his preliminary reductions yielded a probable error for a single plate in the range of ±0.05 seconds of arc, which was well within limits required for competitive parallax work.[72] Boss was most impressed. Not only was Russell's program "judiciously chosen" but his efficiency of labor was remarkable. "I should say that he has improved on his contemporaries in this line very decidedly in that respect."[73] Boss believed that something of "real importance is going on."

> It is too early, of course, to be very sure of the extent to which Mr. Russell will develop as an astronomical investigator of high rank; but I feel so hopeful of good results that I do not hesitate to urge that Mr. Russell's application for an extension of his term as Research Assistant for another year should be granted. Last year I felt somewhat doubtful about his chance of making a good thing of what he had undertaken, and I am free to say that those doubts have been removed.[74]

H. H. Turner was fortuitously visiting Boss just as Boss was reviewing Russell's first year of work. Turner knew Russell as a man of "uncommon ability and efficiency" and evidently spoke glowingly of the young American who was now being noticed at meetings of the Royal Astronomical Society.[75] Boss assured the Carnegie that the quality of the man was at least as important as the quality of his work. Russell had contributed theoretical weight to Hinks's and Turner's techniques, and was now showing that his parallaxes possessed a high degree of internal consistency.

Russell demonstrated that thrift worked. Hinks had long wanted to overthrow Loewy's plan, to "supplant the French method by the Anglo-Saxon," but he made little headway until Russell arrived on the scene. His strategy had been to make himself useful to people like Campbell, who as Lick director was embarrassed by heavy backlogs reducing photographic observations. Campbell's staff was stretched to the limit in their meridian circle and spectroscopic radial velocity departments, "in which connection some two thousand plates are awaiting measurement and reduction." He was open to any economies that Hinks's method had to offer to meet his Eros obligations but had to be convinced that they were reliable.[76] Campbell's backlog was typical of observational astronomy at the time, when the rate of accumulation of photographs vastly outstripped the ability of most observatories to process the data. The efficiencies that Hinks, Harold Jacoby, Turner, Schlesinger, and Russell sought out, therefore, must be appreciated in this light. When Russell announced that he could reduce his plates as fast as he could take them, astronomers took notice.

Eros's passage in 1900 led in time to more than a refined knowledge of the astronomical unit. It was an important milestone in the adoption of photography in positional astronomy. The challenges facing Turner, Hinks, Russell, and Schlesinger as they fought to establish photographic astrometry were all embodied in the nature of the Eros Campaign, the *Carte du Ciel*, and the quest for stellar parallaxes.[77] Although Russell became a player in the process that marked the close of the visual era, his personal goal went beyond establishing a new technique.

Russell's Cambridge Experience

Russell's ultimate goals were never defined by, nor constrained by, proving the efficacy of photography. His work for Hinks was a singular event in his career, and the parallaxes were not an end unto themselves; they would soon lead to more interesting astrophysical problems. This was a goal he carried to Cambridge from Princeton, where he became fascinated by binary star systems and the secrets they held about the lives of the stars. His labors for Hinks, however, also brought him to think about other ways to obtain the data he desired, either through neat and crafty shortcuts, or through depending upon the industry of others. We will find both strategies emerging in his last months in Cambridge, as Russell began to take the steps that would make him an agent of change in astronomy.

Russell's professional outlook was influenced by his years in Cambridge, through contact with Darwin, Eddington, and Jeans. He savored the many different points of view he found in such marked contrast to his life in Oyster Bay and Princeton. It was in Cambridge that he obtained a fresh appreciation for the interpretive powers of mathematical physics.[78] But it was also in Cambridge that he was introduced to the theory and practice of photographic reduction techniques and found that he could make a contribution there as well. It was rigorous training that he would soon use to great advantage as he completed his parallax reductions. But that would happen only upon his return to Princeton and after several more years of intense labor.

• CHAPTER 5 •

Return to a New Princeton

"THERE IS UNDOUBTEDLY ROOM FOR A PROFESSOR OF ASTROPHYSICS IN PRINCETON"[1]

When Russell's second year at Cambridge was approved in September 1904, he returned home to vote and to attend a cousin's wedding in Toronto. After the wedding, back home in Oyster Bay, Russell was stricken with typhoid fever and was knocked out for months. "First I had a headache," he told his children years later, "then I had a worse headache, then I had the worst headache I ever had. And then I didn't remember anything for three weeks."[2] Russell was laid up in Oyster Bay until the end of the year. Typhoid was a serious matter: "No disease requires more vigilant attention or greater medical experience."[3]

Russell could not even write his columns for *Scientific American*.[4] He was allowed to return to England at the end of the year only after he agreed with his doctors not to return to Cambridge immediately: "[O]rdered to a warmer climate," as he later reported, he tried Sidmouth, but he soon found southern Italy and the island of Capri more to his taste. He did not return to Cambridge until May 1905.[5] In the meantime, Hinks filled in with the observing, but reductions and analysis lay fallow.

Just before he left the United States, Russell had met with Young and Lovett to talk about his future at Princeton. In March from Capri Russell wrote the new Carnegie Institution president, Robert S. Woodward, asking for an extension to his term as research assistant and for funding to cover a third year. His illness had delayed progress, and Russell needed a full academic term to finish his work, which, he claimed, "has compelled me to postpone the acceptance of another position, which I had hoped to occupy for that year."[6]

Such a job offer, if one was actually made, could only have come from Princeton. But Russell wanted to stay in Cambridge for at least another year, maybe two, though he did not map out his plans coherently until he was recuperating on Capri. A glimpse of what he was thinking about comes from the record of his contact with the prominent geologist and archaeologist Raphael Pumpelly, who was then exploring Turkestan with his son and had looked in on Russell on behalf of the Carnegie Institution. Although his report has not been found, years later Pumpelly's son recalled their meeting vividly. Evidently Russell captivated both father and son with his scenario of the past life of the Sun as a giant star. Their discussions had ranged widely, from archaeology to free will and God's Design, but the past history of the Sun and the fate of the universe proved to be the most memorable vistas Russell chose to paint for his callers.[7]

Russell employed these vistas to actively sell himself. In what he had evidently taken as a job offer, Young and Lovett had urged him to write to Woodrow Wilson, now president of Princeton, to place on record his desire to return as a member of the faculty. From Capri, Russell told Wilson that he wanted to finish his parallax reductions and to "devote my spare time to some of the admirable opportunities which Cambridge offers for the study of mathematical and spectroscopic Astronomy."[8] Mindful that Young was about to retire and Lovett was going to replace him as chair of the department, Russell told Wilson that what Princeton really needed was a good astrophysicist:

> There is undoubtedly room for a professor of Astrophysics in Princeton . . . It is in this branch of Astronomy that the greatest advances of recent years have been made . . . and the most important unsolved problems of the science are at present astrophysical;—for example, the constitution of the Sun, the explanation of the different types of stellar spectra, . . . All these problems lead up to the greater one of stellar evolution—the answer, as far as may be, to this question[:] How did our universe reach its present form, and what will become of it?[9]

Russell declared that he had learned "modern methods of research" at Cambridge that would lead to the solution of these problems, and that a vast store of knowledge lay untapped in America that he was now ready to exploit:

> Thousands of photographs of the stars are being taken in America every year . . . But only a very few of the plates are being measured. In this respect European astronomers are far ahead of us . . . Plenty of people in America can take good photographs, but there are not more than half-a-dozen men in the country who are familiar with the modern ways of measuring them, and of reducing the result to a useful form.[10]

Russell knew how to reduce photographic data now, but he was also armed with Darwin's celestial dynamics and Jacobus Cornelius Kapteyn's statistical methods, which had caused great excitement at Cambridge.[11] Kapteyn, who with Gill and Turner had first demonstrated the superiority of photographic astrometry, developed statistical formulae for determining approximate distances to groups of stars based upon their mean proper motions (motion across the line of sight), and Russell had toyed with them while in Cambridge (see chapter 6), basing his $\nabla^2 V$ Club talk on Kapteyn's ideas. Kapteyn was also organizing a worldwide network of observatories to explore the structure of the sidereal universe. Russell admired Kapteyn's ability to organize and co-opt observers, and wanted to emulate his tactics at Princeton. He promised Wilson that he would not demand new telescopes, equipment, or staff and would be content with the means at hand: "In fact Professor Kapteyn of Groningen has made a distinguished reputation for himself and his University in the astronomical world by such work, although he has no observatory, and not even a small telescope."[12]

Russell had set his sights precisely. Just as Russell told Woodrow Wilson that the analysis of astronomical data fell seriously behind its rate of accumulation in America, Kapteyn argued at the 1904 Congress of Arts and Science in St. Louis that the radial velocity data required to test his theories "are even now on hand in the ledgers of American astronomers—alas! not yet in published form."[13] Kapteyn was, most definitely, chiding W. W. Campbell, who, as noted in the previous chapter, was painfully aware of this fact. Kapteyn's plea became Russell's agenda throughout his career; it also spoke to the problem Russell himself would encounter years later when he would follow in Kapteyn's footsteps at Mount Wilson and then at Lowell Observatory, goading the observers to publish their results.

Recovering in Capri, Russell knew just how he would fit in at Princeton. Cambridge had opened Russell's eyes to many things, not least the tutorial system, which he had heard that Wilson was planning to imitate in his "preceptorial system" at Princeton. "I should feel it a privilege to assist in its establishment," he told Wilson, "and to do what I could to make 'reading men' of those whom I taught." Russell asked for an assistant professorship, at $1,500 per annum.[14] Wilson was "entirely in sympathy" with having Russell back, but not at that salary, which he considered "decidedly presumptuous." But Wilson also knew Russell and did not "attach any great importance to such a suggestion."[15] Lovett, in fact, with Young's hearty concurrence, had already written Russell on 18 April, inviting him to join the staff as an instructor, at $1,000 per year, to commence with the beginning of the academic year 1905–1906, one year earlier than Russell had requested.

Russell returned to Cambridge in May to find Lovett's letter and notification from the Carnegie that his request for a one-year extension had not been approved. "I was therefore obliged," Russell told Woodward, "to find other means of livelihood."[16] These were years when competition for such awards was high and most of the Carnegie money for astronomy was being absorbed by one huge project.[17] Russell's fate was sealed when he and Hinks found that the special magnitude reducing color-screen had been ruined by moisture. This put an end to the observations and meant that Russell could complete his parallax measurements and reductions anywhere.[18] It was time, Russell realized reluctantly, to go home.

Knowing that his days in Cambridge were numbered, Russell hastily spent May and June gathering up all the research materials he would need to complete the parallax reductions at Princeton. Russell secured a pledge from Lovett that he would have enough free time in Princeton to complete his parallax work, and that Princeton would provide a measuring machine similar to the one used at Cambridge.[19] Before he left in mid-July, he managed to complete two papers: a long one with Hinks on methodology and a short note on the first two stars that had complete reductions. They wanted to set out their rules and procedures for observing and reducing the plates well before Russell engaged in the bulk of the work, to invite constructive criticism if it existed.[20] The two parallaxes were for the visual binary Gamma Virginis, which was on Lockyer's list, and for the star Lalande 21185, which was one of the closest stars to the Sun. Russell found that the Gamma Virginis stars were "either less dense than the Sun or have a greater surface brightness, which accords well with the fact that their spectra are of the first type." Russell's conclusions supported the idea that these stars were at an "earlier" stage in life. Once again, he was quick to think in evolutionary terms.[21]

Wilson's Princeton

Russell was not the only neophyte instructor to arrive in Princeton in the fall of 1905. There were many others who gained far more notoriety and drew more controversy, and whose appearance was anxiously awaited that fall by the Princeton community. In the three years since Wilson had assumed the presidency, many changes were afoot.

Wilson had reorganized the faculty into formal departments to improve management and control. Henry Fine was now chair of the mathematics department and dean of the faculty. A formal physics department was created under Cyrus Fogg Brackett, but Wilson delayed any action on astronomy, since he knew that Young was soon to retire.[22] Wilson also pushed for a major graduate school and improved undergraduate instruction with his "preceptorial method" that would emphasize discussion of concepts, not facts. Wilson hired fifty preceptors, but Russell was not one of them. A preceptor had to "place teaching ability ahead of scholarly qualifications." Russell was hired as a regular lecturer on the permanent staff. Wilson and Fine knew "the character of the man"; his forte was research, not intimate contact with students.[23]

Wilson's preceptorial system was confined largely to the humanities, but it gave Fine and W. F. Magie the chance to build up mathematics and the sciences. Magie in fact stated outright that he looked first for "capacity for research ahead of promise as a teacher. . ."[24] Thus many were called—George Birkhoff and Oswald Veblen in mathematics, Owen Richardson in physics—who had already established reputations. This was Fine's chance to put Princeton mathematics on the map. He was delighted to capture James Jeans, who, as with Russell, arrived fresh from Cambridge where he had been a lecturer in Trinity College. Though commonly regarded as another of Wilson's preceptors, Jeans actually came as full professor of applied mathematics at a salary of $3,500 per year, three and a half times what Russell received, and $1,500 above that of the highest-paid preceptor.[25]

Jeans's attainments, strong teaching experience, and prominence in professional circles would be, Fine hoped, a real boost for Princeton mathematics. With Jeans, Princeton could plan to produce "first rate applied mathematicians such as are produced nowadays in Cambridge only."[26] In contrast, Russell's return did not even warrant comment in the trustees' minutes, nor was it discussed among Wilson's trusted lieutenants. Russell lacked Jeans's stature and cultivation, as well as his track record. But he was a Princetonian who had now studied and worked at Cambridge.

The Department of Astronomy under Lovett

When Russell reached Princeton, probably by September, there were at least two people waiting for him, one in person and the other in spirit. Aunt Ada once again made 79 Alexander Street his home, and after he dropped his bags at the door and dashed to the observatory, he found a letter from Young, now fully retired to his family home at Hanover, New Hampshire.

Young's last years were sad ones. He lost "the vigor of younger life" and seemed "withdrawn, remote as a star, but brilliant in our firmament."[27] After his wife of forty-three years turned gravely ill and died on 18 January 1901, Young was distraught. By 1903 he could barely muster the energy to teach and suffered from episodes of dizziness and forgetfulness. He was excited by Wilson's reforms and managed to finish his *Manual of Astronomy*, taking note of recent advances in physics, but his energies were spent. He resigned as chair of astronomy in December 1904, after twenty-eight years at Princeton. Anticipating Young's retirement, Hale suggested Edwin Frost as his successor, but Young knew that Wilson and the faculty Committee on Curriculum would confine their search to campus. Opting for a known quantity at Princeton in tune with the administration, and feeling no real pressure to mount a search for an established researcher as successor to Young, Wilson found Lovett to be his man. At the turn of the century there was nothing terribly unusual about a mathematician's taking over a small astronomical department, and it didn't make much difference what Lovett's specialty was, since he was most interested in administration and was popular with students and trustees alike.[28]

In his melancholy, Young wrote to Russell when he realized that he would be gone by the time Russell returned: "It was hard work tearing away from Princeton but it had to be done," Young wrote in August, exhausted by the wrenching move. Russell would find Princeton astronomy under Lovett much changed, Young predicted, and the change would be good. Lovett was already touring the great western observatories looking for ways to revitalize Princeton astronomy. He was also getting a jump start learning the observatory trade himself and was looking to recruit another faculty member because William Maxwell Reed (Taylor Reed's replacement and no relation) had resigned to take a position with the

Roebling Iron Works in Trenton. Lovett had just heard from someone named Dugan, who had studied under Wolf at Heidelberg, "and I should not be at all surprised if he turned out to be the man."[29]

Lovett visited Campbell at Lick and then inspected Hale's new Mount Wilson Solar Observatory. He had not yet moved into Young's house on Prospect Street next to the Observatory of Instruction, because the university had promised to renovate the old place. All this was with Young's blessing, Young assured Russell, and he held out every hope that his former student would do well working under the affable mathematics teacher, though he hoped that Russell's health "is now at least reasonably good, and all the time getting better." Young missed seeing Russell but appreciated his kind words from an earlier letter: "[I]t is pleasant to be appreciated by a pupil like you . . . I hope that when you are settled then you may sometime wander up this way, to see our hills and mountains, and if you do you must visit old Dartmouth, and your old Professor." Young sent his daughter's regards to Aunt Ada.[30]

Young was too spent to be concerned with his replacement. Lovett, six years Russell's senior, trained at the University of Virginia and then in Leipzig, and came to Princeton in 1897 as an instructor. He was made professor of mathematics within three years, rapidly becoming a popular campus personality. Interested in celestial mechanics, he began teaching in the astronomy department in 1900 and was one of Russell's examiners. In time he focused on the complex mathematical problem of the motions of three or more bodies in a mutually interacting gravitational system but managed little more than a review of the literature. Lovett's weak research record was of little concern to Wilson or even Fine, apparently, since their priority for astronomy was to provide instruction to a broader clientele. Accordingly, Lovett's first move was to strengthen course offerings in celestial mechanics and practical astronomy, gearing them for engineers and mathematicians. He added a new course of study on the history of mathematical astronomy but did not touch astrophysics. Quite likely, their knowledge that Russell was in the wings underlay Lovett and Young's decision to leave astrophysics out for the time being.[31]

The prospect of funding from the new Carnegie Institution was then causing a feeding frenzy among ambitious American astronomers. Lovett wanted a major telescope for Princeton, which, reflecting popular sentiment, might be built in the Southern Hemisphere. He visited telescope makers and observatories to learn what a first-class observatory would look like and cost. Options included rebuilding the 23-inch refractor; building a new, highly versatile 3-element 18-inch photographic refractor; building 16- and 18-inch doublets for astrographs; and, most boldly, erecting an 84-inch Cassegrain reflector. He received bids for all of these, including the reflector, which was larger than anything yet conceived.[32]

Lovett learned that photography was the tool of modern astronomy. In a long letter to Wilson in November 1906, he talked incessantly about a big telescope for South Africa, noting that many other colleges and universities were talking about going south. Most of these plans died when Hale walked off with the bulk of Carnegie's funds earmarked for astronomy and built his 60-inch reflector on

Mount Wilson. But Lovett was mostly concerned with Princeton's place in the rush; he counseled for a quick start, to get ahead of the game, reminding Wilson that "[f]or thirty years astronomy at Princeton has consisted of a great man and his reputation, neither of which was made in Princeton." And as if he needed to say it, Lovett added, "That great man is gone."[33]

Indeed, astronomy had been on the wane for some time. Young had done nothing to modernize the observatory after the early 1890s, and the curriculum remained much as it was in Russell's day. A succession of junior appointments had not produced an heir; the last, William Maxwell Reed, had come with the endorsement of the Harvard director, E. C. Pickering, as well as a copy of Pickering's visual photometer that Reed used on the 23-inch to monitor and measure the brightnesses of variable stars, especially eclipsing binaries. In 1903, Reed's work, aided by a Carnegie grant secured by Pickering, gave Young comforting assurance that the Halsted was not idle.[34] Reed enjoyed the aid of an ardent undergraduate named Zaccheus Daniel, and together they were diligent about observing variables with the Halsted telescope. But the bulk of Reed's energies still went into teaching the practical courses; he left the matter of publishing to Daniel.

Of the trickle of graduate students who had both preceded and followed Russell, few amounted to anything. Young had helped Stanley C. Reese secure a volunteer position at Yerkes Observatory in 1899, but Reese soon ran afoul of the director and his senior staff.[35] Following Russell by a year was John Merrill Poor, who had graduated from Dartmouth in 1897 and entered Princeton after two years of teaching. He tried a thesis on a comet orbit and returned to Dartmouth, where he taught until his death in 1933.[36] There was also Daniel of Pennsylvania, who as a promising undergraduate gained access to the 23-inch. Walter Mann Mitchell suffered from a lack of guidance when he was not able to interpret peculiarities in the spectra of sunspots he had observed visually with the Halsted. Even though Mitchell's observation of what he thought were line reversals in sunspots attracted attention, no one could explain them.[37] So he merely reported his descriptive findings, to "obtain as complete a table of lines as possible." As Russell related years later, "if he had gotten it he'd have beaten [George Ellery] Hale with the discovery of the magnetic field in sunspots."[38] Mitchell was the only Young student to attempt a thesis in astrophysics, receiving the Ph.D. in June 1905. And there were a few graduate students who did not complete their course of study.

Mitchell's fate highlights the limitations for research in modern observational astrophysics at Princeton. Princeton lacked equipment and au courant guidance and could not compete with Hale's unparalleled laboratory and observatory at Mount Wilson. Indeed, after Hale invited Mitchell to Mount Wilson for a summer in 1908, Mitchell wrote Russell reporting that "Hale finds that my reversals are due to Zeeman effect! I tried for that at Princeton several years ago, but for lack of proper apparatus did not get any results. If somebody besides Edgar [Lovett] had been director [*sic*] I might have succeeded better."[39] Though Russell thought highly of Mitchell, and at one point tried to hire him, Mitchell never prospered in astronomy and soon left for industry. Mitchell's fate was an important lesson that Russell would use to his advantage in later years. It may have been easy to blame Lovett, but the real problem was Young's, who in his later years emphasized teaching over research. Russell would never make that mistake.[40]

With William Maxwell Reed's resignation, Lovett knew that he needed someone to handle the observatory and practical astronomy. Russell was hardly the one to take charge of these responsibilities. He had been promised that his primary tasks would be to finish up his parallax work, and to teach. Someone else had to be found who could take care of the telescopes and the observatory.

Dugan and the Halsted Observatory

Raymond Smith Dugan, seven months younger than Russell, graduated from Amherst College in 1899. Influenced by the flamboyant David P. Todd, Dugan declared for astronomy and secured a research studentship under Max Wolf at Heidelberg, where he obtained his Ph.D. in 1905 with a thesis on the photometric characteristics of the Pleiades star cluster. During routine observing duties for Wolf, a pioneer in wide-field photographic surveys, Dugan discovered sixteen new asteroids.[41]

In June 1905, Dugan, still in Heidelberg, wrote to American observatory directors looking for a job. He had practical knowledge of photographic photometry, plate measuring, and asteroid hunting. Writing to Frost from Rome during a tour of European observatories on his way to Spain to join Campbell's Lick contingent to view a solar eclipse, Dugan admitted that any work in astronomy sounded good: "As a matter of fact I am not very particular just now." Dugan had offers from Todd, Lovett, and the Morrison Observatory at Missouri but preferred the Yerkes Observatory "for its outfit and its push." Frost tried to hire Dugan, but after Hale's departure, Yerkes was impoverished. So Dugan accepted the Princeton position, telling Frost that he would jump ship if the money came through.[42] Although he knew that little was happening at Princeton, he needed to "get on in his profession . . . ," and Princeton, at least, was an elite school and showed some promise. After meeting Lovett and realizing that he would indeed be master of the Halsted, Dugan felt better about the situation.[43]

Dugan wasted no time tuning up the Halsted. He mounted Reed's double-image polarizing photometer on the telescope, taught himself how to use it, and observed every clear night through the rest of October and well into November, adding the newly discovered variable RT Persei to his growing list of Algol-type eclipsing binaries.[44] How he decided upon this course of action remains unknown, but in large part it was a continuation and expansion of the program Reed had started. It also fit perfectly with Russell's more mathematical interests in binary stars. For Russell, Dugan would be a gift from heaven, but there is no evidence that anyone realized it at the time.

The Halsted Observatory became Dugan's life and the bane of his existence. He was not the first to attempt photometry but was the first, apparently, to sense how architectural features of the old octagonal stone structure with connecting offices, while quaint and homey, did not work well for precision photometry. The eight balconies, tall glass doors, and bull's-eye windows were nuisances. Spring-loaded shades snapped up without warning at night, "bringing one dormitory light after another to bear on the observer's eye." Thin clouds could sneak up without notice and seriously compromise observations, since one had to struggle with the glass doors to gain entry to the balconies.[45] But the most challenging problem for the solitary observer was the fact that the telescope sat on a campus, with its own native culture:

> It is an immemorial custom at Princeton during the examination period, when the college bell rings at 9 P.M., to throw up windows, to yell and fire revolvers. Fresh air comes into rooms and lungs. Bedlam lasts for five minutes and is known as "poler's recess." The dome of the old Halsted was regularly used as a target for revolver shots, because of the delightful reverberations. The thickness of the sheeting alone kept them from coming through.[46]

Dugan was comforted by the protection of the heavy dome and thick walls, and, like many observers, he befriended the night creatures who inhabited the place. "Bats squeaked and scraped as they circled the wall and with a quick flutter of wings dodged the observer's head." Pigeons claimed the nooks and crannies of the shutter housing and scrambled about when he opened the shutters. In the winter, opening the dome sometimes meant that "a half ton of snow was pushed off and came hurtling down" onto the observing floor. "Pieces of hardware were liable to do likewise." A black cat lived somewhere around the pier. He would often streak across the floor, pausing "to make eyes at one in the dark."[47]

Dugan became master of the double-image polarizing photometer. With this machine he could visually compare the brightness of a standard star to his program star. He preferred the design because it did not require a comparison lamp and provided consistent measures of relative brightness.[48] All the equipment, however, required tender loving care. In January 1906, he reported to Campbell on his progress: "Mucho trabajo, Señor, at the Princeton Observatory removing dust and oiling up. The status quo is about as I thought it would be. Frost didn't know when he would have the money raised, I was on my uppers and took what was coming. 'The opportunities are large' but I wish I had someone to help push."[49]

Indeed, Dugan had sole responsibility for the observatory, save for the occasional student like Daniel. After the end of his first year at Princeton, Dugan suffered ill health from overwork and from the climate. Mitchell worried about him but was relieved when, after a particularly bad episode, Dugan started to gain weight and strength and was "sensible enough to stay home and rest, instead of trying to do some work in his weak condition."[50] By the end of his second year, however, the illnesses returned, and Dugan longed for a better climate. He knew he had made a good impression on Campbell, who found him "well-trained, resourceful, and efficient . . . ," so in September 1907, he asked Campbell for a job. Campbell, however, needed spectroscopic specialists at Lick Observatory.[51]

Dugan's dissatisfaction with Princeton was not only health-related. Mitchell, who had hovered about the observatory a bit after graduation, found Lovett as much of a nitpicker as his present boss, Frank Schlesinger, who created a cumbersome bureaucracy at Pittsburgh that was a "!!!!!!! nuisance."[52] Schlesinger reminded Mitchell of Lovett: "I suppose that Lovett will be much relieved when the clocks are out of the room. I never knew that he worked so hard that he needed absolute quiet. Has he had the white labels on the books removed yet? That must worry him alot."[53]

Lovett labored over details and failed to grasp larger issues. Dugan complained more than once that Lovett had mislaid work and correspondence that was of great importance to his research.[54] Through 1907 he became aggravated with the fact that he was carrying the entire responsibility of the observatory while teaching two full-year courses (theoretical astronomy and practical astronomy) in his first two academic years. His teaching load equaled Russell's, but Russell did not have to run the observatory. Lovett taught only half Dugan's load the first year and spent more time on campus politics.

Dugan and Russell remained as instructors at $1,000 per year through 1907. When Lovett resigned in January 1908 to take the presidency of the Rice Institute in Houston, Texas, Russell and Dugan were advanced to assistant professorships, with a doubling in income. Dugan did not seem to mind that Russell initially got a larger raise, to $2,200. What was important, he noted later to Campbell, was that "[m]y health is capital this year, and our headless department is quite enjoying itself. I have twice as much teaching as formerly, but other things have doubled, too, so I do not repine."[55]

In addition to his doubled salary and increased status, the University granted Dugan free housing adjacent to the Observatory of Instruction at 14 Prospect, next to the director's residence at 16 Prospect where Lovett had lived. Dugan's life also brightened in early 1909 when he became engaged. This brought him a $200 raise, and parity with Russell, because, as a married man, Dugan could no longer live at the observatory.[56]

Lovett's Departure

Lovett recalled years later that the inspection tours he took in 1905 and 1907 helped to make him the founding president of the Rice Institute in Houston, Texas.[57] When Lovett's resignation was announced, Frost wrote W. F. Magie, gratuitously hoping "it is not true."[58] Others pressed Lovett to hire their students as his replacement. A. O. Leuschner at Berkeley pushed his protégé Russell Tracy Crawford "as a man eminently qualified to head the department."[59] But Fine and Wilson preferred to "take care of the department here for the immediate future by the promotion of Dugan and Russell to assistant professorships."[60]

Years later, Russell recalled that when Lovett resigned, Dean Fine "stepped into the breach and protected my inexperience" for several years until the time came for him to take over the chair and directorship.[61] This is a likely scenario, given that Princeton was reluctant to look beyond its borders for its administrators. Neither Dugan nor Russell was ready to assume the directorship, nor was Princeton willing to hire a third faculty member in astronomy. The graduate school still lacked support, even though other areas of Princeton science were flourishing with the opening of the Palmer Physical Laboratory in October and, a year later, of the massive Guyot Hall, which housed the growing geological and biological sciences.[62] Lovett had apparently secured some funding to explore possible sites for a Princeton observing station in South Africa, but his initiative was stillborn when he resigned. Also by then a portion of Wilson's bubble had burst. Expenditures to hire the fifty-odd preceptors had outstripped income, and there were other troubles on the horizon over the governance and siting of the graduate school.[63] In addition, graduate enrollment was nil in astronomy. But graduates mattered far less than undergraduates, as there was a small but steady enrollment in the elementary courses.[64]

Russell's and Dugan's promotions were taken out of the remains of Lovett's salary. The annual budget of the astronomy department for the past two academic years stood at $5,400, just a bit more than Jeans's salary. Out of this, starting in the fall of 1909, Russell and Dugan received $2,200 each, leaving $1,000 for all operational expenses, such as the upkeep of the observatory, clerical assistance, and new equipment.[65]

With Lovett's departure the department of astronomy ceased to be a formal unit. Although course catalogs and colleagues referred to it casually, it was now part of the mathematics department. Dugan and Russell came under Henry Fine's wing; as dean of the faculty and dean of the Departments of Science, Fine naturally took charge of loose ends. The observatory residence was taken over by the physicist Malcolm MacLaren, incoming chair of the Department of Electrical Engineering.[66]

Dugan seemed not to worry about their loss of status. His light-hearted report to Campbell in February 1909 implies strongly that the freedom he now experienced without a boss like Lovett over his shoulder was, in fact, worth more than the loss. Dean Fine, one of the most powerful allies anyone could have at Princeton, kept watch but rarely interfered. He let his two young assistant professors concentrate on astronomy with no evident agenda for the observatory other than to keep it open as cheaply as possible.

• CHAPTER 6 •

Parallaxes, Pedagogy, and the Lives of the Stars: Russell's First Years on the Princeton Faculty

Nondepartment Organization

After Lovett's departure Dean Fine periodically called Dugan and Russell to his home for meetings to discuss course content, minor purchases, policies for student access to the observatory, publishing Dugan's observations, and other matters typical to the running of an observatory.[1] Starting in March 1910, Fine asked Russell to act as secretary, recording decisions. His sparse notes indicate that Fine did not question research programs or second-guess needs. The department thus continued on much as before, though fewer courses were taught.[2]

Dugan maintained the telescopes and instruments, and Russell handled the photographic measuring equipment and the classroom. But Russell had the inside track. In December 1909, he, not Dugan, was asked in confidence to update Lovett's plan for overhauling the Halsted. Though conservative compared to the visionary Lovett, Russell called for "radical change." The telescope had to be remounted and rebuilt as a photographic instrument. It had to be removed from campus lights, smoke, and the "shivers of the ground" caused by the railway head less than one hundred yards away. The observatory still had to be within a mile of the university for the students, and so the best place was on the golf links behind Russell's home, the open ground between campus and the new site for the Graduate College. Golf, he insisted, could coexist with astronomy.[3]

Russell wanted to convert the Halsted into a large version of the Sheepshanks. The mounting would be expensive, but the building would be much smaller and cheaper to maintain than a huge rotating dome. The entire renovation would cost $50,000, assuming that the present library and offices would be housed elsewhere, possibly in the new Graduate College. The latter would be safer than the present ramshackle wooden structure on Prospect, which was in danger of fire now that the area was heavily residential and overrun by students. Finally, wherever the observatory was placed, the observer had to be housed nearby.[4]

Knowing that his advice was sought out for political reasons, Russell said nothing about spectroscopy or a southern site. He knew that the golf links choice supported the winning side of the controversy over the location of the Graduate College, favored by A. F. West, the graduate dean. West convinced powerful alumni and trustees that a separate campus was more conducive to serious study and research than was Wilson's idea of an integrated campus quadrangle. This issue nearly tore the faculty apart, since it carried deep emotional and philosophical implications concerning the role of graduate study in Princeton's future. Russell's suggestions, no matter how private, indicate a break with Wilson and with those who fiercely supported him, including Lovett, Veblen, James Jeans, and most of all Henry Fine.[5] He strongly supported West's vision of a strong program for graduate research-based instruction at Princeton. Although the Graduate College became a reality, the Halsted Observatory stayed put on University Place amid the light and smoke of campus. Dugan drew the shades.[6]

Dovetailing Research Programs

Russell's vision of converting the 23-inch was impractical. The same conversion could have been accomplished, as Frank Schlesinger had demonstrated at Yerkes, through the use of a double-slide plate holder and yellow filters. Although the meager record of departmental meetings does not reveal that Dugan ever suggested modifications to the equipment, at one point Russell did propose adding a yellow filter for photography with isochromatic plates.[7] Dugan seemed settled using the visual photometer: after all, as he claimed to Frost in 1905, he was not too particular about what his research might be.[8] Therefore on the surface Dugan and Russell's research interests complemented each other, but this was more a matter of circumstance than choice. There was no photographic equipment at Princeton such as that found at Heidelberg, and Dugan had no training in spectroscopy.

Dugan's observing program was defined by his instrumentation. Visual observations of eclipsing variables like RT Persei were also desirable because they exhibited fascinating variations that indicated the presence of a third body. As Dugan later argued, understanding classes of stars like RT required observations over the whole of their light curves. He wanted "to cover the entire mean curve thickly with observations" and accomplished the feat in the two observing seasons 1905–1906 and 1907–1908. The results were compiled as the first volume of the *Contributions of the Princeton University Observatory*, in 1911, which was subsidized out of the operating funds of the department.[9]

Russell's first priority was his parallax program: he had to finish measuring but was stalled in the fall of 1905 without a new promised measuring machine, modeled after one at Cambridge. Russell therefore concentrated on developing course materials and turned to parallaxes only when the machine arrived from the Cambridge Scientific Instrument Company, Ltd., in March 1906.[10] Thereafter, Russell spent weeks checking and calibrating the machine, tearing it apart and determining the periodic error in the screw. He looked for all known types of systematic error and developed procedures for measurement, his "rules of practice." Consistency was everything in a measurement-and-reduction process that was likely to take years, since he was working alone.[11]

After his summer vacation in Oyster Bay, Russell started measuring plates and by the fall started to produce preliminary parallaxes.[12] His lectures, aside from elementary courses, reflected his research, centering on the theory and practice of reducing photographic star plates and the theory and practical use of the spectroscope for astronomical investigations. He offered them sequentially in the 1905–1906 and 1906–1907 academic years to both undergraduates and graduate students, if there were any.[13]

Russell continued to study Darwin's papers on rotating figures in equilibrium and on the stability of a liquid satellite in orbit around a large disturbing mass. He also explored the theoretical characteristics of rotating bodies that contained random surface markings or spots, revisiting an old idea that had been suggested to explain variable stars but applied as well to starlike planetary satellites and asteroids. This study, stimulated in part by his desire to reconfirm that eclipsing binary light curves could not be understood in this manner, was completed in April 1906 as he was calibrating his new measuring engine.[14]

His monthly *Scientific American* columns continued unabated and often reflected his own research interests. He highlighted special events and was especially proud of Zaccheus Daniel's discovery of a new comet, which dominated his column from June to September 1907. It brought honor to the university and to this exceptionally active undergraduate.[15] Russell also collaborated with Daniel in a reanalysis of Venus's luminous ring, and they revisited a statistical distance-determining technique Russell had first tried out at Cambridge.

Hypothetical Parallaxes

Russell never failed to look for shortcuts to acquire astronomical data, and it sometimes got him into hot water. Frustrated by the tedium of his parallax work while laboring away in Cambridge, Russell had applied Kapteyn's new mean parallax technique to determine the distances to a set of visual binaries with known orbits. Between 1900 and 1904, Kapteyn showed that the proper motions of stars, their apparent motions across the line of sight, could be evaluated statistically to obtain useful distance information, and he derived a relationship linking distance (parallax), proper motion, and apparent magnitude that became widely accepted. Russell applied Kapteyn's technique to show that he could determine, "to a very fair approximation," the average mass of stars in binary systems. He tried it out on a small sample with trigonometric parallaxes, finding an average mass for the individual stars of 4.5 times the Sun. He then reversed the procedure and used this mean mass to rederive parallaxes to all systems with known proper motions and magnitudes. He found good internal agreement for visual systems and for spectroscopic systems, which convinced him that a new statistical way to determine parallaxes was at hand. His method exploited the fact that the observed orbit of a binary star system, and the system's combined mass, period, and parallactic distance were all related through Newton's law of universal gravitation and geometry. Russell called his technique "hypothetical parallaxes," crediting Kapteyn and William Doberck.[16]

Now at Princeton, with Daniel's help, Russell applied the technique to nineteen suspected visual double stars surrounding the Orion Nebula. They all were of similar brightness and spectrum and had extremely small proper motions. None had definite orbits, but Russell convinced himself that their common proper motions indicated orbital motion, which he extrapolated into rates of motion and approximate periods, making statistical corrections for foreshortening. Using some rather weak criteria, assuming that the masses of these stars were similar to those he studied in Cambridge, he assigned a mean mass of ten solar masses for each system and with this calculated that the Orion Nebula group was some six hundred light years distant, give or take 50 percent. At this great distance, "the principal stars in Orion must be of enormous brightness. Rigel in particular being probably more than 10,000 times as bright as the Sun." Russell presented these results at the December 1906 meeting of the Astronomical and Astrophysical Society of America (AAS) at Columbia University and wrote them up for the *Astrophysical Journal* and *Popular Astronomy*.[17]

Edwin Frost, editor of the *ApJ*, rejected the paper on the advice of the venerable double star observer Sherbourn Wesley Burnham. "There is no more eminent authority" than Burnham, Frost told Russell. The problem was not with Russell's mathematics, but with the data, which were "so totally inadequate that the conclusions must be wholly illusory." Burnham could find no evidence of orbital motion, so Frost counseled Russell that "ten years from now you will be glad if you should withhold it from publication altogether."[18] Russell, however, did not give up, knowing that George C. Comstock, the well-respected director of the Washburn Observatory at the University of Wisconsin, was then applying a similar technique to faint binary systems. Comstock was one of the few senior Americans who respected Kapteyn's methods, though he found fault with many of his assumptions and felt his results were vitiated by space absorption. Comstock's keen criticism convinced Russell that his assumed stellar masses were too high. Reducing them to 2.4 using early parallax results, Russell found that the internal accuracy of his hypothetical parallax technique improved significantly.[19] But he hesitated to resubmit his paper. Even though Russell wanted to develop statistical methods that could draw as much as possible out of the data at hand—what he would later call his "tissue of approximations"—and even though he wanted to show that stars of enormous intrinsic brightness existed, he also knew that he was pressing too hard against the standards of acceptable practice in astronomy. Lacking Comstock's prestige, Russell had no recourse but to keep his nose to the grindstone and produce stellar parallaxes that were beyond reproach.

Russell's parallax reductions between 1906 and late 1910 fill some twenty-three thick folders and reveal an enormous amount of work, excruciating attention to detail, and an insatiable curiosity for ferreting out correlations that might reveal systematic error, or real physical effects of astrophysical interest. Through 1907, Russell measured and remeasured his parallax stars, paying additional attention to various "problem" stars whose probable errors were large. He had to account for any perceptible motions in some three hundred comparison stars.[20] He ultimately determined that the probable error of a single exposure, measured once, was a mere 1/10,000 of an inch. "It is simply amazing that the gelatine film, subjected to so many processes, shows such great stability."[21] Russell was finding, along with others at the time, that the reseau system championed by Hinks and Turner was not really required after all. Glass-backed photographic plates were very stable.

Russell searched for errors caused by color differences, brightnesses, and spectral types. He also needed these data to convert his relative parallaxes into absolute parallaxes. Such information, Russell knew, lay at Harvard, where the Henry Draper Memorial had accumulated over ten thousand classifications of stellar spectra and visual magnitudes for stars. The classification project alone, the product of a corps of dedicated female assistants working doggedly for decades, would be acclaimed at midcentury as "the greatest single work in the field of stellar spectroscopy."[22] When Russell told Wilson in 1905 that thousands of photographs were being taken at American observatories every year, he most assuredly had in mind what was going on at places like Harvard.

Pickering's Offer and Its Context

Edward C. Pickering, Harvard College Observatory director and president of the AAS, had encountered Russell at society meetings, hearing him talk about his parallax work and his statistical schemes. They spoke on one or two occasions, which emboldened Russell, on a personal trip to Boston in April 1908, to write Pickering asking for an audience. They talked of parallaxes, and Pickering asked to read Russell's papers. Within a week of their meeting, Russell sent Pickering his lists of parallax and comparison stars, asserting their astrometric accuracy: "[T]here is not one chance in ten that such a result is in error by one-twentieth of a second of arc." Pickering soon agreed to provide unpublished magnitudes and spectra, which years later Russell wanted to remember as "an unsolicited contribution to the work of a young and unknown instructor!"[23]

Pickering had good reasons to help Russell. He had invested decades establishing new systems of stellar brightness and classification, and was anxious to have them accepted by astronomers worldwide. Even though Pickering enjoyed great personal influence and had often tried to achieve consensus, there was as yet none and no assurance that it would come. Thus he championed any use of his systems, especially if someone offered the possibility of making a fundamental discovery that might inform the so-called sidereal problem, statistically analyzing the form, structure, and extent of the sidereal system.[24] Accordingly, Pickering suggested that Russell's parallaxes "would perhaps be sufficient to determine which were the most distant, stars of Class A or Class K." Russell was confident that if a difference in mean parallax between stars of classes A (hot stars showing a predominance of hydrogen) and K (orange-red stars showing metallic lines) of so little as 0″.01 of arc existed, it "would be clearly indicated" by his ultimate analysis.[25]

What Pickering asked of Russell lay at the heart of the sidereal problem, the "major (traditional) research program of stellar astronomy" at the time, which not only rationalized the great effort needed to produce stellar parallaxes but also prompted various shortcuts that had both ardent supporters and conservative opponents. Pickering was keenly aware of these various shortcut methods, developed by Kapteyn and others, and in fact had been trying his hand at statistical modeling to infer a density law directly from his data. But he knew that hard evidence from trigonometric parallaxes would offer the best test.[26]

Pickering also had a deeper motive. An obscure Danish astronomer named Ejnar Hertzsprung had been pestering him about the importance of a competing classification system, one also from Harvard but created by Antonia C. Maury. Maury, the only woman on Pickering's staff trained in science and mathematics, noticed that some stars exhibited especially sharp hydrogen lines and enhanced metallic lines. She called them "c" stars and speculated on their evolutionary significance. Pickering was never happy with Maury's system of classification but acquiesced in its publication as she was the niece of Henry Draper, Harvard astronomy's major patron. Pickering did his best to ignore her classification, but now this young Dane was stirring it all up again.[27]

Whales and Fishes: The Parallel Work of Ejnar Hertzsprung

Working within Kapteyn's framework, Hertzsprung confirmed earlier hints that not all stars were alike, "one being dull and near us and the other bright and remote like the Sirians."[28] In 1905 he found that the brightest stars in the sky tended to exhibit Maury's c-characteristic and also had the smallest proper motions. Stimulated by this remarkable correlation, Hertzsprung then extended his examination to redder stars where the c-characteristic was not visible. Using Kapteyn's mean parallax technique, he found that the brightest red stars had the smallest proper motions as a group, the faintest red stars had the largest proper motions, and the resulting luminosity gap between the two increased with advancing color, or redness. Maury's c-class stars were therefore of a wholly different luminosity than common stars.[29]

In March 1906, Hertzsprung sent Pickering his findings to demonstrate the value of Maury's subclassification.[30] When Pickering refused to follow Hertzsprung's advice, Hertzsprung was deeply puzzled, partly because others, like especially belongs to the stars with relative great reduced proper motions.

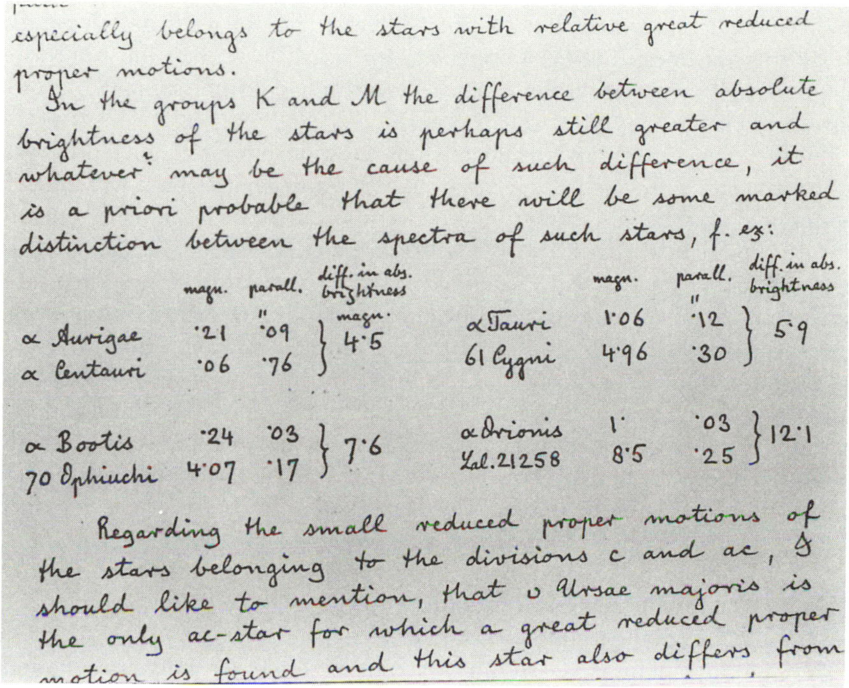

In the groups K and M the difference between absolute brightness of the stars is perhaps still greater and whatever may be the cause of such difference, it is a priori probable that there will be some marked distinction between the spectra of such stars, f. ex:

	magn.	parall. ″	diff. in abs. brightness magn.		magn.	parall. ″	diff. in abs. brightness
α Aurigae	·21	·09	4·5	α Tauri	1·06	·12	5·9
α Centauri	·06	·76		61 Cygni	4·96	·30	
α Bootis	·24	·03	7·6	α Orionis	1·	·03	12·1
70 Ophiuchi	4·07	·17		Lal. 21258	8·5	·25	

Regarding the small reduced proper motions of the stars belonging to the divisions c and ac, I should like to mention, that υ Ursae majoris is the only ac-star for which a great reduced proper motion is found and this star also differs from

Fig. 6.1. Pickering knew of Hertzsprung's results after receiving his letter of 15 March 1906 with a clear tabulation showing the two sequences of stars. Hertzsprung reminded Pickering of this relationship in subsequent correspondence through 1908. Reproduction from the E. C. Pickering Papers, courtesy of the Harvard University Archives.

Hertzsprung's new mentor Karl Schwarzschild, were very excited about his work, and Antoine Pannekoek of Amsterdam was finding the same luminosity effects in spectra.[31] In July 1908, Hertzsprung wrote to Pickering again urging that Maury's c subclass be part of the Harvard system:

> It is hardly exaggerated to say that the spectral classification now adopted is of similar value as a botany, which divide the flowers according to their size and color. To neglect the c- properties in classifying stellar spectra I think, is nearly the same thing as if the zoologist, who has detected the deciding differences between a whale and a fish, would continue in classifying them together.[32]

By then, Hertzsprung knew that giant stars lurked among the common stars of space. This idea was not entirely new; in the words of the chronicler Agnes Clerke in 1903, "[t]he existence of a luminary so vast as Canopus, although bewildering to imagination, need not appear incredible when we consider the immense scope of creation."[33] But Hertzsprung's analysis relied upon an unpopular classification system and used indirect statistical techniques, ones Pickering had little faith in, since he tried them himself without success.[34] Pickering believed that Maury's classification went beyond the data. He knew that a few high-quality slit spectra showed "great differences in the width and sharpness of the spectral lines," but her sample was too small to warrant the wholesale adoption of the subclassification. Thus Pickering would not support Maury's classification.[35]

But Pickering never answered the obvious question: if the c characteristic was an instrumental effect, why did Hertzsprung find strong correlations with proper motion and brightness? Hertzsprung said as much in rebuttal and continued to press his findings. In August 1908 he prepared a summary paper for the *Astronomische Nachrichten* and sent Pickering his draft manuscript. Hertzsprung told Schwarzschild, "I am annoyed with the Harvard spectral classification since one cannot determine how the spectral identification is to be taken. Thus the entire classification is inhomogeneous."[36] Thus Hertzsprung initiated a complaint that would be echoed by astronomers for the next four decades. And he had good reason to feel this way. His statistical studies had confirmed the existence of high luminosity stars among the yellow classes. Moreover, he had earlier applied his training in photochemistry under Ostwald at Leipzig to be one of the very first to use Max Planck's new radiation formula, which equated the energy distribution of a body (its color, roughly) to its temperature, to calculate the angular diameter of a star, the c-type star Arcturus, showing that it was truly a giant, over one hundred times the diameter of the Sun.[37]

Hertzsprung sent his papers to American and British observatory directors and influential astronomers.[38] They are preserved today in the libraries at Yale and Yerkes, but there is no indication that they were read. Karl Schwarzschild did what he could to make Hertzsprung's discovery known, lecturing on it and calling the highly luminous stars Hertzsprung had discovered "giants." In 1909 Schwarzschild invited Hertzsprung to Göttingen and then took him to Potsdam after Schwarzschild was named director of the most prestigious astrophysical observatory in Europe.[39]

Russell's Final Reductions and Revelations

Against this background we can appreciate Pickering's motives and his suggestion that Russell determine the status of the K stars, but there is no evidence that he ever told Russell about Hertzsprung's letters. In August 1909 Pickering sent Russell the magnitudes and spectra he needed, and, upon return from his usual vacation in Oyster Bay, Russell set himself to the task of incorporating them into his reductions. By the end of September, after a flurry of work, Russell had made some striking observations based upon a sampling of less than half of his parallax stars. Removing the brightest stars from his rough sample, he found "very strong evidence that the fainter stars average *redder* than the brighter ones." Russell then added, to establish his priority, "I do not know of any previous *direct* evidence on this question."[40] Indeed, Russell's "direct" parallaxes confirmed Hertzsprung's

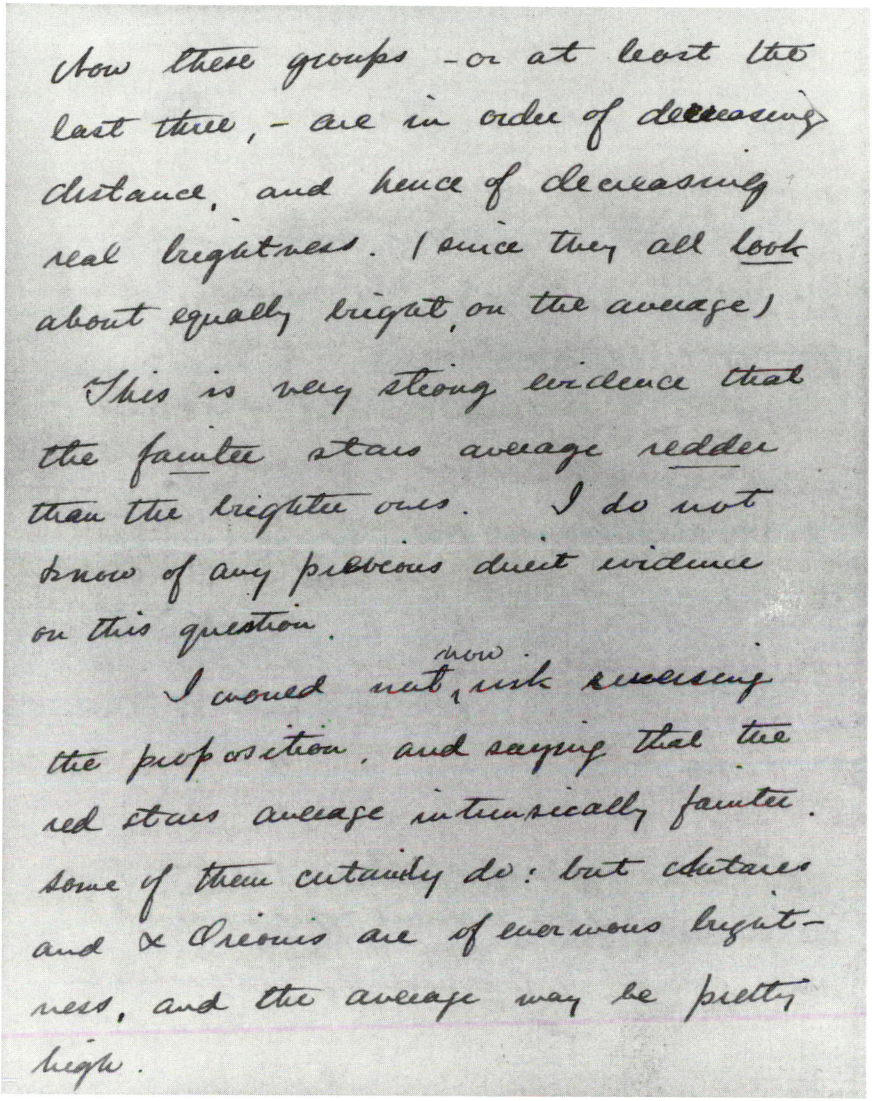

Now these groups — or at least the last three, — are in order of decreasing distance, and hence of decreasing real brightness. (since they all look about equally bright, on the average)

This is very strong evidence that the fainter stars average redder than the brighter ones. I do not know of any previous direct evidence on this question.

I would not now risk reversing the proposition, and saying that the red stars average intrinsically fainter. Some of them certainly do: but Antares and α Orionis are of enormous brightness, and the average may be pretty high.

Fig. 6.2. In a 24 September 1909 letter, p. 4, Russell reports to Pickering on his finding a correlation between brightness and color, and the existence of highly luminous stars, stating that he knows of no previous work on the subject. Reproduction from the E. C. Pickering Papers, courtesy of the Harvard University Archives.

statistical inferences. He even came to the same conclusions: "I would not now risk reversing the proposition and saying that the red stars average intrinsically fainter—some of them certainly do: but Antares and [Alpha] Orionis are of enormous brightness, and the average may be pretty high."[41] Russell outlined the steps he would take to firm up his discovery, but in surviving correspondence

Pickering never mentioned Hertzsprung's identical findings. He only encouraged Russell and promised more help if needed.[42]

Why did Pickering remain silent? He claimed to be skeptical of objective prism spectra and Hertzsprung's indirect statistical methods. Possibly he did not want to bias Russell, or demoralize him, since Russell was evidently confirming the effect Hertzsprung had found in a way that did not promote Maury's complex system. Any of these motives could have developed from Pickering's overarching need to ensure that his favored alphabetical scheme survive an ultimate shakeout, which by then he knew would happen in the summer of 1910 as the world's elite astronomers gathered at Mount Wilson for the fourth meeting of Hale's International Union for Cooperation in Solar Research.

Consensus Building: Planning for the 1910 Solar Union

Hale had been planning this meeting for over a year. He convinced Pickering to participate by suggesting that it would be a safe venue for the assessment of competing systems of spectral classification and magnitudes. They both were anxious to establish the Union as a new forum for standardization that had been heretofore the sole responsibility of European-based organizations such as the Astronomische Gesellschaft and its contentious offshoot, the *Carte du Ciel*, which was about to meet again in an Astrographic Congress in Paris.[43] Pickering's best hope for his magnitude and spectral classification systems lay with Hale's organization.

Pickering planned it all out. He invited the AAS to meet in Cambridge just before the Union meeting to set the stage and invited the most prominent foreign astronomers to attend as honored guests. Hale also agreed to Pickering's suggestion that he hire a special train to enable participants to "go west together holding meetings on the way."[44] This gave Pickering the opportunity to lobby for his systems with people like Kapteyn and Schwarzschild. Pickering must also have known that Schwarzschild would carry news of Hertzsprung's work to America, and that this would rekindle interest in Maury's system. But now he had Russell's results, based upon his own system. Anticipating the inevitable showdown, Pickering decided to test Maury's system himself. He contacted both Frost and Campbell a month before the meetings, asking them for examples of high-quality slit spectra, saying only that it would help clarify "small differences in the relative intensity of the lines."[45]

Russell's Parallaxes: The Last Lap

Meanwhile, Russell worked feverishly, beating down all sources of error to obtain an impressive probable error for his parallax stars of ±0.005 seconds of arc, equaling or eclipsing any other technique. Russell had sent his results to Kapteyn and received his enthusiastic approval and endorsement. With Kapteyn in his corner, Russell proclaimed that he had broken new ground and was now in a position to draw important conclusions from his data.[46]

Russell was extremely aware of the fact that he had uncovered something really important. Not only did his discovery of giants justify his years of laborious searching for systematic errors, but he was telling the most influential astronomer in America that he claimed ownership of a new technique and its first fruits. Accordingly, Russell signed and dated every separate page of correspondence with Pickering that contained a result, and later asked Kapteyn for an endorsement of his achievements, which Kapteyn gladly gave to promote Russell at Princeton.[47]

Constantly worried about being scooped and seeking more data that might be useful, Russell poured over the literature, looking for any relationship between spectrum and luminosity that could influence the one he had found. In more than one instance, he found articles of interest in the *Astronomische Nachrichten* and read them faithfully. He even read an article in the same issue that contained Hertzsprung's 1908 summary paper but failed to note Hertzsprung's work explicitly to Pickering.[48] By the end of the year Russell must have known that he was in a race for priority. Still, surviving correspondence with Pickering was silent on this point.

By 1910, along with his teaching duties and continuing interest in the origin and evolution of binary systems by fission, which he took up again following Darwin's lead, Russell was in the final stretch of reducing his parallaxes. His undated working notes reveal that while he was still calculating parallax factors and equations of condition in early 1910, there were only a few stars that still did not behave properly. By then he had collected data for forty binary systems and had compiled a visual "progress chart" listing observations, some covering four epochs. He also organized long listings of "Astrophysical Data" on magnitudes and proper motions arranged by spectral type that allowed him to derive crude average luminosities. He derived the space density of stars of different types and hoped soon to report "positive results as to the relative distances of the different groups"—just what Pickering had asked and which Hertzsprung's analysis had purported to solve.[49]

In the spring, Russell was ready to announce his finding "that increasing redness goes with increasing p.m. [proper motion] or parallax."[50] Pickering made sure that Russell could air his findings several times before the critical August meetings. In late April, at Pickering's behest, Russell spoke "[o]n the distances of red stars" at a meeting of the American Philosophical Society. "With the latest determinations of temperature and surface brightness," Russell suggested that the faint red stars in his sample "are somewhat smaller, and presumably denser, than the sun, while the brighter ones are very much larger than the sun, and presumably of very small density. The latter class probably represent an early stage of evolution, and the former the latest stage that can be observed."[51] Much of Russell's research effort over the next three years, with the aid of his graduate students, was devoted to getting rid of the qualifiers "presumably" and "probably" in what became his theory of stellar evolution. Indeed, from this point on, Russell emphasized how his parallaxes informed this perplexing problem.

At the AAS meeting at Harvard, Russell presented two papers: one on his theory of stellar evolution and another on a new highly efficient means of calcu-

Fig. 6.3. Detail from group photograph of the 1910 American Astronomical Society meeting. *Left to right*: Annie Cannon (*left front*), Mrs. Russell (*black hat*), Dugan (*back row, to Russell's right*), Russell, and Pickering (*right front*). Photograph courtesy Yerkes Observatory.

lating the orbital parameters of eclipsing binary systems like Algol to determine actual brightnesses, dimensions, and ultimately stellar densities, just what he needed to rid himself of the qualifiers. Of the two papers, "Some Hints on the Order of Stellar Evolution" was the more spectacular, for here he finally linked his theory to Lockyer's old idea of a double-valued temperature history for the life of a star. He began, however, not with parallaxes, nor with Lockyer, but with the theories of Lane and Ritter, and basic assumptions shared by many astronomers, "that a star grows denser as it advances in evolution; that it is in equilibrium under its own gravitation," that it is an isolated system, and that the gases composing its volume behave according to the familiar gas laws until they reach a critical density, beyond which they cool.[52]

Russell said nothing of Hertzsprung in his long summary report "Determinations of Stellar Parallax," which he had completed just before he left for Harvard and the Solar Union meetings. But in the abstract of his Harvard address, written after the meeting and after his first encounter with Schwarszchild, Russell finally acknowledged Hertzsprung's work.[53] By the time the group had moved on to Mount Wilson and the meetings of the Union, Russell and Schwarzschild had plenty of opportunities to talk. Russell saved a dinner menu from the Union banquet. In addition to the notable signatures Russell had gathered on it—as was the custom—one finds next to Schwarzschild's name a cartoon of a rapidly rotating star about to fission. No doubt they talked of many things.[54]

Russell drew positive attention to Pickering's classification scheme. Thus the Boston meetings and the train west were all Pickering could hope for. Along the way, Schwarzschild agreed to adopt the Harvard magnitude system, and strategy meetings were held with Schlesinger, Russell, and others to plan out how they would decide on standardization for spectral classification. Hale also did his part, appointing Pickering chair of the ad hoc committee that would deliberate over the classification question. Pickering became the center of attention on the train

Fig. 6.4. The fourth meeting of Hale's International Union for Cooperation in Solar Research in August 1910, at Mount Wilson. Among those important to Russell at this time are: (*front row, left center*) W. S. Adams (Russell is behind his right shoulder), E. C. Pickering, and Hale. The third man from Hale's left is J. C. Kapteyn (*white mustache, dark suit*). Behind Kapteyn and to the right is Frank Schlesinger. Karl Schwarzschild is behind Schlesinger (*to the left*), standing to W. W. Campbell's left. Courtesy Huntington Library and Art Gallery, Manuscripts Division.

west: there were "informal committees wherever I sit," he wrote in his diary. "My part in this will be regarded as one of the most important things I have ever done." Thus by the end of the year, as a result of these meetings in Boston and Pasadena, as well as the train ride west, Pickering's systems of photometry and spectroscopy gained strong endorsement. And of course he also had Russell's confirmation of Hertzsprung's findings.[55]

Russell, for his own part, was single-minded. As J. S. Plaskett, director of the Dominion Astrophysical Observatory in Victoria, British Columbia, and a major supplier of radial velocities, recalled the events that summer:

> I will never forget the trip of the Solar Union in 1910, how hot it was through the desert, what a fine time we had at the Grand Canyon and particularly how Henry Norris Russell worked so hard to persuade his fellow astronomers, what a conservative bunch they were, that the course of stellar evolution had an ascending as well as a descending branch. What a beautiful and satisfactory theory that was . . .[56]

Beyond his political motives, Pickering's resistance to Hertzsprung highlights what were then very different methodologies practiced by astronomers. Russell knew that if his binary data were not impeccable, he would be rejected outright. This experience both chastened him and led him to test Kapteyn's mean parallax technique using his trigonometric parallaxes. Hertzsprung, however, based his conclusions solely upon statistical techniques, or physical theory. Russell was, in fact, one of the very few in America sympathetic to Hertzsprung's type of analysis. There was greater acceptance among Europeans like Kapteyn, Seeliger, Schwarzschild, and Eddington.[57]

Some astronomers have suggested that Hertzsprung's obscurity kept his work from being appreciated in the United States.[58] But Hertzsprung was no longer an obscure worker in 1909; after he published in the *Astronomische Nachrichten*, his work was obscure only to those who did not read German.[59] Both Russell and Pickering read and cited papers in the *Nachrichten* during this time. Russell indeed must have read Hertzsprung's papers not later than 1909, and the abstract of Schwarzschild's address too. By then, obscurity was not a factor in this astronomical triangle.

Russell's Goals and Motivations

Russell's abiding interest in the dynamical histories of binary star systems had taken him to Cambridge. But his aborted theoretical study of the cooling of a sphere of gas and his study of the density of stars in Algol systems confirm that he was also seeking ways to study how stars form and age: what astronomers then and now have called the process of stellar evolution. Although at the turn of the century astronomers in the United States, if pressed, would argue that their chosen specialties informed such ultimate questions, few tackled them head-on, since such matters required a facility for physical theory and a willingness to speculate. It was far safer merely to collect data, as one observer wryly put it in 1900: "American astronomers were more the 'stonecutters' rather than the 'architects' of their profession."[60]

Although stellar parallaxes were a means to an end for Russell, they were what people wanted. He could indulge in speculation only if his parallaxes were impeccable and Kapteyn certainly thought so. He was delighted with Russell's exhaustive error analysis; the "*sine qua non* of really first class work." His "great thoroughness and completeness" vindicated the spectra and magnitudes in Pickering's vault.[61] Of course, Kapteyn had much to gain from Russell's success, not least the warrant it gave to photography and his statistical methods. Kapteyn's imprimatur, however, also gave Russell license to exploit his parallaxes and the relationships they revealed between the various physical characteristics of the stars.[62]

Since Russell always set forth his thoughts on evolution in addenda to his parallax reports, astronomers assumed that he was led to his theory of stellar evolution after he had discovered the fact that giant suns indeed lurked among the dwarfs.[63] The actual process was, however, just the opposite; Russell was, in every respect, theory driven. He had Lockyer's *Catalogue* in front of him as he planned his parallax campaign, and was constantly thinking of how stars formed and lived their lives. He had tried his hand at developing Lane's law, stated his interest in evolution in his 1905 letter to President Wilson, discussed the general question of "The Creation of a Star" in an unsigned November 1906 *Scientific American* article, and was remembered by disciples around the observatory for talking such "things over in the library on Prospect Avenue" as he worked out his theory.[64] The closest he ever came to admitting that his parallaxes could be a test of any theory of stellar evolution, before the fact, was in the spring of 1907 as he prepared lecture notes for the second semester of "Senior Practical Astronomy," which he dovetailed with his graduate course in spectroscopy.[65]

After lecturing his students about the major spectral classification systems, he noted that "[i]t is therefore extremely probable that these types represent stages in the history of a star probably intimately related to its temperature but also to other factors—mass, density, surface gravity."[66] He voiced the common opinion that the spectral classes represented stars at different stages of their lives, but, unlike his contemporaries, Russell left uncertain the order of things: "It is not certain which stars are the youngest," he added. "A cool star may be getting hotter or colder . . ."[67] The answer would come, someday, Russell lectured, from physics, where important advances were being made every year. The old qualitative estimates of color temperature, which from Stefan's law gave wildly different values for the temperature of the Sun's atmosphere, could now be placed on a stronger theoretical footing with Max Planck's new radiation law. In 1907, there were few astronomers outside Germany who believed that stars could be described in this manner. In the absence of a rational basis for spectra, the best one could do, Russell argued, was make empirical correlations. Hale's studies of the relative temperatures of sunspots and red stars was an excellent example, as were

Lockyer's laboratory studies of the ionization ("dissociation") of elements by high temperature. He "was a pioneer in this field."[68]

During these years when he was a new instructor, Russell's notes reveal that he followed very much the character of Young's lectures: he was open, speculative, and anxious to bring in the latest news. Unlike Young, however, who introduced "astro-physics" with humility, Russell spoke of the newest branch of astronomy as an established force: "[I]t has grown so rapidly that its students are already treating it as an equal of all the other branches of the science combined. (Witness the name of the Astronomical and Astrophysical Society of America)." Even though the "old *Astronomy of Position* and the newer *Astrophysics*" were considered to be at opposite poles of the discipline, Russell argued that in fact, "the two are closely allied." Naturally he would say that, considering how he was interpreting his parallax work then.[69]

"Because one star is hotter or cooler than another it is not certain that it is older or younger," Russell lectured his senior class in 1907. Turning to the slate board, Russell drew two possible temperature histories for a star: he first drew a gentle arch to depict one possible course of evolution. He placed young stars at the left-hand base and old stars on the right. Then, moving up the arch from the left to the right, he labeled regions Type III, II, and, at the apex, Type I. Descending down the right-hand branch, he added I, II, and III once more. With that completed, Russell then moved on to draw another diagram, a simple descending line this time, which he labeled Type I at the upper left, Type II in the middle, and Type III at lower right. Turning back to his students, he concluded "we cannot be sure at present though some things look as if the first hypothesis is correct."[70] As of March 1907, then, well before he could have had confirmation from his parallaxes, Russell was squarely in Lockyer's camp, envisioning a double-valued temperature history for stars.

Russell was also very much in Pickering's camp. In 1910 and 1911 he celebrated the work of the Harvard College Observatory in his *Scientific American* column. Pickering, obviously delighted with the publicity, flooded Russell with data and pictorial examples of how variable stars were discovered. Russell always reassured Pickering that he appreciated Harvard's kindnesses: "[M]uch as I admire the photographs as illustrations of the work done at Harvard, I am most inclined to admire them more as examples of the generous policy toward other astronomers of which I have often been a beneficiary."[71] Pickering's patronage had come full circle.

Initial Reactions to Russell's Theory

Aside from his 1907 lecture notes Russell never admitted any influence from Lockyer's meteoritic or dissociation theories; that would not help his case. He knew that Lewis Boss and other senior astronomers who had heard his remarks at Harvard were not ready to roll over and abandon their cherished ideas about how stars aged.[72] Boss was impressed with a striking correlation that Campbell

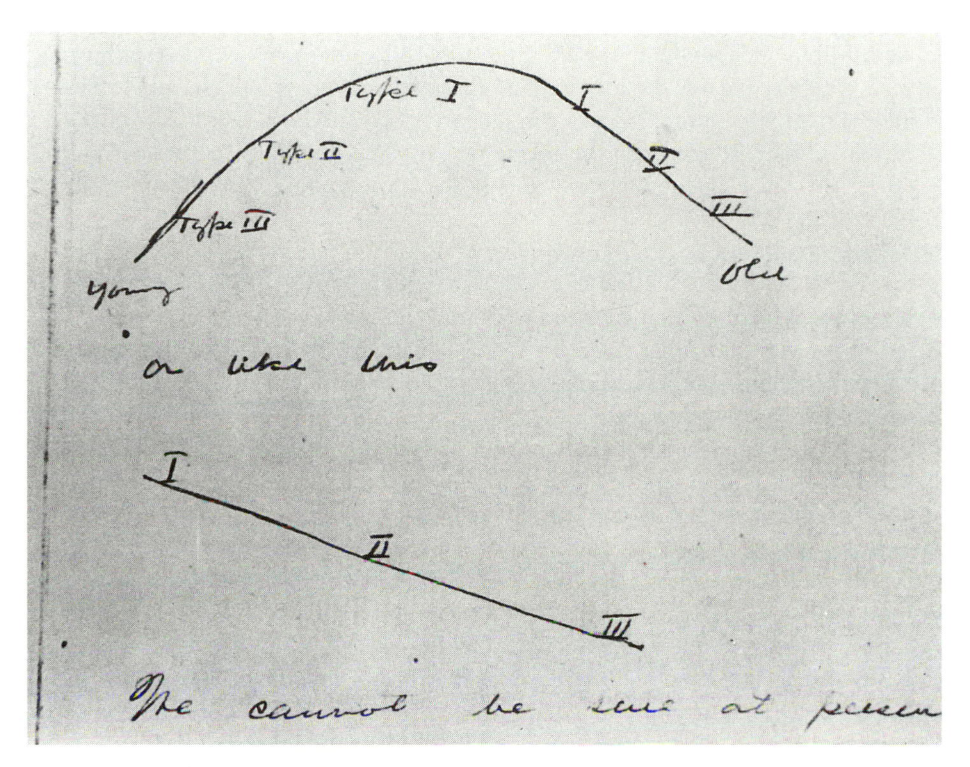

Fig. 6.5. Detail from Russell's lecture notes of 14 March 1907 illustrating Lockyer's theory of stellar evolution (*upper curve*) and traditional view (*lower*) of a linear cooling sequence. Russell preferred Lockyer's view, which he ultimately confirmed with his parallax stars. Reproduction from the Henry Norris Russell Papers, courtesy of the Princeton University Library.

and Kapteyn had just found between the linear sequence of stellar spectra and space motion. Russell, as we shall see, soon had his hands full facing this perplexing observation, which made it all the more important to demonstrate, beyond doubt, the reality of giant stars, and to demonstrate as well that they were giants because of extreme low density and not because of high mass. Binary stars were his only path beyond his parallax work; indeed, as he would say years later, they were his "Royal Road" to the stars. But it would take several years for Russell to marshal these data. In the meantime, he used his parallaxes to their fullest.

Lockyer was, of course, elated to read Russell's abstract in the June 1910 issue of *Science* and evidently heard news from the meetings of August and September that had been attended by some of his closest colleagues, but from which he had felt, and in large measure was, excluded.[73] He wrote Russell in October asking for more: "I have had to wait some years for such a clear cut support of my views & am delighted that it is afforded by researches of a different order from my own."[74] Lockyer was pleased that confirmation came from Princeton, since he and Young had long been cordial correspondents and had often discussed his views. Russell congratulated Lockyer but wanted to be clear that his methods were wholly independent, as were his results: "I should however differ with you very widely as regards the assignment of individual stars to the classes of rising and falling temperature," he wrote in January 1911, and he also emphasized that on his scheme, based upon the theory of Lane and Ritter, not all stars experienced the same evolutionary paths; only the most massive would become the hottest stars.[75]

By then Ritter's theory of convective equilibrium had been revived and extended in the influential textbook *Gaskugeln* (Gas spheres) by Robert Emden. James Jeans believed that Emden's work was premature because there were still too many imponderables influencing the behavior of a hot sphere of gas, but Emden's treatment of Ritter's work was far more defensible than Lockyer's speculations and did provide Russell a theoretical framework for assigning critical density: the point where a star would no longer adhere to the gas laws.[76] Thus Russell told Lockyer, "In my opinion the real test of place in the evolutionary series is *density* (which must increase) we cannot get at this directly, but low density means extended surface . . ." Extended surface area increased luminosity, which, Russell added, was "the best evidence of [a star's] place in the series."[77]

Russell acknowledged how Lockyer's speculation about temperature classifications underscored the importance of looking for a relationship between luminosity and spectrum, as Hertzsprung had shown. "[T]he relative thickness of the hydrogen and 'enhanced' lines must depend on some quite independent set of physical conditions."[78] At the time, Russell thought only in terms of density and concluded that something like Hertzsprung's interpretation of Maury's spectroscopic line criteria might reveal luminosity, but that it required an independent check. In the 1920s Russell would congratulate Lockyer on his daring use of spectral criteria for luminosity and density information, but he did so only years after the spectrum-luminosity relation discovered by W. S. Adams and A. Kohlshütter in 1914 was shown by Meghnad Saha and others to have a strong physical as well as empirical basis.[79]

After Schwarzschild met Russell at Harvard in August 1910, he dashed off a letter to Hertzsprung urging him to send Russell all his papers immediately. If Russell had not already seen them, they were waiting for him when he returned, exhausted, from his month-long transcontinental junket that took him through Flagstaff and the Lowell Observatory, to Pasadena, then to Lick, and finally home via Chicago and the Yerkes Observatory. Russell was looking forward to his family vacation, but Hertzsprung's papers demanded his full attention. In his brief reply, Russell did not indicate whether he had seen them before, but admitted that he had come to much the same conclusions about the two classes of red stars "some time later than you did." After outlining his own theory of stellar evolution in a few sentences, and noting that he had already published a bit of it, Russell added that he hoped to publish a fuller account soon, and he would take "much pleasure to refer to your work, and to see, so far as I can, that you receive the credit which it richly deserves."[80]

Russell never published what was titled "A Provisional Theory of Stellar Evolution," which explored the spectroscopic assignment of relative age. It was a hesitant step, was far from clear and confident, since he admitted then that theories of stellar evolution "appear to be in considerable confusion," and that no one theory fit all the observations, especially the recent kinematical studies of Kapteyn and Campbell.[81] He felt that his theory rested "upon a few relatively simple and plausible theoretical assumptions whose results are in close agreement with the principal observed facts," though it was, he admitted, "seriously inconsistent with some of those at present established." It was, at best, only "a basis for future discussion and criticism" offered as a "forward step in the solution of the great problem with which it deals."[82] He knew that Hertzsprung had now disavowed any evolutionary interpretation in his own work, though he initially harbored such speculation in his early papers.[83] Thus Russell was alone in his speculation and knew that Boss, Kapteyn, and Campbell all were skeptical. He put his paper on hold lest it draw strong criticism at a time when so many questions remained. Hertzsprung had made the situation quite clear:

> On my opinion one of the most important problems of present spectroscopy is to find the spectral equivalent of the great physical difference between bright and dark red stars. The object is to indicate the absolute brightness of a star only from the quality of its spectrum.[84]

Within a year, Hertzsprung would make this suggestion to Mount Wilson astronomers far better equipped than Russell to make the test. Giant suns had to be established beyond a doubt, however, and Russell wisely chose not to deal with spectroscopic criteria but concentrated upon getting real densities from his binary stars.

After the first outings of his theory, in Philadelphia and at Harvard, Russell quieted down. Accordingly, in June 1911, knowing that Pickering was then planning the paper roster for the August meetings of the AAS, Russell offered to talk on "this matter of giant and dwarf stars . . . it is new, has important consequences and is not controversial if the bearing on stellar evolution is not emphasized unduly."[85] Thus within a year of his first announcements of his theory, Russell knew that he was walking on thin ice. He had to have unassailable evidence in the form of direct proof of how density varied with spectral class.

• CHAPTER 7 •

Building a Life at Princeton

Marriage and a Family

There is no evidence that Russell ever considered living elsewhere than at 79 Alexander. During the summers he would visit Oyster Bay. These were melancholy visits now after his mother's death in 1903, but summers in Oyster Bay soon took on a new interest.[1] There was a woman involved.

"I am neither married nor engaged," Russell claimed in his September 1907 class report, which, by the time it was published, was "[Ancient History!]" to the class editor.[2] But he was corresponding with one girl, a summer visitor in Oyster Bay from New York City whom he had known for some time. They had met through a mutual friend and saw each other at church, lingering to talk. They even tried to play tennis, which was more for courtship than for sport. Their relationship remained tentative until her grandmother, "Gammie" Cole, "got definitely impatient." She pushed: "May, why don't you marry Lord Henry?" to which May replied, "Well, I don't know—maybe because he hasn't asked me!"[3]

Lucy May Cole was born into a family of prosperous New York City lawyers on the last day of the year, 1881. She entered the elite Brierly School in October 1892, one of the first prep schools for women in the city, and graduated with the class of 1899. "It was one of Mother's great regrets that she had to 'come out' in New York, instead of going to Smith the way her cousins did," her youngest daughter recalled. Lucy was not comfortable with this prospect: "[S]he was rather a wallflower type and very serious." The Coles found "Lord Henry" serious to a fault, which was fine for Lucy. He presented an air of aristocratic stiffness and knew he was "marrying the daughter of somebody who was much better off than he was." Only with his position secure at Princeton did he feel ready to ask her father for her hand.[4] The marriage took place on a foggy and wet 24th of November 1908, in New York City, at the Coles' St. George's Church on Stuyvesant Square. The St. George's pastor and Henry's father led the ceremony, Lucy's cousins were bridesmaids, Henry's youngest brother Alex was best man, and his Princeton colleagues Luther D. Eisenhart and Oswald Veblen were ushers. The *New York Times* highlighted the event.[5]

They honeymooned in Europe. Always feeling "at home on the sea," Henry, just like his father, insisted upon walking the deck during a bad storm. Unlike his father, Henry fell, broke his ankle, and had to endure a rough setting by a ship's officer. There was a lot of water between Henry's ankle and a proper doctor; it would be another week before they arrived in London, and by then the leg was poorly set. Henry lived with his injury for the rest of his life, requiring high-top therapeutic shoes. When the weather turned cold, he would limp, though the affliction hardly kept him from walking about. At home, his gait was unmistakable: his children recall the " 'thump-bump, thump-bump' as he came down [the stairs] favoring that leg." When Henry and Lucy returned home from their honeymoon, Aunt Ada, "conscientious and self-deprecatory to a fault," moved out for eighteen months because "she wasn't about to have her Henry put out of his house."[6] She eventually returned and lived with the couple in an apartment on the third floor until her death in 1914.

The "genteel country life in Princeton" appealed to Lucy as she learned to run the house, managing two servants who were already well trained, and likely paid, by Aunt Ada.[7] In matters of religion and life, Lucy maintained her independence. She was a devout Episcopalian, less Calvinist than Henry, so they joined their own churches in Princeton, alternating between them week by week. She was interested in Henry's work, and, with Pickering's permission, attended the August 1910 meetings at Harvard where he presented his new ideas on binary reduction techniques and on stellar evolution. In these early years Russell rehearsed his talks by "trying [them] out on my long-suffering family."[8]

By 1911 there were children on the way, so Lucy stayed home more and more. On 26 March Lucy gave birth to twin daughters, Lucy May and Elizabeth Hoxie. Within two months Lucy accompanied her husband once again to Cambridge, taking one of the infants and leaving the other with Aunt Ada.[9] But as the family grew, so did the parents' responsibilities. Russell passed up an AAS meeting in August 1912 in Pittsburgh because Lucy was pregnant again and due anytime. They were resting at Bay Head, on the New Jersey shore, and all seemed fine for the imminent birth.[10] Russell never ceased work but stayed closer to home. Knowing he would miss the Allegheny meetings, Russell sent his graduate student Harlow Shapley to talk about the problem of limb darkening in Algol systems. Russell asked Pickering to look after Shapley: "He is the best student I ever had, and a very modest fellow."[11] Henry Norris Russell jr. was born one month later, on 13 September 1912, but there were complications. At home and in labor, with family and a local doctor acting as accoucheur, Lucy went into violent convulsions, which continued after she gave birth, and were calmed only by morphine. Henry faced mortal crisis for the first time. He would often recall his terror at the prospect of being left with three children to raise.[12]

Russell made his growing family the centerpiece of his life, and his classmates responded by featuring Russell holding the twins in their *Quindecennial Record*. Pickering sent a letter directly to the infant son, and Mitchell quipped: "As far as I can see it you have now a concrete example of the problem of three bodies on your hands. If you have any troubles with perturbations or other inequalities I would suggest that you place them at the vertices of an equilateral triangle. This arrangement I believe is to be recommended as stable."[13] As proud and delighted as the Russells were with their children, they soon realized that all was not quite right with Elizabeth. She required a back brace because "vertebrae in the small of her back are not yet ossified, and if she was allowed to try to sit up, she might get a crooked back." Lucy sought out the best specialists for her twenty-month-old child, but the doctors found "no evidence whatever of disease." Russell hoped the condition would heal itself in time, he told Pickering. "In fact she is amazingly good and happy throughout it all. All this has kept us pretty busy, and for a time caused us much anxiety."[14] Outwardly he expressed confidence that she would improve to normalcy, so there is no way to know whether he and his wife were aware of the challenge they were to face as Elizabeth's problems persisted and multiplied.

This was a time of profound change for Russell, a time of joy and of crisis. In the summer of 1911, his father was suffering from Bright's disease, spending the fall in a sanitarium, failing slowly in acute and painful distress. In October the pastor visited Princeton to see the twins, and even at that distance he tried to tend to his congregation, constantly ministering to their needs.[15] A month later he was dead. Alexander Gatherer Russell's life was his work, and evidently he worked to the end.

None of his sons spoke at the funeral in Oyster Bay on 13 November or at the memorial services the following week. Russell's surviving correspondence reveals nothing of his father as he discussed his binaries and their implications for stellar evolution; one letter written just two days before his father's death bore no indication of any distress, nor did he mention his father after his death, nor were there letters of condolence. One cannot distinguish between denial and ambivalence here, but there is evidence that the pastor's sons had complex feelings about their father. Henry "identified himself with the Norris's rather than [with] his father and his father's background," his own son recalled, and also "talked a lot more about Auntie than he ever talked about his mother." "Though he looked upon his father as a very righteous and sober man," his son added, "he hadn't very many words of commendation or praise for him at all." It did not help that the pastor had remarried, picking a childhood sweetheart from Truro. And he also chose to be buried with her in Nova Scotia (she died barely a year after their marriage), rather than with the mother of his three sons in Princeton. This was tantamount to a "desertion of the Norris house and Norris family of Princeton."[16] Only Gordon MacGregor Russell, Henry's middle brother and by now a fledgling pastor, attended the unveiling of a memorial tablet for his father in June 1913, reporting back to Henry, in England at the time, that it was very difficult to find words.[17]

The most devastating jolt was Aunt Ada's sudden death of liver cancer in January 1914. Nothing is known of the details, but Russell was clearly hit hard. He had just returned from Atlanta, where he had delivered a major address on his theory of stellar evolution, to find his aunt ill. At first he and Lucy thought her condition was not serious. They did not know that she was in danger until two or three days before the crisis, "so that it was a great shock to us, as well as a very grave loss." Now Russell's correspondence was filled with emotion, and his work was severely disrupted. Lucy, pregnant again, was in her ninth month when Aunt Ada died, and with the uncertainties of labor and childbirth again imminent, Russell relived her traumatic near-death experience. His old childhood anxieties and insecurities welled up from the depths of memory as he once again contemplated a lonely and bleak future. Although he was inwardly shaken, outwardly he remained positive.[18] Emma (later called Margaret) was born on 17 February 1914 after suitable precautions had been taken to handle a possible recurrence of the eclampsia. But this time it was a normal birth. It would be the last Lucy would endure.

With his aunt's death and a fourth child at hand, Russell faced the world with "much anxiety, but I think I never was so busy," he told Pickering in March. He was trying to keep the house together, "which, when one takes it up after it has been organized by competent hands, seems to be not much more labor than the famous Mrs. Boisserain assured the suffragists that it was!" "Nevertheless," he admitted, "my work has gotten a bit behind."[19] Indeed, Henry stayed at home a lot now, but he could still think about work. As he filled his notebooks with calculations under the rubric "Effects of Tidal Evolution upon Visual and Spectroscopic Binaries," he used the backs of the pages to copy out requests for coal and made shopping lists for "Veal, Beans, spinach, Bells, oatmeal, chicken carcass, beef soup, veal soup, grapefruits, bananas, prunes, radishes, dried apricots, peaches, apples," which made up the standard summer fare of the family.[20] Servants and cooks managed by Lucy would soon take over these chores, but Russell had his first, and probably last, taste of domestic responsibility.

Changing Fortunes at Princeton

As Russell's family grew, so did his reputation, but he retained the stature of an energetic youngster, even as a thirty-three-year-old assistant professor in 1910. He was not among the fifty-odd astronomers identified as "starred" that year by the American Association for the Advancement of Science; this elite body averaged fifty-six years in age. Most were either observatory directors or prominent staff from the Yerkes, Mount Wilson, and Lick Observatories.[21]

But Russell, groomed by Fine, rose quickly at Princeton, even though the campus was suffering through hard times as Wilson's preceptorial system slowly bankrupted the university, draining its $4 million endowment, far smaller than Columbia's, Harvard's, or Yale's. Princeton also lacked a systematic fund-raising strategy, and the issue of the siting and character of the Graduate College was tearing the campus apart.[22] Wilson charged that this controversy was no less than a struggle "between the forces of privilege and those of democracy," but he lost the battle for his Quad plan in the winter and spring of 1910. By the end of the summer, as Russell attended Hale's International Solar Union meetings in Pasadena, Wilson knew that his days were numbered. He resigned in the fall.[23]

Some of Princeton's brightest stars had also left. James Jeans resigned in March 1909 to return to England, blaming the climate and the failure of the graduate department to develop in the manner in which "we had hoped it would . . ." He felt that he had accomplished nothing at Princeton.[24] Russell recalled that few Princeton undergraduates were able to take advantage of Jeans's brilliance. Stories about Jeans at Princeton focused on his rudeness, his intolerance, and his penchant for sarcasm and irony. Princeton did not suit Jeans, who missed the challenge of his intellectual roots. Russell seems to have had little contact with Jeans in their years together.[25]

John Grier Hibben, Stuart Professor of Logic, succeeded Wilson as president. Hibben had been close to Wilson but was more conservative and broke with him over the Quad plan. He openly and honestly opposed it, siding with those, like Dean Andrew West and Oswald Veblen, who believed that it would prevent Princeton from becoming a real research university.[26] Russell certainly agreed with Hibben, although his only direct involvement in the fray was his confidential report on the renovation of the Halsted. Accordingly, one month after his installation, Hibben proposed that a Department of Astronomy be established, to be headed by Russell, "and that, in accordance with the practice at other Universities, Professor Russell, as head of the Department, be made Director of the Observatory." He was made professor in April 1911 and director of the observatory in June 1912. Hibben even found $3,000 from "Extra-Departmental" funds to give Russell a raise. Over the next year or two there would still be some confusion as to the governance and independence of the department, but there was no question that Russell was in charge of astronomy at Princeton.[27]

Word spread quickly about Russell's ascendancy, and the usual congratulatory remarks flowed in. Schlesinger read the news in the *New York Evening Post* and was concerned: "This appointment is as it should be, but I must say that I should be sorry to see you drawn away from the theoretical work you are now doing—equipment for which is, in my opinion, much rarer than the equipment that makes a good observer. But I understand that there is no prospect of any such pressure in the immediate future."[28] Schlesinger knew Princeton well, and he was coming to know Russell. The department was very small and would remain so. Dugan retained the bulk of the observing load and paid attention to students and infrastructure. When Russell took over, Dugan remained an assistant professor.

Dugan's salary had been raised to $2,200 when he married, but he was given no relief from teaching, and the operating budget of the observatory was static at $1,000. There was no growth in sight, and none was requested.[29] There was a continuing trickle of graduate students and special students around to help out, however, and they tended to work with Dugan, who made sure that the 23-inch refractor was used for visual photometry every clear night. Russell rarely observed and had no routine or purpose beyond bringing visitors in for a look through the glass. On one night, 9 December 1910, Russell and Alfred Joy observed together; this was the first record of Russell's presence in the dome since 1906.[30]

Within one month of Russell's promotion to professor, Dugan received a call from the University of Missouri inviting him to become the director of the Laws Observatory. After F. H. Seares left in 1909 for Mount Wilson, the Laws Observatory drifted. Dugan would be perfect to continue the photometric programs Seares had established.[31] Dean Fine gave Russell the problem, and Russell sought out advice from Pickering and Frost. He was certain that Dugan would accept the offer because, in all frankness, "Dean Fine has advised him to do so." The autonomy and status of a directorship had its obvious advantages. But Dugan may well have chafed at Russell's rapid advancement; he had notified Fine, not Russell, of the offer.

Russell needed to find an observer able to maintain the Halsted.[32] Pickering argued that the person had to fit the instrumentation at hand, and Dugan's photometry had to continue: "You have the best equipment for that, as the telescope is large, and the photometer I believe to be unsurpassed."[33] Russell thought of Mitchell, who knew the Halsted. But others like Joel Stebbins, Philip Fox, and Eric Doolittle were good double star observers. No matter his specialty, the new person had to take up to eight hours of undergraduate teaching of descriptive and practical astronomy, as well as one graduate-level course of his choosing, and be willing to manage the Students' Observatory and to work for $1,800 per year to start.[34]

Dugan, however, eventually turned down the Missouri offer, which was bad news for Mitchell.[35] Dugan took a close look at the Laws Observatory and decided it was too small, and the university too poor, to make this move a prudent step for him. Indeed, the position went to a younger man, who quickly became bogged down in bureaucracy and teaching, with no institutional support.[36] Dugan, furthermore, was no empire builder, and though he would be in Russell's shadow, at least their research was compatible. By now, eclipsing variable stars loomed highest on Russell's list of objects to study. And Dugan got a $300 raise to $2,500 to match the offer from Missouri.[37]

Enter Shapley

The Laws Observatory loomed large in affairs at Princeton. In mid-July 1911, with no decision from Dugan forthcoming, Russell, on vacation at the Coles' summer home in Southport, Connecticut, a rambling three-story house with plenty of room, with shoreline and pier for family sailboats, felt "at a standstill, awaiting events." But at least the telescope would not be idle "as our Fellow will be Mr. Harlow Shapley who has already done good photometric work at Missouri."[38]

Fig. 7.1. Harlow Shapley ca. 1913.
Photograph courtesy Owen Gingerich.

Up to the time of Harlow Shapley's appearance, the few graduate students in astronomy at Princeton had been quietly competent but not exceptional, except for Daniel, who was notable for his "attachment" to astronomy, and Alfred Joy, who came for only a year to learn photometry under Dugan.[39] Shapley had been one of two astronomical assistants under Seares at the Laws Observatory, and Seares thought highly of him: "His ability is far above the average; his personality is good; he is faithful and conscientious; and he works incessantly—much too continuously for his own good." As an undergraduate, Shapley managed reductions of Seares's photometric observations: he was unusual, Seares reported, because "he *thinks* about what he is doing." When Seares left Missouri, he tried to get Shapley to a better place. But Shapley remained to complete his A.B. in June 1910 and then stayed for graduate study. He turned down a fellowship at Lick because he wanted to remain in the East.[40]

Shapley ended up at Princeton because Oliver Dimon Kellogg, a Princeton man now on the Missouri faculty, urged him to apply in January 1911. Kellogg wrote to Deans West and Fine and secured a Thaw for Shapley. Kellogg warned West to expedite his application lest Shapley be grabbed by the University of Chicago, or elsewhere: "[H]e is the kind of man you want." So Shapley, twin son of a rural Missouri hay producer, trained in journalism and astronomy, with a sharp wit but eager to make good, arrived in Princeton that September. Dugan put him right to work.[41]

Shapley was only eight years younger than Russell, but they were worlds apart. Shapley found Russell "a high-class Long Island clergyman's son and very high hat," recalling how Russell expected his students to open doors for their betters: "He intimated that I was, after all, a wild Missourian of whom no one should expect much." But Russell already knew of Shapley's worth as an observer and computer, and soon accepted the man; by the fall they strolled the campus together, and Russell used his cane to sweep the undergraduates out of their path. Russell, along with Dugan, also started advising him on where to find rare wildflowers.[42]

Shapley worked for Dugan at first, lived in the observer's room at 14 Prospect, and took meals at Merwick, the Graduate College. He became the primary observer on the Halsted, working continuously throughout the academic year and summer of 1912. It was a tough winter for observing; on 9 January Shapley expressed relief, and more than a bit of glee, at having to forfeit the night since the dome was frozen tight.[43] His courses were closely tied to his observing: astronomical photography, celestial mechanics, and theoretical astronomy under Russell, practical astronomy with Dugan, and a year of philosophy with Hibben. Later he took spectroscopy from Russell and photometry from Dugan, and another year of celestial mechanics. Shapley excelled in all of this and recalled his Princeton years as a time when he became convinced that he could succeed in astronomy.[44]

Russell had his own way of remembering what Shapley meant to him. He later recalled, facing the monumental challenge of obtaining density information from as many binary star systems as possible, to confirm the existence of giants: "But then I was up against it. There were more than 50,000 observations of eclipsing variables in the literature, all now clamoring to be observed now that tables were available for doing them quickly [which meant] years of work. And at that moment the Lord sent me Harlow Shapley!"[45] Russell knew just what he would do with the man from Missouri.

• CHAPTER 8 •

Building a Case for Giants

Assaulting Binary Stars

As Russell advanced on the Princeton faculty, married Lucy May Cole, and reduced his parallax data, binaries became central to the confirmation of both the reality of giants and his theory of stellar evolution. He approached the problem from two directions: first, he wanted to know how binary systems changed with time, which could give him another link between density and age. Second, he needed stellar masses, radii, and surface temperatures to confirm the existence of low density giants.

To achieve his first goal, between 1909 and 1912 Russell applied Darwin's fission model to binary systems and tested it against actual orbits and stellar dimensions to argue for his view of how stars evolved. He completed his theoretical analysis by January 1910 in a long study, "On the Origin of Binary Stars." Here he defended the Darwinian theory against recent claims by T. C. Chamberlin and Forest Ray Moulton that spinning gaseous masses did not fission into binary systems but would either evaporate or disintegrate.[1] He also countered Moulton and Jeans's contention that Darwin's theory was imprecise about the conditions for dynamical stability just prior to fission.

Chamberlin and Moulton presented a new model of the origin of the solar system and the Earth-Moon system. Their "encounter theory," which described the solar system as the result of a disruptive tidal interaction between a passing star and the Sun, was "a shock treatment to the 19th century idea of a stable, slowly evolving universe." Russell had to resist them if he wished to continue to use binary statistics in his evolutionary theory. He did so by appealing to "the facts of observation."[2]

Characteristic of a style that would later emerge among theoretical astrophysicists, Russell cautiously left wrangling about the mathematics to physicists like Jeans or celestial mechanicians like Moulton. He concentrated instead on characteristics predicted by Darwin's model that he could verify by observation. He calculated the orbital properties of binary systems with a range of masses in which one or both of the components had subsequently fissioned into secondary bodies. Tabulating his results for mass ratios and orbital size ratios, Russell compared his models against observations drawn from a wide range of sources such as Burnham's standard *General Catalogue of Double Stars*. Among some eight hundred systems he found seventy-four that were triple or multiple, and for those, the ratio of secondary to primary orbital size was "in harmony with the fission theory."[3]

From statistics Russell moved on to specific systems. The exotic and complex light variations of star systems like Beta Lyrae, he argued, were best explained "on the hypothesis that they are composed of two ellipsoidal masses of very small density revolving practically in contact . . ." just at or barely beyond the point of fission. Their very existence, Russell argued, gave sufficient warrant to the fission theory. Even though the theory itself was not yet complete, he was sure that when the subject was further explored, "some series of figures of equilibrium of a *compressible* gas, ending in fission into two comparable masses, will be found to be stable."[4] He had more faith in his analysis of observational data than in Moulton's mathematics and adopted the fission theory as a "working hypothesis" until compelling contrary evidence was produced.[5]

Russell knew he might have trouble with Moulton, an established mathematical theorist and colleague of Edwin Frost, so he tried to head him off by sending him drafts before he sent his paper to Frost for the *ApJ*. Moulton at first felt that Russell had overused the data. But Russell stood by the observations: It was the strength of these data, Russell argued, "and not my own work, that leads me to say that there is a very high degree of probability that stars can exist in stable, approximately spheroidal forms, nearly in contact." Even so, he was "ready to abandon [the fission theory] in favor of any theory that can give a distinctly better *quantitative* account of the phenomena." Russell knew that Moulton was more concerned with preserving the encounter theory than with banishing the fission hypothesis. Accordingly, Russell modified his text to satisfy Moulton, showing how Moulton's stability arguments in fact were confirmed by observational evidence. This tactic also satisfied Frost, who, as Russell suspected, deferred to Moulton: "I agree thoroughly with the policy of having differences of opinion settled out of print in so far as possible."[6] Russell would carry on all future negotiations in this manner, to the extent that he was able to control the astrophysical database, wielding it to beat theory into submission.

Russell's appeal to observational evidence and his ability to use that evidence to make a theoretical point, even when formal proof was not forthcoming, mark his style. He knew that his audience was of like mind, and that any appeal to observation weighed heavily in the assignment of credit or priority for a discovery. Russell was also less attached to rigorous mathematical analysis than was Moulton, who let his concern for mathematical stability override any physical considerations. Russell relied on intuitive arguments, based upon observations of what seemed to be happening in the real world. The extent to which he maintained a theorist's outlook, therefore, depended more upon his pragmatic ability to reproduce astronomical reality than it did upon any desire for rigor or elegance. Russell came to personify this pragmatic brand of theoretical astrophysics, one that was welcomed by observational astronomers.

Having protected the fission theory, Russell then used it to link the evolution of binary systems to his theory of stellar evolution. He accepted the prevailing view that tidal friction would cause the separation of components of double stars to increase with time, which was based loosely upon Darwin's model for the Earth-Moon system. But Russell also believed that stars contracted over timescales that were short compared to the time required for any gross increase in separation after fission. Thus presently observed separations represented, on the whole, separations at the time of fission. Hot and massive systems like Algols were close binaries and so fissioned at a later stage in their lives, whereas widely spaced pairs fissioned early and were now observed at a point where contraction had made their radii small compared to their orbital separation. Russell used arguments like these to predict the frequency of orbital periods and separations expected and, appealing once again to observational data, found what he was looking for. Among spectroscopic binaries, the shorter period systems, less than two days, were massive early type (young) B and A Harvard spectral types, whereas the longer period systems exhibited later G, K, and M (older) types.[7]

The Russell Method

While he was applying Darwin's theory, Russell was also developing a reduction method to turn observational data on eclipsing binaries into useful information, like radii, the shapes of the stars, and their mean densities. The last was of critical importance to the demonstration that giants existed, and that they contracted into dwarf stars.

At the turn of the century, astronomers suspected that close binary systems (systems not resolvable by visual or photographic inspection) did not contain spherical stars, nor did they move in circular orbits, nor were they equal in size, brightness, or color. By analogy with the Sun, the surface brightnesses of these stars were not constant but darkened at their limbs. Moreover, each star could illuminate the other in reflection, as Dugan had shown in 1908 after careful observations of RT Persei. One could not assume that the orbital plane of the system lay exactly in the line of sight, a fact that further complicated any analysis. Russell knew well that a complete description of the light curve of such a system in terms of its orbital elements was so complex that a general theory was not practical.[8]

Eclipsing binary light curves were very messy things, highly resistant to simple geometrical description. In short period systems the light curve was rarely flat in or out of eclipse, and analytical functions were very difficult to develop for them.

Distinct from others at the time who searched for specific solutions, Russell recalled that "I got interested in that side of the problem and realized that it was possible to tabulate those functions and very much to simplify the thing" for the general, ideal case.[9]

Although he was not the first to employ simplifying assumptions, Russell was the first to create a practical method that reduced computations to a set of standardized mensuration formulae and numerical tables. His Algol studies in 1910 eventually resulted in a series of six papers in the *Astrophysical Journal* between 1912 and 1914, the last three coauthored with Shapley. The first papers dealt with the simplest cases, of spherical stars and orbits with total and partial eclipses. Russell then generalized the analysis to include ellipticity and reflection, and finally, with Shapley, limb darkening.[10] Within a very few years these papers embodied what came to be known as the "Russell Method" or "Russell model," which was widely adopted and applied, since it required little in the way of mathematical expertise beyond arithmetic manipulation. In his 1918 text *The Binary Stars* Robert Grant Aitken based his chapter on eclipsing variables "entirely upon this investigation."[11]

Russell knew that exact solutions were neither possible nor necessary. First of all, he argued, "[T]he limited accuracy of photometric observations" justified making approximations. Next, he knew that astronomers were not likely to use complicated mathematical reduction techniques. Therefore he created a means to reduce data that he knew would be acceptable: a set of tables that described the brightness of an ideal eclipsing star system as a function of the phase of the eclipse for a range of orbital elements.[12] First, Russell produced a light curve, plotting observed brightness as a function of time on graph paper and drawing a rough curve drawn through the scattered points to best fit the observations. From that curve, Russell identified phase points where the primary eclipse started, reached maximum depth, and then ended (these are the phase points he calculated for his ideal systems). He did the same for the secondary eclipse if one distinguished itself. These phase points then defined his orbit, and he employed his tables to convert them into the orbital elements of the actual system, using those in turn to reconstruct a theoretical light curve that he could then compare to the observed curve. The object of the process was to minimize the differences between the reconstructed and observed curves.

He returned more than once to his hypothetical parallax technique to expand his database. In late December 1910, studying Aitken's new listing of visual binaries, Russell noticed immediately that their masses fell within a narrow range, which he felt justified his technique. As before, in a binary system, the combined masses of the stars, their orbital parameters, and the distance to the system are all related. If he could assume an average mass for a binary system, then an analysis of its orbit would yield the distance to the system. Thus he could use hypothetical parallaxes to expand his sample to over 350 systems, and, with Harvard data, to broaden it to examine relations between the spectra and masses of the systems.[13] He quickly found two sequences that differed dramatically: "one marked by high luminosity per unit of mass, nearly the same for all spectral classes, and the other by small luminosity per unit of mass, diminishing very rapidly for the redder stars."[14] In January 1911 he felt that he had enough Algol masses and dimensions to show a progression of density with spectral class. As he often said when he thought he had made an important discovery, "I think that this is the first direct evidence bearing on this important question."[15]

Russell wanted to report his findings quickly, but Pickering advised him to collect more evidence. The next AAS meeting would not be appropriate; it was the first to be held in Canada, and Pickering wanted the meeting to be uncontroversial.[16] Russell did speak about visual binaries, however, and his hypothetical parallax technique. He reported that the mean densities of giant stars increased steadily "with *decreasing* redness," and, equally striking, dwarf star densities increased "with *increasing* redness . . ."[17]

Russell was in a rush for good reason. Hertzsprung was then visiting Mount Wilson and was working on visual binaries too. Reading Russell's abstract in *Science*, he wrote Russell that their work was so similar "that interfering might easily have occurred." Hertzsprung's insinuation heightened the fact that, even though they worked with different data sets, their research had now overlapped twice, and would again.[18] So Russell kept the pressure on during these years, owing to competition with Hertzsprung. But Russell had several projects going at once too; he sometimes gave more than two papers at AAS meetings. He was then collaborating with Pickering and E. W. Brown at Yale to improve knowledge of the orbit of the Moon using photography, an extension of his thesis, his first multi-institutional cooperative effort, and a collaboration that helped bring him into the hierarchy of the American astronomical community.[19] But other than this and his stream of *Scientific American* columns, during the period 1910–1912 Russell's main attention was on binaries and how they would ultimately confirm his theory of stellar evolution.

Shapley Accelerates the Process

Without doubt Shapley appeared at the right time. His competent and enthusiastic assistance made it possible for Russell to acquire the evidence he needed quickly. Shapley also had assistance from Martha Betz, his mathematically talented fiancée at nearby Bryn Mawr, who quietly helped him with the laborious computations on the weekends.[20]

With Shapley hard at work reducing as well as making observations under Dugan's guidance, Russell revisited some of his computational shortcuts and started to realize that Pickering's cautions were well advised. For instance, Dugan and Stebbins had taken care to observe secondary minima as carefully as primary minima, which Russell at first thought was unnecessary. But in November 1911, he found from his theoretical analysis that a single minimum "without constant phase (i.e., for a partial eclipse) does not give satisfactory data for the determination of the elements." The entire light curve had to be known and had to be accurate to 0.02 magnitudes along its entire course to be really useful. Russell

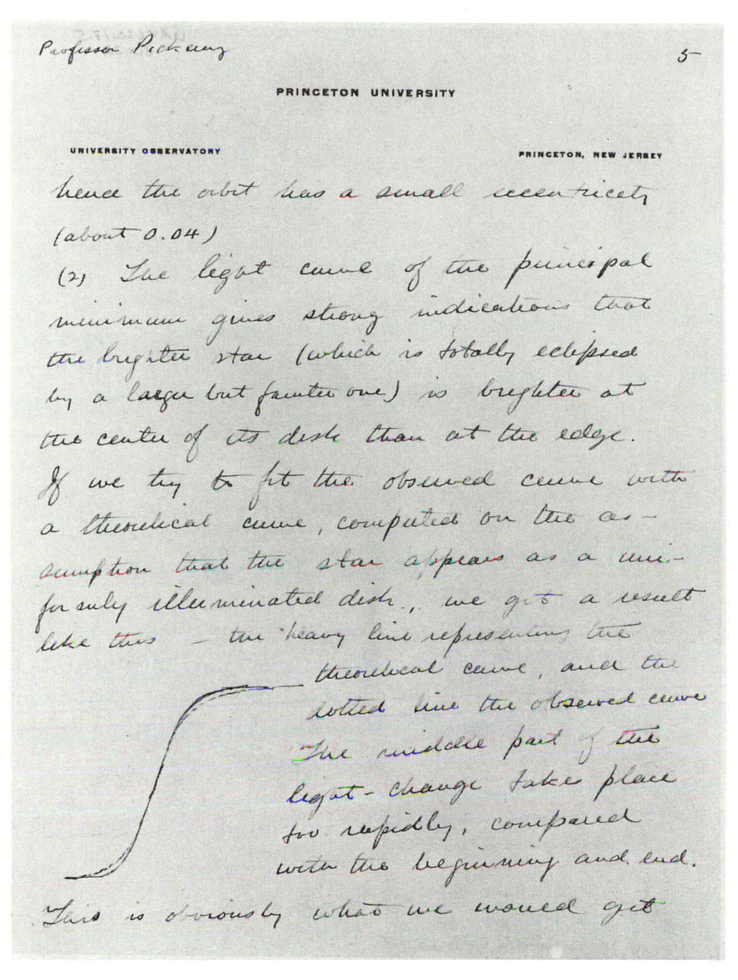
Professor Pickering 5

PRINCETON UNIVERSITY

UNIVERSITY OBSERVATORY PRINCETON, NEW JERSEY

hence the orbit has a small eccentricity (about 0.04)

(2) The light curve of the principal minimum gives strong indications that the brighter star (which is totally eclipsed by a larger but fainter one) is brighter at the center of its disk than at the edge.

If we try to fit the observed curve with a theoretical curve, computed on the assumption that the star appears as a uniformly illuminated disk, we get a result like this — the heavy line representing the theoretical curve, and the dotted line the observed curve. The middle part of the light-change takes place too rapidly, compared with the beginning and end.

This is obviously what we would get

Fig. 8.1. Russell describes his iterative graphical method for assessing limb darkening in eclipsing binary systems. Detail from Russell to Pickering, 9 December 1911, p. 5. Reproduction from the E. C. Pickering Papers, courtesy of the Harvard University Archives.

made sure that Shapley was fully aware of this change in policy. In December Russell and Shapley decided to make Algol variables the subject of his doctoral thesis. He would use Russell's method to determine elements of previously observed systems, and then focus in on those which proved to be "specially interesting," Russell told Pickering.[21]

As he had for his parallax stars, Russell again looked to Pickering and the "great Harvard library of photographs" to fill in the gaps in light curves and spectra. Russell now had some control over limb darkening, he reported in December; if the theoretical curve remained brighter longer and dimmed to minimum faster than the observed curve, then the star was brighter in the middle.

Russell's iterative procedures required more physical intuition than mathematical rigor and lent themselves to visual inspection.[22] Although Darwin's wranglers were mathematicians, Russell was more interested in adapting the tools of mathematics and physics without being bound by the formalisms inherent in those domains.[23] And he enjoyed considerable freedom to do so at Princeton, where there was no one to test him. After all, it was a place James Jeans had found intellectually uninteresting.

Shapley's industry rapidly yielded critical information about density. One of the solar-type components in a system he was examining could not be "more than 1/160 as dense as *air* under normal conditions," whereas the mean density of other solar-type stars was 2.5 times that of the Sun. Russell was delighted with this evidence, echoing contemporary thought equating density with age: "I don't think anybody will claim that these two stars are in the same stage of evolution!"[24]

Russell Marshals the Evidence

By the end of 1911, Pickering felt that Russell had made sufficient progress to make a stand. He convinced the American Philosophical Society to sponsor a "symposium on the spectroscope" to highlight Russell's work and, not incidentally, Harvard's data. Pickering suggested that they bring in Albert Michelson, America's first Nobel laureate in physics, to discuss the history of producing spectra, and Edwin Frost to talk on radial velocities. Pickering would speak on the use of the objective prism in astronomy, and Russell would talk on "Relations between the Spectra and Other Characteristics of the Stars," including, Pickering promised, "the results of Schwarzschild, Hertzsprung, and others." Campbell might open the session and make comments, but Russell would be given most of the time.[25] This was Russell's first chance to put together in one place all the clues he had been gathering that supported his theory of stellar evolution. Thus in April 1912 he spoke on ways to identify giants and dwarfs, on ways to determine density, and on means to determine the real brightnesses of the stars. Russell compared absolute stellar magnitudes to stellar spectral class, made his now familiar conclusions, and lavishly acknowledged Harvard and Hertzsprung.[26]

By now Russell had seen the graphical plots Hertzsprung and Hans Rosenberg published in the *Astronomische Nachrichten* depicting the relationship of brightness to color (or color-equivalent derived by various means) for stars in the Pleiades and Hyades star clusters.[27] Even though he had verbally "plotted" spectra against absolute magnitude in his lecture, Russell at this time did not publish a diagram as Hertzsprung did. This was because their goals were different: Hertzsprung wanted to demonstrate cluster membership through physical affinity and used the plot as a visual filter. Russell, on the other hand, wanted to make his arguments as general as possible. Thus he would not have wanted to confine his attention to clusters, did not need the cluster technique as a filter, and hence did not need to construct such a diagram. Russell wanted to establish the consistency of the relationship he had found first, rather than look for deviants, and so he tabulated his numerical data and expressed his results as a simple first-order linear equation. Here Russell followed Kapteyn's style of persuasion, even though he well knew that Kapteyn, and Campbell, were not comfortable with his theory.[28] Since Kapteyn used numerical arguments to make his points, Russell did the same, for the moment.[29]

Although he drew no diagrams, Russell, like Hertzsprung, utilized the observed relationship to derive the distance to the Pleiades star cluster by fitting its apparent magnitude sequence to the absolute magnitudes of stars in similar clusters whose parallaxes were known.[30] Russell thus put into practice what Hertzsprung had already suggested, deducing the absolute magnitudes of stars from spectra alone.[31] Clearly the two were again working along the same lines, using the relationship they discovered for a wide range of astrophysical applications.

Russell went still further, marshaling his binary data to show that giants differed from dwarfs mainly in luminosity and density, not so much in mass. He now searched for a rational basis behind the relationships he was finding, linking differences in stellar spectra to differences in brightness efficiency that he knew had been confirmed empirically but were also "in good agreement with those computed by Planck's formula."[32]

In Philadelphia, Russell was still circumspect about his stellar evolution theory, although he now explicitly discussed it in terms of the gas laws and Lane's law: that temperature was an inverse function of radius in a star in the perfect gas state. The mean temperature of a giant star rose as it contracted. Compressibility diminished with age, however, slowing down the process; at some point where a critical density was achieved, the star no longer behaved according to Lane's law, and upon slight further contraction, it cooled to extinction. "The highest temperatures will be attained at a density for which the departures from the gas laws are already considerable, but probably long before the density becomes as great as that of water." He once again identified B stars as very close to the critical range of departure, with densities 1/5 that of water. They were also the hottest stars, which he argued fitted his theory nicely. But, without Shapley's evidence, he was not yet ready to stand firm: his was still a "good working hypothesis."[33]

Russell did not talk much about Shapley's work in Philadelphia. He was only softening up the audience, knowing he would get mixed reviews. Lewis Boss had trouble conceiving of how one recognized giant suns, and Pickering, though impressed with the relation between spectrum and magnitude, remained silent about his theory. W. S. Adams and Aitken were more receptive; Aitken found his work on short period variables "the most interesting objects and they hold, I think, the key to many important laws governing the evolution of stellar systems."[34]

Shapley incorporated limb darkening into his solutions during the summer of 1912, working on it well into the autumn. Russell sent Shapley to the AAS meetings in Pittsburgh to announce that the forty-four systems he had studied thus far showed "remarkable agreement" between prolateness and distance between centers, which supported Darwin's fission theory for equal masses and homogeneous incompressible fluids.[35] Russell meanwhile was becoming more interested in applying Planck's law to derive stellar diameters using John A. Parkhurst's highly accurate "actinometry"—the relationship between the spectrum of a star and a photometric indication of its color, its color index.[36] Russell soon found "excellent internal agreement" between his formulae and the measurements of Wilsing and Scheiner at Potsdam, who used visual spectrophotometric methods. Russell had again crossed paths with Hertzsprung, who had performed a similar calculation in 1906.[37] By the end of the year, Shapley's efforts started to pay off, especially an exhausting study of five thousand observations of the eclipsing system Epsilon Aurigae, whose huge invisible component had a density barely one-millionth that of the earth's atmosphere and a diameter comparable to the orbit of Jupiter.[38] Shapley had analyzed almost ninety eclipsing systems, and Russell felt that his work was far enough along to warrant a new assault.

The summer 1913 meeting of the AAS had been canceled in favor of the fifth meeting of the International Solar Union, to be held in Bonn that August. For Hale these meetings were a means to standardize practice through coordination of the research workforce.[39] For Russell it was a new venue in which to air his theory, and a chance to show Aunt Ada a bit of the world. Russell wrote Hale hoping that they would meet along the way, if not in Bonn then possibly in London, for he wanted a chance to "talk about stellar evolution." Russell seemed most concerned to ensure that he would travel not alone but in the company of astronomers. He ached for an audience. Russell managed to book two staterooms next to Pickering's on the S.S. *Laconia* and wrote his Harvard mentor, "We must arrange for a table together in the dining saloon . . ."[40] At the very least Russell knew that he had to campaign for his ideas with those who could be kingmakers in the astronomical community.

Preparing for the meetings, Russell tried to persuade those he knew could make a difference, leading them step-by-step through his evidence, from Shapley's analyses of eclipsing systems to how his results supported Darwin's theory of binary fission and tidal evolution, to how all this supported his overall vision of stellar evolution. He impressed upon Adams the importance of Mount Wilson radial velocities, which provided solid evidence for the absolute dimensions of stars in spectroscopic binary systems, "free from all conjecture." Mount Wilson data also provided "striking accordance with Darwin's theoretical results." Writing to Frank Schlesinger, Russell realized that he had better not press too hard: "[Y]ou have heard me ride this hobby before," he admitted, but "[b]efore long I hope to convince everybody that it is a real live horse."[41]

Russell found Adams sympathetic but Campbell intractable. Campbell resisted Darwinian fission, the low densities Shapley was finding, and the fact that stars rotating slower than the Sun could be so highly distorted. R.T.A. Innes, director of the meteorological department of the Transvaal and a curious but respected freethinker, often poked holes in fashionable ideas. Innes had his own ideas about stars: "What evidence is there," he challenged Russell, "that stars *do* contract?" Campbell's resistance, however, went to the core of the problem: "My point of view is a feeling which is not supported by direct evidence," he admitted to Russell, "and hence is not entitled to much weight. Pretty definite ideas as to the course of stellar evolution are at the bottom of my trouble."[42] Campbell's investment in his radial velocity statistics lay at the root of his feelings.

Shapley defended his thesis just before he left for Bonn. Gathering one morning in the library of the Students' Observatory, Dean Magie, Russell, and Dugan presided over the traditional public inquisition. Dugan, Shapley recalls, asked him arcane questions about the proper adjustment of a telescope, rather than questions about the stars he knew more about than anyone else alive.[43] But all went well. Shapley packed up his magna cum laude and headed off for Europe and the Bonn conference, spreading Russell's gospel far and wide as he sought out the great centers of astronomical knowledge he had tapped as a graduate student. Hertzsprung wrote Russell to say how much he enjoyed Shapley and his many stories about Russell's exploits. "It is a curious thing," Hertzsprung added, "that we seem to have for the second time made something similar, as I just am finishing a small paper on the absolute brightness and distribution in space of the δ Cephei-variables, and I see from a meeting report, that you have read a paper on absolute magnitudes of variable stars . . ."[44] Hertzsprung missed the fact that at the same meeting Russell used Planck's radiation law to predict diameters, just as he had done. Instead, Hertzsprung went on to critique Russell's theory of stellar evolution, on the basis of not only Shapley's relating of it but also what he had been reading. He reminded Russell that he too had dabbled in evolutionary speculation in his first papers but now could not justify it. Hertzsprung hoped that Russell had good evidence, because he predicted that there "may be some serious objections to that view."[45]

Hertzsprung was concerned that gaps existed in his cluster diagrams. "Where are all the connecting stages?" he wondered. He also favored Kapteyn and Campbell's demonstration that "peculiar" space velocities (the motions of the stars after the solar motion is subtracted out) increased with spectral type. How could they be reconciled on Russell's evolutionary model? "How to explain," he wondered, "that the peculiar motion is first great, then small and then great again in the course of star life?"[46] Hertzsprung mused that each star had its own evolutionary path, but Russell had already thought of this. At least Hertzsprung agreed that giants existed, whereas other astronomers remained skeptical. Most seriously, Russell had to fill in the gaps Hertzsprung worried about—he had to find the connecting stages. Russell believed that he had the data at hand in his field stars and knew that the best way to demonstrate continuous change was to create a visual image. This is what led Russell to construct his first diagrams.

The final weeks leading up to Russell's departure were very hectic. In his excitement, apparently unable to think of anything else beyond his work and the trip, Russell prepared his June *Scientific American* column, "How the Navigator Lays Out His Course."[47] He also rushed to complete several chapters of a revision of Young's *Manual of Astronomy*, finish his teaching, and put the final touches on the talks he hoped to give in London and, possibly, have ready at hand for anyone who might listen in Bonn.

Russell's Presentation at the Royal Astronomical Society

H. H. Turner had arranged an RAS meeting on 13 June 1913 for American astronomers en route to Bonn, and Pickering made sure Russell was on the agenda. Russell carefully shepherded a valise that included his talk and several lantern slides of graphical illustrations he hoped to show around that would galvanize all thinking into one powerful image.

The *Laconia* docked at Liverpool, and its passengers fanned out across England. Some were attracted to Turner's golf links at Oxford; others like Pickering headed straight for London. If Russell had his mind on anything, it was his presentation before the RAS on the 13th. It was a very crowded meeting that evening at the Royal Astronomical Society's rooms in Burlington House near Piccadilly Square, and as usual the pace was glacial. All the normal formalities had to be dealt with, and then reports were read into the record from outlying observatories of the realm. Eddington spoke on instrumentation at Greenwich, and a physicist spoke from the National Physical Laboratory. Finally, the visitors got their turn. Pickering spoke of work at Harvard, describing their two million measures of starlight as "a beautiful problem in 'scientific management.' "[48] Then Annie Cannon spoke on the spectra of nebulae, novae, and Pickering's fifth type of stellar spectra. Russell must have worried that as the dinner hour approached, his audience would be fatigued. Expecting to be called when Annie Cannon stepped down, he must have been dismayed when the society president called on S. S. Hough from the Cape. Hough was not brief, and by the time he had finished, the president, noting the hour, deferred discussion and called on "one other visitor you would like to hear, Prof. H. N. Russell, the Director of Princeton Observatory."

Russell leapt into action, calling immediately for his first of four lantern slides. There was no time to waste: he had to hit them hard. With his first lantern slide projected, he told the audience that he wanted to talk about "some studies bearing upon stellar evolution." He plotted the spectra of his stars on the vertical axis against their absolute magnitude, drawn across the horizontal coordinate, mimicking Hertzsprung's and Lockyer's original orientation and his own 1907 lecture notes. Directing attention to how stars were distributed—where they clustered and regions they avoided—Russell spoke at top pitch about his parallaxes and credited Hertzsprung as the first to detect giants among the dwarfs: "All I have done in this diagram is to use more extensive observational material."[49]

It was this overwhelming evidence, however, that made all the difference. Each successive lantern slide revealed the same relationship from visual binaries, hypothetical parallaxes, Shapley's results, and data from four moving clusters. And each improved the diagram.[50] With his last slide on the screen, Russell hammered home: "All the previous relations," he argued, "are reproduced with remarkable exactness."[51] Now he was set to discuss stellar evolution, using his final slide to trace out the lives of the stars. Now possibly because of Hertzsprung's worry about gaps, Russell argued explicitly that only the most massive stars would achieve the highest temperatures. Not all stars followed the same evolutionary path, which explained the observed fact that the average masses of stars decreased down the dwarf sequence.[52]

Russell's "hurried account of things" was highly schematic, crammed as it was into the last twenty minutes of a very long program. Lacking time, Russell not only dropped many of the details that supported his observations and theory, but dropped as well, at least in his written remarks, any indication that his ideas still had to be regarded as a "working hypothesis," the term he had used to characterize his theorizing at the American Philosophical Society in 1912. By now Russell felt he was on firm ground, so he pushed ahead with a certainty that, evidently, was more attractive to his English and European audiences than it had been back home.

No doubt Russell kept talking at the club dinner that evening; Pickering asked him to speak about the Harvard-Yale-Princeton lunar orbit program. After the RAS meeting, Russell and Aunt Ada toured St. Paul's and other ecclesiastical masterpieces, covering London and its environs to the limits of her endurance.[53] Pickering visited with Sir Norman Lockyer and left a note at Russell's hotel that the aged spectroscopic pioneer was very anxious to meet the person who had revived his discredited theory. When they met, Russell promised Lockyer a full accounting of his research for his journal *Nature*.[54]

Even though Russell's talk came at the end of a long evening, and vague accounts of it confirm its hurried nature and the distracted audience, there was one person who paid close attention and was captivated by the lantern slides. Eddington was preparing a summary monograph on the kinematics of the stars, emphasizing the ramifications of Kapteyn's star streaming, which had captured the attention of so many astronomers since it implied that two or more great classes of stars existed, distinguished by their dynamics. He was particularly anxious to discuss what these studies revealed about the structure of the universe and its history, especially the provocative but unexplained correlation of space motion with advancing spectral type that Kapteyn and Campbell had announced in 1910. He was probably thinking about this problem when he listened to Russell's talk at the RAS, and decided there and then that he needed to incorporate some of what he was hearing in a later chapter he was developing on phenomena associated with spectral type. Above all Eddington wanted to publish one of Russell's fascinating diagrams.[55]

Leaving England, Russell and Aunt Ada headed to the Scandinavian countries to see the midnight sun.[56] They reappeared in Bonn toward the end of July, and, after taking tours of Bonn and Cologne, Russell settled in for the Solar Union meetings, which started on the evening of 30 July at H. Kayser's Physikalisches Institut, a legendary center for spectroscopic studies and experimental physics.[57]

The primary purpose of the Union was to act as a clearinghouse, as a forum for setting standards and for coordinating the work of observatories worldwide in areas ranging from sunspot counting and solar radiation to spectral classification. Pickering's classification system was quickly ratified, "Ein, [*sic*] zwei, fertig!" Turner reported, along with a universe of other matters. Between committee meetings and plenary sessions, astronomers talked incessantly about their research. Karl Schwarzschild invited Russell to the Potsdam Observatory and showed him around the place. But here, as among the many other festive and social invitations Russell received, the purpose was to talk. The talking continued to the end; it stopped for a moment only during a "never-to-be-forgotten" party on a Rhine steamer when a Zeppelin flew over them. Just as Kayser knew, "[m]any of us," Russell recalled, "had never seen an aircraft in flight before!"[58]

After the Bonn meetings, everyone moved on to Hamburg for the Astronomische Gesellschaft. The Gesellschaft meetings were very different. Instead of astrophysics, the Gesellschaft concentrated on classical technique. Russell wrote into his program, "Lunar Theory is like a Chess problem in 3 dimensions blindfold[ed]."[59] He did, however, take more notice during a rump session led by Schlesinger. Among the twenty-odd Americans at Hamburg were the majority of Schlesinger's AAS committee for determining fundamental star positions by photography. It was a great opportunity for the assembled members to meet and declare, once and for all, that the photographic determination of the relative positions of stars was the highest priority of the society, the Union, and, they hoped, of the Gesellschaft, "in the belief (founded upon results already obtained) that the time has come to supersede the plan of observing them visually with the meridian circle."[60] Their proposal met with frowns, but it carried the day.

At end of what was, to all intents and purposes, almost three months of meetings, caucuses, and lobbying, Pickering's spectroscopic and photometric systems had remained intact, and the long campaign by Turner, Gill, Pickering, Russell, and Schlesinger to establish photography as the medium of choice in astrometry finally won out.[61] The Americans returned home confident that astrophysics was being established finally along lines that Hale and Pickering had initiated and defended.

While observatory directors haggled over standards of practice, Hertzsprung and Russell compared notes about recalcitrant stars that refused to fit neatly onto their diagrams. One of the biggest puzzles was a faint white stellar companion in the triple star system o^2 Eridani, dominated by a fourth magnitude orange K type star. The white companion, classed as an A type star by Pickering's staff, fell many magnitudes below the dwarf sequence for its spectrum and had long puzzled Russell. Hertzsprung wondered about it too, and at Bonn the two spent their time trying to explain it away. Clearly, any stars that did not fit their sequence weakened the whole picture. As Hertzsprung reported later to Adams, asking for more spectra, "[t]his absolute dark white A-star, so far off the rule, is very interesting."[62] It made him search for other deviants, and, curiously, they all seemed to fall into the same part of the diagram. These were dwarf stars and eventually were called white dwarfs for want of a better term. They would be a nagging puzzle for over a decade.

Hertzsprung and Russell knew that they harbored different views over what constituted scientific evidence. But Russell was surprised that Hertzsprung did not accept his conclusion that dwarf stars decreased in mass with increasing redness. Hertzsprung accepted that the blue helium stars were the most massive, but did not feel that Russell's hypothetical parallaxes for the redder late-type dwarfs were reliable. Where Hertzsprung demurred was not over the correctness of the statistical technique but over the extent to which Russell went to demonstrate orbital motion and to derive masses. In some ways Hertzsprung could be as conservative as Burnham.[63]

En route home Russell wrote several *Scientific American* columns celebrating the Union and Gesellschaft and the many resolutions that were made to standardize practice in astrophysics worldwide.[64] He was most excited about standardizing spectral classification: for even though the members of the Union spoke many different languages, "[i]n future, therefore, all astronomers, when discussing stellar spectra, will (so to speak) be talking the same language, and much confusion will be avoided."[65]

Beyond Bonn

Russell arrived home in late August, and the family packed up immediately to take their annual vacation at the Cole farm in Southport, where they stayed for a month before the fall semester started. Relaxing on the Connecticut shore, to the extent that such a thing was possible with his three "very little people" to contend with, as he lovingly called them, Russell must have felt a certain sense of accomplishment. The children were a cheerful puzzle to him, a never-ending source of astonishment, and, in the case of Elizabeth, a growing concern.

Russell had come a long way in a very few years. At age thirty-six, a family man, a full professor and observatory director at an elite East Coast university, he was now pushing one of the hottest and provocative new theories in stellar astronomy and had caught the attention and interest of a small but growing circle of colleagues. His first intellectual issue, Harlow Shapley, was a success. The offers he received from observatories across the land reflected well on Russell and Princeton. Shapley also proved his worth as a pitchman, heralding Russell and his theories across Europe.

Despite Shapley's rhetoric and Russell's years of concentrated marshaling of evidence, Russell could not know, while puttering around with the Cole family sailboat, or playing with his children, reciting limericks he had learned in Cambridge and practicing the craft of origami, that his theories and techniques were actually going to take hold. One thing was sure: he would not succeed if he left them alone. Many people, not least Hertzsprung, Eddington, and Campbell, had raised many questions. And these questions had to be answered. As he had said more than once to Pickering, "There is room for years of work still on the eclipsing and short period variables." Russell knew that he was in for the long haul.[66] To show that Darwin's fission theory and his own giant-to-dwarf scheme were compatible, that they applied to real star systems and ultimately to the origin and evolution of the solar system itself, Russell had to marshal more evidence that every stage in the fission process could be found among the stars. Shapley had helped immeasurably in this process, as he spread Russell's reduction techniques far and wide among the observing astronomers of his generation.

Russell jumped right back to work when he returned from Bonn, even though he was with his family vacationing in Southport. Without breaking stride, Russell picked up where he and Shapley had left off in June, examining Shapley's latest calculations, correcting numerical errors he found there, and cautioning that they had to remove all discrepancies that had "seemed to me to give rather a black eye to our methods."[67]

In 1921 Alfred Fowler observed, as he awarded Russell the Gold Medal of the RAS, that in 1913 "Russell was fully alive to the danger of making false deductions from data which might be biased by the principles of selection of stars for parallax determinations . . ."[68] Although Russell really worked in the opposite direction, it was certainly accurate to say that in 1913, he left no stone unturned in his search for systematic effects. Indeed, he continued to fill in all possible holes. Now that his parallaxes were finished, he turned to approximate methods. Russell was still a man in a hurry who lacked the patience of the "born astronomer."

• CHAPTER 9 •

At the Theoretical Interface: Defending His Theory

The Department under Russell

At the fifteenth reunion of his class in 1912, Russell was no longer the awkward fellow from Oyster Bay but the Princeton don reporting on "The Progress of Science in the University." His bearing, stiffened by high starched collars, formal studs, loose-fitting but expensively tailored fine suits, and high-top therapeutic black shoes, exuded authority and formality.[1] His mission was to assure his classmates that Princeton was in good hands under John Greier Hibben, and he took the occasion to make a pitch for astronomy. Alumni had supported physics, biology, and geology, Russell asserted; Hibben was "a friend of science," and Dean West was his agent, securing the Proctor Fellowships from William Cooper Proctor, the soap manufacturer. But what of astronomy? There was little to show but ancient telescopes, a tiny staff, the loss of "Twinkle" Young. Russell felt the weight of the "giant's robe" and called upon his brethren for an endowment to make Princeton a leader once again. He explained that his own success was chiefly made through "the liberal policy of other astronomers," and that Princeton needed to repay the debt.[2]

Russell's solicitation could have been taken directly from Henry Rowland's dictum that supporting science was a means of "moral elevation and public service." He made no mention of utility or application, which were the buzzwords of that day.[3] He also had no evident game plan to establish an endowment, though in his yearly reports to Hibben, Russell complained about the loss of the director's residence and the outdated equipment, and he voiced the need for a publication fund and his own demands for a raise in salary.[4] When Pickering, as chairman of the American Association for the Advancement of Science (AAAS) "Committee of One Hundred," asked observatory directors to state their needs in 1915, Russell asked only for two people to perform computations ("computers") for a period of five years. Pickering felt he could do better. On Russell's behalf, Pickering petitioned a Princeton trustee, asking for an endowment large enough to provide an annual income of $20,000 to $50,000. Russell, Pickering claimed, "has a brilliant mind and is an astronomer of great originality [and was among] the twelve most eminent astronomers of the United States." The trustee was unmoved, though he did support a second computer for the department.[5]

Under Russell, astronomy remained one of the smallest departments on campus. He and Dugan shared the undergraduate teaching load and had few graduate students to worry about, usually one or two per year who lived by assisting the research needs of the department. As he had promised Wilson, Russell made no effort to make a larger footprint. He did hope to regain what had been astronomy's domain under Young and to refurbish the telescopes Young had built.[6]

Losing Shapley

In the fall of 1913, Shapley was back for the year, supported by a Proctor Fellowship, and Richard John McDiarmid was the Thaw fellow. By now astronomy graduate students took more physics than mathematics. And physics students looked into astronomy: K. T. Compton, Shapley's contemporary, and later John Quincy Stewart, would be attracted to astrophysics by Russell, especially after watching Shapley's meteoric rise.

Since late 1912 and his presentation at Pittsburgh, Shapley was a hot prospect. Schlesinger wanted to hire him immediately, and Seares, now at Mount Wilson, reminded him that "you are wanted here." But Shapley elected to finish his degree, expecting that after he completed it in the spring of 1913, he would end up at Mount Wilson that summer, where Seares promised him access to the 60-inch reflector to study binaries. Russell, however, wanted Shapley to stay for a year beyond his thesis, to finish up the binary work. Shapley parlayed Russell's demand into a salary raise from Mount Wilson, but when Russell secured one of the first Proctor Fellowships, Shapley decided to stay.[7]

Shapley knew that Princeton could not provide a permanent slot, unless Dugan were to leave, "which is not likely." But he was also worried about leaving too soon: "I have a rather serious aspiration to be an authority on eclipsing binaries, but still am not at all sure that I would want to know nothing else," he told Seares. "The bigness of Mount Wilson somewhat confuses my desires, and I fear sometimes that variable stars look so interesting and important merely because I am ignorant of other things." Shapley worried that he would "spread myself out too thin" at Mount Wilson. He also knew, because "Russell says it is impossible," that he could not finish everything up in time to get to Mount Wilson by the summer of 1913.[8]

Seares was less concerned about a delay and more concerned that Shapley see the astronomical world that summer, visiting observatories and absorbing what they had to offer. He could arrive at Mount Wilson as late as the summer of 1914. Shapley would have his share of observatory chores, Seares admitted vaguely, but "[t]he work of the Observatory as a whole is pretty closely knit together in its bearing upon the general problem of stellar evolution." So Shapley would be in the mainstream.[9]

Seares's promise that he would be contributing to stellar evolution was not hollow. Shapley's analysis of the eclipsing binary RR Draconis had caused Seares to revise some of his own thinking about the order of stellar evolution, especially Shapley's unequivocal result that the redder component was also less dense than the brighter component. What was needed, above all, was improved spectra of the components, and this was best done, Seares felt, at the 60-inch on Mount Wilson, which would be available to Shapley.[10] At the least, Seares let Shapley and Russell know that the best place to test Russell's theories was Mount Wilson.

The extra time Shapley spent at Princeton sufficed to enable him to finish his binary work. He finally headed west in April 1914, stopping off in Kansas City to marry Martha Betz, who had completed her studies at Bryn Mawr and was waiting for him there. As driven as his mentor, Shapley wrote Russell on the eve of his wedding discussing the state of their collaborative work, hoping to have a paper in Russell's hands for the spring meeting of the American Philosophical Society.[11]

Shapley's first reconnaissance of Mount Wilson left him breathless. Reporting to Russell on 5 May, he was delighted with the interest the staff was showing in his binary work. They were all talking about applying photoelectric detectors to improve light curves, and observing the spectra of the fainter components of his binary systems as telescope time allowed. Mount Wilson therefore seemed to be the next step in the attack on binary systems, where the combination of telescope and climate were unknown in the East, or anywhere else in the world.[12]

In his yearly report to Hibben for 1914, Russell made much of losing Shapley: "This Observatory could not at all compete with the attractions of the position offered him there, and we are fortunate to have had him here so long."[13] What regrets Russell may have had were countered by the value of having Shapley at Mount Wilson. His ever-loyal student was now sitting on top of a growing cache of astronomical data that Russell dearly desired, and Shapley proved to be an ardent disciple and emissary. He soon reported back after his first meeting with Hale that "[a]mong other things he took up the matter of your theory of stellar evolution. He is keenly and deeply interested in that—said that it was one of the things we had to consider seriously from now on until it is established or disproved." Hale wanted to know what could be done at Mount Wilson to confirm the existence of giants and dwarfs. Shapley reported that he had made several suggestions, and that they would continue the conference in a week or two after Russell made his own suggestions: "I feel here like an ambassador for that hypothesis. Seares is quite in favor of the theory, but he says that Kapteyn is extremely skeptical."[14]

By the time Shapley left, Russell had only partially overcome the many hurdles facing the acceptance of his theory. Most important was proving that giants and dwarfs existed as separate evolutionary classes, and, for some, that giants existed at all. Russell, with Shapley and McDiarmid, using Dugan's observations, had honed the evidence he had marshaled in June 1913 before the RAS. In the interim, Pickering also gave Russell another chance to present his views at the December 1913 meeting of the AAS in Atlanta. This time he would have forty minutes and could answer the criticisms of his work by Eddington that had just appeared. If he could win over a man like Eddington, it was likely that the rest would follow.

Eddington's Criticism

In July 1913 Eddington left Greenwich to establish residence at Trinity College as George Darwin's successor. Within the year, with Ball's death, Darwin's Plumian Chair and the directorship of the Cambridge University Observatory were combined into one. Passing over a sadly outdistanced Hinks, Eddington took the reins, a decision that years later Russell, ever loyal to those who had befriended him, attributed to university politics.[15]

For Eddington, the defining problem of sidereal astronomy was the spatial and kinematic arrangement of stars in the visible universe. He was among a small group of astronomers who searched for order, trying to understand Kapteyn's two star streams as the motions of particles in a spiraling vortex. Some treated the problem as "an extension of the kinetic theory of molecules" based upon the equipartition of energy. Eddington preferred mathematical modeling.[16] Along with understanding star streaming was the equally enigmatic correlation of space motion with spectral type that Kapteyn, Campbell, and Lewis Boss had interpreted as an increase of space velocity with age. Though it seemed to fit nicely with the traditional linear view of stellar evolution, Eddington worried that the correlation lacked a rational basis. He, along with Kapteyn, looked for other systematic agents, such as a resisting medium in space, which might restrain larger stars more than smaller denser ones.[17] Therefore giant suns came as a shock.

Eddington took Russell's theory seriously because it drew on so many lines of evidence. But as he listened to his talk at the June 1913 RAS meeting, and then read his manuscript, he realized that Russell failed to account for the observed kinematics of the stars: the increase in velocity with advancing spectral type. The old theory gave a neat view of stars forming in the galactic plane and then diffusing away from it with advancing age. But Russell's alternative theory challenged the dynamical history of the stellar universe. The youngest stars, giants, were randomly dispersed and somehow collected in the galactic plane at midlife. How could stars start life with high velocities, then decrease, and then increase again? Hertzsprung had wondered about this, and so did Eddington, who also noted

Fig. 9.1. The astronomical contingent at the Atlanta AAAS meeting, 29 December 1913–1 January 1914. *Left to right, front*: Comstock, Moulton, Pickering, Russell, Fox. From *Publications of the American Astronomical Society Volume* 3 (Northfield, Minn.: AAS, 1918), frontispiece. Reproduction courtesy of Carleton College.

that there seemed to be a steady increase in the periods of spectroscopic binary systems with advancing spectral class, which he concluded was a "strong argument for the usual order." These were the questions Eddington raised to challenge Russell's theory.[18]

The Atlanta Address

Russell was not ready to reply to Eddington in his address to the December 1913 Atlanta meeting of the Astronomical Society and Section A of the AAAS, though he acknowledged the problem, and in his later write-up he crafted a strong defense.[19] Before he had read Eddington's criticisms, Russell's goal for Atlanta was to have more time: "[T]he stuff I tried to get into twenty minutes at the meeting of the Royal Astronomical Society last June will just about make a respectable forty minute talk to Section A." He actually spent little time revising; updating chapters of Young's textbook once again demanded his time.[20]

Not many astronomers showed up at the AAAS meeting, but Russell found his audience very receptive. Comstock and Moulton were impressed, and Pickering was elated. On the return train Russell and Pickering were full of hyperopic visions for new lines of investigation to answer Eddington's questions.[21] These occupied him as much as Aunt Ada's death and Margaret's arrival allowed. By the spring, he managed to complete his lengthy summary argument, write an explicit answer to Eddington, meet his classes, work on revising Young's textbook, say goodbye to Shapley, and devise a more efficient means of calculating the orbits of spectroscopic binaries. His submissions to *Scientific American* continued unabated.

The printed version of Russell's talk appeared three times—first in *Nature*, then in *Popular Astronomy*, and ultimately in the *Publications* of the AAS.[22] The four diagrams appeared here for the first time, and their orientation now took the familiar form of what is today called the Hertzsprung-Russell Diagram, but for years was known as the Russell Diagram, an issue I take up in chapter 21. He gave no reason for his reorientation. Maybe he did it to distinguish his work from that of others. Or, since he used only discrete Harvard spectral types, the reorientation provided a clearer representation of his view of evolution. Years later when asked by James Cuffey how he came to the "Russell Diagram," Russell replied: "It wasn't so named when I invented it. I invented it because it represented the phenomena I wanted to describe in such a way I could get them on a page."[23]

As he would throughout his career, Russell structured his rhetorical argument carefully, starting with what he knew would draw in his listeners and readers. Thus he began with an appeal to the physics of spectra and moved quickly to supporting astronomical and laboratory evidence. Next, he used radiation theory to show that the Harvard classification was a temperature sequence. He derived luminosities from trigonometric parallaxes, then mean parallaxes, then moving group statistics, and finally his own hypothetical parallax technique. At this point Russell displayed his diagrams, the first relating spectra to luminosities for some three hundred stars based upon trigonometric parallaxes alone. Only when he had exhausted all other lines of evidence did Russell bring out Shapley's densities to demonstrate that the distinction between giant and dwarf stars was not due to differences in mass. Presenting his evidence in successively higher waves of certainty, Russell then thrust home, compelling a "theory of development to which it appears to me that they must inevitably lead."[24]

Here, Russell appealed more to physics than in his earlier presentations. Thermodynamics demanded a single direction to evolution: contraction, converting potential gravitational energy into heat. Some sort of "radio-active" heating source might momentarily cause a star to expand, but in the long run "the order of increasing density is the order of advancing evolution." The existence of red giant suns of low density, therefore, demanded that the old view of "early-type" hot blue stars' becoming "late-type" cool red stars now made no sense and had to be abandoned. "This is a revolutionary conclusion; but, so far as I can see, we are simply shut up to it with no reasonable escape."[25]

From thermodynamics he turned to Emden's convective gas spheres and then used Planck's black body radiation law to calculate surface brightnesses, temperatures, and radii of stars. He also employed Lane's law to show that the temperatures at homologous points in two spheres of perfect gas will be directly proportional to their relative masses. Thus more massive gas spheres achieved higher maximum temperatures when they reached critical density. This broke the back of the argument that all stars had to pass through the same stages.[26]

Russell's successive presentations of his theory, from 1910 through 1914, drew upon greater stores of evidence and revealed a strengthening grasp of relevant physics. With the appearance of the Bohr model of the atom, Russell considered

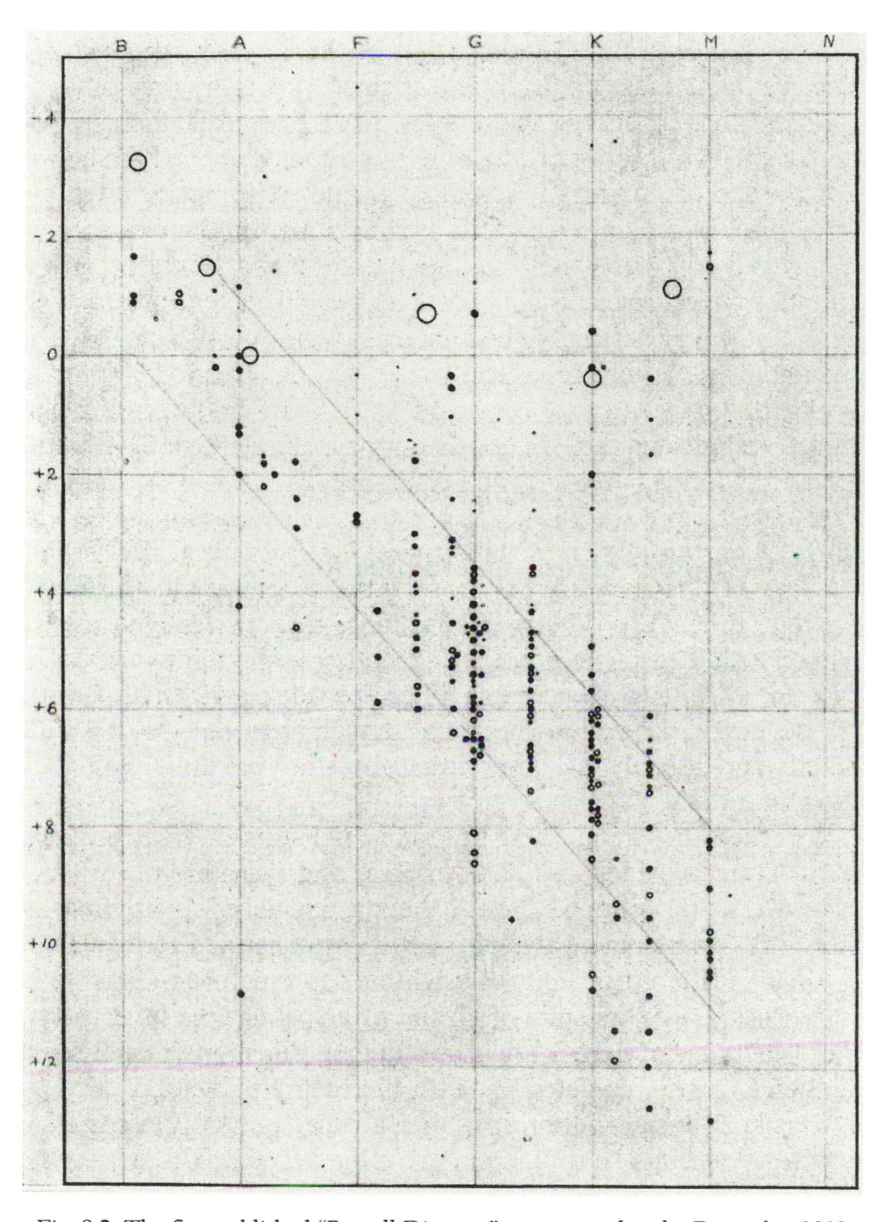

Fig. 9.2. The first published "Russell Diagram" as presented at the December 1913 meeting of the AAS in Atlanta. Based upon direct parallaxes, the sizes of the dots are proportional to the size of the probable error. The open circles are stars with only one set of parallax observations. Russell noted that he constructed this diagram in the spring of 1913 (p. 284), but the record of his talk at the Royal Astronomical Society in June suggests otherwise. Reproduction of fig. 1 from Russell, 1914b, 285. Reproduction courtesy of Carleton College.

the influence of ionization ("dissociation") upon density in the hot stellar interior, and the presence of "radio-active or similar atomic" sources, neither of which, he conjectured, could alter the course of evolution.[27] As had Hertzsprung, he explained why giants were so rare in space by arguing that the initial stages of contraction of a star must be very rapid, whereas later stages, when the bulk of the gravitational potential had been expended, proceeded slowly. "This reproduces, in its general outlines, just what is observed."[28]

Russell also strengthened his statistical arguments. No red giants had been found in spectroscopic binary systems, which meant that they had not yet fissioned, "in agreement with the view that they are in earlier stages of evolution."[29] Russell did not try to explain why velocity should increase or decrease with age, preferring to emphasize that there was every reason to argue that a dense star was in a late stage of evolution. Density was his age determiner, not velocity, but he admitted that his theory left " 'star streaming' as puzzling as ever."[30] At least his theory, Russell concluded, did not require unknown forces of acceleration or retardation.

Russell paid homage to the labors of astronomers who made his synthesis possible. He hailed the generosity and industry of observers, "who, through long and weary nights, accumulated bit by bit, and, through monotonous days, prepared for the use of others, the treasures of observational knowledge with which it has been my pleasurable lot to play in the comfort of my study."[31] This would be his oblation throughout life.

Answering Eddington

After Atlanta, Russell took the offensive, showing how the traditional view created more problems than did his alternative view.[32] He started with his conclusion that orbital periods were determined at the point of fission and did not change much over stellar lifetimes. Since most spectroscopic binaries were hot stars, and the longer period systems were among the later classes (and fissioned earlier), this lent "strong evidence against the conventional view of their relative ages." He tried to make this point with Campbell at Lick, who refused to listen, but with Eddington he invoked Darwin: "This proposition is implicitly contained in Darwin's original equations, and attention has been specifically called to it by Moulton . . . and by the writer."[33]

Conventional theory could not explain how a Harvard type B star lost some two-thirds of its mass as it grew older, but Russell's double-valued temperature scheme allowed a way out. Since only the most massive stars achieved the highest temperatures, they formed the only pure sample, whereas all the cooler classes were admixtures of lower mass stars at their maximum temperatures and higher mass stars still heating or cooling. Further, if the stellar system was dynamically relaxed, the most massive stars would have the smallest velocities. So if a few high-mass giants were mixed into a sample of low-mass dwarfs, this admixture of stars of different masses would skew any velocity distribution, and the observed distribution, Russell argued, could be a selection effect caused by mixing and not an age effect. In this manner, Russell explained Kapteyn and Campbell's correlation of peculiar velocities with spectral type. Russell knew that applying equipartition theory to the stars was becoming popular with some mathematical types. J. Halm had suggested it, and Eddington soon followed suit. At the least, Russell argued, "a correlation between mass and velocity, especially in the sense here assumed, seems more probable than one between temperature and velocity, or velocity with age."[34] With arguments such as these, among others dealing with the spatial distribution of stars with respect to the Milky Way plane, Russell thus was able to throw considerable doubt on the conventional theory as he bolstered his own. It was a masterful rebuttal.

Eddington was impressed: "I see your point about spectroscopic binaries," he admitted in March. Even though he worried about angular momentum and tidal friction, he harbored an "instinctive feeling (in vulgar phrase—prejudice)" that Darwinian fission worked. He was also swayed by Russell's admixture argument. It was a "very interesting problem," he added, "Really, I can't make up my mind about it; except that there is something more to be found out."[35]

Russell's new way of looking at the stars in the aggregate, and especially his arguments that showed giant stars to be of extremely low density, helped to draw Eddington into the study of the constitution of the stars. Eddington would take a circuitous path, trying to explain the mechanism of Cepheid variation and the dynamics of globular cluster systems in terms of the kinetic theory of gases. But eventually he would construct a theory of giant stars as condensing gaseous bodies where the inward gravitational force of a star's mass was offset by radiation pressure caused by the outward flow of heat. In particular he would show that Russell's interpretation of the relationship he and Hertzsprung had found could be supported theoretically.[36] This rough harmony of theory and observation would strengthen over the next decade, but Russell still had more convincing to do.

Continuing Resistance

Hale had already identified stellar evolution as a major motivation for his research; in 1908 it was his rationale for building larger telescopes.[37] Though Hale was keenly interested in Russell's ideas, he was not an early convert, even with Shapley at his side. He was not ready to jeopardize his relationship with two very important astronomers who were highly skeptical: Campbell and Kapteyn. As a result, when Hale searched for someone to speak on stellar evolution at the National Academy of Sciences, as part of a series on evolution in 1914, he turned to Campbell and not Russell.[38] Campbell still hoped that something would explain away giants. Not a mathematical theorist, he was therefore not able to parry and thrust, only to resist stubbornly, hoping that new evidence would emerge to preserve the status quo. For Campbell, the order of stellar evolution was not a theory but a belief.[39]

Russell labored hard to sway Campbell, at least to give his theory a fair hearing during his academy lecture.[40] Instead, Campbell marshaled evidence for the conventional view. He did mention Russell, admitting that not all astronomers held to the conventional view of hot blue bright-line stars descending from nebulae. But he lumped Russell and Lockyer together, and he never mentioned giant stars, nor Russell's diagrams, nor the fact that more and more astronomers, even some on his own staff, were beginning to take Russell's arguments seriously.[41]

Even Alfred Fowler, Lockyer's old assistant, was on the fence. He organized a discussion on stellar evolution to debate Russell's theory at the British Association in September 1915 where he agreed with the Astronomer Royal and with Eddington that "Russell's work demands careful consideration, but I am not yet prepared to acknowledge that the order of evolution suggested by the Draper classification has been overthrown."[42] By the end of the year Russell's theory could not be denied, nor could it claim victory. But Russell had gained some powerful allies: one in particular was Hale's right-hand man, Walter Sydney Adams.

Giant Stars Confirmed

While Campbell prepared his address at Lick Observatory, three hundred miles to the southeast, Adams, Hale's chief stellar spectroscopist, engaged in work that eventually proved that giants existed. Adams had long tried to find temperature and pressure effects in the fine structure of stellar spectra. Working with the laboratory physicist Henry Gale, he found that variations in line strength ratios in stars could be correlated with changes in temperature and pressure in laboratory spectra, but the correlations were weak.[43]

As Adams worked away on his high-dispersion spectra, in 1911 Hale asked him to conduct a study for Kapteyn, who was looking for evidence of an absorbing medium in space. Kapteyn was concerned that systematic effects like space absorption could alter many of the statistical arguments he had been making about the distribution and kinematics of stars. Adams chafed at the assignment and Kapteyn's direction, but since Kapteyn was a Carnegie Research Associate, he had no choice. Over the next year he compared the continuum intensities of stellar spectra in different wavelength regions between stars of high and low proper motion to look for differential effects of absorption between these statistically close and distant groups of stars. In mid-1913 he enlisted Seares's help with the 60-inch reflector and was assigned a young German, Arnold Kohlschütter, who had worked under Karl Schwarzschild at Potsdam.[44] By early 1914, Hale excitedly wrote Kapteyn that they had indeed found evidence of interstellar absorption, giving Kapteyn all credit for his prescience. Adams, however, was not so sure. What he found instead was a link between his line intensity ratios and the average luminosities of the two groups of stars Kapteyn had asked him to isolate. Influenced by Hertzsprung's analysis of Maury's c stars, Adams had been looking for luminosity effects all along but was having trouble convincing Hale.

He sent his evidence to both Russell and Hertzsprung; Hertzsprung believed he was seeing a real luminosity effect.[45] Bolstered by this, Adams and Kohlschütter confirmed that there was a strong correlation among luminosity, line intensity ratios, and continuum brightness. Adams soon managed to convince Hale, who relayed the news to Kapteyn, finding the Dutchman none too happy with the outcome.[46]

What is now called the "spectroscopic parallax" technique became a valuable new tool for estimating the absolute magnitudes and hence the distances of stars, to depths in space far greater than trigonometric parallaxes and even proper motions could match. But the technique also confirmed the existence of extremely luminous giant stars.[47] Both Russell and Eddington realized this immediately. Russell used his *Scientific American* column to hail the method a "most valuable instrument of investigation."[48] Eddington congratulated Adams, marking the deeper significance of his work: "[I]n particular I think the question of giant and dwarf division of the red stars must now be considered finally accepted."[49]

Hale now believed in giants and told Norman Lockyer that it was Adams's work that made all the difference: "Russell's adoption of a similar curve, derived from a very different series of investigations, is striking evidence that the old idea of a single branch is not likely to survive." Hale added that Adams, Seares, and other members of his staff were busy refining the technique to explore Lockyer's theory of stellar evolution.[50] Although the spectroscopic parallax technique was a great discovery, as with most empirical relationships in spectroscopic astronomy, no one had a clue as to why this one existed. This bothered Adams. What was causing the luminosity effect? Adams asked Eddington whether he had any ideas, wishing that "we had more physical knowledge regarding the interpretations of stellar spectra." He also confided his doubts to Russell in 1917, concerned that the laboratory evidence he and Gale had collected was not an explanation. Why did reduced pressure favor the strengths of some lines and not others? Neither Russell nor Eddington could shed any useful light on the subject at the time, but they all knew that spectra harbored clues to varying conditions of temperature, density, and pressure in stellar atmospheres.[51]

Kapteyn never accepted Adams's work, mainly because the spectroscopic parallax technique, and giants, cast serious doubt on the calibration of his mean parallaxes.[52] Kapteyn's continuing resistance surprised other early skeptics, like Benjamin Boss: "Many astronomers do not accept Russell's ideas on the evolution from giant M through B to dwarf M, but I did not know that anyone refused to admit the existence of two divisions."[53] More than scientific evidence influenced Kapteyn's views. He had originally liked Russell's parallax work and the relations he was finding between parallax and spectrum. But in the wake of spectroscopic parallaxes, Russell's theory challenged one of the cornerstones of Kapteyn's lifework.[54]

One can only speculate on how Pickering must have felt as Russell's theory, bolstered by the spectroscopic parallax technique, led to renewed calls for a two-dimensional classification. At one level Pickering could take pride (and he did) that Russell's work was made possible by Harvard data. But at a deeper level, Russell found, as did Adams and Kohlschütter, that the linear Harvard classification sequence (OBAFGKM) had to be modified. As Hertzsprung told Adams in 1915, no longer would the Harvard system suffice, tied as it was to the hydrogen lines and to the color of a star. "In the same way that the old classification is connected with the color it was most desirable to get a new one connected with absolute brightness."[55] Eventually the pure Harvard system did not survive; at the Rome meetings of the International Astronomical Union (IAU) in 1922, three years after Pickering's death, it was altered to include Maury's subclassifications.

Eddington Models Giant Stars

In the spring of 1916, Russell was made a foreign associate of the Royal Astronomical Society, one of his earliest honors. The Astronomer Royal Frank Dyson, among others, was convinced of the rightness of his theory, especially after Eddington found theoretical support for it.[56] Eddington was certainly not won over quickly, however. At Alfred Fowler's September 1915 British Association soirée, Eddington still had trouble accepting the densities of the giants, claiming that he was "not an out-and-out supporter of Russell's theory." He agreed that Russell had overthrown the "mystic order," but he knew that Russell's theory could not manage any relation between nebulae and stars as the old one could.[57] Nevertheless, Eddington believed that giants and dwarfs were here to stay and provided a valuable clue to the probable direction of the stellar aging process.

In 1916, with Cambridge disrupted by war, Eddington, a Quaker pacifist, found more time to think about stars. He told Norman Lockyer in March that he was warming to Russell's views, with reservations. The Harvard classification no longer satisfied Eddington: now he agreed with Lockyer that subdivisions were required.[58] Eddington had been rethinking the apparent relation between spectral class and velocity. Assuming that "[t]he feebly luminous stars move with much higher average speeds than the bright stars," he realized that some factor beyond spectral class and brightness must be at play. Without evidence to the contrary, he argued, "the natural suggestion is that it may be *mass*."[59]

Three factors moved Eddington in Russell's direction. First, his statistical criticisms had been answered. Second, although he must have encountered it before as a student of the physicist Arthur Schuster, he reread Emden's 1907 *Gaskugeln*, the most complete description of the behavior of gas spheres in convective equilibrium. Emden developed polytropic relationships between pressure and density in a gas sphere in the perfect gas state, and *Gaskugeln* gave a coherent basis for Ritter's theory of stellar evolution based upon Lane's law. It convinced Eddington that there was strong theoretical support for Russell's ideas.[60] Third, Eddington had already used Emden's polytropic models to explore the distribution of stars in globular clusters, an outgrowth of his exploration of Kapteyn's theory of star streaming.[61] Thus by the end of 1916, Eddington's interests in the dynamics of the sidereal system and his attempt to criticize Russell's theory of stellar evolution brought him to examine in depth Emden's theory of polytropic gas spheres and to think about how a star works as a thermodynamical system. One more step was required before he actually sat down to build his stellar models, and that step involved something Russell and Shapley had found.

In his rebuttal to Eddington, Russell referred to a short study he had made of the distances to variable stars in 1913, where he found that Cepheid variables were very distant as a class, very luminous, and hence were giant stars. In fact, they were so large that if their light variations were due to eclipses, their diameters were far larger than their orbits. They could hardly be binaries, and their light variations could not be due to eclipses.[62] To make matters worse, their light variations did not agree with their observed velocity variations if both were due to orbital motion. What, then, caused the light of these stars to vary? By August 1914, Shapley was convinced that Cepheid variables pulsated, causing changes in luminosity, color temperature, and radial velocity. He offered no rationale for these stars' pulsation, though he argued that "such oscillations are both possible and probable."[63]

Shapley suggested that the pulsation mechanism could act as a probe into the structure of a star's interior, and Eddington followed it up, later stating that it was by this path that he was led to the study of stellar structure. Ritter had also explored pulsation mechanisms, treating them as harmonic oscillations, but Eddington preferred a thermodynamical explanation: the opacity of the gas in the stellar interior acted as a governor, regulating the engine. To make this work, in 1916 and 1917 he developed a new type of stellar model where radiative transport dominated. He was the first theorist to develop a comprehensive theory of stellar structure on that basis. He did this assured by the fact that he was modeling real stars; giants had densities so low that they had to behave as a perfect gas.[64]

Eddington used opacity as a brake to balance the outward pressure of light and the inward force of gravity. He found that a perfectly gaseous body where energy was transported by radiation alone could be described by an Emden polytrope, and that the total radiation of a giant star was "independent of the stage of evolution."[65] Thus Eddington found a theoretical basis for Russell's observation that all giants had nearly the same luminosities. He also examined the relationship between the effective temperature of a star and its mean density as a function of mass, and found that the maximum temperature attainable by a star was determined by its mass, just as Russell had argued.[66] And finally, Eddington found a density law that fit the giant branch of Russell's 1914 diagrams beautifully.

Eddington found that the effective temperature of a giant star was six orders of magnitude more sensitive to density than to mass. As a giant contracted out of the interstellar mists, it would heat up, turn from red to blue, but not change much in luminosity because as its surface area diminished, its luminous efficiency increased as it heated. Eddington's new theory for giants was in rather impressive agreement with what Russell had found. Eddington excitedly wrote Russell in January 1917 telling him all this, especially how his density law could reproduce Russell's giants.[67] Even though Eddington's first models were little more than deft dimensional analysis, he felt that they confirmed Russell's theory. When he found that major modifications were required to account for a much lower mean molecular weight due to ionization and the presence of free electrons, the agreement was even closer: "[T]he beauty of the radiation pressure hypothesis is that it makes things vary very little with the molecular weight—the ionization to a large extent simply substitutes electron pressure for radiation pressure." What Eddington did worry about was that such highly luminous stars could not last very long. Gravitational contraction alone could not provide more than 100,000 years.[68] Further, dwarf stars did not work out very well. They were not in the perfect gas state, so Eddington substituted the van der Waals equation of state. He found little agreement with observation but still concluded that the Sun had to be close to the critical density where deviations from the perfect gas laws became serious. He placed the critical density for all stars at about 1.5 times that of water, the mean density of the Sun. Among many assumptions, Eddington required that the opacity of stellar material be constant. This would plague his theory for years.

As Eddington refined his theory, bringing it into greater coincidence with Russell's, he prepared forceful reviews for *Nature* and the *Astrophysical Journal*. These were debated equally forcibly by his first and chief critic, James Jeans. In a flurry of letters to *Nature* between mid-June and mid-July 1917, the two emerging titans of astrophysical theory jousted over rhetoric, dimensional analysis, and even a bit of physics.[69] Throughout their debate, however, the existence of giant stars was not an issue; Russell's theory, based upon giants, now formed the background for the debate itself. Even if Jeans had shown Eddington's models to be without basis, which he did not do, this would not necessarily cast doubt on Russell's theory, since other means could still be found to make sense of Russell's deductions. The debate soon focused more on the nagging problem of timescales and the source of energy of the Sun and stars, one of the most perplexing problems astronomers faced.[70] Whatever the stars' source of energy, gravitational or atomic, Russell's theory was not impacted differently from any other. Expansion could exist only as a temporary instability, as in a Cepheid. This amused Ernest Rutherford, who chided astronomers for their conviction that "evolution only proceeds in one direction . . . I see no reason why we should not have some stars condensing and others diverging."[71] But no one listened.

Nevertheless, Russell agreed with Eddington that Jeans was "barking up the wrong tree." His conviction was based solely upon how Jeans handled observational data. Russell followed the discussion "mainly on the side approached from the statistical data." He was fascinated by "theoretical astrophysics" as practiced by Eddington, but as yet did not identify personally with the new and competitive arena. He recognized, however, that the debate between Eddington and Jeans was "rapidly opening up a new branch of mathematical astronomy."[72]

Russell did employ his "statistical data" to make minor adjustments to Eddington's argument. He found that maximum temperature varied more rapidly with mass than Eddington had suggested, and that Eddington's critical density did not "fit the facts." He did not touch the mathematics, however, and was relieved when Eddington agreed with his modifications for maximum temperature and critical density. "I think the most important thing," Eddington declared, "is the explanation of why the stars have the particular mass that they [do]. It seems to me that the indication that this is settled by radiation-pressure suddenly rising to importance is very strong." Eddington believed that the convergence of theory with observation was evidence that his radiative models made physical sense.[73] His theoretical reconstruction of the giant branch gave Russell great confidence, so much so that Russell overstated what Eddington had done. Writing to Hale in September 1917, Russell believed that Eddington's theory

> accounts for the whole character of the system of giant and dwarf stars. With certain alterations in the numerical constants—for which I have data available—it will give an extraordinarily good *quantitative* agreement with almost all we know about stellar masses, absolute magnitudes, densities and temperatures . . . It is naturally not displeasing to me, too, to find that Eddington's theory, developed from physical grounds, agrees almost absolutely with that which I advocated on astronomical grounds four years ago.[74]

Russell had approached the problem from the astronomical side, although it was physical intuition that led him to it. Russell began with an unpopular theory of stellar evolution, amassed an enormous amount of observational evidence in support of it, and then showed that it did not do violence to known physical law, thus demonstrating not only his power but the growing importance of physical theory to problems in astrophysics. Russell soon tried the same method to search for the source of stellar energy, convinced that the stars themselves would provide the evidence needed to reveal the physical process responsible. He did not meet his goal, but along the way his intuition uncovered a simple self-regulating thermodynamical process that showed why stars built upon Eddington's models were stable.[75]

With Eddington's models Russell's theory entered the mainstream. Popularizers and theorists alike who speculated about stellar energy and cosmology, no matter their particular ideas, tended now to think of the lives of stars in his terms. Attracted to the stellar energy problem, the physicist William Duncan MacMillan, professor of astronomy at the University of Chicago, contemplated a steady state universe in 1918, arguing that "[t]here is no necessary limit to its [a star's] age, and though the star itself may rise and fall, the universe as a whole is not necessarily altered."[76]

It is clear, however, that more people believed in giants than in Russell's theory. When Russell was awarded the Gold Medal of the Royal Astronomical Society in 1921, Alfred Fowler made the presentation. Fowler had come a long way since 1915, stating baldly that "[t]he question of the existence of giant stars is no longer a matter for discussion." Any theory of stellar evolution now had to account for them, and "Russell's theory represents the only attempt which has yet been made to give a rational explanation of the part they play in the general scheme of stellar development." Hardly a ringing endorsement, though Fowler's reticence really reflects the fact that Russell won the medal in a compromise vote when the council was deadlocked over Einstein.[77]

Even so, Fowler appreciated the moment. Giant stars were a fact, forced on an unwilling world by Hertzsprung and Russell, proven beyond a doubt by spectroscopic parallaxes and, in December 1920, by the spectacular measurement of the angular diameter of Betelgeuse by Michelson's stellar interferometer at Mount Wilson. Rising to some eloquence, Fowler proclaimed Russell's theory "almost as wide as the universe itself. The theory has already had a profound influence on astronomical thought, and seems destined to be of great value in directing the future course of astronomical research."[78]

Indeed, Russell himself had already played a major role in plotting "the future course of astronomical research." In the ten years prior to Russell, all medalists had been either orbit theorists or observers. Russell was the first to be awarded the medal for theory-driven astrophysical research. He would be followed rapidly by Jeans and Eddington, and this signaled a shift in the nature of astronomical practice to what one astronomer later called a "very good mix of traditional and avant-garde astronomy."[79]

That Russell eventually was chosen as the gold medalist in 1921 speaks to the growing importance of the theoretical interpretation of astronomical evidence. Russell, Fowler argued, had demonstrated that "the study of stellar evolution calls for the co-ordination of observational data obtained among many different lines of astronomical research." No one person could accomplish such a task, even though a theorist could well devise a model without performing any observations himself. But what marked Russell as special for Fowler was that he not only developed a theory but planned out an observational agenda to test it.[80]

Russell's ability to assimilate and manipulate observational data and direct them to the solution of theoretical questions, first seen in full force during this period, marks well his style of research. In following decades he would do the same to study the atmosphere of the Sun, to explore complex spectra, and to refine knowledge of binary evolution. His role at the interface of theory and observation would change, however. Up to 1920, he made no excursions into physical theory itself. Indeed, as Fowler had noticed, the beauty of Russell's theory of evolution was that it "violates no known physical laws."[81] But after 1920, the success he had enjoyed applying his broad intuitive powers emboldened Russell to make such excursions, convinced that the solution of the major questions in astronomy, such as the source of energy or the chemical composition of the Sun and stars, lay in the creative interpenetration of astronomy and physics. Russell made these forays with the help of physicists, though he would always do so heavily influenced by the data available, or, more characteristically, by the data he asked others to provide.

How Russell developed and defended his theory of stellar evolution reveals how he forged links between the old observational astrophysics and the new theoretical astrophysics at a time when the latter specialty was hardly yet formed. His position at the interface of the two specialties would shift back and forth a bit as opportunities arose.

Above all, Russell was an opportunist, always looking for the surest path to achievement. In February 1919, delighted with his recent success showing that Eddington's stars were stable, Russell predicted to Shapley that "Eddington and Jeans have shown us that the time is ripe for fundamental work in theoretical astrophysics."[82] The time was also ripe to shift allegiance, and Russell had already made that shift, from Pickering to Hale, as he became an agent of change in astronomy.

• CHAPTER 10 •

Shifting Allegiance

RELATIONSHIPS

Princeton faculty typically summered in the mountains of Vermont or New Hampshire, but the Russells took to the shore and the rambling Cole home in Southport, Connecticut. It was reminiscent of his Oyster Bay roots. What started as short visits to show off the children became, by 1914, month-long stays "with our whole tribe."[1] John H. Cole was a widower now and loved to have his grandchildren around. They had over a dozen acres to play in, with outbuildings, beachfront, and pier. Russell spent his days as active as ever, but with family, sailing and writing.

In Southport, Russell would focus on his essays for *Scientific American*, which kept him in close contact with observatory directors who would feed him late-breaking news. The *New York Times* kept him in touch with the world, and he kept in touch with Pickering and Shapley. His wife managed his life and the children, with the help of two nurses, since Elizabeth still needed constant attention. She was unable to control her little body and, most noticeably of concern, had difficulty with speech. Her condition failed to improve over the years and was eventually diagnosed as one of the cerebral palsies, a "[c]onjunctive defect of development in cerebellum . . . ," as Russell described it for a 1935 survey by the National Academy of Sciences. The Russells were hardly immune to the guilt, anger, self-doubt, and distress attendant on having a cerebral palsy child as they continually sought a diagnosis and cure from doctors locally and in New York City. That he referred to the condition as a "defect" suggests that he thought of it as congenital rather than as acquired through some insult to the fetus such as an umbilical cord injury, maternal illness, a birth injury, or some other intrauterine condition—a question that was debated then as now and has been the source of much psychological destruction. Their son keenly recalls that Elizabeth's condition "was the cause of anxiety and a lack of understanding and tension" in the family. Her condition was not severe enough to warrant institutionalization, but her options were far from clear at a time when there were few established treatment centers and even less comprehension of the causes. Elizabeth would be their greatest challenge.[2]

Yet, outwardly, these were years when his family was young and growing and a source of joy; they were generally good years. The astronomy department had stabilized; Dugan's double star industry meshed nicely with Russell's need to fill in gaps in his evolutionary scenario.[3] As always, Russell had his distractions. He found ways to make his commitment to revise Young's textbook complement his research. He was as productive as ever, but his focus flirted and his papers were mercurial. Between 1914 and 1919 Russell produced some twenty-six astronomical papers in a wide range of problem areas. His textbook work led him to dabble in a qualitative analysis of the abundances of the elements in the Sun's atmosphere, performing a comparison with the abundances found in the earth's crust and in meteorites. He continued analyzing eclipsing binaries with Shapley, who spent part of his first year at Mount Wilson firming up the relationship of densiity and spectral class.[4] Russell also derived a new and highly efficient method of computing the orbits of spectroscopic binary systems; explored interesting visual doubles; discussed the masses of the stars; investigated the parallaxes of high proper motion stars; studied the albedoes of the planets and the Moon; commented on relativity and on the origins of periodic comets; realized why stars obeying Eddington's theory of radiative equilibrium and powered by radioactive sources remained stable; and predicted the apparent angular diameters of red giant stars barely in anticipation of their measurement at Mount Wilson.

Many of these works proved seminal. His insights about stellar stability aided acceptance of Eddington's theory and Shapley's pulsation model for Cepheid variable stars. And his brief study of albedo provided the basis for the physical investigation of the nature of planetary surfaces by linking astronomically observed albedoes with the reflectivity of terrestrial substances. What separated Russell's work on albedo from all earlier efforts, according to a contemporary, was his willingness to abandon hypothesis "in favor of experiment." Russell always checked theory against data and modified theory accordingly.[5] This style would more than repay the additional effort in years to come as he migrated into the analysis of complex spectra and, with F. A. Saunders, sorted out the alkalis.

In these years Russell established strong relationships with a few key people. Pickering had been his first, and strongest, ally. Going through family records in July 1914, Russell found a deed for a house that his "great-great-great-great-grandfather Gabriel Holman" had bought in Salem in the year 1746. The house and its lands were bounded on the south by land belonging to Timothy Pickering. "Perhaps our ancestors were neighbors," he told Pickering. "I hope our ancestors did not quarrel about the name of the street," Pickering responded.[6] The two astronomers were now collaborators in many projects, not the least the lunar orbit program. Russell often sought out Pickering's advice, and visited Cambridge frequently, especially during the summers when the family was in Connecticut. But just as so many scientific bonds, national and international, would be broken "into hostile political camps" during the next few years by the passions and aftermath of war, theirs would strain to the breaking point as American astronomy found itself divided over its needs and its future.[7]

When the world plunged into war, organized science in America took the opportunity to assert its importance to both national sovereignty and economic competitiveness. But American science was far from monolithic. Old institutions like the American Association for the Advancement of Science and the National Academy of Sciences had very different power bases and created different mechanisms to manage science. The more egalitarian AAAS had just emerged from an extensive horizontal reorganization to accommodate the growing forces of specialization in American science, and the elite National Academy took steps to reassert its historical role as the adviser to government.[8] Pickering was a major power broker in the AAAS, whereas Hale was the architect of the academy's new activist arm, the National Research Council. They, and their organizations, had very different views concerning the organization of science in America and drew Russell into the debate over how his profession could participate in the present emergency and what its needs were in the future.

In this chapter we explore how Russell entered the debate, and how he came to realize that his view of the needs of astronomy differed from Pickering's, reserving for the following chapter his parallel experiences in the war. The two discussions are closely linked, however, as his war work confirmed his growing conviction that astronomical practice had to incorporate physical theory at its core in order to break the bonds of its naturalist roots.

Proper Practice in American Astronomy

Russell's obstreperous 1905 letter to Wilson contained the seeds of his view of how astronomy should be practiced: astronomy had to take a problem-oriented approach, following the examples set by Kapteyn and Darwin. Astronomical observations had to be planned with problems in mind, developed within a framework informed by physical theory. Russell's actions reflect these convictions; they are clear in his approach to stellar evolution and his application of physical arguments between 1910 and 1914 to buttress his theory. They are most evident in his work on binaries and his hypothetical parallax technique, which depended upon iteration. Overall, Russell believed that progress could be made through iteration between observation and theory.

He said as much in a series of a dozen lectures on the "Scientific Approach to Christianity" delivered on campus in 1916 and 1917.[9] He hoped to reconcile a deterministic, mechanistic universe governed by divine law with the existence of free will and personal responsibility, but succeeded mainly in revealing his view of proper scientific practice. He depicted science as a "trial and error" process, eschewing dogmatism and belief. He described the process of theory formation as based not upon observation alone but, true to his Princeton heritage, upon "vastly simplified and . . . incomplete working models of the phenomena, devised by the human brain." Iteration brought the phenomena of Nature into a coherent theory; ultimately, if the theory turns out to be in "harmony with the observed facts throughout the range of its applicability it attains the [degree] of a natural law." Careful to limit the power of science, Russell reminded his listeners that "these laws do not explain, they merely describe," and, moreover, were not "an agent of spiritual and moral degeneracy." In all of this Russell echoed mainstream thought.[10]

The difference with Russell is that he put his rhetoric to practice. The creative shortcuts Russell delighted in developing, using statistical inference and intuition, were not yet accepted by traditionalists.[11] He was impatient with the common view, expressed by J. S. Plaskett, director of the Dominion Observatory in Victoria, that "[a]stronomy is, perhaps more than any other, a science which requires long continued and systematic investigations to be carried through with faithfulness, unselfishness, and untiring perseverance before any definite results can be obtained."[12] For Plaskett the heart of astronomy remained the act of observation, unsullied by theory or approximation. Any "born astronomer," in the sense in which Gill had used the term, shared this view.

Russell was reminded of this once again in 1914 as he fought to have his hypothetical parallax technique accepted, and found continuing resistance not only from Frost and Hertzsprung but from mundane double star observers like Eric Doolittle, an ardent Burnham disciple.[13] Doolittle, a functionary in the American Philosophical Society, seemed to view Russell as an interloper and criticized his papers. In an angry response, Russell released his usually well-controlled sarcasm:

> I have not Professor Burnham's serene and self-denying willingness to let the derivation of results from my work wait until fifty or a hundred years after the probable date of my decease. I am always trying to best Father Time with the aid of what mathematical weapons I can bring to bear on things. A hundred years hence all this work of mine will be utterly superseded: but I am getting the fun of it now. The altruistic nature of the work that you double-star people are doing arouses my lively admiration. But I fear that I am too selfish to emulate it.[14]

Russell was clear about his motivations. He strove for computational efficiency and shortcuts because he was intensely impatient. Although he was no stranger to deliberate and careful work, it always had to be a creative and insightful exercise and had to be designed to lead to the solution of a problem. Since he could think and work at speeds frightening to others, he regarded the rest of the world as proceeding at a glacial pace.

Publicly, Russell extolled the selflessness of the observer, but he knew that the continued authority of people like Burnham and Doolittle had to be challenged if American astronomy was ever to become a modern physical science.[15] Soon he would have his chance, but he would have to break ranks with his strongest ally.

Pickering

By 1915, at age seventy-one, with people like Newcomb, Young, and Langley gone, Pickering was the most influential astronomer in America. He was president of the AAS and was a strong voice in the American Philosophical Society, where he was instrumental in gaining membership for Russell in 1913.[16] He was also a gatekeeper for numerous funds for astronomical research administered by Harvard, the American Academy, and the National Academy of Sciences. His popularity and power derived both from his generosity in sharing data and from his commitment to support the rank-and-file astronomer. He believed in apprenticeship programs rather than formal training and did his best to support observatory directors in need of routine assistance.[17]

Pickering had devoted decades to establishing Harvard College Observatory as a central source and defining authority for spectroscopic and photometric data and practice. Since his installation as Harvard College Observatory director in 1877, he had created a plan to support astronomical research by disbursing funds from philanthropists through Harvard.[18] By the 1890s Pickering had established himself as an international broker for both the management and the support of astronomy. He would first ask astronomers what they needed, and then approach benefactors with those needs, brokering as he saw fit, with Harvard as funnel. His view of what deserved support did not change from the 1890s to the end of his life. The continued health of the astronomical enterprise, Pickering argued in 1895, centered upon "the undertaking of large pieces of routine work, and the employment of numbers of inexpensive assistants whose work is in a great measure mechanical, such as copying and routine computing." The financial assistance Pickering could provide was moderate, though it gave him great power.[19]

Pickering's greatest competition came from Hale. They competed over the 40-inch glass disks that Hale eventually secured for Yerkes. They competed bitterly over how astronomy should be funded by the new corporate philanthropies, such as the Carnegie Institution of Washington, and again Hale won handily. Pickering tried to adjust. He proclaimed that observatories were "great factories, which by taking elaborate precautions to save waste at every point, and by improving in every detail both processes and products, are at length obtaining results on a large scale with a perfection and economy far greater than is possible by individuals, or smaller institutions." But his Harvard-centrism put him out of step with the great industrialists who were then transforming the landscape of corporate philanthropy.[20] He joined forces with Columbia University psychologist and *Science*

Fig. 10.1. Edward Charles Pickering. Photograph courtesy Dorrit Hoffleit, Yale University Observatory.

editor-publisher James McKeen Cattell in a new AAAS venture, a "Committee of One Hundred" to raise and dispense funds and give the AAAS a dominant role in American science.[21] In 1915, Pickering chaired both the AAAS committee and its "Subcommittee on the Needs of Astronomy." As he had done before, he canvassed observatory directors, asking their needs, and then collected their responses into a proposal to the new Rockefeller Foundation. Ignoring the advice of those who provided nuanced views, Pickering's synthesis of the needs of American astronomy did not change: the greatest single need "was for assistants—either an experienced observer or one or two computers."[22]

The Rockefeller Foundation turned him down flat, but the committee had garnered some funds, so Pickering asked his astronomy subcommittee, including Campbell, Frost, Schlesinger, E. W. Brown, and Russell, to once again "notify me in what way you think we could best promote the progress of Astronomy." Russell asked for two computers and, as noted in the last chapter, eventually secured them through Pickering's lobbying. Some, like Schlesinger, were growing weary of the constant drill, whereas Russell was still very much under Pickering's wing and beholden to him for his reputation. But the escalating war, and his self-confidence, shifted his attention to a new form of collective action.[23]

Shifting Allegiances to Hale

Because of his upbringing, Hale spoke the language of the industrialist and knew how to develop broad coalitions to build institutions and attract patronage.[24] Hale was comfortable with corporate philanthropists and knew how to craft his pitch in their language. Rather than naively blustering ahead, as Pickering seemed to do at times, Hale worked systematically. Instead of launching frontal attacks, Hale preferred to work through golf partners.[25] Unlike Pickering, who tied himself too closely to Harvard as a broker, Hale argued that control should be reserved for representative committees and societies. "No *single institution* control for me!" he told a confidant. By 1904 Hale knew that he had to break with Pickering: "I suppose I could survive the loss of the advantages of the *salon*!"[26] Hale walked off with the majority of the Carnegie Institution's funds for astronomy, building his 60-inch telescope and the Mount Wilson Solar Observatory.

Hale always knew he had to co-opt Pickering, and did so for his Solar Union. And as Hale worked to create the National Research Council of the National Academy of Sciences, he knew too that he must deal with Pickering, who was already chair of the Section on Astronomy and had served on the Academy Committee on Research in 1913.[27] Hale's NRC would be a national agency to promote and coordinate organized science in service to the nation. After a difficult birth in 1916, the NRC created a series of committees not unlike those the AAAS had formed to assess disciplinary needs and capabilities. As the NRC gained power, it weakened the parallel AAAS effort.[28] It was here that Hale and Pickering once again butted heads, and where Russell emerged as a player.

Hale had, in characteristic fashion, wanted Pickering to head an NRC central committee for astronomy, hoping that this would align Pickering's AAAS Committee of One Hundred and the American Astronomical Society with the NRC. Envisioning that a "Committee on Cooperation" would keep everyone happy under the aegis of the NRC, Hale charged the committee with two tasks: first, to appraise the present needs of the discipline, and later, when it was obvious that the nation would soon enter the war, to assess how American observatories could be marshaled for the war effort.[29]

Pickering agreed mainly because Hale gave him a free hand to form a coalition committee and prepare a draft report on the needs of astronomy. Since NRC committees were not limited to academy members, Hale suggested that Pickering invite Kapteyn, Campbell, Eddington, Frost, Russell, Schlesinger, Adams, Abbot, Moulton, and Leuschner to write essays on "the larger problems and the best methods of attacking them, and indicating opportunities for further cooperation in research." By January 1917 they had compromised on a committee representing equally the NRC, the AAS, and the AAAS. Pickering, of course, had his

10.2. George Ellery Hale in 1908, working in his Mount Wilson office. Group photograph no. 2875, reproduction courtesy Huntington Library.

own game plan.[30] He managed to retain most of his original AAAS committee intact, although Hale had inserted himself too. And instead of calling for a series of reports, Pickering merely repeated, once again, his old familiar pattern of polling. In other words, what he had been doing for Harvard and the AAAS he now did in the name of the NRC.

What Pickering did not count upon was where Russell stood. In November 1916, Russell called for "co-operation between different observatories . . . largely under your inspiration." Cooperation for Russell, however, meant organizing to solve specific problems; and, with Hale, he believed that "each problem should be attacked by the men most competent to advance the solution." He called for more attention to physically meaningful calibration efforts; for example, the Harvard magnitude system had to be put on a physical basis through the calibration of photographic emulsions against laboratory standards.[31] Russell's suggestions differed from those Pickering expected. They were fresh ideas, not "more of the same" and certainly not calls for routine computing assistance.

When Russell read Pickering's first draft report, he was alarmed not to see any of his own ideas. Pickering only repeated his familiar conclusion that "[i]n almost every case, the demand was for more assistants to aid in extensive routine observations. A relatively very large increase of output could thus be secured . . ."[32] Russell reacted quickly but carefully, writing Pickering that "[i]t is too often still the case that the routine work is initiated *too early*, before the methods are fully perfected . . ." He gave examples, like the *Astrographic Chart* program. He also felt that Pickering had slighted Hale's programs at Mount Wilson, which were "emphatically *not* routine work." What distinguished the Mount Wilson style from all others, Russell argued, was that

> [i]t has involved rather the discussion of specific, very carefully chosen problems, with the intention, not mainly of accumulating a mass of data for the use of some astronomers in the future, but of collecting data that are believed or known beforehand to be of critical importance in the attack on some specific problem.[33]

Not wanting to appear too belligerent or disrespectful, Russell stated that though he had these reservations, he still endorsed Pickering's report. But when Pickering issued a second draft report that was essentially unchanged, Russell became annoyed. Still trying to be diplomatic, but emboldened by increased contact with Hale, Russell stated his points in more specific terms. He now openly attacked the report's emphasis on routine work; observational work might have a better chance of getting funded, he admitted, but good observational work required creativity. "[O]ne good man like Shapley, who would have to be paid a good salary to keep him, is worth more than ten ordinary computers," Russell argued. It was wiser to increase the salary of "one good and experienced man, rather than to pay smaller sums to two or three ordinary girls." Routine work was now better left to amateurs. What was really needed were new physical techniques to quantify astronomical observations. He called for an open discussion of the issues at hand when the committee next met. In effect, he was ready for a showdown.[34]

Comstock and Schlesinger had their own concerns with Pickering's report, mainly its lack of emphasis on wartime needs. Stebbins called for more international cooperation. With Russell he also criticized Pickering's system of largesse, which reinforced decades of entrenched observatory practice: "If we get into a system of a few directors with large numbers of routine assistants, no bright young man will undertake [theoretical] work, and the natural outcome will be a lack of material for even the attractive positions."[35]

Hale was most critical. By the spring of 1917, shuttling back and forth from New York City to Washington on NRC business, Hale stopped in frequently to talk with Russell. In terms remarkably close to Russell's, Hale, in correspondence with Pickering, urged "[t]hat at least as much emphasis be placed upon the formulation and solution of the chief problems of astronomy, as upon the collection of data."[36] In rapid fire Hale laid out the Mount Wilson agenda, based upon Kapteyn's plan: nothing less than a coordinated attack to gather data to answer specific questions about the structure of the sidereal universe warranted support, or his endorsement. Cataloging was still important but had to be done with specific problems in mind. "Since Darwin's time, the indiscriminate methods of Linnaeus have become almost obsolete in Biology. Darwin himself was a great collector, but he had a definite end in view, which guided him in all his work." In astronomy, Hale added, one could not completely ignore purely inductive processes, but one had to be judicious and promote those that stood the best chance to solve "specific sidereal problems." Above all, Hale argued, responsible observatory directors had to find and train staff "capable of thinking for themselves."[37]

The person who discovered magnetic fields in sunspots after pondering Pieter Zeeman's papers was understandably concerned that new lines of work in physics "should not be ignored in preparing schemes of astrophysical research." Hale wanted nothing less than a "radical change" in outlook. Astronomers needed to pay more attention to physical studies: how radiation could be influenced by electric fields, how radioactivity might power the stars, and how Bohr's theory of the atom might explain stellar spectra. "[T]hese and many similar advances warn us that the empirical procedures of the recent past must continue to yield place to studies based upon a physical foundation . . ."[38]

None of this moved Pickering, who issued his final report intact, mentioning only that there were some on the committee who had made suggestions that "could not readily be included among the routine investigations, since they relate rather to principles and social problems." Clearly the committee was at an impasse, and Pickering was imperious.[39] He would consider support only for routine work, not for radical change. Pickering also paid no attention to the preparedness question. He merely passed along the suggestions of others without editorial comment.[40] Throughout April, committee members wrote Pickering with many suggestions for how the committee might coordinate, or at least identify, war work among astronomers. Russell, Hale, and others identified areas of application: optical design, teaching navigation, general computing, and instrument development. Observatories could calibrate compasses, chronometers, sextants, and other navigation devices. Russell felt that "the intellectual training and breadth of outlook [of an astronomer was] fully comparable in value with the more immediate benefits conferred by the more 'practical' studies." Astronomers could make versatile and effective researchers in many areas. E. W. Brown lobbied strongly for "the organization and efficient carrying out of large masses of computation" by astronomers.[41] Pickering remained silent but shaken.

Pickering submitted his report in April, urging Hale to publish it promptly in the council's *Proceedings*. Hale, totally involved in the war, let it through, mainly because he knew that Pickering wished to resign his post with the publication of the report, and there was no use fighting. Pickering did resign, suggesting that Hale appoint Russell as his successor, which had been Hale's plan all along. But Pickering was clearly bitter. He reminded Hale of the many "large routine investigations" that deserved support, and that the majority of the committee had originally sided with him.[42]

Pickering was, indeed, deeply embittered. This was only the last of many skirmishes with Hale that ended poorly. He spoke his mind about Hale and the Carnegie Institution president, R. S. Woodward, in August 1917. Confiding in Charles W. Eliot, retired Harvard president, Pickering chastised Woodward for his myopic support of Hale's wastefulness at Mount Wilson. Pickering asked what of use had come from Mount Wilson thus far, after it had enjoyed an annual income which "probably exceeds that available for research in all other American observatories combined" (excluding the Naval Observatory). The Harvard plan, by comparison, had yielded an abundance of data that had been freely shared with the world. The Carnegie, he countered, "instead of finding the particular man and aiding him, has established its own institutions without proper expert advice or a knowledge of men and modern methods of research." Citing his forty years of stewardship of the Rumford Fund, his thirty years with the Elizabeth Thompson Fund, and other like services, Pickering felt he had far more experience, and success, than Woodward in such matters. He predicted that Woodward's pending "resignation would probably lead to a great advance in American science."[43]

When Pickering stepped down, Hale took charge and asked Russell to run the subcommittee as its "executive officer" and to take the lead by preparing a general essay entitled "Some Problems of Sidereal Astronomy." Russell eagerly accepted Hale's invitation, even though he had earlier rejected Pickering's invitations to manage committees. Russell rushed a draft to Hale, in confidence, and in July 1917 Hale liked what he read, though he only glanced at it. Hale felt that Russell "accomplished precisely what I had in mind, thus preparing the way for other writers." Russell said nothing about the essay to Pickering, even when Pickering asked.[44] But he still wrote Pickering trying to clear the air, managing only to criticize Harvard photometry, which, he felt, still lacked "definite *physical* definition and meaning."[45] Meanwhile, Hale moved fast. Confident that through Russell he was now in control, Hale wanted Pickering to rejoin the committee as its figurehead. Pickering agreed, because it meant he would chair open sessions the committee would hold at the August meetings of the AAS in Albany. Pickering insisted, however, that Russell do all the work.[46]

At Albany, Hale's plan for a series of reports was ratified, and Russell took the lead, choosing the authors. The subcommittee also discussed how astronomers could aid the war effort, and passed several resolutions. Most contentious was a French initiative on reorganizing science at war's end, excluding the neutrals as well as the Central Powers. Russell objected, as did others, but feelings were clearly strained over Sweden's efforts at obtaining a "peace without victory," which "God forbid!—would make the restoration of friendly scientific relations impossible." Some, like Stebbins, simply hated the Germans; whereas Schlesinger felt that the French were hysterical, and when the war was over, "we will 'come to our senses' " over such matters.[47]

After the meeting Russell visited his brother Alex in Rochester and then took Lucy and two of their children to a lakefront resort near Minnewaska. Happily ensconced in Wildmere House, Russell wrote out the resolutions from the Albany meeting and reported to Hale that all had gone according to plan.[48] Hale liked Russell's report and was delighted that Russell's opinions were "so closely in accordance" with his own. "I agree that it would be a mistake to exclude the neutrals." He was also happy to hear that the question of who would write the topical essays was left completely in Russell's hands.[49]

After the Albany meetings, the war dominated everything. Russell taught a bit of astronomy, but now navigation was his largest class, drawing enthusiastic Princeton undergraduates. He was also drawn deeper into sound-ranging work under Augustus Trowbridge. In October, after Pickering inquired about the status of his report, Russell finally admitted to Pickering that his essay "Some Problems" was now in Hale's hands, but that Hale had not yet written preliminary commentary, so the "matter rests for the moment." Russell still did not offer to send Pickering a copy, nor was he overly anxious to see his essay published for the moment. But he did outline for Pickering what the general series would look like, and finally exposed their deep differences to full view.[50]

Russell planned some nineteen essays by leaders in each specialty defining "the larger problems of sidereal astronomy" such as solar physics, solar radiation, stellar parallaxes, stellar kinematics, binaries, and clusters. Eddington would be tapped for stellar constitution, and Jeans for tidal evolution and "kindred problems." Of all the names he suggested as writers, only Pickering's appeared from Harvard, for magnitude systems, and even here, Russell suggested that F. H. Seares lend a hand. He was only following Hale's wish, Russell claimed, to have the "most competent men" involved, whether they were Americans or not.[51] Pickering felt that Harvard was slighted. Four writers from Mount Wilson, two from Lick, and only one from Harvard? Where was Annie Cannon? Henrietta Leavitt or Solon Bailey?[52]

Russell was blunt: "[T]here are two sides of astronomical research, one of which has to do with the collections of facts, and the other with their interpretation." The former was routine; the latter was not. Pickering's report, he argued, had already covered the routine aspects. Now the other side needed airing: "[I]t is upon studies of this sort that the future advances of any science must very largely depend."[53] Cannon, Russell added, could write any report on method, and Russell hoped she would, once the "routine labor of the Revised Draper Catalogue is through." Cannon, however, was not the astronomer to write on interpretation:

> [T]o be quite frank, it seems to me that Miss Cannon has been more concerned with what the spectra *look like* than with what they *mean*. I do not think that this fact diminishes the service she has rendered to astronomy; on the contrary, her strict attention to the facts, disregarding the current theories, has given her a peculiar aptitude for her great work.[54]

Russell felt that Adams would be most "likely to know rather more than anyone else what the significance of spectra and their peculiarities is, and to make more valuable guesses at the probable meaning of the things that no one understands as yet."[55] In like manner, Russell justified choosing Shapley over Bailey, and Seares over Leavitt. Still, he felt, using wartime rhetoric to convince his old ally, the two sides of astronomical research had to be in close touch. "It seems to me that present-day astronomy is like an army advancing with two wings." Each needed the other, or "the army will not get far." "If, in proposing reports about the other wing, I pay more attention to the officers of that branch of the army than to equally or even more distinguished officers of the other wing, I hope that it will be recognized that I did not mean to slight them."[56]

Pickering knew he had been trapped by his own exclusion of Hale's and Russell's original criticisms and was now being patronized by Russell. He was far from happy with the situation. Reminding Russell of his indebtedness, he responded sarcastically: "I think it would be a good plan to embody suggestions of the routine investigations desired to determine such quantities as absolute magnitude and the distinction between Giant and Dwarf Stars." The time was right "for the dreamer to suggest to the practical man what facts he wants."[57]

Outwardly, little changed between the two astronomers. Russell continued to feature Harvard work in his *Scientific American* columns, though he now gave equal attention to Mount Wilson. Their correspondence diminished as Russell concentrated on war work and put his essay on hold, but he wrote Pickering "just to show that I am still alive, astronomically speaking!" He kept Pickering updated on his family, and Pickering inquired now and then of Russell, mainly to ask about committee progress.[58]

At the end of the war, Pickering was seventy-three years of age and outwardly vigorous and active. His correspondence lacked the constant references that Hale, Russell, and Frost made about their infirmities, and indeed he seems to have been generally healthy until just before his death on 3 February 1919.[59] Genuinely shocked, Russell carefully hid what remorse he may have had over their split as he wrote eulogies for *Scientific American* and *Science*, proclaiming Pickering a pioneer in the new astronomy, the creator of spectroscopic and photometric data that he shared willingly with everyone, Russell most of all:

> He possessed a genius for organization which would have undoubtedly brought him both wealth and fame in the world of business; but he preferred to devote these talents to the service of science, and, because of them, enjoyed work of a sort which most other men would have regarded as drudgery. He once said to the writer, "I *like* to undertake large pieces of routine work."[60]

Russell passed over delicate issues of standardization and heralded the devotion and productivity of Harvard staff working under Pickering's "unifying guidance." And most of all, he celebrated Pickering's generosity and his own indebtedness.[61] "Who among us has not gone to Harvard, enjoyed the delightful hospitality and finished courtesy of the director, and returned, loaded down with data for investigations new or old, and inspired by his experience with new enthusiasm alike for the magnificent research of the great observatory, and for his own humbler work?"[62] Who indeed more than Russell, who with Pickering's passing must have wondered more than once about the future of the Harvard College Observatory and where he would go now for data. As he wrote Pickering's obituary for *Science*, Russell also rewrote and edited his long-languishing essay "Some Problems of Sidereal Astronomy," his blueprint for astronomy in the postwar world.

"Some Problems of Sidereal Astronomy"

Russell's manuscript lay fallow because Hale never was able to prepare an introduction. Hale also felt that it would be unwise to publish an essay on pure science in such difficult times.[63] Russell picked up the essay only after he returned from the war (his war work is discussed in chapter 11), and after other chores had been attended to. He completed "Some Problems" in February 1919 as he was writing Pickering's obituary, noting his delay to Hale: "I have added a good deal of new material, and references to about a hundred original papers." Of the ninety-eight papers he cited, eighty-two were published after January 1914.[64]

Russell prepared his final essay just as the new International Astronomical Union was being talked about, and when it was clear that the German grip on astronomical research and practice was now broken.[65] Though he was in a nationalist mind-set when he completed his NRC essay, his opening volley took direct aim at what had been, particularly, an American tradition in astronomy:

> The main object of astronomy, as of all science, is not the collection of facts, but the development, on the basis of collected facts, of satisfactory theories regarding the nature, mutual relations, and probable history and evolution of the objects of study.[66]

As he had told Pickering, Russell described astronomy advancing on two fronts. Observation, however, should guide theory, although theoretical investigation would be critical "in the solution of the larger problems of sidereal astronomy." The legacy of George Darwin now flowering under Eddington and Jeans was his ideal—the "Cambridge School," he called it—"which combines keen mathematical analysis with a thorough knowledge of modern physics."[67]

At the time Russell regarded the field as physical, but based in mathematics. His lectures to students had long identified theoretical astronomy and celestial mechanics as "really a branch of mathematics—one of the most difficult ones in its higher developments and one of those in which exact science has gained its greatest triumphs." Astrophysics was the application of the spectroscope and "deals with the physical characteristics of the heavenly bodies."[68] But in July 1917, around the time he was drafting his essay, Russell also wrote to Eddington describing "theoretical astrophysics" as "a new branch of mathematical astronomy."[69]

If Russell viewed Eddington's brand of research as theoretical astrophysics, and described it as a branch of mathematical astronomy, he had not yet begun to look at astronomical problems from the standpoint of physics. Indeed, he regarded his own contributions as astronomical, based firmly upon observation, but informed by theory.[70] In like manner he hoped that his essay, and those that followed his in the planned NRC series, would alert astronomers to the many theoretical problems that might be addressed in part through directed and focused observational programs. He wanted more physicists to take an interest in astrophysical theory, since Eddington and others were taking care of the mathematical side.[71] Carefully planned campaigns to collect high-quality solar and stellar spectra were the "master key" to the solution of many of the pending problems he had identified, but, he added strongly, observers had to become sensitive to the fact that astronomy was no longer the mere collection of data. Observing was no longer the raison d'être of research and had to be informed by theoretical questions, like that of the source of energy which drives the Sun and stars, "at present the greatest of all the unsolved problems of astronomy."[72]

Russell's essay was widely read. In January 1920, Robert Aitken, acting director at Lick Observatory, congratulated Russell: "I read it, as everybody here did, with the greatest interest."[73] Kind words notwithstanding, it was abundantly clear that

Russell was moving against the tide. Theorists like Eddington remained keenly sensitive to the whispers among his listeners as he delivered the British Association for the Advancement of Science presidential address, "The Internal Constitution of the Stars," in Cardiff in August 1920: "If we are not content with the dull accumulation of experimental facts, if we make any deductions or generalizations, if we seek for any theory to guide us, some degree of speculation cannot be avoided."[74] The watchword of the day remained, however, as F. H. Seares articulated it in a posthumous appreciation of Kapteyn in 1922: "The most pressing need for further study of the structure of the stellar universe is still the accumulation of observational data . . . until the end it was [Kapteyn's] great concern that there should be no cessation of activity in gathering the facts essential to further advance."[75] Pickering would have said much the same thing.

"Some Problems" produced few if any immediate converts. As Otto Struve observed in 1943, frustrated with the remnants of the legacy of Gill, Kapteyn, and Pickering, "[a] physicist would consider it incomprehensible that anyone should find satisfaction in observing a phenomenon only because it is measurable." Struve had spent much of his life watching Russell, in 1955 recalling that "[m]y own work in astrophysics was stimulated by this article, and even today it forms one of the most inspiring pieces of astronomical literature."[76] The lessons of "Some Problems" were not forgotten, because key people like Struve listened and knew that the author had faithfully lived its message during much of his career.

What Russell strove for indeed emerged during Struve's generation, but there was much he could not foresee. His ideal image of the "Cambridge School" and his promotion of physics were prescient, but he could not anticipate the changes that would come in the 1920s and 1930s as a result first of the new quantum physics and then of the influx of theorists in physics and astrophysics who came to America with the rise of Nazi Germany. Nor could he fully appreciate how physics would eventually inform astronomical problems. He did not cite the growing literature on atomic physics, and there was no call to incorporate Bohr theory in the pursuit of a rational understanding of stellar spectra. He only touched the surface, noting R. W. Wood's studies of bromides and Charles Fabry's speculation on discrete line spectra as useful for a better understanding of nebular spectra.[77] Although Russell faithfully discussed Alfred Fowler's spectroscopic observations, he made no mention of how they had helped to refine spectroscopic line series formulae, or how such interactions revealed the growing potential for synergism between modern physics and astronomy. Russell looked for a connection between physics and astronomy that made sense within his universe, but he hardly could have guessed how profoundly his universe would be transformed. He had not yet assimilated the outlook of the physicist. This would soon change, partly as a result of his wartime contacts.

• CHAPTER 11 •

The Great War: Transformations

Brooding

Russell often turned his lectures on religion and science to the problems of the day. In December 1916, fretting about the war and continued U.S. neutrality, Russell spoke on "Prayer and Action." Christianity is "a world-transforming religion—a stage that is still in process of realization." In a "Call to Action" quite in keeping with the mood on the Princeton campus, where Hibben refused to permit peace meetings, Russell admonished his students to "*Pray* that the [Right] may win—but pray too for strength, courage and guidance, and DO YOUR PART." If God ordained all things, Russell asked, why "should there be evil in the world?" Paraphrasing the words of Sir Oliver Lodge, he answered, "We do not know: but the evolution of responsibility and character is hardly conceivable without a period of struggle such as we are in." In these difficult times, he appealed to his charges to " 'do one's level best.' "[1]

The internationalism in astronomy that Russell had celebrated in his *Scientific American* column after his return from Bonn in the fall of 1913 was now gone. During his European travels, Shapley had reported on the persecution of Jews and the student riots that were closing the schools. Russell now found the war "an incredible abomination . . . all without the least shadow of real reason."[2] By early 1917, he had given up completely on Wilson, fearing that his administration was in hopeless torpor.[3] In accord with American clergy and eastern intellectuals, especially those with strong Anglophilic connections, Russell felt that it was his personal duty to serve the "common cause" but had little idea at first what that would be.[4]

What Russell finally did during the war demonstrated, to himself at least, that the techniques and talents he had honed as an astronomer, mainly his ability to find fast and efficient means of reducing observations, served him well during the national emergency. George Ellery Hale would guide him to this realization by orchestrating his participation. For Russell it would be a time when he learned that his professional life was no longer bounded by a campus and that physicists worked in wonderful and mysterious ways.

Searching for War Work

Sure of himself in the pulpit, Russell crumbled at home. Russell's children remember well, even though they were just toddlers, how they would be rushed from his presence, even banished from the first floor of 79 Alexander, when he became angered. His son keenly recalls: "The [*New York Times*] would come and he would read the newspaper and then he would come out of the study where he had been reading and he . . . would pace up and down the floor with his face scarlet and snorting and shaking his hands. Wow was he in this kind of state."[5] "Henry Jr." asked his mother what father was on about, but she said only, "[J]ust leave him alone, just leave him alone." Russell would rant and rave for about ten minutes, really frightening the children. Margaret, a year younger, recalled that the spells could last for what felt like hours. "Oh, he just went to pieces, physically," she remembered.[6]

Worried about Britain as well as restraint of trade, Russell admitted to Hale in January 1917 that "[t]he war has got even a deeper hold on me than ever this year, and I am inclined to believe that the real crisis of modern history may come in this year."[7] Acting on advice from Adams, Hale had just invited Russell to spend the following summer at Mount Wilson, but Russell, who very much wanted to get his hands on Adams's spectroscopic material, now declined, saying that one way or another, he would find war work that summer, possibly in Canada, which, unlike the United States, had been in the war since 1914.

Hale persuaded Russell that a larger mission awaited. He would be of more use on campus, working for the "war research committee[s]" that Hale was trying to form nationwide.[8] Hale wanted to coordinate all wartime research through his National Research Council (NRC). Believing that through such a mechanism science itself would prosper, Hale argued that "the most fundamental form of preparedness lies in the promotion of research in pure science."[9] He asked observatory directors to marshal forces to aid the Shipping Board in securing and calibrating compasses, chronometers, sextants, and other navigational devices. He made sure people like Russell read and distributed his NRC circulars on preparedness and the support of science.[10]

Russell liked Hale's vision, but he ached for action. His most extensive plans were made with his brother Gordon. Together they would find a way to get across the ocean to work in England in the Y.M.C.A. relief "huts," at army training camps and in the larger cities.[11] By early April, Gordon had secured the approval of his church and only awaited word from their family friend, Colonel Theodore Roosevelt, the "vocal Volcano of Oyster Bay": "[U]nless he very strenuously objects the chances are very strong that I'll sail with you," Gordon wrote.[12] Russell took steps to train for the job; he wanted to accompany Princeton students who were volunteering and felt revitalized: "These are great days to live in," he told Frost in March 1917, caught up in the Allied propaganda wave after the czar's abdication and the assassination of his allegedly pro-German ministers.[13]

Frost, Hale, and Campbell all sent their blessings along with lantern slides, but Hale was determined to keep Russell home. At a convocation of astronomers at the American Philosophical Society in mid-April, a week after the United States entered the war, "[e]veryone [was] talking war." Hale made a renewed call to his colleagues to support the NRC's efforts to establish research committees.[14] Hale again prevailed on Russell, who finally decided to stay. He told his Y sponsor that when Hale called upon his services, "I feel that it is the duty of any man of science to follow his advice on a matter of this kind." Gordon went alone, with a considerable infusion of cash from his brother, who remained guilt-ridden over abandoning their common mission.[15] Two weeks later, when the universal draft was enacted, Russell was delighted. He had "vigorously favored this as a public policy" and applied immediately, but at age thirty-nine with four children, he was declared ineligible in the first two draft cycles, and by the third he was working for the NRC.[16]

Sound Ranging

Russell joined his campus war research committee, finding colleagues like physicist Augustus Trowbridge and zoologist Edwin Conklin already deeply involved with Hale's strategies for the National Research Council.[17] Trowbridge felt that the development of poison gas for the trenches was a brilliant "application of chemistry." But in early 1917 he was more interested in finding ways to improve fire control, bombing systems, and sound and flash ranging.[18]

Sound-ranging systems used microphones to pinpoint enemy firing positions. Each microphone would detect cannon fire and send the signal to a central processing station, which recorded differences in arrival time. The intersection of three circles of position located the source of the sound. Trowbridge toured war zones in Europe to examine how present systems worked under battle conditions, and decided to refine a Belgian design into a reliable and efficient portable sound-ranging system.[19] Trowbridge could handle design and construction; what he needed was someone who could create quick and efficient means of triangulating to targets. Russell, of course, excelled in such techniques and found his niche.

The work started slowly over the 1917 summer as Russell adapted and simplified the Belgian system. That fall he still met classes, managed the work of the NRC subcommittee on astronomy, taught navigation, worked on the lunar program, and analyzed eclipsing binaries for stellar masses. He wrote *Scientific American* columns and followed closely Eddington's theoretical modeling of giant suns. He could still be excited by astronomical work, for it was terribly exciting stuff, and Eddington now was a major supporter.[20]

Russell dropped all astronomy late in the fall, telling Shapley and Pickering that he was "devising methods for observing the parallax of hostile artillery. It is quite as interesting as the other kind of parallax work . . . and at present it appeals to me as more immediately useful."[21] By the end of the year Russell was working with a team of Army Signal Corps engineers both on campus and on the Jersey coast near Long Branch. Writing in his rooms at the Hotel Pannaci, Russell seemed happier than he had been in years, more fulfilled and vital at least: "[I]t is a great comfort to be able to do something."[22] Even his NRC essay "Some Problems in Sidereal Astronomy" took a backseat now: he promised to return it to Hale as soon as possible, but the tests of his sound-ranging technique came first; as for his essay, "I never supposed that I would put it second."[23]

As he read Gordon's stream of letters from London, and learned that his brother-in-law, John Cole, was performing medical services at Fort Sheridan Officer's Training Camp, Russell wanted to do more. He eagerly read Cole's glowing reports of being a first lieutenant. His own letters were filled with ideas: aerial photography with box kites, using movie cameras to record battles. He would feel really fulfilled only if he could don a uniform and take his sound-ranging systems to the front: "[N]othing but war work seems worth while."[24]

By the New Year the Signal Corps adopted Russell's methods. After trials on the Jersey coast, Russell wanted to test them under more realistic conditions, in actual combat as Trowbridge's boys were then about to do. "The methods of calculation that I had the luck to hit upon are much faster than anything that we have had sent to us from abroad," he informed Hale; "it is highly desirable that someone should go over there, obtain by word of mouth all that is available, and come back with the information."[25] With Hibben's approval, Russell turned over the department to Dugan. He was raring to go, and Lucy supported him: "If she had been born a man," Russell told Hale, "I think that she would have been 'over there' before this."[26] Trowbridge, however, passed on Russell, preferring to have the Harvard physicist Theodore Lyman by his side. Lyman had experience hunting big game and seemed more fit for the front. He eventually commanded both mobile sound and flash sections that were able to keep up with the shifting fronts. They proved themselves able to work under conditions of shell fire and gas attacks.[27] Russell worried, however, that there was much practical information which was not getting back to him in Princeton.

Russell had also become something of an aide-de-camp for Hale. Beyond managing the NRC subcommittee, he helped to locate talent in applied mathematics among astronomers, calling upon Frank Schlesinger and E. W. Brown to look for problems to attack and the expertise with which to attack them.[28] He traveled often to Washington now to meet with Hale, using his time on the train to catch up on astronomy and correspondence. Writing to Adams in mid-January, filled with admiration for his latest spectroscopic parallaxes, Russell predicted that "you won't probably see many more astronomical papers of mine till we win."[29] Indeed, Russell submitted no astronomical papers in 1918 nor well into 1919.

Russell wanted to be off the Princeton payroll and on government books as an NRC engineer so that he could be ready for action anytime. Hibben was more than willing to give Russell leave with full pay to work on anything Hale wanted.[30] Dugan also dropped his observing routine to teach navigation, and the lone graduate student, Thaw fellow Bancroft Sitterly, left to be a computer at the Aberdeen Proving Ground and then at Sandy Hook, New Jersey. Work on the lunar plates stopped early in the year when both computers left for war work, although plates continued to be taken at Harvard, where Pickering managed to hold his staff together. Indeed, what was happening in astronomy had already happened to physics and mathematics at Princeton. Mathematicians like Oswald Veblen left in late 1917, and physics graduate students were disappearing even faster. John Q. Stewart, who was working under K. T. Compton and H. L. Cooke but had taken extensive training in astronomy under Russell and Dugan, had moved into the Sound Ranging Detachment in the Palmer Physics Laboratory in July 1917. In September he resigned his fellowship for a commission in the Signal Officer's Reserve Corps, leaving for France in January to serve under Trowbridge and Lyman. By July he was chief instructor in sound ranging at the Army Engineering School, Langres, France.[31]

Hale constantly looked for other ways to use Russell's talents on the home front. Possibly sound ranging could be used to locate enemy aircraft, especially slow-flying zeppelins. Russell liked this idea, "barring some natural regrets as the idea of wearing the uniform fades into the distance."[32] Accordingly, at the end of January Russell received word that he was to report to Langley Field near Hampton, Virginia. He made a quick visit to the Western Electric Company in New York to obtain diagnostic test equipment and reported to physicist Robert Millikan that "[t]he principal geometrical problems involved appear to be in a fair way toward solution."[33]

Aerial Navigation

Millikan was then chief of the Signal Corps Science and Research Division and wore a uniform as "a swivel-chair officer." He was based in Washington and set up a small laboratory at the National Advisory Committee for Aeronautics's (NACA) new Langley Field to perform general aviation research, including aerial sound ranging.[34] Much of his time was spent fighting turf battles. The Bureau of Standards wanted primary responsibility for sound-ranging equipment, and the NACA objected when the NRC started experimenting with aircraft systems. Although there was much work to do, many bureaus wanted to be in control; boundaries were vague and often breached. The NRC proved to be more aggressive in its search for aeronautical problems and soon edged out the NACA.[35]

Russell had barely started work in March 1918, however, when President Wilson reorganized army aviation, taking it away from the Signal Corps and creating a civilian Bureau of Aircraft Production and a military Division of Aeronautics. Millikan's group went into the civilian agency. Although the instrument group survived, it was never very large and was continually attacked by the military as a "claque of impractical academics."[36] Russell soon found that with the reorganization he was bound not for Langley but for an "Engineer Research Camp" at Ellington Field, near Houston, Texas, where he worked on aircraft detection schemes.[37] He lived in a tent camp on the prairie, finding the "outdoor life very agreeable and healthy." He returned to Washington in May to perform ballistics calculations as a "Consulting and Experimental Engineer" and in August switched to aerial navigation and trajectory computations. He was soon assigned to an instrument group and had his first flights in De Havilland DH-4 aircraft from Langley Field. While stationed at Ellington Field, Russell was elected to the National Academy of Sciences.[38]

The NRC outpost at Langley was directed by Charles Mendenhall, an amiable physicist from the University of Wisconsin who ignored the typical military directives on deportment and dress.[39] As a result, the atmosphere was rather casual, and spirits were high. It was here that Russell met David Webster, a physicist turned pilot; the two would become lifelong friends. They took rooms in a boardinghouse and shared their copies of the *Physical Review*, the *Astrophysical Journal*, and *Science Abstracts*.[40]

David Locke Webster, son of a Boston leather manufacturer and eleven years Russell's junior, was a Harvard physics graduate who worked on a magnetic bomb project for the United States Navy at the University of Michigan. When the navy dropped the program, Webster stayed on as an assistant professor of physics but soon left to accept a commission as a lieutenant in the Signal Corps Aviation Section, working as an aeronautical mechanical engineer. He ended up at the NRC evaluating magnetic compasses for the Naval Consulting Board. He liked the work, especially since it included testing instruments at Langley in real aircraft. It was "one of the most interesting and instructive jobs I have ever undertaken," he told Harrison Randall back at Michigan.[41]

Life at Langley under Mendenhall suited Webster. At first he flew as an observer in DH-4 and Bristol Fighter aircraft to test and calibrate instruments. But when Mendenhall and Millikan decided that their "scientific experimenters" needed "practical knowledge" of how instruments behaved under flight conditions, they asked Webster to be the first physicist to take pilot training at Gerstner Field, Lake Charles, Louisiana.[42] By June 1918 Webster was back at Langley testing rate-of-climb indicators, compasses, azimuthal bearing plates, course and distance indicators, airspeed meters, and barographs. Only the news that his beloved sailboat, the *Bluebell*, was destroyed in a wharf fire in Cohassett, Massachusetts, dampened his spirits. They were lifted again when he learned that his wife Anna and their children, still in Ann Arbor, would soon be able to join him at Langley Field.[43]

In August Webster started to take a steady stream of observers aloft, and Russell was one of the first civilians on his flight manifest that month; Russell's first flight on 15 August lasted for one hour and forty minutes and reached 3,300 feet.[44] Russell flew twice with Webster the next day, trips totaling almost three hours in the air. Most of Webster's flights were less than one hour, some only a few minutes, and he logged seven flights that day. Russell became his most constant observer, flying thirteen times in August alone. These flights lasted into October, and Russell's were the longest by far and reached the highest altitudes. As he told Pickering a week before the Armistice, flying had become a passion:

> I have had some fine flying at Langley Field. Altogether I have been about 21 hours in the air and have been up to 14,000 feet by day and 8000 by night. [I] have been miles above a sea of clouds and have "acquired an experimental knowledge of loops, tail spins and other 'stunts.' " . . . It has been an invaluable experience and I have more of the same work ahead.[45]

Beyond the common means of map reading, compass reckoning, and the use of primitive air speed indicators and pressure altimeters, what was really needed was some form of accurate, reliable, and efficient location finding independent of weather or time of day.[46] The two most promising techniques were radio direction finding and celestial navigation. Russell tried to adapt the latter for the Bureau of Aircraft Production. Between August 1918 and January 1919 he flew dozens of times with Webster and led a small group of engineers and scientists to devise ways to take celestial sightings from aircraft and reduce them quickly and efficiently. To be practical, the system had to be used by military aviators working in the open cockpits of their machines. It was a long shot.

Russell was assisted by James Percy Ault from the Carnegie Institution. Ault had practical experience navigating oceangoing research vessels and knew that aerial navigation was different from navigation at sea in one significant way: aircraft were faster than boats. Aviators could not employ the second Sumner line of position technique commonly used by sailors, unless they had very accurate knowledge of speed and direction between the two observations. Accordingly, Russell concluded from his months of testing and "from theoretical considerations, that very careful piloting is necessary, if sextant observations of value are to be obtained from airplanes."[47]

The sextant became Russell's primary tool at Langley Field. During his flights, Russell found that when a good physical horizon was available, on land or sea, standard sextant readings were as easy to take from an airplane as from a ship. He also decided that fancy attachments, like microscope eyepieces to magnify the dials, were more a hindrance than a help, and their precision did not translate to substantially greater accuracy of observation.

Since any acceleration of the craft altered the apparent direction of gravity, simple bubbles or plumb bobs were useless as horizon indicators. Russell thought about using some form of gyroscope to maintain true vertical. But, as he and Webster knew, a gyroscope maintained its direction relative to an inertial frame, whereas the direction of true vertical changed as the craft flew over the curved surface of the earth. As Russell wrote after the war, "to maintain the *vertical*, any apparatus whatever must ultimately be controlled by gravity as a directive force, and an instrument which could discriminate between the components of the apparent gravity relative to the airplane which arise from the Earth's attraction and from the reaction due to acceleration of the motion would furnish an experimental disproof of the Principle of Generalized Relativity."[48]

Although Mendenhall championed a damped pendulum mounted in gimbals, Russell and Webster believed that a properly designed gyroscope could damp the differential effects of short period acceleration and act as an averaging agent, but such an instrument was too "heavy and complicated."[49] So they tested a viscously damped Mendenhall pendulum and found that it worked, though everything had to be in perfect adjustment, sightings required long periods of stabilized flight, and, generally, the bulky instrument was hard to handle.[50]

Russell and Webster also tested a device invented by R. W. Willson of Harvard that incorporated an artificial horizon and a sextant, called by him a bubble sextant. Using an unsilvered mirror placed at forty-five degrees in the optical path of the telescopic sight of a standard sextant, Willson was able to optically superimpose the image of a small bubble, constrained by a downward facing spherical surface of glass, onto the field of view of the telescopic sight. Thus when the navigator looked through the telescope sight of the sextant, he would see a distant object superimposed on the bubble; when the bubble was centered in the crosshairs of the sight, the object in view defined the horizontal.[51] Russell felt that "[t]he observations at first appear almost ludicrous—like trying to place an orange in the middle of a soap bubble—but after a little practice readings may be made very easily and with surprising precision." The viscosity of the liquid provided the required damping.[52]

The bubble sextant was tested on five flights, and Russell concluded that it was the best choice, if only because it was small and portable and was easy to use with proper training. On his last night flight, Russell, sitting in the front cockpit of the airplane, made nine separate observations of Sirius. The craft was fully loaded with two heavy bombs and by design was flying north, requiring that Russell make observations to the stern over the wing. His average error was ±14 minutes of arc, and the mean of the nine a mere 2.5 minutes, which located the position of the craft within an error circle less than three miles in latitude. In Russell's opinion, the bubble sextant left "very little to be desired."[53] Repeated observations convinced Russell that in practice, to achieve six-mile accuracy, at least six observations of the same object had to be made when the pilot was flying a straight and steady course.

Russell did everything he could to simplify and accelerate the in-flight reduction process. He created standard tables for the various calculations, simplifying them by precomputing ephemerides for a full day of flight. By using a device designed by Charles Lane Poor that was similar to a circular slide rule, Russell could solve the navigational equivalent of a spherical triangle in barely one minute, accurate to two minutes of arc. The slide rule was cumbersome, however, since it weighed nine pounds, and even at his quickest calculation rate, a full reduction to a single line of position took about ten minutes.[54] The Sumner method required a minimum of two lines of position.

By April 1919, completing his optimistic report for the Bureau of Aircraft Production, Russell declared the problem solved "within the limits set by the nature of the subject, and with sufficient accuracy for the practical purposes of navigation." His colleague J. P. Ault, however, felt that future navigation techniques would employ not Russell's methods but radio. In practice, celestial navigation was too slow and cumbersome for aviation. Still, celestial techniques akin to those Russell developed for Willson-type bubble sextants were used by Richard E. Byrd and by the Royal Aircraft Establishment and were used with radio direction-finding techniques as a basic ingredient of the navigator's tool kit.[55]

Russell's contribution may have had no measurable effect in winning the war, but it was a distinct gratification to him that the highly efficient computational techniques he had honed as an astronomer could solve real-world problems. His work, and that of astronomical compatriots who served in a broad range of capacities, demonstrated at least to themselves that astronomers could be of use during times of national emergency.

War's Aftermath

Russell called Webster his "chum and flying partner . . ."[56] They found that they had many interests in common, including sailing and the sea, and collaborated on a number of ventures apart from aerial navigation. Russell's penchant for calculating bomb trajectories led to a collaborative design for a drift sight, which they tried to patent. Some of their most exciting flights were at night when they dropped bombs with illuminated tracers.[57] Years later, Webster looked back to their wartime experiences: "We were expendable," he felt, because parachutes were not issued to the flyers or observers. Their many adventures were "more fun to look back on than at the moment."[58]

In November 1918, Webster learned that his brother had been killed in action in France, and Russell did what he could to console his friend.[59] But it made him long for his own brothers, who were then overseas: Gordon had been in London and France for over eighteen months, and Alex had just left to take up a post with the army of occupation. Alex spent most of 1919 in France in the office of the General Purchasing Agent, auditing the books of the American Expeditionary Forces while he watched the glacial progress of the peace conferences. At least he could send his brother Gillette shaving blades, which were more plentiful in the A.E.F. than at home. He lived for the day when he would be home for a "definite Russell tribe reunion."[60] Henry said nothing of Webster's loss when he wrote Alex in February 1919, speaking only about the exciting times he had flying with Webster. He even recommended it as fine sport. These were politically charged times, both in Europe and at home, but Henry was preoccupied with flying, watching Wilmer Duff use his "luminous bombs" for trajectory analysis, and reliving the many harrowing moments they all experienced flying through low clouds and making forced landings. All that was at an end now as he and Webster wrote their reports and looked for ways to make good on many of their wartime plans.[61]

Webster and Russell pledged to pool their talents after the war. They talked about studying stellar energy, relativity, and the structure of the universe, and planned to apply Webster's physics to problems in astrophysics. Webster knew atomic physics and electromagnetic theory and during this time clearly was stimulated by their many talks. They both wanted to extend the relationship, at least to the point of summering together for sailing and relaxation at Clark's Island near Plymouth, which was a Webster family haunt accessible only by the Fall River boat. All of this was dreaming in the fall of 1918, but it stayed alive at war's end.

Four days after the Armistice, Pickering reminded Russell that he was an astronomer: "Have you forgotten that you are the active executive of the Astronomy Committee of the National Research Council?" Pickering wrote, suggesting that Russell, when he prepared his ultimate report, highlight all the contributions astronomers had made to winning the war. Astronomy itself had little bearing on war, Pickering added, but "the training of the astronomers has rendered them especially fitted for solving many problems essential to the successful prosecution of war."[62] Russell was not impressed. He was not a Millikan, who would preach widely how "the war has demonstrated the capabilities of science."[63] Nor, for that matter, was Russell like Hale or Pickering, who spoke out mainly on the need for new institutions and funding schemes. Russell limited his war-related writings to technical reports, mainly accounts of his group's development of aerial celestial navigation.

For Russell, the war was a turning point. It drew him into a larger community populated by the likes of Hale, Trowbridge, and Webster. For the first time he worked in close contact with a physicist and became comfortable working in Washington and at Langley, sites far removed in temperament and scale from Princeton. It also was a time when Russell was drawn to issues larger than himself and his community. As anxious as he had been to get involved in the war, he had learned a great lesson from Hale: to deal from strength. On many occasions he remarked to colleagues how fit he felt when using his talents for practical ends. Whether working on sound-ranging computations or flying at Langley, Russell felt alive.

Russell had poured all his energies into these efforts, so much so that by the Armistice and for months afterward he was exhausted, though quite happy with himself. Once he had finished up all his reports and his NRC essay, cleaned up some minor astronomical projects, and caught up with Shapley's progress on determining distances to globular clusters enough to warn him that his results might well be vitiated by interstellar absorption, he and Lucy, on Hale's strict orders, took a steamer to Puerto Rico. They became so uncharacteristically relaxed during the sixteen-day cruise that even a diversion of the steamer to rescue a disabled ship, which their liner then towed for two days, did not ruffle Russell's feathers in the least. It gave him more time to write and think about the future.[64]

As Russell shifted back to Princeton and normalcy, Webster accepted a post as assistant professor in the physics department at the Massachusetts Institute of Technology. Reading Russell's draft of their report on aerial navigation, Webster felt they had really made a contribution: "The great question is, will we ever have a chance to do something more?" Russell certainly hoped so: In June 1919 he offered Webster a joint appointment in physics and astronomy as professor of "astronomical physics," raising his MIT salary by 50 percent to $3,000. Russell had leveraged the new position after he was asked to assume the directorship of the Yale University Observatory (see chapter 12). Princeton physics, Russell promised, was "excellent" under the leadership of Magie, Trowbridge, Adams, and Cooke. K. T. Compton "is a live man," and the place was not bureaucratic. Most of all, "[I] promise you the same spirit that was exhibited in a certain flying partnership at Langley Field." Princeton was an expensive place to live, but not so much so as Cambridge. It was also a good place to raise kids.[65]

Webster was torn but decided to remain at MIT to develop his laboratory and curriculum, and to be near his parents. Although he relished the idea of collaboration, Webster wanted an independent career centered in experimental physics. He also knew that he needed some distance from his impetuous and highly aggressive astronomical friend. He expressed regret that "it is impossible to be in the same institution with you."[66] Indeed, though Webster claimed that he chose MIT to be near his parents, he soon abandoned the East Coast altogether to become chair of the physics department at Stanford, where there was a greater chance for the independence he craved.

Even though Webster was clearly an experimental physicist, Russell wanted him at hand for access to modern physical theory, the new core of astrophysics. He needed a jump start to study what he perceived as a very hot field: the source of energy of the Sun and stars. The existence of giants made the problem even more challenging. In July 1917, writing to Eddington, Russell used Antares as an example: "It has radiated more energy since Ptolemy's time than the whole gravitational supply possessed by the Sun! There *must* be some more abundant source of stellar energy unless the rate of radiation into empty space is different from that measured in a material enclosure."[67] When Shapley made much the same suggestion in print a few years later—that radiation acted only between bodies—Russell reproached him for such wild speculation, fearing that it would damage his career.[68] But this only put the problem into sharper focus. Astronomers were ill-equipped to speculate. Intuition was useless without a firm grounding in the latest theoretical physics. At the very least, Russell hoped, the right physics might be found in the astronomical evidence, and a good physicist knew best how to look for it.

This was Russell's intended agenda for Webster. They had mulled over many questions between flights and apparently made plans to search out the nature of the radioactive source of the Sun after the war. Jeans and Eddington looked at the universe from the standpoint of mathematical physics, whereas Webster,

Russell believed, had "just the physical point of view . . . needed for such work."[69] Evidently, Webster did not share Russell's optimistic enthusiasm for the collaboration. He knew he was not a theorist, and knew too that he would always be the junior partner. If Russell was, indeed, looking for a theorist, he was looking in the wrong place.[70] Most likely, what he was really looking for was someone who was intellectually compatible, someone he could talk to and feel comfortable with, as we shall see when Russell begins looking for his successor (see chapter 22).

Russell nonetheless was confirmed in his conviction that he benefited from contact with good physicists when he realized, after talking with Irving Langmuir, that one of the problems they had batted around had a most elegant solution. If some form of nuclear disintegration was active in the stars, what kept them from blowing up? With Langmuir's help, Russell realized that if "induced radioactivity" was indeed created by atomic collisions, this new energy source, once ignited, would create conditions within the star that would turn it off. Atomic collisions would produce extreme local heating and expansion, which would in turn reduce the chances of another collision's happening immediately. This led Russell to realize that "[a] mass of gas in equilibrium under its own gravitation has, as a whole, a negative specific heat. When it loses heat, it contracts, and rises in temperature, though at less than the adiabatic rate. If heat is supplied to it, it expands, and its temperature falls." From this Russell concluded that a gaseous star could never become unstable from a source of heat that had a positive temperature coefficient, "for the production of an excess amount of heat lowers the temperature and automatically stops the overproduction."[71]

Russell quickly applied his self-regulating mechanism to Eddington's model for Cepheids, and it worked. "There is no physics proper in this," he told Webster in June, "only astronomical considerations of a rather elementary kind." To work it all out properly required real expertise in atomic physics and "some lively work in the solution of differential equations of equilibrium." Russell's goal now was to "derive, from astrophysical data, a sufficient number of facts regarding the nature of the unknown energy-liberating process to make it possible to find out a lot about its physical nature, and perhaps to extend our knowledge of the properties of matter into a field inaccessible to experimental study."[72]

Russell's theory of stellar evolution had rested on physical ideas about gravitating bodies. He had no preexisting concept of what the physical mechanism was that produced induced radioactivity. Possibly the stars themselves would reveal the nature of their energy sources, but it would take a good physicist to interpret the evidence. Before Webster had turned him down, Russell told Hale that "it would be as much service to astronomy as anything I could do to enlist such a man—who knows thoroughly all the modern physics,—radiation, x-rays, atomic structure, quantum theory, etc.—for it is just here that American astronomy is weak."[73]

Russell wanted to create, as Karl Hufbauer has described it, "a genuine interplay between physics and astronomy." With Webster he could find the atomic mechanism responsible for radioactive energy by combining astrophysical data and physical theory.[74] Webster was his key, and Russell did what he could to make his offer compelling. Astronomy, he suggested, was the surest path to rooting out the fundamental properties of matter. Russell's interest in modern physical theory increased as the power of the new physics emerged. He was ready to apply Planck's radiation law in 1914 but remained silent about Bohr's theory in his 1919 NRC essay. Yet he and a growing circle of astronomical spectroscopists faced the problems of rationalizing the spectroscopic appearances of the Sun and stars, luminosity effects in stellar spectra, and the source of stellar energy. Russell could only admit in March 1917 that "I don't know enough about spectroscopy to have very valuable opinions about enhanced [ionized] lines."[75]

Prior to his contact with Webster, Russell did not appreciate the power of modern atomic physics. Eddington's mathematical models, the kinetic theory of gases, and thermodynamics did not deal with physical processes on the atomic scale. But now, in the spring of 1919, the deepest problems in astrophysics seemed to be pointing in that direction. He shared Hale's faith that the answer lay in the integration of theory with the laboratory and observatory, but he knew that the linkages to physics were not yet forged. At the time, he simply did not know what his role would turn out to be, but the way he practiced astronomy, and his increased attention to physics (he started to sit in on Compton's lectures at this time), positioned him uniquely when the opportunity arose.

• CHAPTER 12 •

Russell's Turn to Mount Wilson

"... MY SHIP'S COME IN, OBVIOUSLY"

With the war over, as Russell searched for a collaborator, he found that he had to make major decisions about his own future. Would he stay at Princeton or accept offers of the directorships at Yale or Harvard? Russell parlayed these offers to increase support for astronomy at Princeton and to secure access to the resources of Mount Wilson. In both cases, he turned to Hale for advice and counsel.

E. W. Brown, Yale's Sterling Professor of Mathematics, had been shepherding the Yale University Observatory during the war years with the understanding that it be closed for the duration and that a new director be sought out at war's end.[1] Brown also led the search and looked for someone firmly grounded not only in mathematics but in physics and chemistry. Most of all, he advised the Board of Managers, Yale astronomy needed a major overhaul. The observatory had languished, but it had an endowment that ranked it sixth in the nation.[2]

The observatory had four options, Brown felt. It could continue as a traditional observatory, it could build a mathematical staff capable of analyzing observations made elsewhere, it could limit its expertise to dynamical astronomy, or it could build up its observational facility to take advantage of campus strengths in mathematics, theoretical astronomy, physics, and chemistry to provide advanced training for students in those subjects. Such an arrangement existed between Berkeley and the Lick Observatory. Brown liked the last option and approached Walter Sydney Adams through Hale, asking him if he wanted to leave Mount Wilson. Adams declined, since as Hale's chief assistant, he was in many functional ways acting director of Mount Wilson, as the bulk of Hale's energies were deployed elsewhere.[3]

Brown then looked for a mathematical astronomer capable of exploiting observations made elsewhere, like Kapteyn, Eddington, Jeans, and Russell. He approached Russell in May 1919, and Russell immediately turned to Hale and Webster (who did not turn him down until mid-June) for advice, and to Hibben for a counteroffer. Yale offered Russell an endowment Princeton could not match and a chance to put into action what he had preached to Pickering and had written in his National Academy essay. The observatory needed a total overhaul, however, and Russell was not sure he could find someone as trustworthy and capable as Dugan. An "excellent photometric observer, but not brilliant or inspiring," Dugan took care of details, and their "personal relations [were] excellent." Yale would demand too much of him, Russell worried to Hale; New Haven was an industrial town far different from pastoral Princeton. Bothering about buildings and instruments was to " 'leave the word of God and serve tables.' "[4] And though Princeton required some teaching whereas Yale did not, teaching was preferable to administrative work.

Russell felt that he could not do his best work managing a large staff: "[T]he sort of work I really care about is rather different . . . most of my work, as I see it, will gain more from my having time to think, than assistants to keep busy." Russell wanted an astrophysical colleague like Webster, however, and chided Hale for stealing Shapley: "I had this while Shapley was here: but you got him away, and I have always felt that the change was for the good of science." Webster, he felt, would be equally happy at Yale or Princeton.[5]

More interested in attacking the stellar energy problem than in rebuilding an observatory, Russell soon talked himself out of the Yale job, but not before he managed to use the offer to secure another faculty position, which is the one he offered Webster, and to wrest the director's residence back from the mathematics and engineering faculty. He also received some tentative concessions about teaching load and financial support for maintaining the observatory. When he heard the news, Dugan was ecstatic: "And you even got the house!"[6]

Russell was delighted with the attention: "[M]y ship's come in, obviously," he told Webster as he invited him to Princeton in June.[7] He thanked Hale for his support and talked excitedly about his stellar self-regulating mechanism: "I propose to pull in my line vigorously until I see how big a fish I have caught." But what Russell needed was data: "May I count on you for that?"[8]

Hale had advised Russell to stay put, where his freedoms were assured: "You could never have them again if you were to set out to organize a large observatory, either at New Haven or Princeton." As for data, "You can get plenty of observational material from Mount Wilson and elsewhere, and the freer you are to go on with your present work, perhaps with the collaboration of some such man as Webster, the better the results for science are likely to be." Shapley agreed emphatically with Hale, and the two plotted to bring Russell west for extended visits. Webster's rejection confirmed Russell's decision: "This shake-up has done me alot of good,—and so has the year of war work," he told Hale. "Between the lessons that I have learned from the two, I ought to be able to do a good deal better than before."[9]

Indeed, these years proved to be the beginning of the most significant transition in Russell's professional life. That transition would continue for another several years at least, and much of it would center on Hale's observatory and the resources it provided Russell. Russell was now one of the best-paid members of the Princeton faculty. His future was secure and bright. There were family expenses and concerns, especially for nine-year-old Elizabeth, whose condition required constant attention. Nurses performed that function first, and then, after about seven years, Miss Jean Hetherington took charge. She was imported from England as a baby nurse after Margaret's birth and stayed the remainder of her life with Elizabeth. Miss Hetherington, proud to be connected to America through the Plymouth Brethren, "was the most completely devoted person imaginable" and soon became a member of the family.[10]

"Mother felt that her first responsibility was to her husband and if they could afford to pay for it they paid for somebody to care for Elizabeth," Henry Norris Russell jr. recalled. Considerable inheritances from both sides of the family supplemented Russell's Princeton income, which made it possible to keep a large domestic staff and to provide an outwardly seamless life for Russell. "Dad was an intellectual person," his son added, and though he enjoyed playing with children, especially games and puzzles, "in which you could learn something," he was more often in his study than in the parlor.[11] He always felt it was his duty to be with the children on Sundays, certainly for breakfast and church, but he was constantly driven back to his study by nervous energy and the passion of the moment, especially when Harvard came calling.

"I WOULD RATHER DO ASTRONOMY"

"The King is Dead," wrote Oxford's Herbert Hall Turner to Joel Metcalf, chairman of the Board of Visitors for the Harvard College Observatory. In February 1919, everyone had an opinion about Pickering's successor. Kapteyn suggested a statistical astronomer like Schwarzschild or Dyson to maintain and tap Pickering's vast legacy. Then he thought of Russell, or maybe Shapley, to run Pickering's empire, which by then included two dozen staff members, two major survey programs, two Southern Hemisphere observatories, and a large off-campus observing station. Kapteyn had been impressed by Shapley's inventive style and industry at Mount Wilson, where he was now distinguishing himself with a flurry of strong papers on Cepheids in globular clusters and was using them to show that the Milky Way star system was far larger than Kapteyn had estimated, and that the Sun was far from its center. Although this was controversial work, it was spectacular, and Shapley was gaining broad notice. So Hale agreed with Kapteyn, telling Harvard president Abbot Lawrence Lowell that Shapley would be his first choice, then Schlesinger, and then Russell. Hale knew that Russell's genius was not in administration.[12]

Lowell's adviser and patron of the observatory, George R. Agassiz, however, wanted Russell. No one on the present staff could raise the observatory out of what he viewed as the doldrums: "a reference library of undigested and probably perishable photographs." What Pickering had amassed now had to be exploited, in Harvard's name. Russell at forty-one was his candidate. He had demonstrated a keen ability to make "broad deductions from accumulated data." Unlike Shapley, he was more than an adequate mathematician, despite gossip at Harvard.[13]

In March, Russell snickered to Shapley: "It is suspected that [Lowell's] distinguished brother [Percival, of Mars fame] would have got the appointment if he had not considerately died! Whoever gets that job has my sympathy!"[14] But Lowell made no hasty moves. He quietly asked Solon Bailey, Pickering's senior astronomer, to manage the observatory and to begin collecting names. Bailey asked Turner for advice, Turner suggested Bailey, but the sixty-five-year-old Harvard astronomer declined in favor of a younger person, like Shapley, Russell, or Schlesinger.[15] Turner was equivocal:

> I think Russell is the man of greatest *intellectual power* you have in the States at present; he is in some respects angular, or *was*, but I don't know that his angles are sharp enough (compared with those we all have) to stand in his way. It is a serious matter to forgo taking the best man because life might be pleasanter under a second choice.[16]

Schlesinger garnered high marks as an administrator, but Turner went for intellectual power and so chose Russell, "though my vote might be different if I had to work under him."[17] Frost said the same, feeling that Shapley was too much the maverick. Agassiz pushed for "new blood and real distinction," and Russell remained at the top of his list, although he was warming to Shapley after reading his recent globular cluster papers.[18]

When Schlesinger took the Yale job, the choice was between Shapley and Russell. In February 1920, Lowell heard again from Hale, who felt that Shapley needed more growth and maturity.[19] Russell also did not feel that Harvard was right for Shapley, advising him privately that the directorship would reduce his ability to do good astronomy. The chief problem with Harvard, Russell continued, was that its big entrenched staff lived in "a land of settled habits." Think about the real problems, Russell cautioned, especially Pickering's brother: "consider W. H. P.—and tremble!" "If I had to run the place," Russell advised Shapley, "I think that I would plan to draw in sharply on the large routine jobs . . ." Echoing his NRC essay, Russell would turn the staff to "investigations on specific problems,—large problems, not in extent, but in content." But this would be a daunting task for an administrator who had to worry about a formal visiting committee in "[t]he home of the bean and the cod—Where the Cabots speak only to Lowells—And the Lowells speak only to God." Managing Harvard was not a pleasant thought: "I would rather do astronomy."[20]

After watching Shapley in action in April 1920 at the National Academy of Sciences in Washington—as he argued for a large Milky Way galaxy on evidence from his globular clusters, countering Heber Curtis's defense of Kapteyn's model and the extragalactic nature of spiral nebulae—both Agassiz and Theodore

Lyman, Harvard's ultraviolet spectroscopist, came away more impressed with comments Russell had made from the audience. Although Russell had cautioned Shapley about the possible influence of interstellar absorption upon Shapley's estimates of the distances to globular clusters, and hence upon his estimate of the size of the galaxy, Russell was persuaded by Adriaan van Maanen's dramatic astrometric evidence from Mount Wilson that spiral nebulae showed internal rotation. Thus they had to be local objects and certainly could not be extragalactic. He helped to convince Shapley of this because it strengthened Shapley's own model, and the two remained in staunch opposition to the island-universe theory until Edwin Hubble determined the distance to the Andromeda nebula in late 1924, which made Russell an immediate convert and Shapley a lifelong competitor. In April 1920, however, Russell was in total command of the facts as he knew them, and came to the National Academy fully prepared to bolster Shapley's case, since Hale arranged for the presiding officer to call on him "as a starter."[21]

Agassiz and Lyman listened to Russell explain Shapley's position during an extended two-hour lunch the next day and came away convinced that Russell was their choice: "Russell, besides being more mature, has more balance more force and a broader mental range. He has, however, a some what peculiar and nervous personality."[22] Still, they wanted the best intellect, so Lowell made an offer to Russell, who now had to make a real decision. Again he sought Hale's advice while he visited Harvard for extended discussions with Lowell, Agassiz, and others. And again he used the Harvard call to parlay his position at Princeton.

Russell took the call seriously enough to think of ways it all might work. He could hire Shapley to take care of the details. "Between us we cover the field of sidereal astrophysics pretty fully," he told Hale. "We can both do some theory,—and I might keep Shapley from too riotous an imagination,—in print." But Harvard did not guarantee the freedom and security Russell had at Princeton. Above all, he cherished "that freedom from care which I think, after all, is the greatest worldly good,—as there is much reason to hope for in this weary world."[23]

In June Russell received Lowell's formal offer, and it put him into a spin. He wrote Hale again as well as Stanley Cobb, a Harvard neuropathologist and psychiatrist, one of Lucy's Boston cousins. Hale dashed off a night letter, worried that Russell might actually accept the offer: "[F]ear effects of severe nervous strain also doubtful about Shapley . . ." Hale also reminded Russell that he could probably make Russell a permanent Carnegie research associate.[24] Hibben matched Harvard's offer and sweetened the pie, promising Russell a leave of absence "whenever my professional work, or my health, would be benefited by it." He also promised to raise the observatory endowment, to match Yale's level at the least. These were major concessions, but, as Russell admitted to Hale, these academic courtships were wearing him down:

> I have been pulling out of a rather deep hole, as regards nervous depression, this year, and have made notable progress; but the last ten days of strenuous consideration have tired me again enough to give me pause.

> My symptoms are not pain in the back of the head, but abnormal irritation on finding that a plumber's bill was bigger than I had anticipated, but I recognize them only too well.[25]

Russell had evidently gone through another of his scarlet-faced rages that were so fearful to his children. Harvard was upsetting his "reasonable peace of mind," and he needed the advice and counsel of Cobb, who had done "me a heap of good" in the past. Russell wanted to be done with this pressure so he could take his family to Clark's Island with the Websters. He was delighted with Hale's repeated offer of a Carnegie research associateship, especially since Hibben had now agreed to the plan.[26]

Stanley Cobb was assistant professor of neuropathology at the Harvard Medical School and practiced neurology and psychiatry at Massachusetts General Hospital. Always available to family, he had helped Russell through a number of crises, including the Yale offer. Tortured by this new challenge, Russell asked Cobb for a complete checkup and evaluation before he answered Lowell. Russell needed to know the "probable limits of my strength" because he did not trust himself: "You have looked me over more carefully than any one I know, and I want the benefit of your opinion as to the line of activity that bids fair to be the best for me, physically and nervously."[27] Russell worked himself into a frenzy thinking of ways he might make the Harvard position work. But the very thought of leaving Princeton was enough to make his head spin: "[M]y mind for the past ten days has resembled Stephen Leacock's hero, who 'rode furiously off in all directions.' "[28]

Cobb liked to classify scientists and probably found Russell among his "opportunists"—"the alert, over active men with scattered activity, stimulating ideas and many satellites . . . they are delightful people." He did examine Russell and predicted that he would self-destruct at Harvard. Nevertheless, even after the second interview with Lowell, where Russell tried to say no, Lowell convinced him to think things out a bit longer.[29] To further complicate matters, Hale liked Russell's scheme: Shapley would be an excellent assistant director, and Russell could remake one of the Harvard staff into a business manager. Still, there would be less freedom at Harvard, and there would be less chance to get away to Mount Wilson; Hale had just proposed Russell for the summer of 1921. But overall, "[t]he question of your *physical* ability to do the work is the real problem, and I hope you can solve it safely. A nervous breakdown would be a serious thing." Hale, notorious for his health problems and fragile psychological state, was really looking in the mirror.[30]

Through July Russell continued to deliberate and started to think more clearly, especially when he realized that he would have to raise funds and manage them. Pickering, he knew, often used his personal wealth to make ends meet, and Russell knew he could not afford to do this. W. W. Campbell agreed that Pickering's successor would be hard put if "not so highly blessed financially. It has long seemed to me an open secret that Pickering frequently made both ends meet by drawing upon his private exchequer."[31] Russell could not imagine such a life, and so decided against Harvard and immediately campaigned for Shapley.

Immensely relieved, Hale christened Russell's decision a "blessing in disguise."[32] All attention turned to Shapley now, who coveted the chance to rise, and after a number of twists and turns, deftly managed by Hale, Lowell agreed in January 1921 to take Shapley on a trial basis with Bailey as acting director. Hale gave Shapley a leave of absence to keep appearances up.[33] Hale now could use Shapley's salary to appoint Russell: "I hope you will accept, as I look forward to many interesting discussions, of great advantage to the Observatory, if you decide to join us."[34] Russell, of course, was ecstatic. "Glory, glory, Hallelujah! I 'did my possible' . . . ," he wrote Shapley a few days later, reporting that he too had worked to "induce President Lowell to appoint you." It was a win-win situation for Russell: "I shan't be sorry to go to Mount Wilson, either! Dr. Hale has written me, making a definite offer, and I have just written to him and President Merriam, accepting."[35]

Russell spent that winter and spring clearing his plate. He sailed for England to accept the Royal Astronomical Society's Gold Medal for his researches into how stars lived their lives, and then tended to various matters at home, planning his family's second summer at Clark's Island with the Websters. And now Clark's Island had another attraction: it would place Russell two hours from Harvard and Shapley: "I am at your service at any and all times, with such aid and counsel as I can bring. I have learned a good deal in the past three years,—principally what a fool I used to be. But I was no fool when I tried to get you at Harvard."[36]

Clearly he was no fool, for with Shapley at Harvard, continued access for Russell was not only assured but destined to increase. Quite pleased with all this, and with himself, Russell planned to leave for his first visit to Mount Wilson in the late summer: "This is a magnificent opportunity," he told Hibben; "I think it is really quite as much of an honor as the gold medal and professionally it surpasses any other offer which I have received." His Princeton world would not change, "other than my obtaining leave of absence for a visit of a few months to Mount Wilson to work there."[37] Russell asked to take leave for the remainder of the academic term, noting that Dugan had already agreed to fill in, and that the Carnegie stipend would fully reimburse him for travel and lodging expenses in the West. He also asked Hibben to consider a second leave during the autumn of 1922, when he hoped to take his whole family.

Just after Russell departed for England, Hale reported that he was about to secure Robert Millikan full-time for the California Institute of Technology: "If he comes success will be assured."[38] Hale was now more hopeful than ever that his vision of a coordinated attack on the physical universe would come true. Someone like Russell fit neatly into his plans, but to appreciate why, we have to consider what Hale was trying to do in the West.

Physicists were Hale's heroes; astrophysical research was merely an "applied science" wholly dependent upon advances in modern physical theory.[39] But Hale also hoped that astrophysics could aid physics, since it studied a realm "far beyond the scope of terrestrial laboratories."[40] Hale created physical laboratories at both Yerkes and Mount Wilson to integrate physics and astronomy. He always was on the lookout for astronomers who knew physics and physicists who were intrigued by astronomical problems. As he moved West to establish Mount Wilson, Hale attracted, for short-term projects, physicists like E. F. Nichols and Henry Gale, who with J. C. Kapteyn started spending summers in Pasadena trying out new techniques and stimulating the troops. Hale's own discovery of magnetic fields in sunspots in 1908 strengthened his conviction; more than half of the staff he hired, such as Arthur S. King, Charles St. John, and later Paul Merrill, were physicists, and the astronomers he added, like Adams and Seares, were adept in physical methods such as spectroscopy and photometry.

Hale thought of his observatory as a physicist's test bench: "If you enjoy work of development, under conditions that will enable us to build special apparatus, and use it under the finest conditions, I am sure the place would appeal to you," he told Charles E. Mendenhall in 1904. Mendenhall, however, preferred campus life to an isolated observatory.[41] Others, like King, came after training at Berkeley and Lick and advanced study in Kayser's institute in Bonn.[42] He had been influenced by W. W. Campbell to search for relationships between spectral peculiarities and temperature, and had corresponded with Hale on building a furnace spectroscopy laboratory, using designs he had learned while in Kayser's institute. King would eventually produce a temperature classification scheme based upon line spectra.[43]

But Hale's vision did not permeate down to daily life at his observatory, especially after 1916 when he became more and more distracted by National Academy activities. The isolation Mendenhall feared became reality for Walter F. Colby, a student of Harrison Randall, who arrived at Mount Wilson in the fall of 1914, fresh from the University of Michigan, to be King's assistant. But Colby left after a year, frustrated with the isolation, the routine, and the lack of interest in physical theory: "To my mind the need of it here is obvious."[44] Colby was not interested in building instruments or observing programs. Others were and stayed, among them John A. Anderson, who trained under R. W. Wood at Johns Hopkins and was hired in 1916 to be in charge of instrument construction.[45] Paul Merrill arrived in 1919 after training at Berkeley and Lick, research for H. D. Curtis at Michigan, and war work in laboratory spectroscopy under William Meggers at the National Bureau of Standards.[46] Merrill had hoped to take advanced training before the war, but when Hale called, he answered: Mount Wilson was "a very wonderful opportunity for *systematic* work in the observation and interpretation of the less understood types of stellar spectra."[47] Merrill devoted himself to radial velocities of long-period variables even though he knew that so much more could be learned from the spectrum of a star. He said as much in 1920 to one of his Lick mentors, J. H. Moore, who agreed that "the radial velocity business has been somewhat overdone here at the expense of a more thorough study of the spectra themselves. As time and opportunity offer we are trying to do a little of it on the side, but not as much as I would like."[48] But as Merrill well knew, the Lick staff was not capable of more than radial velocity work.

Despite his convictions, Merrill was typical of Hale's staff in avoiding atomic physics. Writing to William Meggers in March 1920, Merrill spoke in heady terms of the great achievements in Pasadena, now that Anderson was obtaining higher temperatures, bringing the physical laboratory closer to the stars. But theoretical matters were less welcome; "the Einstein stuff" was known as the "accursed theory" around Mount Wilson. Merrill felt there were "dozens of other things which are more worth working on." Merrill hoped the astronomical tests of general relativity might dispose "of the Einstein foolishness." He placed quantum theory in much the same category: "Single line spectra, so-called, have had my animosity because of the non-descript theory which has been used to account for them," he told Meggers. For Merrill, quantum theory was "rather made to order." Empirical work was safer; the "recorded facts and the analogies between different spectra are of the highest importance, and the work is noble if not Nobel."[49]

Many factors kept theoretical physics out of Hale's observatory. As much as he wanted to fuse physics and astrophysics, Hale was really building observatories hungry for observational technicians: builders and users. Not only was Mount Wilson a new, very large observatory, but it was an experimental test bed upon which new forms of observational devices could be built and evaluated. People like Anderson and A. A. Michelson, the Chicago Nobel Prize–winning physicist and "Master of Light," came to Mount Wilson to build new devices, and Adams's highest priority, as Hale's chief lieutenant, was to keep the instruments happy and used to their fullest degree. To this end, Adams managed the daily observatory operations along traditional lines where the local specialty was learned "on the job." Even after Hale's creation of Caltech, therefore, he still needed a link between the physicists on campus and his Mount Wilson staff, who were both physically and temperamentally isolated.[50]

Hale's hope that Kapteyn would stimulate Mount Wilson was killed by the war and by the changing face of astronomy. Although the statistical techniques Kapteyn championed remained important, the real push after the war was to harness knowledge about how matter interacted with light. Here was the key to the nature and the evolution of the Sun and stars. Neither Kapteyn nor Michelson could satisfy what Hale felt was the new direction his staff had to take. And most of all, Hale knew that he was inadequate to the task. He had fallen too far behind.[51]

One episode taught Hale a critical lesson. In 1919, Hale's staff was building a gigantic twenty-foot stellar interferometer for his new 100-inch reflector, the largest in the world, to measure the angular diameters of stars for Michelson. But Hale found that neither Michelson nor his staff, including Shapley, were capable of predicting which stars could be observed, or if any could be observed, with the device.[52] Shapley thought he knew how to predict angular diameters using the Planck formula to estimate brightness efficiency (amount of energy per unit area), but he incorrectly applied various photometric constants and fell off the mark by an order of magnitude. Hale, spending most of his time on the East Coast, became concerned when Adams claimed the project was doomed. Only when Eddington showed that the measurement was worthwhile, to "make sure that our theoretical deductions are starting out on the right lines," and demonstrated that the measurements could be made with the twenty-foot beam, did Hale take action.[53] Eddington, as Hertzsprung did in 1905, calculated the apparent angular diameters of some of the brightest red stars, finding that Betelgeuse would subtend 0.″051 seconds of arc in the sky, 1/36,000th the angular diameter of the full Moon. Eddington's predictions spurred Hale's staff to finish the beam and make the difficult measurements. They soon confirmed Eddington's prediction, sensing a disk "about the same size as a halfpenny fifty miles away." The measurement made the front page of the *New York Times*.[54]

One American had also made the right prediction, and that was Russell. In January 1920, after Hale told him of his concerns, Russell provided estimates that were close to Eddington's some nine months later.[55] Russell had even advised Shapley of the proper photometric constants to apply to obtain surface brightnesses, but Shapley, overly sure of himself, failed to heed his advice. Finally, in October 1920 Russell submitted a paper giving a detailed set of predictions for over three dozen stars.[56] Michelson's interferometer was a spectacular success, but it also confirmed that Hale's staff was ill-equipped to adapt and apply physical theory. Even with Caltech's dazzling accumulation of talent, Mount Wilson remained isolated. Hale therefore needed someone like Russell to bring modern physics within reach of his staff. He needed someone to act as interpreter and as catalyst, someone able to bridge the gap between his physicists on campus and his astronomers on the mountain. Coming to know Russell through Shapley, as well as through the NRC committee, sensing Russell's desires as they both tortured themselves over the offers from Yale and Harvard, and then, finally, through the stellar diameter episode, Hale knew that Russell was the right choice. As a result, starting in 1921 and continuing for over two decades, Russell devoted much of his energy to spectroscopy and split his institutional affiliation between Princeton and Mount Wilson. Thus began a new phase in Russell's professional life, the Mount Wilson years.

Russell turned down the two most distinguished positions that had opened in American astronomy in those years. He did so in order to preserve a lifestyle that he felt would allow him maximum freedom to engage directly in research. As he told Hale in June 1920, "The real question, as I see it, is: what is my duty to the advancement of science in general, and of American astronomy in particular?" Clearly, he decided that his duty to science was to do science, to focus on his own research. For Russell, duty and self-fulfillment were synonymous.[57]

If he learned anything from these calls from Yale and Harvard, it was that he could deal from a position of power with his Princeton administration. This is quite possibly what he meant when he described himself as a "fool" to Shapley, and when he told Hale that he had learned an important lesson after the Yale offer. Russell's annual petitions to Hibben, trying to secure a promotion for Dugan and regain the observatory residence, had fallen on deaf ears. Now he realized he was in a position to demand things. He not only regained the residence for Dugan but secured his promotion to professor in June 1920 with a very substantial raise.[58]

His most significant demand, however, was for his own freedom, mainly his freedom to remain absent from the campus for a considerable fraction of the year. Three months in the West, added to a Princeton professor's already substantial summer vacation, meant that Russell could not fulfill all the functions normally expected of an observatory director. But Russell had no intention of doing so. His plan, as it always had been, was to leave the details to Dugan. His obligation was to see that Dugan was properly compensated. Now that he had met this responsibility, he felt truly free.

• CHAPTER 13 •

Rationalizing Stellar Spectra

"I would rather analyze spectra than do cross-word puzzles or do almost anything else."

—*Russell to William F. Meggers, 1927*[1]

Filling a Gap in the Logical Argument

When Russell was ruminating over the offers from Yale and Harvard, he was auditing Karl T. Compton's course on atomic physics and the electron theory of matter and attending a special series of lectures by E. P. Adams on quantum theory. He was also reading an early edition of Arnold Sommerfeld's *Atombau und Spektrallinien*. In February 1920 he lobbied the NRC to publish Adams's lectures as "an admirable account of this difficult subject . . . for the benefit of American physicists."[2] At age forty-two Russell was rapidly absorbing the new physics and would adopt many of the tools and techniques of the physicist. As he did, he also changed in more fundamental ways. Russell's astrophysical style had always been radical and opportunistic, but upon entering the world of the physicist, though he retained these traits initially, he eventually turned into a very different sort of worker, as I show here and in the next chapter. He had entered a very different world.

Sometime in December 1920 Russell picked up the October issue of the *Philosophical Magazine* and was struck by what he read. A paper written by a young physicist from Calcutta seemed to hold the answer to a question that had long plagued astrophysicists: why stars exhibited different spectra. For over two decades, astronomers suspected that there was a definite relation between effective temperature and type of spectrum, and since Adams's and Kohlschütter's detection of luminosity-sensitive line criteria in 1914, the puzzle only got deeper. Some, like Adams, inferred from experimental evidence that pressure could alter the appearance of a spectral line. Transition probabilities—how easily atoms absorb energy—were not known then, and only a few atoms had measured ionization potentials. Theoretical physicists knew that these properties held the key to linking stellar spectra to the atom; as long as the relationship remained empirical, E. A. Milne recalled a few years later, "[t]here was a gap in the logical argument."[3]

Russell sensed that Meghnad Saha had found a way to fill that gap by linking thermodynamics and Bohr atomic theory to the stars. Saha showed that line strength in a star's spectrum was primarily a temperature effect and secondarily a pressure effect, and both were artifacts of thermal ionization, the loss of electrons through the application of heat. Russell found Saha's work "extremely suggestive," he wrote Adams in December. "I believe that within a few years we may utilize knowledge of ionizing potentials, and so on, to obtain numerical determinations of stellar temperatures from spectroscopic data—or at least of relations between temperature and pressure." Within a few more months Russell was reading other Saha papers, which provided a temperature calibration for the Harvard spectral sequence and demonstrated that temperatures and pressures in the atmospheres of the Sun and stars were for the first time determinable by physical theory.[4]

Saha was able to explain why spectral lines appeared, grew in strength, and then diminished in the Harvard spectral sequence, from the cool red stars to the hottest blue stars, in terms of how temperature influenced what energy states electrons occupied in the atoms in the stellar atmosphere, and also how electrons could be lost to those states through ionization as temperature increased. He used his equations to construct theoretical visibility curves for each major element and compare them with observed visibility curves from the Harvard sequence; Saha made his calibrations by fitting the curves at their wings.

Saha's fitting procedure represents the first successful link between the structure of an atom and the appearance of its spectral fingerprint in the atmosphere of a celestial body. Saha applied techniques and concepts from areas as diverse as physical chemistry, atomic physics, thermodynamics, and astrophysics and fitted these together into an amalgam that framed a general picture. His achievement had immediate impact in astrophysics.[5]

Although Russell realized that Saha had hit upon something really important, it took him several months to absorb enough of it to be sure that he could make a positive contribution. There was really not much awaiting Russell at Mount Wilson that would help him make progress on his theory of stellar evolution or on stellar energy, his main enthusiasms, but there was a treasure trove of material ripe for testing the Saha ionization theory. So during the spring 1921 term, as he prepared for his first visit to Mount Wilson, Russell refined Saha's theory to account for a gas whose electron component came from an admixture of elements, not from a single element, as Saha's model presupposed. He showed that his modification altered ionization levels "to an appreciable extent" and fit nicely with A. S. King's experimental furnace data.[6]

Russell's first stop in Pasadena was to visit King's Santa Barbara Street laboratory, and then he raided the plate vaults. He soon found features in solar spectra that Saha's theory predicted, such as rubidium in sunspot spectra. Just as he was setting his sights on further confirmations, Hale asked him to answer a letter from none other than Saha himself, who, on a temporary fellowship in Walther Nernst's laboratory in Berlin and due to return to India and oblivion, had pleaded with Hale for help.[7] Russell excitedly told Saha how his prediction of rubidium had already been confirmed, and how he was now directing Mount Wilson staff to look for other elements Saha had predicted. Russell was already linking King's furnace spectra and temperature classifications to Adams's stellar work, all based upon Saha's suggestions. Regardless of how Saha received this news of what the well-equipped Americans were doing, it is clear that his papers had caused a fundamental shift at Mount Wilson, and that the shift was being orchestrated by Russell.[8]

In his NRC essay, Russell had claimed that physical processes underlying the behavior of stellar spectra held the "master key" to the major problems confronting astronomers.[9] Now he found himself in the best of all possible positions to utilize that key and was soon on the lecture circuit telling the world how the Bohr atom and the Saha ionization theory were central to a combined attack on the physics of stars and atoms. Echoing Hale, Russell predicted, "It is not too bold to hope that, within a few years, science may find itself in possession of a rational theory of stellar spectra, and, at the same time, of much additional knowledge concerning the constitution of atoms."[10] Russell did his best to make Bohr theory accessible to lay audiences as well as to his peers. His excitement was more infectious than his rhetoric was effective, as Robert Aitken reported to Paul Merrill at Mount Wilson in August 1921. A recent Russell speech at Berkeley "went over the heads of many of his auditors, although we thought he made everything as clear as the subject would permit. One lady was heard to remark that she was glad she went for now she knew the difference between a cell and an atom!"[11]

Mathematical theorists who recognized the potential of Saha's breakthrough reacted differently from Russell. E. A. Milne publicly hailed its importance but rejected his thermodynamical arguments and, with Trinity College colleague R. H. Fowler, rederived the equations using statistical mechanics. They soon found that maximum line strength was a better physical marker of temperature than Saha's wing-fitting, and upon recalibrating the spectral sequence, Milne realized that atmospheric pressures were three to four orders of magnitude lower than Saha's estimates. A-type stellar atmospheres were only one-ten-thousandth the surface pressure on Earth. And among the giants, pressures were vanishingly smaller. Once they felt that they had set Saha's theory on a proper footing, Milne and Fowler invited "investigators who have access to extensive collections of stellar spectra" to take up their lead.[12]

Closer to the great astronomical plate vaults in America than any other theoretically minded astrophysicist, Russell found himself in the right place at the right time to take that lead, exploiting the theory, testing and extending it against the spectral testimony available from Mount Wilson and Harvard. During the next decade Russell spent a good fraction of each year steering members of Hale's spectroscopic staff, which was headed by Adams, toward areas he thought useful to examine. True to his rhetoric, he worked just as hard to employ astrophysical data to unravel the structure of atoms as he did to apply the latest theories of modern physics to the stars. In Russell's mind, astrophysics and physics were synergistic: astrophysicists needed data from physics on ionization potentials, but astrophysicists could also provide physicists with ionization potentials from astronomical evidence. Working from both ends, by 1924, Russell had published over a dozen papers extending Saha's work, and whenever he required data, most of it came from Mount Wilson. By the end of the decade, his multiple lines of attack would produce convincing evidence for the compositions of stellar atmospheres, a new form of transition rule that could account for anomalies in the spectra of alkaline earths, and new procedures for determining the structures of atoms from an analysis of their spectra. All three of these contributions demonstrate Russell's new commitment to spectroscopy and his impact at Mount Wilson. In this chapter we look at Russell's efforts to contribute to physics; in the next we turn to astrophysics.

RUSSELL AT MOUNT WILSON

Russell, at the interface of observation and theory, chose problems that could be addressed by Hale's observers. His influence was palpable: "You want to know what Russell's interests were at any one point?" a Mount Wilson staff member later recalled: "Look up what the topics were in the Mount Wilson *Contributions* for the next year." Younger physicists at Caltech regarded him as "one of the keenest minds at that time" in theoretical spectroscopy, as he was always full of ideas.[13]

After Russell's first summer at Mount Wilson, Hale knew that his plan had worked: "Nothing ever stirred the staff to action so effectively," he wrote Russell.[14] Hale persuaded Hibben to let Russell return often, and during the regular term: "[N]o one has ever shown so intelligent and active an interest in every branch of our work, or done so much to stimulate the members of the staff to wider thought and greater effort." Russell's "encyclopedic knowledge, and his extraordinary ability to grasp a new idea and apply it effectively," made him "invaluable to us." Hale invited Hibben to share the responsibility of fostering and preserving "Russell's rare talents."[15] Most of all, he argued that Russell should spend winters at Mount Wilson, so as not to have his vacations curtailed: summers were for rest. In September 1921, Hale, suffering from nervous stress himself, demanded that Russell rest too, and was happy to hear about Clark's Island, where there were no telephones: "Your high-strung nature and insatiable love of research will surely break you down unless you get completely away from all thoughts of astronomy for several months of the year." Hale said as much to Hibben, adding, "[N]o alliance could be more promising than this between Princeton and Pasadena."[16]

Russell had spent July, August, and most of September at Mount Wilson. At the end, Hale wanted Russell back in a scant three months to help his staff appreciate the lectures of H. A. Lorentz, who was due to visit Caltech. Russell would also be a bridge to the lectures of Caltech's new theoretical physicist, Paul S. Epstein.[17] Although Hibben had some obvious reservations, the Curriculum Committee of Princeton's Board of Trustees approved an extended leave, mainly because John Q. Stewart was now on the staff (hired to fill the position Russell obtained for Webster) and could take over Russell's courses.[18] So Russell returned in January 1922, lectured on the theory and application of atomic physics in astrophysics, and provided not only a bridge to the world of Lorentz and Epstein but introductory materials that would help the Mount Wilson staff become conversant with the new physics. These lectures were formal, intensive, and mandatory and continued on through the 1920s; one staff member reported later, "[W]e have an hour each day with Russell who talks on spectrum analysis."[19]

Russell stimulated Mount Wilson spectroscopic staff to think first in terms of how the latest advances in modern physics might help in the interpretation of their astrophysical data, and then how their observations could advance the theory of spectra. Adams and King were receptive, and even Merrill was cautiously warming to it: "As to quantum theory both Lorentz & Epstein agree that the foundations are unsatisfactory, inconsistent and practically incredible," he told William F. Meggers. "Nevertheless great truths are contained in the theory and it is very useful."[20] What convinced Merrill was Russell's argument that one could apply the tools of the new physics without worrying about their foundations. He was concerned less with "great truths" than with the expediency of approximations to the truth. King liked this. He had been laboring for over a decade studying the furnace spectra of the elements and had done a wonderful job separating out how spectra changed with temperature. Now for the first time he was able to appreciate that the spectra of elements changed with changes in temperature because electrons were permitted to jump around from energy state to energy state only in certain ways. He did not have to worry about the physics per se, since Russell acted as interpreter. Adams said as much in February 1923 when he wrote Russell about Arnold Sommerfeld's "immensely interesting" lectures. They seemed to have all sorts of "suggestive ideas which you are one of the few persons who could utilize fully." He had also listened to A. H. Compton lecture on scattering and hoped that Russell could explain it all: "Our older conceptions seem to be going by the board at the rate of about one a week," Adams groaned.[21]

Russell's own research ranged broadly at Mount Wilson. From 1921 through 1929, building upon Saha's theory, he published some forty-eight research papers, reviews, and monographs on atomic structure and stellar atmospheres. He sought out the underlying atomic structures of atoms that were important in illuminating the nature of solar and stellar atmospheres. He gave special attention to the metals and to the alkaline earths because they hinted at the existence of new transition rules. He knew that physicists had been developing a set of rules based upon the observed behaviors of elements in the laboratory, but Russell had the added advantage of hints from celestial spectra, where temperatures outstripped terrestrial laboratories. Russell also became interested in connecting the observed strengths of spectral lines in stellar spectra to the number of atoms in their atmospheres that were capable of producing that line. Driven by a complex set of issues, he among others groped toward a quantitative analysis of the abundances of the elements in the Sun and stars. Along the way he studied the intensities in families of lines called multiplets, producing in 1925 long papers on theory and how the theory agreed with observed data. At the same time, he also examined, with his computer Charlotte E. Moore, the intensities of "winged lines" in the solar spectrum, lines that had been broadened by a number of mechanisms, including abundance. This led him finally to redirect a long-standing project at Mount Wilson to photographically recalibrate the standard visual Rowland scale of line intensities in the solar spectrum.

In addition to working along the dual paths of atomic structure and quantitative studies of stellar atmospheres, Russell also proselytized for the integration of physics and astronomy, speaking sometimes as a physicist, and at other times as an astronomer. Boundaries became artificial for him, so when speaking about the relationship between physics and astronomy at the inauguration of a new physics building at Vassar in 1927, on stage with Millikan and F. A. Saunders, he quipped that "the question one might ask is not so much what the relation is as what the distinction is." Indeed, by now, for Russell, the relationship was "simply the relation of the part to the whole, or of one part to another, and the boundary between the fields of the two sciences now is practically non-existent. It is simply impossible to tell where astronomy ends and physics begins." During the 1920s Russell spoke out many times, and in many venues, on the need to recruit physicists to the problems he thought important. Most of his appeals were for collaboration in what, he believed, was the greatest task facing both physicists and astronomers. He often spoke of the "delightful times" that he had had with Saunders, "working out together" the structure of atoms from spectra. "Everybody is doing it now, and we are gaining information very fast indeed . . . Literally dozens of men all over the world . . . are working at the problem."[22]

Russell's Collaborations

Russell engaged a wide array of complementary talent to get his work done. His collaborations took three forms: first, his early associations with K. T. Compton and the Harvard spectroscopist Frederick A. Saunders were stimulated by a need to sort out and explain anomalies between what the Saha theory predicted and what was observed in stellar spectra. Second, his collaborations with Moore and the Mount Wilson staff were to interpret astrophysical spectra. And third, he worked with Meggers of the National Bureau of Standards (NBS) as the nucleus of a widening circle of spectroscopists on the analysis of complex spectra, not to explain anomalies, but to fill in the picture of the structure of atoms from their characteristic spectra as completely as possible. The first type of collaboration was intense and brief, resulting in a significant contribution to physics; the second led to a wholly new view of the composition and structure of stellar atmospheres; and the third was long-term, plodding, and thorough. It would last the rest of Russell's professional life and result in Moore's monumental tables of spectral lines and associated atomic transitions.

New Regularities in the Alkaline Earths

Saha could not explain why barium was stronger in the solar spectrum than sodium, even though the two had similar ionization potentials. To explain barium, Russell looked for an *atomic* explanation as to why one element would absorb energy differently from another. At first he freely speculated; possibly the presence of an excited electron might inhibit recombination, but, talking with Compton, he also wondered whether an "atom could emit two quanta at once from different electrons . . ." He decided that this latter possibility was a "more delicate question" because it raised many problems with known transition rules.[23]

As Russell pondered barium in late 1921, he sought out Saunders at Harvard, whom he had known from the war when Saunders worked for Trowbridge on sound ranging. Saunders, who had received his Ph.D. under Henry Rowland at Johns Hopkins in 1899 and had worked with Friedrich Paschen, was an expert on alkali spectra. He not only held a firm grasp of the latest literature in quantum mechanics but displayed the "essentially pragmatic" philosophy of the American physicist, finding ways to exploit theory in the laboratory without worrying about the basis of the theory itself. Russell shared this common trait among American spectroscopists, so the two meshed nicely. Saunders was easygoing, devoted to his laboratory and students, and was ready and willing to collaborate with a needy theorist.[24]

Russell drew upon a wide range of talent to solve the barium problem: he needed data from Adams and A. S. King for temperature classifications, and advice on theory from Compton. He and Saunders also found that they had to examine the alkaline earths as a class, but they had little theoretical guidance. Reading more widely now, Russell began to realize just how many obstacles stood in the way of a full elucidation of multiple valence electron atoms; the Bohr-Sommerfeld model worked well for hydrogen-like atoms but could not handle complex spectra because they embodied multiplet terms.[25] This problem attracted the best physicists of the day; in Max Born's words, quantum theory had to be able to deal with "the motion of more than one electron."[26]

Over the next two years, with Saunders's guidance, Russell found that Alfred Landé's vector model of the atom and Werner Heisenberg's earlier and cruder Rumpf model, also known as the core model, were very useful interpretive tools, even though the latter was never fully accepted by physicists. What Russell liked about the vector model was that it was an easily visualizable geometric representation of the atom which rendered most aspects of complex spectra comprehensible, if one were willing to overlook its theoretical frailties. The vector model, "so dear to the heart of the atomic spectroscopist," was accessible because it reduced problems of theoretical spectroscopy to trigonometry: puzzles that could be worked out in a visualizable and conceptually mechanistic manner.[27]

As Russell absorbed the theory, Saunders studied the appearance of neutral and ionized barium spectra subjected to a magnetic field. The resulting Zeeman patterns of the multiplets would reveal the energy levels involved. Thinking in terms of multiple electron interactions in late 1923, Russell used the vector model to figure out how to assign quantum numbers to the new terms (energy states). Their breakthrough came when Russell realized that the vector sum of the valence electrons also had to be quantized with respect to the angular momentum vector of the Rumpf. This additional quantum requirement brought order to the observations that Saunders had made, and allowed them to construct a new transition rule, or "coupling scheme," to account for the anomalous behavior of the barium atom. They had the basic structure for what was later called Russell-Saunders coupling (later still "L-S" coupling) in hand by the end of November, but there were still many obstacles to overcome, including the need for more data, the growing problem of consistent spectroscopic notation, and, most daunting, skepticism from Niels Bohr.[28] Overcoming these obstacles took more than a year of effort, and both Russell and Saunders knew they were in a race: "I think that it is worth while to hurry up the publication of the present paper anyhow," Russell advised, "because I am afraid that some German will get out the theory and steal a march on us." Saunders worried whenever Russell became distracted with some other spectroscopic passion, like the structure of aluminum and titanium, and advised him to publish "PDQ if I were you."[29]

There was competition: Gregor Wentzel published a very similar explanation as Russell and Saunders worked away. But their delay was prudent because Saunders's exhaustive observational confirmation, together with a new notation scheme, constituted strong persuasion. The theoretical physics community in Germany as well as in the United States accepted and used Russell and Saunders's work, citing them generously.[30] Heisenberg, in particular, found it useful in his last defense of the Rumpf model and rederived Russell and Saunders's coupling mechanism by completely independent means. And there were others who extended the approach, using the vector model, Pauli's exclusion principle, and Russell-Saunders coupling, to build up new ways to manage the analysis of complex spectra. Chief among them was Friedrich Hund, who created a set of rules to predict the patterns seen in complex spectra. Thus as Russell and Saunders's coupling scheme became an accepted tool in spectroscopic theory and analysis, Hund's theory appeared and propelled Russell's spectroscopic agenda for much of the rest of his life.[31]

Although their work enjoyed a quick and positive reception, Russell and Saunders found the analysis of line spectra sufficiently challenging and far more rewarding than frontal assaults on theory. This choice was in fact typical of American spectroscopists. As one historian has shown, modern German theoretical physicists like Sommerfeld looked to experiment mainly to illuminate the fundamental framework of quantum theory, whereas Americans were more interested in applying quantum theory as a set of rules to learn more about the objects of experimentation. Millikan, the Compton brothers, and most definitely Saunders, Webster, and Russell all preferred "practice over principle."[32]

The Analysis of Line Spectra

Starting in late 1923 and continuing through much of 1924, Russell prepared and revised constantly, always with Moore's aid and Mount Wilson data, a preliminary listing of prominent spectral lines for which the various spectroscopic and energy relations were known. His goal was to produce a standard listing of lines that arose either from the lowest ("ultimate") or first excited ("penultimate") states, which he believed were most important for astrophysical application because they predominated in stellar spectra. An orderly display of ultimate and penultimate lines of neutral atoms and then of ionized atoms helped to create an astrophysically meaningful temperature classification. For instance, ultimate and penultimate lines from neutral atoms were strong in sunspot spectra and in the cooler stars, whereas ultimate lines from ionized (or, as he still called them, "enhanced") lines, which were strong in arc spectra, also dominated in solar-type stars. "Progress in the study of complex spectra is at present very rapid," Russell told an astronomical audience at Vassar in December 1923. "With the work of a year or two more (especially on enhanced lines) it should be possible to put the classification and interpretation of [the] stellar spectrum on a really firm rational basis."[33]

The work of "a year or two" lasted for decades, culminating in Moore's monumental multiplet tables. But what Russell set in motion ultimately became an important element in the research agenda of a significant fraction of spectroscopists at American observatories and physical laboratories. His suggestion a year later, that "the astrophysical importance of the results [of the analysis of line spectra], appear to justify a collection from . . . widely scattered sources," defined Moore's lifework.[34] What Russell was after, however, was not only astrophysically useful data but clues to the basic structure of the atoms that produced complex spectra: his work with Saunders. As such his work on the analysis of line spectra may be seen less as a retreat from theory than as an application of the dictum "practice over principle" to elucidate theory.

Russell worked as an analyst, tending to complete what others had started. His first active forays were based upon published data. In the summer of 1923, he extended Paschen's study of the spark spectrum of aluminum and then turned to titanium and iron. Soon he found that he needed more data from spectroscopists, and so Russell took steps to become a coordinator, asking others for specific data. Just as Harvard botanist Asa Gray created a network of collectors, knowing that he could not personally gather all the plants he needed to create a representative classification of the flora of North America, Russell enlisted physicists at the National Bureau of Standards and at Princeton, and finally, through the National Research Council, he created a working committee that included the most active laboratory spectroscopists in the nation. Russell became a self-styled "headquarters scientist," acting as a clearinghouse in the development of an atomic taxonomy.[35]

Titanium drew Russell to Meggers, eleven years Russell's junior, who had taken a degree from Johns Hopkins in 1917 and was now chief of the NBS's four-person spectroscopy laboratory. Meggers was an ambitious and tireless networker, intent upon establishing a major role for spectroscopy at the bureau that would relieve him of the more mundane chores that government work entailed. Meggers already had a wide circle of correspondents, but Russell became his first active astrophysical collaborator.[36] In Russell, Meggers found not only an emerging master who could significantly aid his mission to analyze spectra, but an elite academic who could champion the cause of spectroscopy at the bureau. What Meggers was looking for, above all, was cooperation. As he suggested to Sommerfeld in June 1923: "Let experimental spectroscopists do the portion of the work in spectral regularities which requires observations and trained judgment in the description of the spectra, and we will reciprocate by asking the theoretical physicists to interpret the results. Isn't that fair?"[37]

To appreciate what Meggers saw in Russell, one must look at how he viewed his own life as a spectroscopist. From diary entries at about the time he encountered Russell, one gets the image of an unhappy drudge: "There are so many things to do and I leave many of them undone. Start off the year [1924] with regular routine. To the Bureau every day to do what I can do. Spend most of the time measuring spectrograms, searching for regularities in complex spectra." One can see how he needed help. After weeks and months of observations, adjusting the equipment, calibrating the photographic plates, and preparing the samples, one then had the data to sort out. Struggling with the vanadium spectrum the previous June, Meggers lamented: "I plug along without making much progress. Lack brains and have no insight. It comes very slowly." One member of his small group, F. M. Walters, sometimes could offer insight, helping Meggers see the path to a solution. When that happened, Meggers would be transformed: "This work is so exciting that I cannot leave it without thinking about it or sleep without dreaming about it. All day, when visitors, letters, etc., do not interrupt and every evening at home I make subtractions and subtractions and try linking up various lines." When it all worked out, he could be euphoric: "Begin to look for more regularities. A puzzling job to unravel a complicated spectrum but success in such a case is rewarded by sublime sense of satisfaction."[38]

These diary entries reveal a man in need of an interpreter, as he admitted to Sommerfeld, and as he sighed privately: "Wish I had more help or less tedious work."[39] He found his interpreter when Russell started asking for titanium information in July 1923 and then wrote glowing letters complimenting his staff's industry. Meggers knew Russell by reputation and was touched by Russell's enthusiasm for what was a terribly intricate and labor-intensive field. Having been deeply impressed by Russell's first publications, he suggested that "[t]here will be a great deal to do in correlating our new knowledge of spectra with stellar phenomena."[40] Thus they entered into a collaboration that lasted over a quarter century.

Russell helped Meggers's staff convert their spectroscopic data into energy level diagrams and the structures of the elements, and promoted NBS activity whenever possible. He soon found that he could work much faster at analysis than any of Meggers's men, so he was careful to report progress both to Meggers and to Adams, and always to cite where the data came from and how each new observation brought new revelations. His identification of a set of titanium triplet terms in August 1923, during a vacation at Clark's Island, moved him to rapture: "It is beautiful."[41] Meggers also looked to Russell to marshal astronomical expertise. Although King was always "especially useful in analyzing complex spectra," not all Mount Wilson staff were so forthcoming, Meggers told Russell. "[I] trust that you will encourage its extension."[42]

Russell had good reason to be interested in titanium. Beyond hydrogen, iron and titanium lines were sensitive indicators of spectroscopic parallax, as Adams keenly knew. Titanium presented a most intriguing spectrum rich in multiplets and was a dominant feature in late-type stellar spectra, which made possible the determination of its ionization potential independent of laboratory extrapolations.[43] He attacked titanium from two directions. First, donning the physicist's cloak, he solicited support from Mount Wilson, from the NBS, and from the University of Alberta to analyze its structure. At the same time, from the direction of astrophysics, he guided a graduate student, Donald H. Menzel, to the Harvard plate vaults to examine the celestial behavior of titanium and other elements to deduce their ionization potentials and the temperatures in stellar atmospheres.

The pattern Russell established with titanium, making himself the hub of a growing cadre of laboratory spectroscopists, was to be repeated time and again in Russell's mounting campaign to sort out spectra. He worked very rapidly and so was able to reward his collaborators with preliminary analyses that helped them plan out subsequent observations. He strengthened his network by providing elite intellectual patronage. After his first visit to the NBS in December 1923, Russell wrote George K. Burgess, the new bureau director, stressing the "great importance of the work on regularities in complex spectra" being done by Meggers, Walters, and Karl Kiess. "This work is of fundamental importance, both to Physics and to Astronomy," he argued, suggesting that Meggers's group be released from their routine testing responsibilities. Russell warned that there was much competition in the field at present, and that "if the results of American investigations are delayed in publication," then probably some German will scoop them and "secure the greater share of the credit." Meggers thanked Russell for his support, noting that his letter pleased Burgess because it gave him the political ammunition he needed when faced with growing demands for applied work.[44]

During Russell's first visit to the bureau, he lectured the staff on "Atoms and Stars," and deeply impressed Meggers, who like Hale, knew he had found a "[v]ery brilliant man," as he recorded in his diary that night. "What a profound impression you made on us," he wrote Russell after his visit. "We had never thought of you as an intensive spectroscopist and were surprised at the grasp you have of the subject and the marvelous facilities you have for finding the laws of spectra."[45] Russell in turn praised the industry of bureau staff, and over time they began to rely upon his powers of analysis more and more. In January 1928, Meggers noted that he needed to "[w]rite to Prof. Russell for help on yttrium spectra," and in exactly one month he cheered, "Russell has cleaned up Y[t]II spectrum for me!"[46]

As much as he praised the bureau publicly, in private Russell griped about them. Moore frequently sensed Russell's impatience when they did not deliver on time. Theodore Dunham, a graduate student, remembered how often Russell would take over the proceedings if someone got bogged down: "[W]henever he went to Washington, the Bureau of Standards, and talked to the people there,—and back and forth to other universities—he would often come back with all sorts of partially worked out spectra." Russell worked very fast. The perception of those around him in the 1920s was that he kept an enormous amount of information on tap in his head and could see relationships and new possibilities before anyone else. Meggers was no exception. During the 1928 Leiden International Astronomical Union meetings, Meggers was working away on an IAU commission report when "Russell happens in and we talk over some spectroscopic problems . . . He is as busy as ever or more so—how can he keep so much in his head?"[47]

In the network Russell created, Meggers, Saunders, King, and Allen Shenstone in the physics department at Princeton, among many others, fed Russell data, and Moore kept it all straight. They were all compelled to work with Russell, as Dunham suggested, because he could perform the analyses "in an extraordinary way. He could do it ever so fast, and he knew exactly what to do so he didn't waste any time on it." Russell barely managed to tolerate the slow methodical pace of men like Shenstone. The two had very different approaches yet were "amazingly effective as a pair."[48]

Outwardly to his colleagues beyond Princeton it was all fun and games. Russell disingenuously told Meggers, "I think you will be wise to take whatever grasp I may have of spectroscopy as an evidence that the subject is not a very difficult one, but it is certainly a most delightful field in which to work."[49] No matter how hard Russell would work on the analysis of line spectra in the following decades, he would not flinch from this view: that there was nothing inherently complicated in such work and it was all great fun. Meggers, of course, thought otherwise about the complexity of the task but loved the game just the same. There was a certain intoxication to working out the structure of physical reality; and, one must never forget, for Russell "fun," as it had been for his father, had to be a most serious, vigorous, and meaningful challenge. Just as his father had enjoyed strenuous hiking, Russell never took anything easily or passively, always acting as if he had to prove himself against all odds, whether he was pondering the odd crossword puzzle or divining the secrets of nature. In his zeal for accomplishment and conquest, Russell could be rather insensitive to the limitations of others.

Russell's Rhetoric and Technique

Russell consistently argued that term analysis was easy and fun, which set him apart from those, like Meggers, who always marveled at how Russell could carry "the theoretical details as well as the empirical facts in his head."[50] Russell's techniques changed as new transition rules, multiplet theory, and Hund's sum rules appeared, so to an extent, they remained ad hoc until the theory stabilized in the mid-1920s. Typically, Russell directed Moore to lay out wave numbers from their spectroscopic line lists to search for term intervals arising from similar transitions and to compare strengths of each line, an indicator of the number of transitions causing it. She plotted the wave numbers on long strips of coordinate paper and made detailed horizontal comparisons of patterns to sort out the spectra.[51]

His working notes from the 1920s reveal an iterative process. His first rough estimates and tabulations based upon Moore's labors revealed gaps in data, and so he would write and cajole his laboratory colleagues to fill in the gaps. He typically applied graphical fitting procedures loosely based upon Walther Grotrian's diagrammatic apparatus to illustrate the energy levels in an atom (see fig. 13.1).[52] He wanted to test how well his theoretical intensities, in sets of quartets, triplets, and sextets, fit King's observed intensities. Filling numerous thin notebooks, blue exam books, and foolscap pages with detailed calculations to arrive at theoretical intensity ratios, organized by the complexity of each group, Russell constructed triplets, quartets, and septets on up to highly complex multiplet structures. He was always elated when he found some agreement with observation, but was no less intrigued when there were discordances, because in the early years not only was he trying to create workable maps of the atom but he was always on the lookout for exceptions that could reveal new rules about how matter behaved on the atomic level.[53] This pattern changed in following years, as Russell realized that he was really filling in the details. But it never seemed to dampen his enthusiasm.

The image Russell liked to convey more than any other, starting in the 1920s, was that the whole process was akin to solving a crossword puzzle. He, like many scientists, liked to describe scientific problems as "puzzles."[54] No fewer than sixteen of his *Scientific American* columns contained the word in the title or subtitle, five of them appearing in an eighteen-month period during 1928 and 1929, his most intense spectroscopic phase. In spectrum analysis, Russell tended to carry the analogy to the limit, as in his 1935 George Darwin Lecture before the Royal Astronomical Society:

> [T]he analysis of a complex spectrum is curiously like the solution of a glorified cross-word puzzle. There are usually a few places where a start is easy. Once a word or two has been written in (or a couple of low-energy levels found), many other words (or spectroscopic terms) can be fitted in, successively, each affording a clue to the other. One must of course know the pattern of the puzzle (the Hund theory). The various spectroscopic data (temperature class and Zeeman effect) may be compared with the definitions of the words of the puzzle (though they are often more similar to the picturesque hints of an English puzzle than to the elementary directness of many in American newspapers). There are even unkeyed letters in the puzzle, or lines in the spectrum, which belong to one word only (or to a level having no other combinations) and must be identified by "making sense." The main difference is that the spectral puzzle takes longer to solve, and that, when you are through with it, you can publish the result and call it research.[55]

Russell typically described his methods in terms of a mechanical "astronomical atom," a tiny solar system whose structure could be revealed by analytical tools like the Zeeman effect, "of great aid in grouping the numerous levels in a complex spectrum into multiple terms . . . They often tell the investigator where to start . . ." He used Hund's Rule "as a main guide, telling the analyst what terms to expect, and how many, and of what nature, remain to be found."[56] The same imagery appeared in his 1933 Halley Lecture, "The Composition of the Stars," delivered in Oxford. He used it again in 1943 in one of his last *Scientific American* columns, where he provided clues for the reader to figure out his own spectroscopic puzzle.[57] Such rhetoric reveals that the techniques Russell developed were hardly controversial; rather, they embodied the essence of normal science: based on an agreed-upon set of rules, directed toward a clear solution, set within the context of an accepted theory, and directed to improve the generality of that theory. Most typical of normal science, term analyses led to definite solutions.[58] To make it all work, the observations had to be very accurate, and for Russell, this was the onerous part of the problem. The analysis was a piece of cake, but getting to the interpretive stage was tedious.

Recruiting Labor

Term analysis was a labor-intensive art, and Russell turned many of his graduate students into practitioners. Most, like Theodore Dunham and Donald Menzel, were exposed to it only as a portion of their course studies, but others, like Louis Green and Robert King, went through to the thesis and stayed in the field. Russell gave Dunham the spectrum of neutral manganese to sort out, based upon Catalán's early multiplet work, which was not complete. Russell insisted that every line be identified and nailed down, and led Dunham to his sources. It was here, in 1925 and 1926, that Dunham began appreciating Russell's growing network

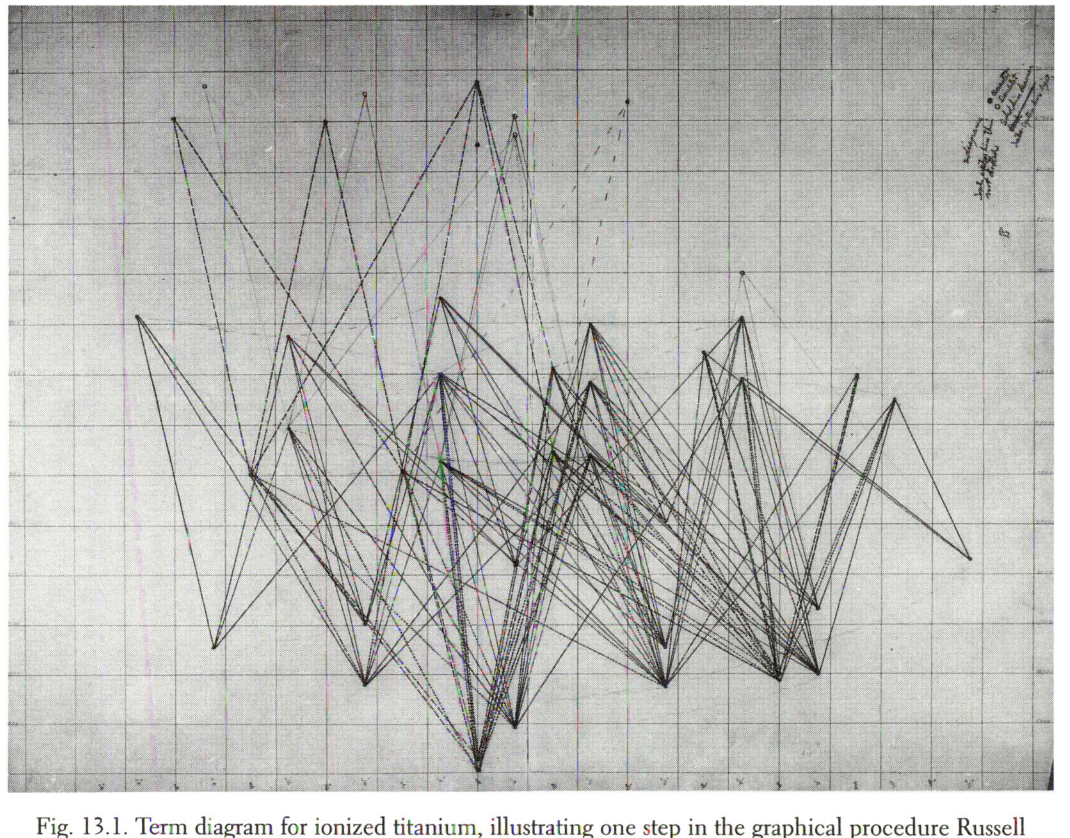

Fig. 13.1. Term diagram for ionized titanium, illustrating one step in the graphical procedure Russell employed to sort out the energy levels in the atom. Reproduction from the Henry Norris Russell Papers, courtesy of the Princeton University Library.

of collaborators and the critical importance of maintaining Princeton as a central clearinghouse. Data flowed in from "all sorts of people all over the country," as well as from Canada and Europe.[59]

Russell rewarded his students with degrees and his collaborators with praise and political favors. He was especially sensitive to Moore's role; he may have possessed phenomenal powers of intuition, concentration, and memory, but it was Moore who kept it all together and performed the preliminary reductions. Russell deeply appreciated her work, pushing for her promotion by telling the Princeton Controller in 1936 that she undoubtedly would have obtained professional standing on the faculty by then "were she not a woman."[60]

Different students at different times had very different experiences, and their differences can be reconciled with changing times. Dunham's laboratory work took place when Russell's excitement, and the promise of spectroscopy, were fresh and at their peak. By the time Louis Green was recruited in the mid-1930s, spectroscopy was on the wane and Russell had given over much of the graduate student guidance to Stewart, Allen Shenstone, and others in the physics department, as well as to Moore. Green worked with Shenstone on the iron spectrum as an observatory assistant, and this led to his thesis, in spite of Green's own interests: He approached Russell several times with at least three suggestions for thesis topics, and each time Russell ignored the alternatives and insisted that he continue work on ionized iron. No matter how vociferously Green objected, "in the end, of course, I worked on the Fe II spectrum."[61]

Russell could be insensitive about how he assigned research to graduate students; his priority was to put his best students onto his most cherished problems, and he thought particularly well of Green.[62] In addition to Green, Russell also trained A. S. King's son, Robert B. King. King learned the practical aspects of the enterprise at his father's side in the Santa Barbara Street laboratory, and both King and Russell were anxious that Robert follow in his father's footsteps. The elder King hoped that Moore, who was then at Mount Wilson, would be able to help Robert unravel his laboratory data. Robert missed out on a regular fellowship in his first year, as they were all taken by continuing students such as Serge Korff, John A. Merrill, and J. A. Bennett. Dugan found research assistantship money and invited Robert to live at the observatory.[63]

Neither Green nor King felt that the experimental work they were doing was astronomy. King had a legacy to maintain, but Green chafed. Laboratory work and plotting term diagrams constituted the bulk of the mental and physical effort that went into the routine of spectrum analysis, but this was not challenging to Green.[64] Green's case nicely highlights some of the recruiting problems Russell and Meggers faced in their spectroscopic campaigns. There was not an overabundance of large vacuum spectrographs available for the work, and not a large pool of sufficiently talented and dedicated youngsters to draw upon. Thus when a new player like George Harrison, a student of Webster's, did appear on the scene, both Russell and Meggers could be very supportive. Otherwise, like Asa Gray, they twisted the best arms they could find.

The NRC Committee on the Line Spectra of the Elements

As Russell watched Meggers in action, he found that despite his call for openness and collaborative work, Meggers was fiercely competitive. Russell confided in Shapley in October 1923 that he felt "somewhat embarrassed by the fact that the Bureau of Standards people seem to regard the field as a good deal their own property." Russell by then was collecting data from a wider circle, including his students, younger spectroscopic faculty at Princeton, Mount Wilson staff, and personnel at a growing cadre of university laboratories. He felt sensitive to Meggers, knowing that he could be offended "if I publish . . . and skim the cream."[65] He needed another venue that shifted focus away from the NBS and expanded his talent pool.

In 1924, Russell found Meggers willing to let the National Research Council act as both a coordinating agent and a legitimizing champion for the analysis of spectra. Meggers had been trying to form a partnership with Arnold Sommerfeld, but the plan bogged down because Sommerfeld's goals—as expressed by Otto Laporte, who had been sent from Munich to help coordinate the effort—were very different from those Meggers thought were important.[66] Laporte and Sommerfeld needed only a partial survey, whereas Meggers and Russell wanted to be exhaustive. And they could never agree on notation.

Meggers and Russell admitted that Sommerfeld's plan would have immediate benefit for atomic physics, but it would be inaccurate, incomplete, and superseded quite soon, whereas their option would likely be longer-lasting. Laporte's needs, however, were not so different from those which had driven Russell to create his hypothetical parallax techniques some eighteen years earlier. The difference now was in the purpose to which Russell wished to apply his term analysis efforts: in the early 1920s he wanted to advance physical theory, but at middecade he turned more to astronomical application, which required exhaustive analysis given the vast number of unidentified lines found in stellar spectra. There were no shortcuts now, nor did Russell possess the intuitive powers to create them. Thus, as is evident in his puzzle-solving rhetoric, this goal eventually turned Russell into the type of worker in spectrum analysis that he had criticized in observational astronomy. He would use no "tissue of approximations" here (see chapter 18).

Sensing Meggers's frustration with the Germans, Russell suggested that they formulate "Anglo-American plans for a complete report." Russell had good connections with the NRC, having been a council member since June 1918. Joseph Ames of Johns Hopkins was then head of the new Physics Division and in January 1925 agreed with Russell, Meggers, and Paul D. Foote to convert an existing committee on ionization potentials into a new committee, headed by Russell, for the analysis of complex spectra. Assured that the NRC would publish the committee's results, Russell forged ahead. By then, Sommerfeld's plans had died, so the NRC Committee on the Line Spectra of the Elements was now the only game in town.[67]

Russell quickly enlisted over a dozen physicists and, with Meggers, laid plans to standardize notation. He also asked his members for status reports, hoping to issue a summary on the first long series on the periodic chart within a year or two.[68] Russell found that he liked being chair. He had a good working core team at Princeton and believed that most of the coordination could be done by correspondence. "Moreover," he told Ames, "as I happen to travel around a good bit, I can act as liaison officer."[69] Indeed, Russell liked to travel; it gave him time to think and plan, and to work out terms in spectra. Depending upon his route, Russell's yearly journeys to Mount Wilson gave him an excuse to stop in Washington to see Meggers and his staff, Pittsburgh to visit K. Burns and F. M. Walters (who had left the NBS in August 1924), as well as Ann Arbor to see Harrison Randall. His westward pilgrimages typically included stops at the Lowell Observatory in Flagstaff, the Lick Observatory just south of San Francisco, and the Dominion Astrophysical Observatory in British Columbia. These stops were hardly casual. A Russell visit was an event usually arranged far in advance, linked to a lecture or two, informal conferences, and inspections of plate vaults and laboratories. At Harvard the call would go out, "Russell has come!" which meant that one dropped whatever was at hand to be attendant to his presence.[70] From the early 1920s through the mid-1940s this pattern would repeat itself on innumerable occasions as Russell attempted to assume intellectual responsibility for larger and larger chunks of American astronomy and laboratory spectroscopy.

Using the NRC committee as a platform, Russell and Meggers led a small army of spectroscopists across the periodic chart. They started out optimistic that a first summary report would not be more than a few years in coming. But as Russell surveyed the landscape, gaping holes kept publication at bay, and, ultimately, no formal comprehensive publication ever appeared from the committee, although individual members continued to issue their own results, and Russell provided annual reports on the state of the field. Russell remained chair of the committee for twenty-three years, never completing what he had set out explicitly to do. When he finally stepped down in 1949, Meggers succeeded him, but it was clear from where Meggers stood that the real power in the process had been with Moore:

> This committee was organized more than 20 years ago for the purpose of compiling and publishing monographs on atomic spectra. After training a perfect compiler in the person of Charlotte E. Moore, and accumulating material for several monographs you generously give everything to me. Under these circumstances I would not dare disappoint you.[71]

Two aspects of Russell's spectroscopic networking in the 1920s inform his subsequent work and life: his continual confirmation of the series relationships embodied in Hund's Rule, and this work's impact upon his staff, especially Moore. First, over the years, as if to encourage his troops, Russell always reminded physicists of how their work confirmed Hund's recipe for how to deduce the structure of atoms exhibiting complex spectra. His 1931 report to the NRC argued this point; by then the theory was on such a firm basis that it fully justified the preparation of a major report, even though much work had to be done actually preparing the tables for publication.[72] One factor behind Russell's optimism was that by the end of 1928, he had, at last, in conjunction with Allen Shenstone and Louis E. Turner, standardized the morass of spectroscopic notation into a reasonably consistent system. The American Physical Society had asked them to perform this task only when theory finally seemed stabilized, and so they canvassed spectroscopic workers worldwide to be sure their system reflected the majority opinion. They knew that agreement on a notation describing energy states and transition rules required consensus on the theory behind those rules, and the rapidity with which they achieved consensus was further proof of the stability of the theory. By the early 1930s, the notation had "been accepted internationally by many investigators."[73]

Charlotte Moore

Even though by 1932 the NRC had provided funds to hire computers to assist Moore and Russell in the production of the tables, Russell's continued concern over completeness and the sheer enormity of the task kept a formal publication beyond reach. If anyone was getting frustrated at this delay, it was Moore. Although she found herself hardly in a position to do anything about it directly, her persistence ultimately paid off, not in the publication of the tables Russell and his committee had envisioned, but in an adjunct project, something she had been developing since the late 1920s, which quickly proved to be extremely valuable for the astrophysicist and did, in fact, embody what Russell had been trying to do. These were Moore's *Multiplet Tables of Astrophysical Interest*, which grew out of her thesis work at Lick and Mount Wilson and expanded as she continued to collect and organize the data flowing from Russell's contributors.

Moore's career trajectory helps to illuminate the world Russell built, for she was in many respects both an essential ingredient in, and a reflection of, his scientific life. A fresh graduate in mathematics and astronomy from Swarthmore, Moore arrived in Princeton in September 1920 as Russell's general computer and assistant. She performed calculations for his lunar program, for double stars, and for spectroscopy. She eventually spent the second half of her first decade with Russell working mainly on the West Coast, first between late 1925 and 1928 at Mount Wilson, helping out with the revision of the Rowland Atlas (see chapter 14), and then another two years at Berkeley, Lick, and Mount Wilson completing her graduate studies and thesis while the Russell family toured Europe. She wanted to get away from Princeton because she was in poor health owing to the climate, and, as well, "[w]orking for Dr. Russell was really a strenuous job and you had to be on your toes all the time and I think it was just too much for me." She would have liked to stay at Mount Wilson, but no one dared take her away from Russell.[74]

Her thesis, "Atomic Lines in the Sunspot Spectrum," was suggested by Russell. Adams, St. John, and A. S. King "offered me a gold mine of data[;] it was one of their pet subjects." More to the point, Moore proved to be indispensable to the

Mount Wilson staff. They became deeply attached to her, not only for her dedication and drive, but for her ability to manage Russell's enthusiasms: "He needed somebody to stabilize him, he really did," she recalled.[75] In 1927 and 1928 alone, Moore had collaborated with St. John, King, Babcock, Dunham, and other Mount Wilson staff in no fewer than eleven separate research publications.[76] Out of this she amassed an enormous amount of information about the atomic lines in the solar spectrum, as well as lists of molecular spectra and laboratory data from the NBS and other sites. She was then compiling an extensive catalog of both laboratory standards and solar wavelengths grouped by element, multiplet, and term designations, including intensity observations and other data of astrophysical interest that were at hand. Building upon Russell's 1925 study "A List of Ultimate and Penultimate Lines of Astrophysical Interest," she wanted to create a multiplet table that was comprehensive: "Work on spectrum analysis was proceeding so rapidly that an extension of his list was imperative."[77]

They agreed on most matters, but Moore and Russell did not agree on when was the right time to publish her compilation of multiplet information. After he returned from Europe in 1930 and inspected her work, it took another year for Russell to agree to print a trial edition, to test the waters; some 250 copies were "exhausted within a few months."[78] It was a gold mine that included the new standardized notation with references to all older systems, as well as a large bibliographic database indicating the sources of information that had been used, something usually provided in only the most sketchy manner. Although it was obviously in demand, her *Table* was not reissued for over a decade.

Throughout the rest of the 1930s, she added to her listings as she coordinated much of Russell's NRC committee work. Nevertheless, he never felt that the time was right to issue a revised multiplet table. When Russell delivered the Darwin Lecture at the Royal Astronomical Society in June 1935, he expressed concern that there were still many faint lines in the solar spectrum not identified that were probably due to elements like the rare earths which were not yet studied. These all had to be cleaned up before a revised table was justified. He was also keenly aware that much data of astrophysical interest lay beyond the ultraviolet cutoff of the earth's atmosphere. Although he lamented this fact, he was powerless to do anything about it, rationalizing that sufficient expertise was not available to handle the accessible regions of astrophysical spectra.[79]

Moore's relations with the Russells remained very close even after her marriage to the delightful and lighthearted Bancroft Sitterly, a Princeton graduate and astronomer at Wesleyan University. In the summer of 1938, Moore and "Banny," house-sitting for the Russells—who were in Salonika, Greece, to see their eldest daughter, Lucy, and her family—reported on new data flowing in from Bengt Edlén in Uppsala, Sweden. She wrote personally to the family and reported professionally to Russell, but Sitterly added commentary that poked a bit of fun at the enormity of the effort, requiring, he alleged, "[t]hree tons of paper, 1700 gallons of ink."[80] There was a lot of drudgery underneath Russell's accomplishments, and Moore assumed the burden, even after she left Princeton for Middletown.

By the end of the decade, so much new material had accumulated that Russell finally decided that a revision of the *Multiplet Table* was in order. Even though no formal report had been issued by the committee, Moore's work would go a long way to satisfy their initial goals. Adams agreed wholeheartedly, even though this meant more work for Moore. "You will," he told her, "have the satisfaction of knowing that few astrophysical monographs are published which are used to the extent that yours is."[81]

Russell always marveled at the power of Hund's theory; in his mind, the lifelong task of the committee was to demonstrate that every element ultimately succumbed to this philosopher's stone. Only around the mid-1940s did he feel a degree of closure. The most important lines in the solar spectrum had been heavily covered; now was the time to support Moore's revision. It was published as Moore resigned from Princeton and took a position with Meggers at the NBS. As with everything in her professional life, this decision was determined by Russell. He was soon to retire and agreed that the bureau was the best place for Moore to continue her work.[82]

Analyst Turned Accumulator

We have followed Russell's migration into spectroscopy, noting how it changed his feelings about observational completeness and sufficiency. Here for the first time he did not seek out shortcuts, nor was he interested in developing elaborate statistical schemes and extrapolations to exploit unripe data as he certainly had done earlier. Now we find Russell willing to devote decades to what was, to be sure, a repetitive and accumulative task; a single theme with infinite variation confirming Hund's Rule and exploiting it to accumulate data of astrophysical importance. Russell now sided with laboratory workers like Meggers, who dedicated themselves to exhaustive analyses and attention to detail.

In this sense, we find that Russell was far from averse to normative practice in science, when he saw a right and proper need for it. As we shall see in the next chapter, Russell had good reason to side with those who opted for exhaustive open-ended analysis: astrophysical applications (such as those arising from his use of multiplet theory to derive theoretical line intensities) required more than the hit-or-miss selections that the Germans preferred. But at a deeper level, Russell's approach to the accumulation of spectroscopic data reveals how his sense of proper practice changed as he confronted physics: he moved from being an avant garde nonconformist early in his career—from being an opportunistic "fox among the hedgehogs," to use Isaiah Berlin's expression—to being one who engaged in normative practice prudently accumulating data. In this regard, the type of puzzle solving that had started Russell out on his groundbreaking work in atomic structure and spectroscopy in the 1920s became something very different for him over time, once the theoretical framework was sufficiently in place. By the late 1920s physics had moved on, and Russell lacked the power or the drive to make intuitive leaps to break new ground. He no longer fit the image of the revolutionary scientist, according to one historian's description, of someone looking "for entirely new scientific innovations in his research area."[83] Now, Russell looked for completeness, not novelty.

Over the three decades of Russell's work in spectroscopy, physics changed in profound ways. In April 1924, Meggers could record that it was "[r]emarkable how spectroscopy has overbalanced the programs" at a meeting of the American Physical Society. In January 1926 he chuckled to himself that "Laporte [has come] back from New York. Physics entirely changed again!"[84] In Meggers's view, theorists like Laporte had to struggle with the new quantum theory "based upon matrixes." Meggers, however, could spend his time computing wave numbers in righteous ignorance of Heisenberg's new theory, because spectroscopy was the key to the code. As long as spectroscopy was king, Meggers was happy. But the number of papers published on laboratory spectroscopy in the *Physical Review* and the *Astrophysical Journal* peaked in the early 1930s.[85] By the end of the decade it was clear that the heyday of spectroscopy was coming to an end. Spectroscopy now had to compete with newer, hotter fields. Yet for Russell, Meggers, and other astronomical spectroscopists, there were still many important elements crying out for full analysis, like the doubly ionized metals and rare earths. By the late 1940s, Russell supported Meggers's efforts to "attract students away from cyclotrons and also cosmic rays and get somebody working on spectra." But he had little advice, other than his conviction that the game was " 'such good fun.' " In 1949 he turned once again to his crossword puzzle metaphor, as he did in 1935: where else could one have such good fun and then "publish the solution and call it research?"[86] Unlike physicists who justified their work as morally uplifting, Russell always described term analysis as "such good fun" when trying to encourage others to keep the data flowing. But it is quite clear that over time, Russell's message weakened.[87]

Russell's ability to keep large amounts of logically related data in his head most likely was a result of the way he approached problem solving. Donald Menzel, who followed Russell's lead in bringing quantum mechanics to bear on astrophysical problems, viewed Russell's approach to spectra as "more by intuition than by precise theory" in his work on the relative intensities of lines in multiplets and in his search for regularities in spectra. Moore maintained that Russell "always had a wonderful ability to get an overall view of a problem rather than to start from the smallest to the largest. He could always picture the problem as a whole." Naturally, if, as his contemporaries suggest, Russell had the ability to carry the entire picture in his head and to make linkages intuitively, the problems must have seemed more tractable to him than to those who labored to keep thousands of wavelengths and line designations straight on paper. Certainly Meggers's early recognition of Russell's powers fit Menzel's view. Over the years they all accepted his role as interpreter, par excellence.[88]

• CHAPTER 14 •

"A Reconnaissance of New Territory"

"[H]is word was law. If a piece of work received his imprimatur, it could be published; if not, it must be set aside and its author had a hard row to hoe."

—*Cecilia Payne-Gaposchkin*[1]

Old Ideas Die Hard

When Lyman Spitzer was a graduate student at Princeton in the late 1930s, the idea that hydrogen dominated in the universe was so entrenched that "it is difficult to visualize now a period when this truth was not realized."[2] Yet barely a decade earlier the possibility that hydrogen was the primary constituent of the universe was not a welcome thought at all. Even though hydrogen was the most persistent line feature in the spectra of the stars, and sometimes the most prominent, astronomers felt strongly that it could not be a major constituent of the stars.

The behavior of the hydrogen lines in stars was troublesome as well for Saha's theory. It was the second anomaly Saha uncovered, and was the most worrisome by far because it implied that a major reversal of long-held beliefs about the chemical composition of the universe was in order. The "hydrogen problem" reveals how Russell reacted to new knowledge that tested entrenched beliefs and challenged a theoretical framework that he believed supported his theory of stellar evolution. How he dealt with hydrogen demonstrates how he used his growing influence and authority to maintain that framework, certain that, somehow, the anomaly could be explained away in a manner that would do no damage. But it also provides another important clue as to how Russell framed arguments that were convincing and promoted consensus. Just as in the case of the Hertzsprung-Russell Diagram (chapter 9), where Hertzsprung may have discovered the relationship but Russell convinced the world of its reality, when he finally came to accept the true message of the anomaly, even though others had provided the evidence, it was Russell who forcibly and tenaciously marshaled the proof that hydrogen was the chief constituent in the atmospheres of the Sun and stars.[3]

The old view, established in the late nineteenth century, was that the composition of the Sun's atmosphere was not so different from that of the earth's crust. As spectroscopic technique improved, more terrestrial elements were found in the solar spectrum. Kirchhoff identified 16 earthly elements among the 600 lines he mapped in the solar spectrum; by 1891, Rowland detected some 36 elements within the thousands of lines on his photographic charts. Russell worked comfortably in this framework. In 1914, stimulated by the need to revise Young's *Manual*, he refined Rowland's lists using recently improved terrestrial abundances and analyses of meteoritic fragments. Using Rowland's own criteria, Russell concluded that elements weak in the solar spectrum were rare on Earth, and that there "is a general similarity in [the] relative order of the two." He felt that he had gone "to the verge of my field," but Russell had not tested new waters.[4] He agreed fully with the view, expressed by Alfred Fowler in 1918, that "as work proceeds, it becomes less and less probable that the Sun contains any elements which do not also enter into the composition of the Earth. It seems natural to infer that the composition of the Sun may be practically identical with that of the Earth."[5]

The Hydrogen Problem

Throughout late 1921 and early 1922, in correspondence and in print, Saha and Russell wondered why hydrogen was so persistent in stellar spectra. Russell hoped that when Saha's ionization theory was refined to account for a mixture of elements, and once better data were at hand for line intensities, hydrogen might be explained. Soon, others joined the search, like H. H. Plaskett and Anton Pannekoek. But they all worked within the same framework: that the relative abundances in the earth's crust reflected the solar composition.[6]

When Milne and Fowler showed that the pressures in stellar atmospheres were exceedingly low, orders of magnitude lower than everyone had assumed, Russell and others speculated that at such low pressures, hydrogen might behave strangely. He and K. T. Compton speculated in 1924 that under such conditions transition rules broke down, causing electrons to remain in excited states longer than they normally would. The lack of collisional de-excitation would promote a "metastable" condition that would overpopulate the first excited state of hydrogen, making transitions arising from that level more frequent and the Balmer series abnormally strong.[7] But Svein Rosseland, a Bohr student who was one of the first of many International Education Board fellows at Caltech in the 1920s, promptly shot their idea down. Metastability required even lower pressures than those found in stellar atmospheres, he argued; conditions might be right in the nebulae. Rosseland suggested instead that hydrogen was repelled electrostatically from the stellar interior and became overly concentrated in a star's outer layers.[8]

Rosseland's arguments held for the moment, but everyone knew that it was just another stopgap rationalization. It was very difficult to see how the Balmer series of hydrogen could be so strong; the lines were comparatively difficult to generate because of their high resonance potential. How could they persist even in the coolest stars? The answer soon came from Harvard, but it was the most difficult to accept.

Donald Menzel and Cecilia Payne

When Shapley became director of the Harvard College Observatory, he put Russell on his Visiting Committee and constantly sought his advice about everything.[9] Russell now took a strong proprietary interest in Harvard's health and started to think of its vast resources as his own. Accordingly Russell sent graduate students like Donald Menzel to Harvard to exploit the plate vaults for their theses.

Menzel was not one of Russell's favorites, but he was one of the most ambitious and capable students to graduate from Princeton. He latched onto Stewart, Compton, and modern physics as he acquired a full knowledge of astronomical spectroscopy and atomic physics in his third- and fourth-term courses in 1922 and 1923. He had studied physical chemistry as an undergraduate at the University of Denver and had been captivated by Saha's techniques. Russell sent Menzel to Harvard in early 1923 to gather new information about manganese multiplets. Menzel also examined the behavior of titanium and yttrium in Harvard spectra and was close to deriving an astrophysical estimate for titanium's ionization potential.[10] Working along lines Russell had defined, Menzel looked forward to exploiting the full potential of the Harvard plate vault the following summer as he submitted his thesis proposal to Shapley. His goal was to confirm Milne and Fowler's theoretical predictions with observational data from the stars, and he recalled that the "prospect was an exciting one."[11]

Another student harbored very much the same goals. Cecilia H. Payne was a young graduate from Cambridge University, where she had studied with Fowler, Eddington, and Milne. Arriving in September 1923, "I followed Milne's advice," she recalled, "and set out to make quantitative the qualitative information that was inherent in the Henry Draper system." When she arrived at Harvard, however, Shapley realized that she had come to do exactly what Russell wanted Menzel to do. Payne thought that she had been first on the scene; evidently she did not know that Menzel had been working in the area throughout the previous summer. The resulting tension was palpable; there was no chance at collaboration, only an atmosphere of hostility: "It was my first taste of professional jealousy, the struggle for priority. I had much to learn."[12]

Russell found a letter from Shapley waiting for him after he and his wife returned from a short trip to Oyster Bay for their fifteenth wedding anniversary, to revisit the site of their first meeting. Shapley outlined the problem and asked for advice on how to split up the work, "before anybody is imposed upon." Russell reminded Shapley that when they had spoken of the importance of following Milne and Fowler's ideas, "[n]o one seemed to be in sight to do it [so] I thought of Menzel." Laying the burden squarely upon Shapley, Russell stated that if he had been told about Payne's plans, the matter would have been simple: "I should have set Menzel at something else."[13]

Russell was concerned that Shapley was not taking care of business. Beyond the Menzel-Payne conflict lay the fact that in three years as director, Shapley had done nothing to strengthen the Harvard classification system (the Henry Draper Memorial) to account for the many spectroscopic peculiarities that the notation could not handle, such as Maury's c-characteristic. At the 1922 Rome International Astronomical Union, Russell urged that physicists like Bohr and Saha be added to the spectroscopy commission to help place the classification on a firm physical footing.[14] His plan required Shapley's active participation, but all he saw coming out of Harvard was more of the same old thing, which would become useless when astrophysics incorporated the latest advances in physics.[15] If Shapley could not orchestrate the work of two bright graduate students who promised to perform this kind of work, he was not paying attention to what was important. It was the director's job to "know at least as much as his subordinates" and to keep on top of them and their work, Russell warned. Above all, the director should be able to handle the "younger people." Russell challenged Shapley to straighten out his priorities:

> [H]ow much time [have] you had to read this recent spectroscopic stuff? It's awfully important; personally I think it is more important than encouraging business men's clubs to contribute, even to worthy causes, but then, you may find the Society of the Rotarians and the Kiwanis sufficiently restful to compensate for travelling and speechmaking. I should not.[16]

Russell's sharp rebuke reveals his priorities, which evidently were not Shapley's. The latter, unlike Russell, had made the heavy commitment to manage and maintain one of the largest and hungriest astronomical observatories on the planet. Shapley also lived in a larger world, one Russell avoided; as an institutional visionary, coordinator, planner, and later social activist on an international scale, Shapley relished a life that Russell respected but rejected for himself. "[T]hink where Shapley is and tremble," Russell advised Lyman Spitzer in 1946, a time when Shapley's political and social activities had diluted his effectiveness as observatory director.[17] For Russell, intellectual responsibilities always came first; from the beginning he knew that he had to keep close watch on Shapley. With Russell's approval, Shapley let Menzel continue to work on spectral lines immediately important to Russell, and Payne limited her work to giant stars that were important for elucidating Maury's c-characteristic. Menzel, however, was already considerably advanced in his work and completed his thesis, "A Study of

Line Intensities in Stellar Spectra," by the spring of 1924, about the time Payne matriculated at Radcliffe. As a result, Shapley's plan did little to prevent either student "from feeling imposed upon."[18]

Payne realized that she had to go farther in her analysis to make her own mark. By the time Menzel defended his thesis, she had already completed a detailed analysis of neutral and ionized silicon, using some two hundred Harvard spectra, and, as had Menzel, she compared theoretical to observed line visibility curves as a function of temperature. But to distinguish her work from Menzel's, knowing that Russell would be the final arbiter of her contributions, she recalibrated the temperature sequence for the Draper classes and provided refined values for ionization potentials. Shapley was initially excited about this work but, true to form, deferred to Russell for a critical opinion of her first manuscript.[19]

Distracted by Menzel's thesis defense, by commencement, and by his own publishing deadlines, Russell ignored Payne's manuscript until, sifting through his papers one morning, he realized "with horror" that it had been on his desk for almost two weeks, which, for Russell, was an eternity. "It looks to me like a first class piece of work," Russell hastily wrote Shapley in mid-June. He liked her derivation of ionization potentials but worried about the continuing odd behavior of hydrogen. As Menzel pointed out and as Payne had confirmed, the hydrogen lines were much too strong in the low-temperature giants, just where the theory predicted that they should be weak or nonexistent.[20]

Payne redoubled her efforts to work out the discordances and laid out an ambitious plan of work, which included an analysis of sunspot spectra to determine absolute abundances. Russell was planning to do the same thing and advised her to limit her work. He did not doubt her ability; he had just written an extremely complimentary and forceful recommendation for her to the NRC. But it was a huge and intricate problem that had intrigued theoretical and observational spectroscopists alike, and was too much for a thesis. Since Einstein's 1917 elucidation of transition probabilities, little had been done to relate the observed intensities of lines to the number of atoms involved in the process. Werner Heisenberg, Arnold Sommerfeld, and other theorists all were looking for a way to solve this critical question.[21]

In October and November 1924 Russell visited Harvard mainly to meet with Payne. To Payne, they were "two alarming, thrilling visits . . . both of which reduced me to a state of prostration—the second one had the same effect upon him!" Finally, he convinced her to concentrate on giant star spectra to determine relative abundances using Saha's wing-fitting method, following Fowler and Milne's suggestion that this would yield "fractional concentrations," the fraction of the atoms available that actually contributed to the intensity of a given line as a function of temperature. Assuming that all elements possessed the same transition probabilities, "in default of a suitable correction," she completed her analysis of the Draper sequence and with her recalibrated temperatures calculated relative abundances.[22]

As Payne completed her thesis chapters, Shapley sent them to Russell, who was impressed but worried over her results. Hydrogen and helium, she calculated, were orders of magnitude more abundant in stellar atmospheres than the rest of the elements she examined. Russell could not accept this. "It is clearly impossible that hydrogen should be a million times more abundant than the metals," he wrote in January. "There seems to be a real tendency for lines, for which both the ionization and excitation potentials are large, to be much stronger than the elementary theory would indicate."[23]

Payne had to make the changes Russell dictated, but she was crafty about it. As she poignantly recalled years later, Russell's influence was irresistible at Harvard. He was a "formidable figure, tall and lean, endlessly voluble, speaking with the voice of authority." For an impressionable young student, "his word was law. If a piece of work received his imprimatur, it could be published; if not, it must be set aside and its author had a hard row to hoe."[24] Her published discussion therefore highlighted not the discordances but the striking similarity between abundance ratios of the elements in stars and in the earth's crust. Indeed, she seemed to belabor this conclusion, drawing heavily upon Russell's ephemeral 1914 study. She made clear what her own results were for hydrogen and helium, but on Russell's and Compton's authority she concluded, "The stellar abundance deduced for these elements is improbably high, and is almost certainly not real."[25]

After a long dinner conversation in Washington during an AAAS meeting, just before Russell had read her thesis, Payne found that Russell could be charming as long as they limited conversation to poetry and ancient Rome. She could not fear such an educated and clever person, but, as she confided to a friend, "His power in the astronomical world is another matter, and I shall fear that to my dying day, as the fate of such as I could be sealed by him with a word." It therefore made good political sense to reject her own conclusions, but to publish them for all to see. This was a very effective tactic. Payne knew that hydrogen was not going to go away; Harry Plaskett had found similar results for ionized helium in the hottest stars but had differed with Payne in his assignment of temperature. If theory was wrong, then she was well advised to acquiesce. But if her abundances were right, then ultimately she would be vindicated. Russell accepted her thesis in April 1925.[26]

Stellar Atmospheres was well received when it appeared. Otto Struve, a young instructor at the Yerkes Observatory working for Frost, reviewed the book critically for the *Astrophysical Journal*, marking it as the latest study in this "newest branch of astrophysics." Mindful of criticisms of her work by Plaskett, Struve admitted that the subject was so unsettled that "it can hardly be expected to represent the views of all investigators." The matter of hydrogen never came up, however; at least it did not distract Struve's attention from what was her most important contribution: the volume was "full of useful suggestions for the practical worker. Nearly every page contains references to problems which are open to investigation by the spectroscopist."[27]

Struve's review vindicated Russell's advice. As Struve observed in his review, the field was yet young, largely untested, and was not robust enough to warrant radical conclusions. It was one thing for Fowler and Milne to show that atmospheric pressures in stars were far lower than previously believed, mainly because conclusions like these did not overthrow any cherished beliefs. It would have been quite another issue if Payne had pushed hard for her findings about hydrogen. She said just enough to offer a hint that something could be terribly amiss with the present picture, but did not go farther. Accordingly, thirty-five years later, with the benefit of hindsight, Otto Struve's opinion of her work increased immensely. In 1962 he hailed her monograph as "the most brilliant Ph.D. thesis ever written in astronomy."[28]

Why Hydrogen Was an Issue

Russell had good reason to reject Payne's hydrogen abundance. As Alfred Fowler pointed out in 1921, Eddington's ability to describe the interiors of the stars in radiative equilibrium produced "results emphatically favorable to Russell's theory." Russell, accordingly, would resist anything that weakened Eddington's models, which had one critical constraint: they had to possess very little hydrogen. They worked beautifully as long as the mean molecular weight in the star was near 2, which was fine since all elements in the hot stellar interior had to be almost completely ionized.[29] A significant amount of hydrogen would lower the mean molecular weight of the gas, as Eddington passionately argued in February 1923. His models were "practically independent of the chemical composition of the star," he added, and beautifully reproduced the observed range of mass for a star, "*provided it is not made of hydrogen*."[30]

Anyone suggesting a high hydrogen abundance in stellar interiors had to face Eddington. Payne recalls that when she discussed her original results for hydrogen with Eddington during a visit to Cambridge in September 1925, he shot back: " 'You don't mean *in* the stars, you mean *on* the stars.' "[31] For Eddington, hydrogen could be abundant in stellar atmospheres if Rosseland's explanation held, but this made little difference to Russell, who had to deal with the issue in the summer of 1925 when asked by the editors of *Nature* to write a review of the status of his theory of evolution. Eddington had just announced his theoretical mass-luminosity law, which showed that dwarf stars were also in the perfect gas state, contrary to Russell's original theory. No longer were the giant and dwarf sequences simple evolutionary paths, as Russell had envisioned; they were now representative of how gas spheres in radiative equilibrium distributed themselves in temperature and brightness, according to their masses. The only way to keep his theory alive was to find a way for stars to lose an appreciable amount of mass as they aged.

Such a way existed. At the time, there were two options for how to supplement the gravitational energy source of the stars to meet both astronomical and geological demands: element synthesis through proton fusion or electron-proton annihilation. Eddington, like Russell, entertained both possibilities at different times and even at the same time. In the early 1920s they preferred fusion, based upon the demonstration by Francis W. Aston in Cambridge that atomic masses were nearly integral (normalized to oxygen at 16), and by Rutherford's success at transforming nitrogen into oxygen.[32] If four hydrogen nuclei were somehow combined into one helium nucleus, Russell speculated, the conversion of the tiny amount of excess mass from the converted hydrogen (a proton's mass was 1.008, whereas helium's was 3.999) would provide an enormous amount of energy, "enough to heat 7400 tons of water from freezing to boiling."[33] By 1924, however, Eddington had turned to electron-proton annihilation, and it seems that he did so to preserve Russell's evolutionary scenario as well as his radiative models. If stars were powered by matter annihilation, their mass would diminish through the conversion of matter into energy: thus, Eddington argued, "diminishing brightness in the dwarf sequence is due to decreasing mass and not to a falling off of compressibility."[34] A star's atmospheric temperature could first rise and then fall, Eddington believed, although its internal temperature would be constantly on the rise. Russell embraced Eddington's astrophysical argument for matter annihilation. Although Russell's physical interpretation of the dwarf and giant sequences had to be altered, Eddington's mechanism let giants still become dwarfs.[35]

Russell's revised scenario depended more than ever on Eddington's models, and now upon matter annihilation as well. But it also made his theory more physical, because he was able to show, using Eddington's radiative theory, that all stars on the main sequence possessed the same central temperature. Russell's dwarf sequence beautifully fit Eddington's theoretical isotherm for thirty million degrees. He therefore had no incentive to doubt Eddington's requirements for mean molecular weight in a star. Indeed, in its modified form, Russell's evolutionary theory worked better than ever. As he concluded his *Nature* essay, "the difference of mass between giants and dwarfs is now explained, and the white dwarfs—formerly most puzzling—now, thanks to Eddington, find an orderly place at the end of the sequence." Russell never dealt directly with Eddington's demand for a low hydrogen abundance, but his own theory of evolution was still alive with such a framework in place.[36]

Facing the Hydrogen Anomaly

In the four years following Russell's rejection of Payne's abundances, time and again the hydrogen anomaly faced him, and he dodged. But just as he advised Payne to discard her findings, he also outlined for her the steps that had to be taken to address the issue. The chief problem was to find a way to relate the intensity of a line to the number of atoms responsible for producing that line. Schrödinger's new wave mechanics would soon make it possible to calculate transition probabilities. But without the new physics, Russell resorted to studying the intensities of lines in multiplets for clues, which put him into competition with European physicists again.

In 1925, sum rules were available to calculate the number of atoms required to produce relative line intensities for simple series relations, but as yet no one had been able to describe more complicated spectra. Sommerfeld, Heisenberg, and others were attracted to the problem, knowing that they had to determine the number of ways an electron might oscillate and therefore absorb energy. By the same reasoning that had led Russell and Saunders to apply the Rumpf model and Landé's vector model, they all knew that the solution lay in determining the permitted orientations between the total angular momentum vector of an atom and the angular momentum vector of its valence electrons. The race was on once again to the next step.[37]

Russell worked out the theory behind determining the relative intensities of lines in multiplets very quickly but, as with his work with Saunders, delayed publication until he was convinced that his theoretical line intensities agreed with A. S. King's spectra. Once again he was scooped. Sommerfeld and H. Hönl, and soon after them R. de L. Kronig, published the theory first. Russell came next in a hasty note to *Nature*, published in May 1925. Though the editors knew of the German work, once again they gave Russell credit for his care in confirming the theory with observation, and in subsequent literature Russell's contributions were listed in parallel with the Germans'.[38] Even so, this was only the first step. Absolute abundances required the formulation of a theory of intersystem combinations between different multiplets. Russell well knew that "the quantum theory, when properly applied, will clear up these questions." But in May, he also knew that "Sommerfeld and his students . . . are probably already deep into it."[39] By then, Russell was satisfied that the theory would be developed one way or another by the German theorists, and his best suit was to stick as close as possible to the astrophysical data.

Russell was wise to let go. He was already overstressed from pressures to complete his long-delayed revision of Young's *Manual*, a project that was then reaching a crisis point at Ginn and Company, Young's publishers. He also had numerous other astrophysical projects and was constantly distracted by term analysis. His family doctor was worried; Stanley Cobb warned him again, and most forcibly his wife insisted that he would break if he did not slow down. He had a perfect pretext for a respite: he had won the Bruce Medal of the Astronomical Society of the Pacific, and so he and Mrs. Russell had traveled west in February for the ceremony in San Francisco. They visited the Websters and then spent two months in Pasadena, escaping the winter, and ended up in Flagstaff for the official prescribed rest. As therapy, the Russells spent May and part of June exploring the backcountry with Lowell staff and then in July met the Websters in Boston and sailed to Nova Scotia aboard the Yarmouth boat. They spent the rest of the summer at Clark's Island and with Lucy's family in Southport. By 1925, Clark's Island was Russell's haven; he took considerable pains to secure an annual lease on a cottage there and took out membership in the Plymouth Yacht Club.[40]

Of course, Russell never stopped doing term analysis. He carried it around with him in his head at Clark's Island and wrote constantly to King, Meggers, and Moore about it as he traveled. This was another form of therapy; puzzles that he could control calmed his nerves. The term analysis was, however, merely a distraction now that cleared his head for his next attack on abundances. All through the summer, Moore was working away at Mount Wilson on another project for which his work on the intensities of lines in multiplets was a preamble: examining the strongest lines in the solar spectrum using the exquisite high-dispersion spectra that Adams and his staff had collected over the years from their unrivaled solar tower telescopes. Russell had set her to the task in late 1924 because of some very provocative results John Q. Stewart had found working on scattering theory back at Princeton.

Stewart and the "Winged Lines" in the Solar Spectrum

The strongest lines in the solar spectrum possessed broad penumbral extensions, or "wings" when their photographic density profiles were traced out on paper with a Koch microphotometer, a device available at Mount Wilson in the 1920s. Solar researchers wondered whether the wings indicated abundance, or pressure, or were scattering or turbulence effects. In 1922 Stewart, recently hired by Russell and Compton as assistant professor of astronomical physics, provided the first theoretical evidence that abundance was at play, but, after several years of trying, could not convince Russell that he was right.

John Quincy Stewart Jr. graduated from Princeton in 1915 with concentrations in physics and astronomy. He had been an honors student and after graduation was awarded a fellowship in physics, which he interrupted to take a commission in the Signal Officers' Reserve Corps in June 1917. Trowbridge sent him to Europe as chief instructor in the Sound Ranging School at Langres, France. Stewart resumed his studies after the war and completed his thesis in experimental physics in May 1919, taking a job with AT&T as an electrical engineer in the development and research department in New York City. He created simulations of human speech and explored mechanisms to enhance hearing, one of Bell's most cherished goals.[41]

After David Webster turned Princeton down, Trowbridge suggested Stewart, and after a series of interviews at the end of 1920 Russell offered Stewart $3,000 to return to Princeton, which was nearly Dugan's salary but well below industry standards. Stewart, a bachelor, could live free at the observatory and so agreed to come, and with his arrival in the summer of 1921 the department of astronomy took on new life. Stewart was the first addition to the faculty since 1906 and would be the only permanent hire to a faculty post Russell would make for another fifteen years.[42]

Among the first problems Stewart attacked, after completing his first semester of teaching, dealt with stellar atmospheres: determining the opacity of gases and the physical causes for line width. Using data from J. A. Anderson's exploding wire experiments at Mount Wilson, Stewart linked opacity, gas pressure, and line width, using classical scattering theory and a new quantum mechanical free-electron absorption mechanism. He could calculate line widths in terms of the number of atoms available to produce a line and concluded that broadening was a manifestation of high abundance. Russell was delighted, quipping to Adams in November 1922 that Stewart was "finding the subject broadening."[43]

As Stewart developed his theory over the next several years, he came under heavy attack from Milne and Rosseland. At Mount Wilson, Rosseland, following up work on opacities he had done while at Cambridge with Eddington, felt that Stewart's theory was unable to reproduce the opacity effects seen in Anderson's data, although he admitted that neither the theory nor the observations were in any sort of settled state.[44] Milne saw useful ideas emerging in Stewart's treatment but felt that he depended too much on old scattering theory and overemphasized the role of abundance. Writing to Russell about this in September 1925, Milne claimed, "[S]orry I am in opposition to J. Q. Stewart so much, but I cannot persuade myself that he is justified in his conclusion about line widths." Milne dismissed his work in a blunt "Council Note" in the *Monthly Notices*, declaring that Stewart's ideas were "mainly based on a misunderstanding."[45]

Milne never clarified the misunderstanding, however, though he rejected Stewart's manipulation of scattering theory. Nevertheless, Stewart managed to establish a foothold in the elusive realm linking atoms to stars. Unfortunately, neither Russell nor Compton was able to defend Stewart adequately in the matter, even though graduate students like Menzel did, who recalled how Stewart was an important influence upon his early training in astrophysics. Menzel found that his laboratory experiments agreed with Stewart's theory to within an order of magnitude, which gave them both faith that the classical theory was valid. Nevertheless, Menzel knew from Stewart and from his readings that "very few of the quantum physicists felt the same."[46]

Opacity was the big issue, and through 1925 Stewart and Russell worked around it, applying ionization theory to examine the influence of a temperature gradient upon line intensities. They explored the border between the photosphere and the "reversing layer" and realized that what had long been believed to be a finite layer was really a broad region; thus ranges of temperature and pressure were at play. As he studied the influence of temperature gradients, Stewart also explored how the quantum equivalent of Rayleigh scattering near resonance might account for the dependence of line width upon relative abundance. In all of this Stewart constantly drew attention to the influence of abundance on line width, finding support in Menzel's studies of Harvard spectra. Milne, however, remained skeptical, even though he liked a parallel study by Stewart and Russell—one that examined all possible sources of pressure broadening in the solar atmosphere—because it supported his discovery of extreme low pressure in the outer layers of the solar atmosphere. Of course, low pressure meant that line width could not be due to pressure broadening, and this result only raised the abundance question again. Stewart wanted to be clear that he was just laying out the physical groundwork and was not posing final solutions, but just as clearly he was arguing that abundances lay at the root of the matter.[47]

Faced with Rosseland and Milne's criticisms, Russell could not depend exclusively upon Stewart's theory, even though both Payne and Menzel supported him. As he would do in such cases, Russell turned to empirical data to look for correlations with other physical characteristics within the atmospheres of the stars.[48] In fact both Stewart and Milne called for more attention to careful analysis of line profiles; after one of his smaller skirmishes with Milne in the spring of 1924, Stewart suggested that Russell "start something in this direction at Mt. Wilson, if they haven't started already."[49] This is how Moore's study of the winged lines began.

In November, Russell asked Moore to examine the behavior of the strongest lines in the solar spectrum. He was in high spirits after the presidential election, elated that Coolidge had won (which reassured Russell that "American people are conservative—and still have a Puritan streak in them"), but sobered up quickly to be sure that Moore had everything she needed to get the large job done. She set about classifying the lines by character: their strengths, "haziness," and "sharpness." By February 1925 she found that the 138 most broadened (or "winged") lines came from only eleven elements. Russell was most excited by this finding, and though he was exhausted and was preparing to travel West to receive his Bruce Medal and then rest, he pushed even harder so that he could get to Mount Wilson as quickly as possible to have a look at her results. With all the tabulations in hand, and with a strong correlation between energy level and line strength acting as a guide, Russell instructed Moore to sum the intensities for all the widened lines of each element; the sum then became a measure of relative abundance. By June, Moore and Russell found a correlation between the energy levels that gave rise to these strongest lines and degree of widening. Most conspicuous was iron, which also possessed the widest lines in sunspot spectra. Thus they concluded that the most widened lines arose from the lowest-lying levels, which meant, Russell told Adams, that "except for the hydrogen lines, solar widening is entirely a question of abundance."[50]

Here was strong support for Stewart's theory, and Russell now defended Stewart. He also reaffirmed that the relative abundances of the elements in the solar atmosphere and in the earth's crust were alike. His conviction seemed unshakable. To make other elements fit, Russell argued—in the case of silicon, for example—that it appeared underabundant because it had a high excitation potential and had only one strong feature in the visible region of the spectrum. If he had applied this rationalization to hydrogen, however, with an even higher excitation potential and only five lines in the visible region of the spectrum, it would have become the most abundant of all. He did not do this, arguing that its strength was anomalous and possibly due to other causes, possibly the Stark Effect, anomalous Doppler broadening, or Rosseland's electrostatic repulsion scenario.[51]

Moore's work highlighted the hydrogen problem once again, but the weak link this time was not theoretical. Russell knew that one of the long-standing programs at Mount Wilson was to revise the venerable Rowland Atlas of the solar spectrum, which was based on visual estimates of intensities for thousands of lines to the limits of Rowland's photographic maps. Improving observed line intensities was the next step, Russell reminded Adams in September 1925: "I feel the need of it frequently."[52] Russell kept watch on the Rowland revision, headed by Charles St. John, making suggestions time and again for ways to improve their work, to speed it up, and also to add more astrophysical information. Adams had to mediate with the grizzled St. John, who disliked Russell's intrusions: "It seemed to me good policy to make it not too easy for Russell to go too far," St. John complained. Adams assured St. John that Russell's requests only meant more work for Moore, "but this is no great matter . . ."[53] If Russell wanted something done, Moore would do it, no matter the scale of the effort. Her presence not only gave the project a great boost but made Russell's suggestions tolerable.

Russell wanted to calibrate the new Rowland intensities in terms of the number of atoms involved, using his multiplet theory and Moore's work on the winged lines. By the summer of 1927 Russell realized that the task would be enormous: he had to make a number of simplifying assumptions that confined the calibration to small chunks of the spectrum.[54] Russell directed Moore to work from multiplet to multiplet, exploiting Russell's finding that the relative numbers of atoms in the lower level out of which each multiplet arose could be predicted from the quantum theory. Moore calculated the relative numbers of atoms involved in the production of each line and then compared each to the observed Rowland intensity that had been redetermined with a photoelectric densitometer by the Mount Wilson staff. She performed these calculations for over 1,200 lines in about 230 separate multiplets for some nine elements and ions to find a relationship between the number of atoms producing a line and its Rowland intensity.[55] Their calibration yielded a simple logarithmic curve that reduced to a set of calibration constants.

Russell brought Adams into the game to examine how the intensity of a line grew as the number of atoms available to produce that line grew, a rudimentary "curve of growth" analysis in modern terms.[56] Unlike later calibrations based upon transition probabilities, line profiles, and equivalent widths, their technique would be useful only for relative abundances. Even so it did not take long for Russell to again appreciate "the enormous differences in the number of atoms which are involved in the formation of the stronger and weaker Fraunhofer lines."[57] In fact, as they worked through the summer, Russell and Adams became so excited by this finding that they moved beyond the calibration to apply it to seven bright stars, using high-dispersion spectra from the Coudé focus of the 100-inch telescope.

Russell, however, had again overextended himself, and though he continually promised his wife and doctors that he would behave and slow down a bit, his pace reached successively higher and higher peaks of intensity. The completion of the textbook the previous spring unleashed an enormous amount of nervous energy that led to over a half dozen major studies including calculations of spectroscopic terms, three huge papers on the titanium spectrum, a new study of the spectra of Cepheid variable stars, and the first of several long papers on the spectra of the elements of the iron group, the last two completed within a day of each other in June in Pasadena, and all of them completed within a three-month period.

Exhausted, Russell left Adams, Moore, and the Mount Wilson staff working on the details of the Rowland calibration while he and his family moved on to Flagstaff for a long vacation. When they arrived home in late September 1927, Russell found an invitation from Eddington to prepare yet another article on stellar evolution for a revision of the *Encyclopaedia Britannica*, which Russell found very much to his liking.[58] Throughout the fall, therefore, Russell kept his ideas about stellar evolution very much in mind as he worked with Adams and Moore to complete the calibration and their applications of the calibration. Russell hoped that this work would highlight "several important things, such as the verification of Boltzmann's formula, the determination of the relative abundance of different elements in stars compared with the Sun, and the questions of pressure and temperature."[59]

Russell devised a differential method to relate the line strength ratio in two stellar spectra to the ratio of the number of atoms producing that line, its excitation potential and state of ionization, and the electron pressures and temperatures of both stars. When ratios were taken, transition probabilities dropped out and yielded some fascinating but also very perplexing results. Although most abundance ratios were in close agreement, there were a few glaring anomalies: calcium was some sixty times weaker in Betelgeuse than in the Sun; only one hydrogen atom in a billion was available to produce the Balmer series in the Sun; and this factor was eight orders of magnitude less in red giants like Betelgeuse! These vanishingly small ratios demanded, once again, either that hydrogen be invisible in the spectra of the Sun and stars, or that it be far and away the dominant element in stellar atmospheres.[60]

Once again, Russell and Adams refused to accept such a conclusion. Admitting that hydrogen's great intensity in the cooler stars and its enhancement in giants "have long been a puzzle" they treated these findings still as an anomaly, suggesting that the theory was somehow at fault. Russell speculated that deviations from thermodynamic equilibrium in giant stellar atmospheres could produce such spurious results, but no matter how he adjusted his equations and ratios, hydrogen always came out on top. Unable to address critically the theory itself, Russell and Adams called for refined line intensities and profiles. Adams committed Mount Wilson to the task. Dunham, recently arrived at Mount Wilson, watched Russell at work that summer and was much impressed with the final Rowland calibration. But by October, Russell realized that his summer

Plate 1. Alexander Gatherer Russell and Henry in 1878.

Plate 2. Eliza Hoxie Norris Russell and one of her sons, ca. early 1880s.

Plate 3. Russell and Aunt Ada Louise Norris, ca. 1910.

All images on this page and the next two pages are courtesy of Margaret Edmondson Olson.

Plate 4. Families gather for the christening of the twins on the side porch of 79 Alexander Street, ca. April 1911. *Left to right*: Ada Louise Norris, Russell, Alexander Gatherer Russell, John H. Cole, Lucy May (Cole) Russell, Josephine Hewson "Gammie" Cole, "Aunt" May Hill (Lucy Russell's maid of honor).

Plate 5. In late 1911, Russell's classmates celebrated the two new joys in Russell's life, highlighting this image in their Quindecennial Record.

Plate 6. The Russell family ca. 1916: Lucy May, Margaret, Henry, Elizabeth, Russell, Lucy.

Plate 7. The Russell brothers (Henry, Alex, and Gordon) gather for a Princeton reunion ca. 1920 with (*left to right*) Margaret, Lucy, and Alice Russell (Alex's daughter) joining in the revelry.

Plate 8. Charles A. Young in the classroom in the Students' Observatory on Prospect Street, much as Russell would have encountered him in the 1890s. Courtesy Princeton University Archives, Seeley Mudd Library.

Plate 9. The heavily fortified Halsted Observatory. Author's collection.

Plate 10. The Students' Observatory at the northeast edge of campus.
From a postcard, author's collection.

Plate 11. The Halsted Observatory at the southwest corner of campus, ca. 1906, surrounded by residence halls and Nassau Street, and near the smoke and vibrations of the railroad head. From a postcard, author's collection.

Plate 12. Russell and Dugan at the Observatory residence, Stewart on his bike, late 1930s. Photographs by Dorothy Davis Locanthi, courtesy E. Segrè Visual Archives, American Institute of Physics.

Plate 13. Russell and Svein Rosseland at the September 1941 AAS meeting at Yerkes Observatory. Frank Edmondson is at rear center. Photograph by Dorothy Davis Locanthi, courtesy E. Segrè Visual Archives, American Institute of Physics.

Plate 14. The Princeton Department of Astronomy, 1949. *Clockwise from left*: Stewart, E. L. Schatzman, Pierce, Schwarzschild, Russell, Spitzer. Orren Jackson Turner photograph, courtesy the Department of Astrophysical Sciences, Princeton.

work really made “the famous ‘hydrogen problem’ much more acute than ever. Perhaps we can get that cleared up some day,” he wrote Dunham. “I hope so.”[61]

Russell's speculation about deviations from thermodynamic equilibrium drew objections from Eddington and others. Such anomalies would make the hydrogen problem worse, not better, Eddington argued, and in any event departures from thermodynamic equilibrium were not necessary if, in fact, there was a deficiency of excited hydrogen atoms in the stellar atmosphere. Russell was intrigued by Eddington's suggestion and wanted to test it. Eddington agreed: “I am much more satisfied that the idea has got to the right quarter, where if it has any value, it will be discovered.”[62]

What was needed was a theory of opacity, which was no simple matter, Eddington reminded Russell. The best way to proceed was to look for hints in observations. Everyone knew that some deviation from local thermodynamic equilibrium had to exist. Otherwise, “there would be no absorption lines at all.” One complicating factor to which Eddington was now sensitive was that the old idea of a discrete reversing layer had to be modified, as Stewart had shown. “Thus it seems impossible,” Eddington added, “to predict even the simplest facts about stellar absorption lines without following the events through a considerable depth of the atmosphere.” Milne, Eddington knew, had a long discussion of the problem in press; Payne had told Russell of it when he had visited Harvard that fall. This was the great problem of the day among those interested in the structure of the atmospheres of the Sun and stars. But in Russell's mind, the important point established by Eddington was that deviations from local thermodynamic equilibrium could not account for the strength of the hydrogen lines.[63]

An Influx of Younger Theorists

Evidence was piling up. Stewart all along had been finding additional clues and feeding them to Russell: the lack of nonmetals in the solar spectrum was likely due to their high ionization potentials and to the fact that their ultimate lines were in the ultraviolet. Stewart also continued to use classical statistical mechanics to show that the number of neutral hydrogen atoms in the solar atmosphere was enormous. He announced his conclusion at an AAS meeting in late 1927.[64] Russell and Payne were there, but Russell paid close attention only when European theorists like Albrecht Unsöld started to say the same thing.

Unsöld completed his thesis on line profiles in the solar spectrum under Sommerfeld and in 1927 published his findings, which Russell believed had “some very good stuff on winged lines.”[65] Although his was similar in form to Stewart's treatment, Unsöld used Karl Schwarzschild's scattering model where energy transport was by radiative processes only. He also applied a new technique for calculating oscillator strengths, based upon a sophisticated quantum mechanical training that fully incorporated the new mechanics. With data at hand in Munich, Unsöld confirmed many of the abundances Payne had determined in 1924. First, hydrogen was extremely abundant; on the order of five million times more abundant than calcium. As had Payne, he called this a “seemingly meaningless” result that might be explained by Rosseland's electrostatic repulsion mechanism.[66] Although some young theorists like W. H. McCrea were convinced it was an abundance effect, Unsöld later recalled that "[a]ll the astrophysicists were quite confused; was it a matter of abundance, a matter of line broadening, or strange excitation conditions?"[67] Indeed, although Eddington applauded Unsöld's work, he felt sure that the time was not right to commit completely to his methods:

> The subject is too dangerous to tackle lightly, since it involves in a rather unusual degree a mixture of the classical atom, the Bohr atom, and the Schrödinger atom in their most (apparently) contradictory aspects. From those physicists whom I have consulted I have been unable to extract a coherent view.[68]

Despite Eddington's resistance, Russell was delighted that Unsöld had confirmed Stewart's work: "John will be glad," Russell wrote Dugan in March from Pasadena. "I wish John would publish more," Russell added. "He has very sound physical intuition." Stewart, however, had taken over the bulk of teaching and advising duties left by Russell and was then absorbed by another delegated task: the "Campaign for Princeton." He had little time for research, a fact Russell seemed not to appreciate.[69]

While Dugan and Stewart minded the store, Russell spent a good part of the summer of 1928 touring European observatories as he made his way to the International Astronomical Union General Assembly in Leiden. At the center of a large traveling party including the Hubbles, W. H. Wright, St. John, and others, Russell took care to listen to the latest findings from younger theorists.[70] At Leiden, he heard reports by Milne, Marcel Minnaert, and E. Finlay Freundlich on high-dispersion solar spectrophotometry. Minnaert, from Utrecht, had successfully derived intensities for the Balmer series using Schrödinger's methods, and Freundlich, from the Einstein Tower in Potsdam, reported on how Unsöld's theoretical line profiles fit his data.[71] All of these efforts, using the new mechanics but still limited by many theoretical and observational obstacles, were nevertheless showing that the relative order of abundances Payne had detected was the correct one.

"On the Composition of the Sun's Atmosphere"

Russell must have realized that Minnaert's and Unsöld's line profiles, which made use of the new transition probabilities just appearing, represented the way quantitative studies would proceed in the future. These methods were laborious, however, requiring highly detailed observations that were as yet available for only the strongest features in the solar spectrum. Russell therefore looked for a way to extend Unsöld's theory using approximate methods based upon the recent calibration of the Rowland map and his study with Moore of winged lines.

Upon completion of his studies Unsöld won an International Education Board fellowship to work at Mount Wilson to apply his theory to the line profiles in high-dispersion solar spectra from the tower telescopes. Russell told Eddington on the eve of his own departure for Pasadena in December 1928 that he intended to talk to Unsöld about Eddington's suggestions, and he hoped to write up a report on abundances. By the time Russell arrived, Unsöld was lecturing, and Russell took the time to attend as many sessions as he could. Of course Russell had long known of Unsöld's work, but now firsthand contact brought him back to hydrogen, and it is likely that, after he obtained Unsöld's lecture notes, he finally decided that there was no other course than to submit to the evidence.[72] Thus as he continued to fill his *Scientific American* columns with what was happening at Mount Wilson, including how the site for Hale's planned 200-inch telescope would be chosen, he also celebrated Unsöld's ability to calculate the number of atoms needed to produce a line using a "quantum theory of spectra."[73]

Russell later described his residence in Pasadena, from mid-December through the second week in February 1929, as an "orgy of work."[74] As usual Russell had too much on his plate. He collaborated with Ira S. Bowen to show that argon was not responsible for the coronal lines; he gave no fewer than five lectures at Mount Wilson's Astronomy Seminar and at Caltech's Astronomy and Physics Club; he also finished collaborative work with Meggers on yttrium and explored other elements with King.[75] By January, however, as he had promised Eddington, he was well along with his paper on abundances. He was still at it late in the month as he traveled to Berkeley to arrange for Moore's graduate program and visited Lick Observatory to see Menzel, who had further evidence.

In 1926 Menzel was hired by Lick Observatory and, at Russell's urging, focused on the enormous cache of eclipse spectra accumulating there, which had not been properly examined by anyone adequately trained in physics. Examining "flash" spectra from the instant of totality, Menzel unraveled the structure of the chromosphere and found that the mean molecular weight in the region of the reversing layer was far lower than expected. When Russell visited, Menzel showed him the results and recalls that Russell "became convinced of the correctness of one of my conclusions, that hydrogen was the dominant element of the solar atmosphere."[76]

If this was the process of conversion for Russell, starting with Payne and ending with Menzel, it certainly did not come out in the structure of the monumental paper he published in 1929, "On the Composition of the Sun's Atmosphere." Russell's rough draft manuscript did begin with astronomical arguments, but his published paper reversed course and started with physical arguments, wherein he identified the great "energy of binding" of hydrogen and how this was an inhibitor of its visibility in spectra. This alteration reveals Russell's method of persuasion: arguments based upon the structure of the atom were stronger than astronomical evidence.[77]

To make the case for hydrogen, Russell marshaled an exhaustive pile of evidence, physical and astrophysical, making the latter support the former. Now he showed how hydrogen's high abundance explained away all the anomalies. He deftly made everything fit into the new scenario, especially a new calibration of the Rowland scale using Unsöld's evidence from a few line profiles, which he now used to derive the relative abundances of some fifty-six elements in the solar atmosphere. Russell made various simplifying assumptions, but in the end he felt that the magnitude of his findings dwarfed any possible sources of error or oversimplification. "The conclusion from the 'face of returns,' " as he put it, was that "the great abundance of [hydrogen] can hardly be doubted."[78]

Only at the end of his seventy-one-page *ApJ* paper did Russell give full credit to Payne's 1925 conclusions, saying nothing of his original rejection of them. Now he showed how his results agreed with hers, which was "very gratifying" because she had used very different methods on her giant stars. A giant star evidently had an outer atmosphere of nearly pure hydrogen "with hardly more than a smell of metallic vapors in it."[79] In an addendum, "Astrophysical Considerations," Russell presented Menzel's and Eddington's arguments, revisited Unsöld's conclusions, and reevaluated his own work with Adams in the light of hydrogen's high abundance. Though it had been these episodes that brought Russell originally to his conversion, they were now presented as confirmations.

Russell's strategy placed physical theory at the core of astrophysical practice. Physical theory was not merely a convenient way to *explain* phenomena, it was a way to frame and guide the evaluation of those phenomena. In structuring his arguments in this way, Russell reaffirmed his 1917 commitment, expressed in "Some Problems in Sidereal Astronomy," to a fundamental shift in astrophysical practice away from open-ended astronomical observation toward well-defined programs informed by physical theory. His 1929 paper made astrophysical practice look more like physics.

The "Russell Mixture" and Its Aftermath

Russell's strategy did not satisfy everyone, but it was widely cited and had its intended effect. Menzel and Robert d'E. Atkinson incorporated his abundances into their work immediately. H. H. Plaskett felt it was the best one could do with outdated methods. L. S. Ornstein, reflecting upon how Russell's approximate methods could yield important results in his derivation of the intensities of lines in multiplets and their application to abundances, felt his methods were "marvelous but rather dangerous." And eight years later, talking about the value of approximate methods in astronomy, Russell knew that his 1929 "method of discussion [was] a tissue of approximations; nevertheless it has given results of great importance."[80]

Russell's results still differed from Unsöld's, and so the old worries about thermodynamic equilibrium and electrostatic repulsion mechanisms lingered.[81] This bothered Milne greatly, partly because Menzel, defending Russell's results for hydrogen, challenged Milne's recent study of the solar chromosphere that depended upon a higher mean molecular weight for the region. Milne only responded, "At present we may suspend judgment."[82] At a deeper level, however, Milne was becoming rather frustrated with the ad hoc flavor of astrophysics. In November 1929, as the newly installed Rouse Ball professor of mathematics at

Oxford, Milne penned an indignant letter to the *Observatory* complaining about how astrophysicists treated the interplay of theory and observation. He objected particularly to researchers' invoking "anomalous phenomena" to account for otherwise "unexplained *observations*" rather than searching for the answer rigorously in known physics. He did not address Russell's 1929 abundance paper directly but did criticize his 1928 study with Adams as an example of the "difficulties of a partial application of thermodynamic methods."[83] Explanations by recourse to anomalies (in this case departures from thermodynamic equilibrium) were, for Milne, no explanations at all: "[T]hough we must recognize anomalous *observations*, there is no need whatever for astrophysicists to accept anomalous *explanations*." Milne hoped that his fellow theoretical astrophysicists would see the light: "If theoretical astrophysics is to retain the respect of its parent sciences, it must have nothing to do with anomalous explanations."[84]

Milne had personal reasons for his frustration, no doubt owing to growing tensions with Eddington, who favored astrophysical opacities over physical ones. But he did put his finger on a significant tendency among astrophysicists to search for unusual explanations to explain unexpected observations. His essay resonates with Russell's conversion, though few, including Russell, refrained completely from invoking anomalous explanations when confronted with discordance between observation and theory. Milne's critique and Russell's rhetoric are important indicators that theoretical astrophysics was then in a state of rapid flux.

Within a very few years the hydrogen abundance Payne and Russell had established was confirmed both for stellar atmospheres and for stellar interiors, though the opacity problem was not solved until the late 1930s when Rupert Wildt demonstrated at Princeton, influenced by the physicist John Wheeler, that the presence of negative hydrogen ions (a hydrogen atom with two electrons, one very weakly bound) accounted for the bulk of the continuum opacity in solar-type stars. But as Russell had predicted, and as Stewart always knew, the established method of choice for abundance work soon came from Unsöld and Marcel Minnaert's use of line profiles.[85]

Even though Russell used outdated techniques, ad hoc assumptions, extrapolations, and approximations, the "Russell mixture" for the solar atmosphere remained a standard measure among astrophysicists for many decades, even though everyone knew it was provisional. In 1940 Bengt Strömgren noted that it was the "main source of information" on solar abundances. A decade later it was still the conventional choice for the heavy elements, and as late as 1965 Minnaert observed that Russell's "admittedly rough determinations . . . are still now, 35 years later, the only ones available for 17 out of the 64 elements, for which abundances have been determined in the Sun!"[86]

After his "orgy of work" at Mount Wilson and a still rather hectic spring, Russell was again seriously close to his mental and physical breaking point.[87] He was up against a personal deadline: his family had long planned a fifteen-month odyssey to England, Europe, and Egypt. The children were growing up and were about to enter college, and it would be the last opportunity they would have to be together for such an adventure. From the summertime and through most of

1930 they toured, at times in a manner reminiscent of Russell's own seven-month voyage with his parents in 1892. Russell desperately needed this break. Not only his research, he confided to Webster ca. 1929 but his professional duties seemed out of control.[88]

During his months of travel, reading sporadically in observatory libraries, Russell well knew that Milne, Unsöld, and Minnaert were now the leaders in quantitative stellar atmospheres. Russell never abandoned spectroscopy but limited his work to the analysis of complex spectra. He continued to read and comment on stellar atmospheres but knew that the wave front had passed him by. In 1941, preparing an address on the Sun's atmosphere, Russell read Unsöld's new textbook: "There is a good deal for me to pick up," he told Adams, "but I am enjoying it."[89]

Russell's contributions to quantitative stellar atmospheres and to term analysis were enormous, but his resistance to change and his failure to appreciate Payne's prescience or defend Stewart's theory mark him as a transition figure, one who held to traditional values not only in atomic physics but in social relationships. He was an authority in a world that venerated authority. He attracted good talent and usually made good use of it, but he could also ignore it or neglect it. His opposition to hydrogen demonstrates how he was able to control how others presented their work, and provides a poignant vision of the high degree of centralization in American astrophysics, which made it difficult for new ideas to emerge and be given a full airing.[90] The factors which ultimately brought him to accept hydrogen reveal, moreover, that America's chief astrophysical theorist was only too willing to let his views be governed by Eddington's models. Russell listened more to European theorists than to domestic ones, which weakened the chances that a strong domestic theoretical infrastructure would soon appear.[91]

Russell as Theorist

To an orthodox theorist like Milne, Russell's use of outdated techniques, ad hoc assumptions, extrapolations, and approximations in his stellar atmosphere and interior modellng was questionable, though Milne envied his freedom of action (see chapter 16). Ornstein regarded his techniques as "marvelous but rather dangerous," as we know, but the question remains, what was it in Russell that allowed him this freedom of action? We can find hints in some of his musings about quantum physics and reality, and as before we find them in his writings for a religious audience.

Late one Sunday evening in October 1935 Russell prepared a short essay for the Clergy Association of New York on how the events of the past generation in science could be likened to the age of the great navigators and their first reconnaissance of the globe. Responding to a general unease that relativity, indeterminacy, and the probabilistic laws of modern quantum theory challenged the basic postulates of Christian thought and the ability to perceive God directly, Russell elucidated his view of scientific practice and the limits of perception.[92]

Echoing James Jeans, Russell explained that the great lesson of modern science was that "we are not in touch with *ultimate* reality." Indeed, with each "new set of images" of reality we might feel we "understand more, and predict more, about Nature than the last; but, from the earliest to the latest, they are all images constructed in our minds and the reality behind them eludes us."[93] Jeans did not worry that reality was elusive, Russell implied; he lived in his own world of mathematical elegance.

Russell, however, preferred to joust with the physical universe; his basic religious nature was compatible with his science, as both recognized limits and never allowed him to expect ultimate victory. Thus there was no law that forbade the scientist to employ outdated models either, if he found them useful in getting closer to the truth. The mathematical physicists might not want to do this, but, ever the pragmatist, Russell felt he could: "[I]n periods of rapid advance, [one could] use two different models at the same time, trusting to some inspiration of genius in the future to invent a new (and probably still more intricate) super-model which shall include the good points of both." For most practical purposes, the earlier models explained phenomena just fine, Russell added, "those which match with immediate experience." After all, "the theoretical foundations of physics may shift radically without affecting in the least the practical applications of cog-wheels or of screws."[94]

Practical application was paramount. At the opening of George Harrison's spectroscopic laboratory at MIT in 1932, Russell celebrated the triumphs of the old quantum theory to explain atoms. He admitted that the hardly "visualizable wave-mechanics or the wholly unpicturable matrix theory" had pushed the boundaries back, but they removed the neatly mechanistic formalisms that characterized Bohr's original model.[95] Anyone who loved eclipsing binaries as Russell did cherished the idea of definite orbits and orientations, so there was some sting to his words as he described how the new mechanics visualized atoms:

> When a modern lecturer tries to draw an atom on a blackboard, he uses not chalk, but an eraser, and constructs a smudge illustrating the relative probability of finding a unit charge in different regions. But as a means of calculation,—interpreting and, on occasion, predicting, the results of precise observation, the new theory advances from conquest to conquest.[96]

And those conquests were immensely useful. He could chuckle with David Webster over the oft-repeated jest, attributed to William Lawrence Bragg, that "God ran electromagnetics on Monday, Wednesday and Friday by wave theory, and the devil ran it on Tuesday, Thursday and Saturday by quantum theory."[97] Even if the new mechanics was incomplete or unfathomable, even though it lacked a "completely satisfactory" theory of the photon, Russell, writing his "Spiritual Autobiography," believed that "we do not have to wait for such a theory to use effectively the knowledge that we have. Within a specifiable range, the particle image represents the facts satisfactorily; within another range, the wave image does so; and both images have been satisfactorily applied to practical problems."

For Russell, and for many laboratory spectroscopists like him, the mechanical atom worked just fine; whether or not it existed was another matter.[98]

Russell shared an attitude toward theory that was common among American laboratory spectroscopists, who agreed with Albert C. Candler's 1937 view that "the physicist asks for something he can touch and see, or as he often says, for something he can understand. This the vector model gives him."[99] There was also a shared concern, as we have seen, that the practice of laboratory spectroscopy be accessible to a wider range of workers because it was so labor-intensive. R. C. Johnson noted in the introduction to his 1946 textbook on spectroscopy that "[t]he experimental physicist and chemist inevitably cling to this wreckage lest they be drowned in the mathematical sea."[100]

In his mature years as chair of the NRC Committee on Line Spectra, Russell can be located squarely within this framework. He thought no longer of breaking new ground but of covering known territory, as he told Webster in 1942; there was much "mopping up that needs to be *completely* done."[101] Much of Russell's later spectroscopic work was indeed, "mopping up." The projects he chose and the conceptual tools he used to attack them, his reluctance to prematurely release Moore's *Multiplet Tables*, his adoption of Meggers's zeal for exhaustive analysis, and his lifelong devotion to that goal all place him firmly within the realm of the classical spectroscopist.

• CHAPTER 15 •

Princeton Astronomy in the 1920s

Russell's Priorities

The out-of-control feeling Russell expressed to Webster in 1929 came from his insatiable appetite for research, not from his administrative duties. Even though he felt that his professional duties were onerous, in fact he had limited them as much as anyone in his position could do. He was the director of an observatory, but Dugan tended to the details; he was a leader of a profession, although he held few offices and no editorships in the 1920s; and yet he was a person with considerable power and influence, made possible by the close-knit hierarchical nature of the astronomical community in the United States.[1] His priorities for Princeton astronomy, in light of his need to maintain his freedom to travel to Mount Wilson whenever he wanted, bear this out. These priorities reveal Russell as atypical among directors of leading American observatories, such as Hale, Campbell, or Schlesinger. His influence was based much less upon the resources he directed as an administrator than upon the resources he commanded intellectually.

Russell always felt that the Princeton University Observatory should not grow. During a series of public lectures in Toronto in the winter of 1924, where he lectured on "the applications of modern physics to astronomy," Russell was asked by a campus reporter whether Toronto should build a great observatory. Russell bluntly said no: "As an astronomer I would not like to see large amounts of money spent on expanding the equipment to a great plant at either Toronto or Princeton, because the climatic conditions do not warrant it." No amount of equipment, however small or large, could compete with the West if housed under poor climatic conditions. Support for research equipment was also useless without "enough assistance given to the man in charge." Russell would give such unwelcome advice elsewhere, but he imposed this view most strenuously on his home campus, not only in the 1920s but, as we shall see in chapter 22, right up to his retirement. This worked for Russell because Mount Wilson satisfied his needs.[2]

After the war Russell never asked the Princeton administration and alumni to support more than the renovation and move of the Halsted Observatory and the restoration of observatory property lost after Lovett's departure. Even the renovation was not a high priority when Princeton, like many major universities, harbored ambitions for expansion. In 1919, Princeton mounted a $20 million fund drive and asked its senior professors to participate by stating their hopes and dreams. Russell, however, along with other Princeton scientists, felt that the campaign had little to do with their priorities.[3] His call to Harvard netted a promise from Hibben to raise an endowment adequate to maintain the observatory and to renovate it someday, and Russell was quite happy at first with the arrangement.[4] But when the endowment failed to materialize, Russell did join his colleagues, including biologists Edwin G. Conklin and E. Newton Harvey, physicist Karl T. Compton, physical chemist Hugh S. Taylor, and his old mentor Henry Fine, to campaign for a separate "Princeton Research Endowment Drive."[5]

The theme of the drive was cooperative scientific effort. Taking advantage of a new climate that fostered, in Herbert Hoover's words, "cooperative individualism" free from governmental control, in 1925 Princeton scientists asked the Rockefeller Foundation's General Education Board for $3.5 million to strengthen existing interdepartmental programs. Russell and Compton called for an expanded program in spectroscopy and atomic structure.[6] The GEB approved $1 million *if* Princeton could raise $2 million as part of its general drive. Hibben acquiesced, releasing his scientists to raise their own funds.[7] Russell's research was indeed a model of interdisciplinary cooperation. Princeton astronomy, unlike its counterparts at Mount Wilson, Lick, and even Harvard and Yale, was very much part of campus life and was closely bound to the physics department through Russell's interests and Stewart's joint appointment. Through the 1920s and 1930s Russell's graduate students tended to spend more time in the Palmer Physical Laboratory than on Prospect Street.[8]

Astronomy's share would be some $300,000, sufficient to renovate the Halsted Observatory and provide other amenities, like an endowed professorship. Russell curried favor with patrons like Daniel Barringer, who was trying to unearth what he hoped was a huge meteoritic fragment buried under a vast crater in northern Arizona. If Barringer struck it rich, mining the iron, nickel, and precious metals, Russell reminded him, he should not forget his alma mater.[9] What Russell would not do was provide Princeton fund-raisers with a statement on the need for a great observatory facility—for anything more than a renovation of the Halsted. Ignoring the historical fact that people gave money for institutions and monuments first, what Princeton staff needed, Russell argued, was more time for research, which was best executed through faculty exchanges with western observatories. Even after repeated pleas by the executive secretary in charge of the fund, Russell would not budge. Above all else he valued the freedom of action his Carnegie connection afforded. Time and again in following years, he reminded astronomers like Joel Stebbins, Otto Struve, Menzel, and Shapley that research came first, organizations last. Why would anyone want to build an empire when he could visit one (Mount Wilson, Harvard, Lick, or Lowell) and reap the benefits? Russell would not go farther than telling Stewart to write a statement about the needs of the present observatory and about cooperative research for the fund drive. Stewart, as the junior faculty member, also handled the details and attended endless meetings.[10]

Shapley, in fact, more tuned in to the importance of institutional health and vitality, campaigned for Princeton more than Russell did, but he did so within boundaries set by Russell. "Russell's contagious enthusiasm and his universal interest have made him indispensable to American astronomers," Shapley claimed. American wealth had built the world's greatest observatories; "[i]t seems, however, that we have more equipment and ambition than brains—we are stronger on observing than interpreting. That is where Russell's generous thinking on all sorts of astronomical research problems comes in so usefully."[11] Shapley's observation was indeed correct, but he well knew that benefactors were more likely to provide bricks and mortar than anything else, and shared the common view that first-class facilities attracted first-class talent. When he campaigned for Harvard, of course, it was for improved equipment and new observing facilities. But Russell did not care for such things and was not interested in managing a larger world. He thought only in terms of what worked best for himself.

Shapley's praise for Russell was more than matched by rhetoric in Princeton's own proposal to the GEB, which celebrated Russell as "the leading authority in this country in astrophysics." There was little question that Russell was a major jewel in Princeton's crown, and, indeed, in American astronomy, based on contemporary surveys.[12] Thus Russell also had little incentive to build a new empire when he had more than one at his beck and call.

Despite Russell's intransigence, Princeton found the money and the GEB made good their commitment in 1928.[13] For all his foot-dragging, Russell did speak at some alumni dinners and wrote letters, and his classmates responded by endowing "a research professorship in astronomy" for $200,000, which, with an additional $100,000 from the GEB, was sufficient to raise Russell's salary to $10,000 per year.[14] The combined support from the GEB and alumni allowed many departments to grow, but astronomy, true to Russell's vision, remained static. Russell, however, achieved his goal: true independence of action. Benefiting coincidentally from a call to become secretary of the Smithsonian Institution in 1927, Russell received his research professorship chair and was freed from all formal teaching duties.[15]

Although his endowed chair released funds for operations and could have provided a new slot, Russell kept a tight hand on expenses, maintaining the status quo. He rarely demanded anything of the Princeton Research Fund. Even when Dugan asked for an additional computer in 1930, which was denied by the Research Fund Committee, Russell didn't lift a finger.[16] Three funds available to him, including the general university fund, the Research Fund, and departmental funds accumulating from the original Archibald D. Russell Fund, covered salaries and operations and secured the department's future, which was, above all, Russell's goal. All of these funds could accumulate too, so Princeton astronomy, by remaining small, suffered no cutbacks during the Great Depression; the unexpended funds from the accounts held for the department were sufficient, in fact, to finally renovate the Halsted and move it to the edge of campus in 1933.[17]

Russell always knew that the Halsted had to be upgraded and moved. But he was savvy enough to know that astronomy needed visibility on campus. Writing from Pasadena in March 1925, Russell advised Dugan that if the Halsted were ever moved, "we must hang on like grim death to the Prospect Avenue site and the house."[18] Russell did his part by writing a steady stream of articles for the *Princeton Alumni Weekly*. He was reluctant to write on relativity but welcomed any chance to write on astronomical matters that engaged the public, like solar eclipses.[19]

The department received more than the chair, which was ultimately named in Young's honor. Russell's second demand was for faculty exchanges. He wanted Dugan to benefit in the way he had at Mount Wilson. Dugan opted for Lowell Observatory, where he spent the winter term in 1929 so that a Lowell staff member, Carl Lampland, could be in residence at Princeton. This was another form of independence of action. At Lowell, Dugan, in theory at least, would have access to far clearer skies than at Princeton, and he would be free of teaching and administration for the duration. Russell would have liked to make this a continuing option for Dugan but did nothing to make it happen.

Russell's Carnegie connections as well as his continuing prominence at the National Research Council gave him further independence. By 1930, in addition to the freedom financial autonomy provided, Russell enjoyed considerable prominence in the astronomical community, now long established through a growing stream of awards and honors. The rewards were already flowing in 1923 when J. S. Plaskett remarked, "It is getting to be an every day affair."[20] By then Russell was a frequent reviewer for the *Astrophysical Journal*, and invitations to speak to both public and professional audiences were rolling in faster than ever, no doubt stimulated by his *Scientific American* articles, which by the 1920s had taken on the character of a monthly review of the astronomical profession. Russell had, in effect, established a public forum within which he could interpret recent events in both astronomical worlds, that on Earth and that in space.

"Russell, Dugan, and Stewart"

If anything exhausted Russell, and Dugan and Stewart, it was their collaborative textbook ("RDS"), which had started out as Russell's revision of Young's *Manual* and became the only department project that engaged the entire faculty. Russell had provided an updated version in 1912 and was well along with a complete revision of *General Astronomy* when the war intervened. He returned to the project in November 1921 only when the publisher, Ginn, pressured him. But Russell was not happy with what Ginn wanted, which was another marketable *Manual. General Astronomy*, he thought, was by far the more important work and needed a complete revision. Russell, wanting a text to train astrophysicists, expanded topics in solar and stellar astrophysics to the point where he suggested a split into two volumes,"Solar System" and "Stars." After Russell obtained endorsements from Brown, Shapley, Frost, Campbell, and Hale for a new two-volume text, Ginn compromised by authorizing a longer text in a single volume. To sweeten the deal, Ginn agreed to raise the royalty from 3 to 8 percent if the manuscript were delivered promptly.[21]

As Russell worked away on the text among his many other projects, each new chapter raised issues that became distractions which sometimes bloomed into short papers or long abstracts for talks. The textbook framed Russell's most extensive excursion in public lecturing: his set of fourteen public lectures delivered during a three-week period at the University of Toronto in February 1924. There he celebrated the successes of both radiation theory and atomic theory and laid out the nature of his science and defined its boundaries. Although he did not shy away from discussing cosmology—just months before Hubble's revelations Russell favored van Maanen's evidence that spiral nebulae were nearby objects—he identified the limits of astrophysical practice as the pursuit of how the Sun, solar system, stars, and galaxies "got into their present shape," and avoided deeper questions of "how they got there." This was a fundamental limit for Russell, who argued that "[i]t is not the business of science to explain how matter got into space; only supposing there was matter in space, what would be the history of it."[22] Russell said as much to his students during his religious talks in Dodge Hall and in the "Sunday Seminars" he gave in later years.

Russell had promised the full revision within a year or two, but even after Toronto he made little real progress. Ginn's C. H. Thurber started to pester him in the spring. The publisher had already advertised the revision and, Thurber reported, "it was hardly too much to say that the whole country was waiting on you." It seemed as though the whole profession was on hold; other professors contemplating textbooks were hesitant, knowing of Russell's plans: "All united in saying that you are the one man to do it and it is perhaps not too much to say that the whole development of astronomy teaching in the country right now seems to be waiting for you."[23]

Only when Thurber threatened to sever the agreement did Russell take action. In the midst of all his commitments and constant travels, he therefore decided to recruit Dugan and Stewart as coauthors to accelerate the process. Thurber was delighted. So the book became a department project; Dugan ruefully noted in the 23-inch telescope logbook, "Working on Young's Manual," to explain a ten-month break in observations in 1924 and 1925. For Dugan, this was a major sacrifice. Throughout all of this, Russell constantly complained, claiming that his own research suffered, a contention not borne out by his rate of publication. "Job was certainly right when he wanted his *enemy* to write a book," Russell cried to Adams as he searched for photographs. "It is an awful chore, and I have hardly had time to think of anything else this Fall." By December 1924 they had completed 45 percent of the rough manuscript. After many rounds of negotiations with Ginn, in which Russell demonstrated a keen knowledge of binding legal contracts and copyright constraints, the disposition of estates and of responsibilities for revising the book in the future, the preliminary contract was signed in November 1925. One month later Russell reported that fifteen of the twenty-five chapters were finished, though he was completely rewriting the last chapters on stellar structure and evolution, just as he had had to do a year earlier for the chapter on nebulae and the distances to spirals, because the subjects were in rapid flux owing to Hubble's observations.[24]

Dugan and Stewart took care of copyediting details while Russell worried over the frontispiece, choosing a colored plate of the lunar landscape by Howard Russell Butler. He liked the treatment, and Butler was a potential donor to Princeton astronomy. Stewart lost his summer proofreading the manuscript and had a "punch-list" of things to do a mile long. Such drudgery made Stewart restless, and Dugan became annoyed when promotional material from Ginn mentioned only Russell as author. "His reputation is so widespread that there is no point in calling attention to it," Dugan complained to Ginn, confident that Russell would agree. There were also surprises about the ultimate length of the book. Dugan and Stewart had assumed that Ginn would estimate the published length as soon as they received the manuscript, but apparently they rushed the book into production without checking. Russell had agreed to a seven-hundred-page book, but by late May 1926 Ginn calculated that it was running to a thousand pages. Thurber snapped, "The book will weigh so much that anyone who wants to carry it around will have to have a special truck." He demanded that the later chapters be cut by at least 150 to 200 pages, or that the book be split into two volumes.[25]

Dugan did not have to ask Russell's advice. The chapters at risk were at the heart of Russell's own vision of astrophysics and were the latest statement of his theory of evolution. Dugan therefore took Thurber's second suggestion, Russell's original idea, and split the book into two volumes. The first volume covered a first-term course on solar system astronomy, and the second a course in elementary astrophysics. He suggested a new title, *Astronomy*, to distinguish it from its progenitor. Stewart also defended their work, arguing that the pace of astronomy was now quickening like physics and chemistry, and textbooks soon went out of date. But, he argued, increased demands from ever-increasing audiences more than justified Ginn's investment in the larger, more comprehensive work.[26]

Stewart's loyalty flagged during his lost summer. At one point he suggested to Thurber that they take 100 to 150 of the best illustrations they had gathered and prepare a picture book that would sell to a wider popular audience. He knew this was heresy and soon retreated by writing enthusiastic copy to advertise the virtues of the two-volume set, especially how it would provide for a "complete and thoroughgoing course." To cut down on length, the publisher set in smaller type many paragraphs containing speculation or commentary, so the book could be read on two levels.[27]

Ginn wanted volume 1 ready for an early September 1926 publication, but try as he might, Stewart knew that was impossible, since Russell, writing from Clark's Island in August, was still making revisions to their discussion of the earth's interior, based upon some new work by Jeffreys within the past few months. But Ginn had already targeted the fall market, sending the first sections of volume 1 to teachers, free of charge, promising that the full volume would be issued by mid-October.[28] With errors still cropping up in the plates, and with Russell reversing many decisions, however, there seemed to be no end in sight. By the end of September, with Russell soon due back, a multitude of details defied closure. Russell rejected the publisher's rendition of Butler's lunar scene but acquiesced when Butler approved it. Letters turned into night letters and telegrams toward the end of September as the project reached fever pitch. Russell, Dugan, Stewart, their secretary Henrietta Young, and some shanghaied graduate students worked full-time on the book, reaching the index stage by early October and submitting the index within days. As volume 1 went to press, Russell started in on volume 2, with Dugan and Stewart following suit as soon as they cleaned up the first volume's loose ends.

At one point, Ginn asked if they might use as a cover illustration a woodcut of the new large reflector under construction at Ohio Wesleyan University. Russell strongly objected to this: "How much scientific publicity the University will deserve will depend primarily not on its *possession* of a great telescope, but upon the *use* which is made of it." Russell had initially felt that such a large telescope was a waste at a Midwestern site where the weather was marginal at best, and was not interested in supporting what was, to him, an unproven venture.[29] An image of Hale's 100-inch adorned the front cover of volume 1.

Russell was even more sensitive about volume 2: "It would be the only text that covers some very interesting astrophysical matters; and while the rapid growth of knowledge will doubtless make certain parts obsolete very soon, much of the material will remain standard for years to come." Indeed, volume 2 would become the first college textbook, in English, to deal with astrophysics using physics as a point of departure, if Russell could just stop fiddling with it.[30] Through the end of 1926 and well into 1927 Russell, Dugan, and Stewart kept up a maddening pace. Russell overextended himself once again, since he would not sacrifice his own ongoing research. Their spirits picked up when they received their first royalty checks for advance sales of volume 1. Russell was delighted with his share of the profit, which amounted to $33.70. Since volume 1 sold for $2.48, this meant some 450 copies had been sold based upon the 122-page advance copy that had been distributed free.[31]

By early February advance copies of volume 2 arrived at Princeton, and the trio tracked down errors while Ginn distributed advance sets to colleges and a few major reviewers for endorsements. Ginn orchestrated the chorus: in late 1926 the publisher listed over three dozen universities and colleges where the set had been adopted, implying that the list was far longer. Statements from four journal reviews and a letter from Herman Zanstra attested to the fact that it was not only "the most complete treatise on astronomy in the English language"

(*Scientific American*) but was the first textbook in English to provide the recent results of astrophysics based upon "modern physical theories" (Zanstra). By the end of January 1928, Ginn reported brisk sales, and Russell found a check in the mail for $324.64, which meant that at least two thousand copies of the set had been purchased.[32]

Volume 2 of *Astronomy* was the first astronomical textbook in English to present both observational and theoretical astrophysics with enough mathematical detail to provide the serious student and researcher a basic understanding of the physical processes underlying astrophysical phenomena. One astronomer hailed it as "undoubtedly one of the greatest boons to the teachers of astronomy."[33] It was more accessible than Eddington's *Internal Constitution of the Stars* as an introduction to theoretical astrophysics and was readily accepted by those who were in the business of training astronomers. C. A. Chant opined, "In no book with which the writer is acquainted is to be found such a comprehensive, though condensed, treatment of not the New Astronomy, but the New Astrophysics."[34]

But RDS suffered as a popular introduction. Over nine hundred pages in two volumes, it entailed as prerequisite a year of college mathematics and physics, which made it unsuitable for an elementary course. Whereas volume 1 offered little challenge to most teachers, as it was similar to Young's *Manual*, volume 2 was wholly recast and demanded more of its readers. Stewart's fears were realized. Although the set was very popular with serious students and offered the best introduction to modern astrophysics available, it could not compete with popular one-volume texts, such as R. H. Baker's *Astronomy*, which appeared in 1930. In March 1931 Russell reported that all three authors felt they could not improve on Baker's excellent text; Russell regretted that RDS had not been more profitable to Ginn, but in his own mind, they had met a real need of the profession.[35]

RDS proved to be an intellectual success. In January 1928 Russell was delighted to report that one of his suggestions in volume 2, "that the nebular lines may be emitted only in gas of very low density" under the condition of metastability, led Ira Bowen to identify the nature of the chief nebular lines.[36] The book became one of the most accessible introductions to the problems of modern astrophysics for astronomers who trained in the thirty years following its publication.[37] The advanced texts by Unsöld and Rosseland that appeared in the 1930s were too mathematical for the average American astronomer to fathom. Though RDS was adequate training for the average college-level instructor or observing astronomer, it was hardly adequate to train competitive astrophysicists, observational or theoretical, by the late 1930s.[38]

RDS in many ways also defined the department of astronomy at Princeton and at times drew the staff and students together. Lyman Spitzer, Ray Lyttleton, Green, Dunham, Moore, and others poignantly recall many informal hours sitting around the table in the library in the old observatory discussing issues raised in both volumes that needed revision or further elucidation. Those graduate students around in the late 1930s particularly remember long sessions where Russell, Dugan, and Stewart sat with them, contemplating the state of astronomy and its future, trying to find new accessible ways to teach what modern astrophysics was all about.[39]

Russell in Action at Princeton and Mount Wilson

Throughout the 1920s and to a lesser extent after he obtained the research professorship, Russell continued to lecture, but it was on his own terms, without much regard for the needs of his students. "His course in astrophysics we found somewhat frustrating," Menzel remembered, mainly because he taught it early in the morning, which was hard for those who had observed the previous night. And Russell would often be late to class; commonly after about a half hour someone would poke his or her head into the room to announce that Mrs. Russell had just called to say that her husband was on his way by bicycle. "Puffing and panting" and without pause, Russell would rush into the classroom and start his lecture, sometimes with notes, more frequently not, drawing from the "voluminous store of knowledge within his brilliant brain." Russell would continue his lectures without regard to the clock and often overran the allotted time, making life difficult for anyone who had classes after his. "But for the rest of us it was a true experience to have this informal, often bewildering exhibition of knowledge. We arranged our other classes accordingly."[40] Fifteen years later Spitzer, as a graduate student, experienced the same insensitive passions: "[H]e would get carried away and talk for hours. He was just supposed to talk for 45 minutes. He would go on for an hour and a half, sometimes more."[41] Russell could get equally excited about least squares fitting procedures, binary stars, stellar atmospheres, and stellar masses. It all depended upon what was on his mind at the moment.

Russell was never at rest while lecturing. Constantly pacing, waving his arms, Russell frequently collided with objects at the front of the room. Menzel recalled how he often fought with a chair: "Russell would stumble over it, climb on it, declaim for a minute or so from the summit, then descend and continue his pendulum-like swings, stumbling over the chair perhaps a dozen times until he thought of removing it."[42] Far from being a show, such antics were a physical release for Russell, who could drive himself to intellectual exhaustion.

"There was always something weirdly exciting about his lectures," Menzel remembered. Russell mainly talked about his own research and would sometimes take the time to think through his own problems in front of the class. He would "often stop to derive a formula from fundamental principles right before us. He let us see the inner operation of his mental processes, how to select research problems, how to analyze them and how to carry out research." Russell did not take on artificial affectations while lecturing his more advanced seminars.

His research informed his monologue, which often hinged on what new scrap of physics or mathematics he was using.[43]

Watching Russell in action, either at research or teaching, could be bewildering to anyone outside Princeton. In 1929, in residence for advanced training as part of Russell's plan of faculty trades, the fifty-six-year-old Carl Lampland from Lowell was a fish out of water, but he marveled at what he saw:

> If you would see Russell in full stride you will have to come here. I never saw anything like it—the way he turns off work is positively amazing. I have remarked to you that he does the work of several men, and he really does. He is constantly being consulted on difficult problems and he always seems to have everything ready at hand to help out. I never did see anyone so generous with his time and energy towards all who come to him for advice and help. And one must remember that he is carrying a very heavy research program of his own, and giving a series of lectures to graduate students and members of the faculty. His lectures this term are on stellar atmospheres, and it is heavy stuff. One should have much time for study to get the real benefit of such a course, as you can readily understand if you glance through Eddington on the internal constitution of stars. Several lectures have been given outside of the University. The other evening he gave the address at the Sigma Xi club at Swarthmore and the following evening he gave another lecture before some society in New York. Just before that I heard him speak at the Graduate School on the constitution of the sun's atmosphere, one of the finest addresses I have ever heard him give, and that is saying a lot. About that time the Philosophical Club of the University invited him to speak on Eddington's last book [*Nature of the Physical Universe*], and as usual he was there full of the subject giving us a most delightful and instructive talk. I have just read the University Weekly Bulletin for next week and see that for the Physics Colloquium next Thursday Professor H. N. Russell is to give "A Survey of the Ionization Potentials of the Elements" and that is one place where they seem to feed on heavy stuff! At any rate, all the sessions I have attended have been highly technical, but a very lively interest seems to be characteristic of this group of keen research workers. I should judge it is well for the leader to be loaded and upon his toes. The first part of next month Russell goes to the University of Illinois to deliver lectures and I understand he will be there about a week. Russell's correspondence is, of course, voluminous and like everything that comes his way, it is attended to with promptness and thoroughness. He works so hard all the time that I do not see how he stands up under it. I do hope he will ease up a bit.[44]

Russell's doctors were also telling him to "ease up a bit" after he had nearly wrecked himself working madly that previous winter on the abundances of the elements in the solar atmosphere. But Russell could not stop, not at Princeton, nor even at Clark's Island, and especially not at Mount Wilson.

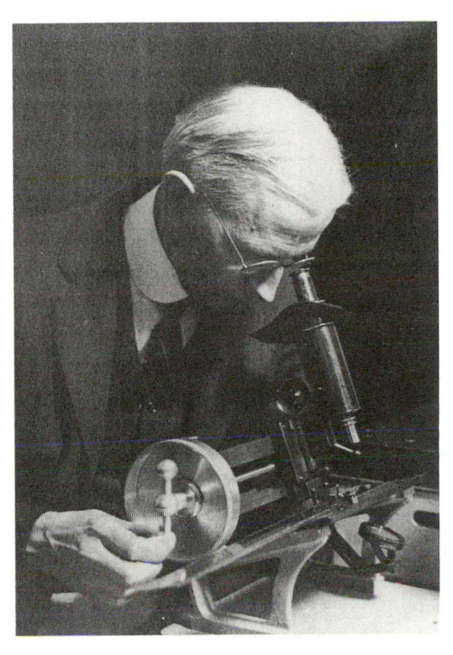

Fig. 15.1. Inspecting spectra at Mount Wilson, ca. late 1920s. Photograph by Margaret Harwood, courtesy Huntington Library.

Among Observers

Theodore Dunham's experience as a graduate student, and later as a staff member at Mount Wilson, illuminates Russell's activity at Mount Wilson. As a Russell student, Dunham found Mount Wilson rather timid about physics. In fact, as time went on, he found that his approach to research there was that of a physicist. He found few on the staff of like mind; they did not think of their work in terms of what physics could do. At Mount Wilson, analysis went only as far as description and some interpretation, "but not too far in the physical direction . . . It was Russell who got this to happen."[45] What Dunham remembered most vividly was how Russell moved among the observers:

> [H]e'd go rushing around with his yellow pads, and people would come in to ask questions, and he'd suggest what would be a wonderful thing if only they could get the data, and then they'd work it out together. He'd talk with them a great deal . . . he really revitalized the place.[46]

Russell's ability to draw out the observers and get them involved in his ideas impressed Dunham immensely. For Dunham, it was not enough to be an excellent observer or to have top-notch equipment. There also had to be ideas. Russell was filled with ideas; he imparted them to his students, to be sure, but was never so full of them as when he was at Mount Wilson, faced with the glorious potential of those fantastic instruments.[47] Dunham keenly recalls that Russell would frequently visit him at the Coudé focus of the 100-inch. It was most like a physics laboratory, and thinking of its potential would drive Russell wild. One night, perched on a chair on the platform next to the slit of the Coudé opposite Dunham so the two could talk in the darkness, Russell forgot where he was. As he talked about spectra, he shuffled his chair back to the edge of the platform until "the back legs [of the chair] went off and he went off." Russell fell about three feet and the back of his head hit iron steps. "So I rushed over to help pick him up. He said, 'Oh, that's all right, don't bother—what about that nebula now?' He really got a blow on the head."[48]

Russell's enthusiasms could sometimes be exasperating: "he just got more ideas every ten or fifteen seconds," Dunham recalled. All this could be stimulating but also "nerve wracking to the observer, who's supposed to make the machine work right, and think, 'what's next?' And all the clouds coming up, and thinking 'what can we hope to get?' " Russell did not have to worry about such practical details or maintain the mental as well as physical concentration that observing with large instruments required. He did not have to worry about strategy, contingency planning, or instrument and observer performance. Russell could concentrate on the intellectual problems, and so "he did most of the talking."[49] Although he may have been an acknowledged intellectual leader, he was hardly willing or able to direct programs that took into account the human needs and limitations of staff.

Russell's talking was not confined to Mount Wilson. Martin Schwarzschild recalls getting similar treatment when Russell found him working at his desk at the Harvard College Observatory in the late 1930s. The experience was indelible: "[H]e understood very fast and asked very fast questions [which gave me] just an enormous moral[e] boost, because when he was enthusiastic you couldn't help but catch it."[50]

Intellectual stimulation was certainly the positive aspect of Russell's presence. Hale found Russell "an amazing fellow" but knew that he had to be watched closely, for his own good. "I have to hold him down to prevent over-excitement in work. One of the best devices is to feed him novels, but if these are too exciting they are likely to be taken too seriously, or read into the small hours when he should be asleep."[51] Adams and King were awed by Russell's energy as well but also found his boundless enthusiasms enervating. Early on, King reported to R. T. Birge that "[s]ummer is not a particularly quiet time here, in fact it never is. Henry Norris Russell is here with his nerves rampant. He really should learn to smoke a pipe. However, it is great fun to talk with him. He has thought over astrophysics so thoroughly."[52] Russell's ideas created needs, and needs created diversions, especially when he was interested in what King was doing; his nonstop banter as he leaned over King's furnace data distracted King from the job at hand. His highly preoccupied walking and talking tours at the Bureau of Standards, Harvard, and Mount Wilson were often helpful, but just as often annoying. Inspecting a spectrum or a photograph, he would make sweeping conclusions on the spot and suggest ways to test them. Sometimes he would stop and provide an impromptu lecture on atomic physics and spectroscopy, but most of his time at Mount Wilson was spent pouring over data and writing, usually in his hotel room where he could concentrate, undistracted by the wonders around him. By Russell's third visit to Pasadena, he had made a convert of King, who enthusiastically wrote Hale about Russell's activities:

> Henry Norris Russell arrived, "sailed in high," and (to continue the disrespectful comparison) with plenty of oil in his crankcase. The talking became a solo and continued unabated during his stay. He gave us three or four talks a week on spectral series applied to atomic structure . . . Most of his time for a while before he left was devoted to working out the titanium series, according to the selections made in the furnace classifications. The complexity proved greater than he expected, and he is still at the job, but the fundamental sorting out was made and it became clear what the character of the multiplets is. Between the furnace and the sun-spot lists, there is now getting to be a pretty satisfactory amount of material for the selection of series groups, and it looks as if the way to go at the many-lines spectra has been found. It is a large subject in which only the barest beginning has been made.[53]

King's observations contrast starkly with Lampland's because Russell had entered King's world, whereas Lampland was entering Russell's. Although he would have preferred more time to himself, and possibly fewer lectures, King wanted to share his data with Russell because he was delighted to have his data exploited. Of course, he expected credit for his labors.[54]

There was a downside to a Russell visit, and more than one staff member at Mount Wilson experienced it, as Dunham and King well knew. Beyond the distractions and insensitivity to human limitations was the question of proprietorship. Adams harbored deep reservations at first, as he confided to Charles Greeley Abbot of the Smithsonian in August 1921:

> We have had a strenuous summer with Russell who has just gone. He is a tremendously able fellow but I sometimes think that the man who secures none of the results upon which theories have to be based has rather an unfair advantage. It is possible to think of many schemes in the time it takes to secure and study a few stellar or solar spectra. However, he was full of suggestions and we enjoyed his visit while he himself had a great time.[55]

Adams had a strong sense of ownership and, common to the breed, harbored some antipathy toward opportunistic theorists. Charles St. John in particular, trained as a physicist, older than the rest, and responsible for the spectroscopic work, chafed the most. In addition to his observing routine, St. John also had numerous programmatic duties, such as revising the Rowland Atlas, which had occupied much of his time and energy since 1919. He was then also searching for relativistic line shifts in solar spectra and was not free to take direct advantage of Saha's discoveries, even though he had a notion to do so.[56]

St. John therefore did not look kindly on Russell's presence at Mount Wilson; at the least he must have found Russell's freedom of action frustrating. Naturally,

Russell became interested in St. John's efforts toward revising the Rowland scale and was not reluctant to suggest how it could be done better. What was originally planned as a refinement based upon photographic intensities became an overhaul, as Russell showed how the recalibration could be linked to King's temperature classifications, and how those in turn could be reinterpreted in terms of the atomic structure of the elements.[57] Even though St. John balked at some of Russell's demands, he also knew that Russell would be his most important patron in getting the *Revised Rowland Atlas* accepted as the new standard. On more than one occasion Russell used his influence as reviewer for spectroscopy papers submitted to the *Astrophysical Journal* to demand that the author abandon the old Rowland designations in favor of the Mount Wilson revision.[58]

The Mount Wilson astronomers certainly respected Russell and followed his lead in designing new research programs in spectroscopy. But Hale's staff was never fully comfortable having Russell around too much, mainly because of his hyperactive nature. Gossiping with Abbot in May 1923, Adams quipped, "Russell is here and will remain until July 1st. We sometimes feel as if we were 'standing under Niagara' when he is in full swing."[59] Russell demanded attention and spoke so rapidly during his lectures that people were sometimes hard put to follow him. Just being with Russell could be exhausting. St. John probably said it best in 1928 when he learned that they would be on the same ocean liner traveling to the IAU meetings in Leiden: "Russell is going on the same boat. I hope we have not been by chance assigned the same cabin!"[60]

Russell's hyperactive nature aside, it was his physical insight and general influence over spectroscopic matters that were of value to Adams's staff. Adams was soon won over as he and Russell collaborated throughout the decade on the analysis of stellar spectra. And King too found in Russell an enormously valuable political ally when he sought new outlets for his spectroscopic data and found that he required new collaborations with spectroscopists on the East Coast, connections that eventually determined the professional fortunes of his son Robert B. King. And Hale also found in Russell a strong political ally in his never-ending quest for ever-greater telescopic power and the theoretical expertise to interpret what was found.

The agenda created by Hale and carried out by Russell proved to be extremely fruitful in bringing much positive attention to Mount Wilson spectroscopy at a time when the Rockefeller International Education Board had made it possible for Europe's best young theorists to take extended fellowships in the United States.[61] Throughout the 1920s, as Bohr's and Sommerfeld's students began showing up eager to apply the newest versions of the quantum theory to the stars, they usually found Russell in residence, ready and willing to work with them at the interface of theory and observation. He helped to explain their ideas to Mount Wilson staff and acted as catalyst and provocateur. But he was also a major beneficiary since he knew best where the relevant data lay waiting in the plate stacks, and had the political clout, through Hale and Adams, to see that it was gathered from the telescopes. One can well imagine that in the presence of young theorists like Zanstra, Rosseland, and Unsöld, ideas came thick and fast to Russell.

Russell introduced the new physics to Mount Wilson staff just as he taught it at Princeton. He applied Catalán's theory of multiplet structure to identify additional lines in the solar spectrum, helping King, Babcock, St. John, and others learn the method even before it was published. He soon acquired a taste for the analysis of line spectra itself and applied it to the study of the structure of atoms, leading his Princeton students and Mount Wilson staff to new ways to exploit theory and, in some cases, to contribute to it. As a result, in the first decade of Russell's tenure as Carnegie research associate, the centroid of publications of the spectroscopic staff, including Russell's own contributions, migrated slowly, from purely empirical studies in the laboratory and at the telescope, to observational studies informed by physical theory, to a growing number of theoretical studies at middecade by Russell, Babcock, and the European theorists in residence.[62]

Just as Saha's breakthrough had transformed Russell's scientific life, Russell extended the transformation to Princeton, Mount Wilson, and his NRC network of spectroscopic workers. In many ways, Russell managed to extend Princeton's astronomical borders to encompass the nation's spectroscopic centers, especially to Mount Wilson. His frequent visits and lectures, and having Charlotte Moore in residence, and later with Theodore Dunham on the staff, made Mount Wilson accessible as a resource for Princeton. Thus Russell's no-growth policy for Princeton was not a no-growth policy for Princeton research. The arrangement Hale orchestrated and that Russell executed freed Russell to do what he did best. As a result, he managed to influence a significant portion of the spectroscopic research at the world's greatest astrophysical observatory and in effect parlayed Princeton's resources clear to the West Coast.

• CHAPTER 16 •

Stellar Evolution

"The problem of stellar evolution is not one of natural history, but of physics."

—*Henry Norris Russell, 1942*[1]

AFTER Russell won the Royal Astronomical Society Gold Medal in 1921 for his work on stellar evolution, he became the chief elucidator of the problem, contributing reviews to *Nature, Scientific American, The American Yearbook,* and *The Encyclopedia Britannica*. He devoted two full chapters to the subject of stellar constitution and evolution in his textbook and invented ingenious ad hoc remedies to extend his theory's life, trying to preserve the view that giants turned into dwarfs. Although most of Russell's energies in the 1920s had turned to spectroscopy, he was continually faced with defending his giant-to-dwarf theory of stellar evolution and the physical meaning of what was variously called the Russell Diagram, the Hertzsprung-Russell Diagram, or, as Russell called it in his textbook, the "relation between the absolute magnitudes and spectra of the stars."[2] The challenge, as we saw in chapter 14, lay in reinterpreting the observed facts in light of Eddington's mass-luminosity relationship. Although he eventually abandoned his theory, he did not abstain from talking about how stars are structured and live their lives. As Russell moved from center stage to the sidelines, while younger theorists schooled in modern physics came on the scene, and the world of theoretical astrophysics changed around him, he not only remained a major gatekeeper in the field but continued to campaign for a specific mode of practice that had proven so successful in his own experience.

When Eddington announced his mass-luminosity relation in 1924, confirming theoretically that these two quantities defining the appearance of a star bore a definite relation to one another, Harold Jeffreys declared the "sudden death of the giant and dwarf theory." If both giants and dwarfs adhered to this relation, they both had to be in the perfect gas state. Therefore both giants and dwarfs would heat upon contraction. But Russell, responding to a reader who inquired, only mimicked Mark Twain: "[R]eports of its death are very greatly exaggerated."[3] Nevertheless, Russell knew that he faced a complex challenge. The major theorists were at war: Eddington and Jeans hardly agreed on anything dealing with stellar structure, and if one of them agreed with Russell, the other would attack. If workers in the specialty could not be made to agree, there would be no hope for consensus on any theory of stellar evolution.

The issue as Jeans saw it was the stability of a star containing a radioactive source. Central to his problem was the nature of that new source, how that source consumed matter, and how it was influenced by a star's structure and temperature. Jeans envisioned a star as a liquid body, which could not "automatically adjust its radius to correspond to any assigned rate of generation of energy within it." This was physically absurd to Russell, who felt that his and Eddington's gaseous models allowed for such adjustments and made radiative equilibrium stable. Appealing to thermodynamics, Russell showed how "the assumption that [the rate of energy generation] increases rapidly with increasing temperature gives a very good representation of the observed facts." In October 1925, far from taking the role of a disinterested mediator, Russell asserted to Eddington that "you and I are right, and Jeans wrong." Eddington and Jeans reacted accordingly.[4]

Jeans believed that if the radioactive source was dependent upon temperature raised to the 2.5 power or greater, this would lead to instabilities so severe that the star would break apart. According to Jeans, Russell's stars could not be stable and remain in the perfect gas state. As Jeans put it in 1925, "If my stellar matter is compared to uranium, Russell's must be compared to gunpowder."[5] Russell knew that there were grave problems: he could not explain the peculiar gaps that Hertzsprung and others had found in the distribution of stars on the diagram, something that had concerned Eddington. Nor could he account for what appeared to be age discordances for stars in binary systems and clusters. Russell mused prophetically, "[I]t may well be that our views of the life-history of a star are destined in the future, as in the recent past, themselves to undergo rapid evolution, and perhaps mutation."[6]

Russell tried to answer Jeans's criticisms, feeling that he "was wrong in his attempt to explode my latest views."[7] But there was little common ground. He was never reticent to admit, of course, that he favored Eddington's views; in the wake of the appearance of Eddington's magisterial *Internal Constitution of the Stars*, Russell felt that "[t]he theory of the internal constitution of the stars, and, in particular, of their luminosity, appear to be at present in a pretty satisfactory state, but ideas regarding their evolution are in a state of chaos."[8] And even after he modified his own theory to accommodate Eddington's criticisms, he admitted in 1928 that "[t]he path of this idea is strewn with the wreckage of abandoned theories," and that the field was at an impasse.[9] Eddington agreed. "I do not think it is too blunt an expression to say that this [theory] is now overthrown," he stated in his book; "at least it has been gutted, and it remains to be seen whether the empty shell is still standing."[10] The frame may have been "gutted," but for Russell it was still standing. No matter how stars were structured, he wrote in 1927, "the distinction of stars of rising and falling temperature has not disappeared, but remains a probable . . . consequence of the new postulates."[11]

After a spring at Mount Wilson and a full recuperative summer spent in Flagstaff with his family, Russell arrived home in late September 1927, armed with an answer to the challenge of the mass-luminosity relation. The situation was "not quite so bad as it looked a year ago," Russell told Eddington. He now had a full answer, an expansion of hasty thoughts he had cobbled together for the last chapter of his textbook's second volume. Russell apologized for using his textbook in this manner, but "the book had to be finished and so I put in all that I had on the subject, though of course in a sketchy fashion."[12]

Russell reexamined the properties of homologous gas spheres to explore the physical meaning of the dwarf sequence, what Eddington had relabeled the "main series," which soon became known as the "main sequence" to reflect the fact that the vast majority of stars were dwarfs. Given a set of homologous stars of the same chemical composition, but a range of mass, how do these stars sort themselves out on the diagram? The stars, Russell found, arranged themselves very neatly, reproducing the dwarf sequence quite faithfully. Eddington's mass-luminosity relation taught him that if the mass and chemical composition for a set of homologous gas spheres were fixed, there were also exact relations connecting mass and radius, luminosity and effective temperature, and, he realized, between "any other pair of their macroscopic properties." Russell believed that this simple relationship indicated that stars acted as a one-parameter family, a fundamental clue to the way stars structure themselves.[13] Russell had come upon the idea as he revised his stellar evolution chapters and regretted later that he "not only buried it in a textbook but camouflaged its grave by 'avoiding the use of the calculus.' "[14] As it turned out, Hermann Vogt had the same idea and published it in 1926. Milne knew this, but it was Vogt who complained to Russell. In 1931 Russell told Milne that Vogt deserved the credit and was "sensitive on the point—and a Nazi to boot!—I want to be scrupulously correct."[15]

The "Vogt-Russell" theorem, as it has come to be known despite Russell's scruples, allowed one to explore how the observed properties of a star changed with mass and composition. It was quickly applied by Milne and Eddington because of its generality and fulfilled Russell's goals completely as an approximate means to study how the structures of stars behaved in the aggregate. Russell always knew that "it would be wrong to conclude that it must be true in all its details."[16] Nevertheless, he used the relationship to revise his theory of stellar evolution to examine the relationship between luminosity and spectral class.

This was a fact of observation, as was the fact that the higher the mass or luminosity of stars, the rarer they were in space. Once again he collected the observed facts into a rational framework, defined by the diagram, to visualize how stars distributed themselves according to mass and to the amount of active material available. Using an idea he had earlier developed for a review in *Nature*, Russell envisioned stars containing two forms of active matter. The course of a star's life was marked by periods of rapid gravitational contraction interspersed with longer periods of stability when one of two types of active matter was being consumed. Drawing evolutionary paths for stars possessing equal amounts of "giant-stuff" and "dwarf-stuff" on a theoretical spectrum-luminosity plane, Russell created a series of curves that allowed him to map out the history of a star. He found that, in general, stars concentrated on the diagram just in those places where active matter was being consumed: "All this is in excellent agreement with the observed facts."[17]

Russell elaborated on his modified theory in an *Encyclopaedia Britannica* entry in December 1927 using plots of star clusters that Robert J. Trumpler at Lick had recently published. They were just what one would expect for a "great quantity of matter" broken into "masses of various sizes which started independent careers as stars." Russell admitted that his energy-generating mechanisms were "frankly invented *ad hoc*, and the best that can be claimed for them is that they account for most of the facts." But they gave him a new set of variables with which he could interpret the evidence. This was the best one could do until physics provided real answers.[18] As he noted privately to Menzel, "The whole theory of stellar evolution is still in a prodigious state of flux." Given the many unknowns aside from the stellar energy problem, such as the great discrepancy between "astrophysical" and "physical" sources of opacity, and the ages of stars, it was quite premature to attempt any bold elaborations on theory, as Jeans continued to do in his defense of his "liquid stars." "My own belief is that it is still most profitable to work on simple theories," Russell argued, and he hoped to have the time, once his spectroscopic work was over, to give this matter his undivided attention.[19]

Milne did not think much of Russell's "giant" and "dwarf" stuff, although R. H. Baker emphasized the heuristic value of Russell's theorizing in his popular 1930 textbook, *Astronomy*: "This is not intended to be a physical picture of what is going on. It is the attempt to find variables enough to represent the observational data, without defining precisely, for the present, what some of the variables mean . . ." Baker agreed with Milne's assessment that nothing was really known about how much of a star's mass was available as active matter, or what the process entailed. Nevertheless, "Russell's 'stuff' expresses the present situation admirably." Finding the exact mechanism that powered the stars "is for the future to solve."[20]

These reactions highlight the major conceptual difficulties at hand, especially the limitations of known physics. But they also demonstrate the heuristic value of Russell's methods. Russell's "giant" and "dwarf" stuff "created a lasting impression that different stellar types might be powered by different processes using different fuels."[21] Here again we see Russell's intuition at work. Gathering in all available evidence and applying theory he could handle, Russell approached stellar evolution as he did the hydrogen problem, where all evidence pointed to a single conclusion. But stellar evolution refused to yield. Once the composition question was settled, however, it made people stop and take stock of notions regarding stellar constitution.

TIMESCALES

As Russell completed his solar abundance paper and headed off with his family for Europe and Egypt for their fifteen-month tour of cathedrals, mountain scenery, and the Egyptian Valley of the Kings, he left behind the realization that hydrogen was dominant in the atmospheres of stars. While he was away, another factor emerged that helped to narrow the playing field. Stellar lifetimes had up to now ranged in the billions to trillions of years, based upon a bewildering array of astronomical clues. The longer ranges fit in comfortably with the time required for stars to lose appreciable mass through matter annihilation, a scenario Russell and Eddington favored. But with Edwin Hubble's marshaling of evidence linking the red shifts of galaxies to their distances, which most theorists interpreted as evidence for an expanding universe, the age of the universe itself, calculated from the observed expansion rate, restricted the timescale to billions of years. At about the same time, physicists were looking for specific mechanisms of nuclear synthesis based upon the fusion of hydrogen nuclei to form the heavier elements. Thus as Russell toured the European world, matter annihilation was slowly losing favor as the mechanism driving stellar evolution. This meant that age effects could no longer be interpreted in terms of mass loss.[22]

The new framework created by hydrogen's dominance also revived the problem of stellar opacities and with it Eddington's theoretical interpretation of the mass-luminosity relation. Thus for many reasons, the field was wide open in 1930, and it remained just as contentious as it had always been. Milne by now had become a major protagonist; as S. Chandrasekhar later lamented, "Milne's enormous originality was frustrated by his attitude of trying to do science which would contradict Eddington all the time."[23]

During the time the Russell family wandered the tombs and temples of Europe and Egypt, Milne made stellar constitution the central theme of theoretical astrophysics; the amount of space in the *Astronomischer Jahresbericht* consumed by "theoretical astrophysics" doubled from 1930 to 1931. Much of the attention was orchestrated by Milne, who mounted a vigorous challenge to Eddington's standard model by drawing in younger European theorists whom he befriended and encouraged. Now established as the Rouse Ball Professor of Mathematics at Oxford, Milne was a man with a mission.[24]

Challenging Eddington: Composite Models

In September 1930, delayed in Paris for a month with sixteen-year-old Margaret looking after him as they both attended to Henry jr.'s recovery in the American Hospital from a frightening attack of tuberculosis, Russell stole time away to read in the library of the Paris Observatory.[25] There he followed the struggle between Milne and Eddington over the physical state of the interiors of stars and realized that Milne was attacking not only Eddington's theories but his methods as well. As Russell read, he found Milne disputing the theoretical basis for Eddington's mass-luminosity relation by the philosophical argument that his theory was untestable since the value of the coefficient of opacity for the stellar interior was not observable, nor was the source of energy known. Milne demanded that a deductive theory of stars started with observed characteristics and then worked inward to clear the theory of all "special assumptions." One had to find a way to reconcile Eddington's model and the Vogt-Russell theorem with the fact that two very different types of stars existed: Giants and dwarfs both were stable configurations with highly condensed, extremely hot cores surrounded by cooler, less dense envelopes. White dwarfs, stars in a highly "collapsed" state, also had to be accounted for. As Russell read on, he soon found Eddington reminding Milne that the wide discordance between astrophysical and physical opacities was the central issue, and that the opacities Milne had chosen for a star's outermost layers were not physically reasonable. The jousting continued as Milne and Eddington fought over boundary conditions for stars and the influence these assignments had on the equilibrium of a star. As he had with Eddington and Jeans, Russell probably thought that an impartial moderator was needed. He was aching to get back in the game.[26] But as he read on, others joined the action, like R. H. Fowler and his students, who were interested in deviations from the perfect gas state. But Russell's focus remained on Milne and Eddington's combat. The "point of cleavage" was clear: Eddington argued that the interior structure of a star determined the state of the photosphere, whereas Milne took the opposite view.[27]

Filled with ideas from his reading, Russell had time on the homeward-bound steamship to reassess the problems at hand. Although the donnybrook was still raging when Russell reached Princeton in October, he had other duties to attend to, such as visiting Flagstaff and Mount Wilson to catch up on spectroscopic work and preparing a set of Lowell Lectures on the "Physics of the Stars." He finally got down to work in spring 1931, organizing his thoughts as a set of lectures. R. B. King recalls that it was his new hot interest then, but his lectures "went over our heads." King sat through all the lectures with Green, Sydney Hacker, Melvin Skellett, and John Merrill. They had a terrible time reading Russell's blackboard writing and were thankful that Russell spoke out everything he wrote down. But Russell also was in the habit of rushing his derivations through as he became more and more excited and caught up in the process. Too often he would fail to vocalize the final steps that led to "some equation that was the moral of the whole derivation." "Flapping his hands" as he turned around to the small group assembled in the Observatory of Instruction, Russell would anxiously await signs of appreciation, but "[n]one of us could read the darn thing. Finally he would turn around and read it to us."[28]

Milne and Eddington, along with Milne's students, had spent much effort rethinking the behavior of polytropes: idealized physical models describing how pressure and density distributed themselves in a sphere of gas according to a specific equation of state governed by an exponential variable called the polytropic index "n." Normally, the value of n determined the stability conditions of a gas, whether it was in radiative equilibrium (for values of n > 3/2) or in convective equilibrium (n = 3/2 or less). Eddington's radiative "standard model" required that n = 3.[29] Milne, however, added a new variable: degeneracy, the state of matter at great densities where Pauli's exclusion principle could be violated. Important work was then being done on the physics of degeneracy: in 1926 R. H. Fowler had shown dramatically how white dwarf stars could be explained as degenerate gas spheres, using the statistical formulae developed by Fermi and Dirac. Milne extended Fowler's lead to describe normal stars as possessing white-dwarf-like degenerate cores, represented by a polytrope with index n = 3/2, surrounded by thin envelopes of gas described by Eddington's standard polytrope of index n = 3. The method Milne used to create his model was developed by Thomas G. Cowling, one of Milne's first postgraduate students at Oxford in 1929. Cowling tested Eddington's assumption that a singularity existed at the centers of stars, what has been called the "point-source" model.[30] Cowling found a new way to derive point-source models integrating inward from the surface of the star and in November 1930 was able to support Milne's contention that such models failed to fit the mass-luminosity relation or reproduce the observed properties of stars.[31]

Russell saw in Cowling's analysis the seeds of a new efficient graphical technique for the solution of Emden's polytropic equations and for the fitting of dissimilar polytropic models on top of one another. Russell elaborated on Cowling's fitting techniques, using Emden's tables, to explore in rapid fashion how a star could be built up of a series of concentric shells, each with its own polytropic index. His goal was to find a set of interlocking mathematical transformations that made the polytropes fit at their boundaries. Within weeks he found a set of homology-independent dimensionless quantities that did the trick; simple adjustment of these variables made discontinuities in pressure, density, and temperature fit at the boundaries.

Russell was in great haste to finish his first polytrope paper but was hampered by his responsibilities as a vice president of the American Philosophical Society in Philadelphia. Thus he had to miss a more exciting AAAS meeting in Pasadena in June 1931 where Fowler would be speaking, so he asked Dunham to read his paper and give it to Fowler. Russell felt that Milne had made some poor assumptions about condensed stars, and was anxious that his paper be heard by Fowler, "both to save him possible trouble and for fear some of these bright Englishmen might get the proof before I publish it."[32] He also sent the manuscript to Milne, who was impressed with Russell's ability to use approximate means to show how Milne's solutions would end up with infinite central densities unless he surrounded his cores with envelopes of polytropic index greater than 5. The proof was so simple, Russell argued, "that it seems quite water-tight." Russell asked Milne to resolve their differences, "leaving no more than necessary for publication at the rate of two papers a point."[33] Milne, however, submitted the paper directly to the *Monthly Notices*, applauding Russell's generality and his ability to deal with a great range of polytropic indexes to test which of three classes of polytropic solutions created real stars. Using neat graphical transformations, which Cowling later described as "a very ingenious method," Russell found that certain types of gaseous envelopes could be fitted around a degenerate core that did not contain a singularity, but they could not be fitted around the type of collapsed degenerate core that Milne had postulated.[34]

Just as Russell had been attracted to term analysis by plugging holes left by earlier workers, Milne's work needed refining, and Cowling's graphical techniques showed Russell the way. Though Russell found the notion of "collapsed" stellar cores very useful, he did not believe they existed in nature. Further, Russell felt that Milne had not fully dealt with the high hydrogen abundance he now knew existed in stellar atmospheres and, he surmised, in their interiors as well. Knowing that Eddington had also criticized Milne on these points, Russell felt he was on the right track, and planned to test his model using the observed apsidal motion of the eclipsing system Y Cygni. But Dugan was having a devil of a time securing better observations, since this star system had an integral three-day period that made it very hard to observe all parts of its photometric orbit.[35]

Russell was not the first to challenge Milne's collapsed core models. Cowling, Bengt Strömgren in Copenhagen, and even Milne's ally Subrahmanyan Chandrasekhar at Cambridge were hacking away at them. Fowler, Dunham reported back from Pasadena, also was aware of Milne's "difficulties with the fitting of polytropes." Russell's neat graphical transformations were "[t]he real achievement," Dunham added, but he felt that stars could not be so static—there had to be mixing going on.[36] Mixing, however, was messy, so people avoided the problem or simply assumed that it was complete in stars. More provocative was stellar energy. Astronomers had been watching Nova Pictoris that year as it turned into a planetary nebulae, prompting Milne to speculate that in a nova outburst, the "collapsed" inner portion of the original star drove off outer atmospheric layers, exposing the core, revealing what looked like a white dwarf. "The collapse theory of novae," Milne told Russell, "accounts for all the facts of observation." Any theory of stellar structure had to deal with white dwarfs as cores of composite stars. Milne was impressed "[t]hat novae are so pregnant."[37]

A young English physicist who had just arrived at Rutgers University, not far from Princeton, found the nova fascinating too. Robert d'E. Atkinson, a former student of experimental physics under F. A. Lindemann at Oxford and James Franck at Göttingen, wrote Russell that they were "just right for the 'corpses' of heavy stars."[38] Atkinson met Russell in fall 1930 when Russell returned from Europe, and Russell was delighted to find someone close by who actually had a theory of element synthesis and energy production. In 1929, Atkinson and Fritz Houtermans explored how protons could penetrate the potential barrier of nuclei to create fusion reactions in stars. He carried these ideas with him to Rutgers and sought out Russell.[39] Atkinson became a major new catalyst propelling Russell to study stellar structure.

Matter annihilation had been very important to Russell as a way to preserve his giant-to-dwarf theory, but he found Atkinson's theory of proton synthesis in stars compelling. Russell regarded Atkinson's theory as the first to provide a realistic view of atom building and energy production in the stars that could reproduce the relative abundances of the lighter elements in the Sun.[40] More to the point, however, Russell's enthusiasm for Atkinson's proton synthesis mechanism marks his quiet realization that his own theory was dead and was best buried without fanfare. In the early 1930s Russell and Atkinson teamed up often on related projects, including an analysis of the central stars of planetaries to test Milne's theory. Atkinson was convinced that these stars had to be generic white dwarfs and had to die in timescales favoring transmutation mechanisms.[41] Atkinson's work had been proposed for the A. Cressy Morrison Prize of the New York Academy of Sciences in 1930. Morrison had created the prize for research that would inform his contention that the Sun and stars are powered by "intra-atomic energy."[42] Russell had been one of the judges for the prize in past years, and Menzel won it more than once. But Russell was abroad with his family when the judges were chosen in 1930, and he left for Mount Wilson before the judging took place. Stewart and Condon were the judges that year, and Atkinson came in second, which, Atkinson recalls, displeased Russell.[43]

As Milne defended his white dwarf theory, debating many points with Russell through summer 1931, Russell took some time to assess the general situation, realizing that the differences among the growing band of workers, led partly by Milne but all working under Eddington's shadow, had to be carefully weighed. Many sensed that there was a need for dramatic change in technique, but neither Russell nor most of the others were sure how to proceed.[44] In such times of doubt, Russell resorted to simplifying assumptions and general approximations, which he believed had great heuristic value; the Vogt-Russell theorem proved that significant advances in understanding came from the most general assumptions, even if those assumptions did not bear up under close mathematical scrutiny. Russell evidently saw more in Cowling's graphical visualizations than Milne had intended, though Milne soon refined Cowling's method in his own development of a graphical fitting procedure based upon homology-independent dimensionless variables. Milne's variables became central to stellar structure calculations in the 1950s and 1960s.[45]

For Russell, graphical fitting procedures were efficient and useful devices as well as illustrative pedagogy. They were most valuable when used to visualize and then build approximate composite models of stars, as he argued in a long review for the Royal Astronomical Society in 1931. He felt that it was hopeless to attempt a "complete deductive theory of stellar constitution" as Milne was trying to do. Too many variables were at play: composition, rotation, mixing, absorption coefficients, degeneracy. Therefore, he told his RAS readership, "It is necessary to work with drastically simplified models, and to test the appropriateness of the hypotheses upon which these are designed, by comparison with the observed facts." Calling the construction of stellar models at best "a difficult imaginative art," Russell admitted that Milne probably would not agree that such an attitude was proper theoretical astrophysics.[46]

Milne was attentive to Russell's concerns because he was still having great troubles making his models work. Maybe Russell had a point, Milne suggested in June 1932. Maybe it was futile "to erect models of actual stars in the way an engineer makes a model of a bridge." It might be better to "examine in great detail the aggregate of properties associated with a simple astrophysical situation in order to gain insight into the nature of stars."[47] In his continuing battles with Eddington, Milne looked to Russell for fresh insight. Stars might or might not be built in the way that either Milne or Eddington wished, Russell mused, so one had to search for new ways to approximate their structure, using every trick in the book.

Russell and Milne

Milne's orthodoxy as a mathematical theorist helps to place Russell's style in context. As Milne well knew, Russell's theoretical strength centered on his ability to draw general clues from a broad patchwork of evidence. Arguing for simplified mathematical models, numerical approximations, and good data, Russell was not concerned with the formalisms demanded by the new physics or by mathematical theorists. He asked only that the physics he did use lead somewhere useful. Milne gained a healthy appreciation of the freedom of action Russell enjoyed when, in 1936, trying to describe the photosphere as a polytrope in terms of the usual physical variables without recourse to a central singularity, Milne failed to find a practicable solution. Russell found it, using the Vogt-Russell theorem as an approximation, neglecting higher-order terms, and solving his equations iteratively by successive approximations. He concluded that proper Emden solutions did exist and could be extrapolated to the interior of the star, although he admitted that this was "only true in a very general and approximate way."[48] Russell wrote to Milne with his solutions, which brought Milne back to stellar structure very impressed that Russell had solved the first-order problem. Grateful that Russell had sent him his note before publication, he wrote back wistfully, "You have very nicely and ingeniously solved the problem I failed to solve elegantly, & applied your solution with a comprehensiveness which I can only admire."[49]

Russell was not bound by the legacy of the Cambridge mathematical tripos and so was free to use any method that satisfied him. As they continued to correspond about how best to exploit the theory of polytropes to fit cores of stars to envelopes, however, Milne became frustrated. Russell seemed to lack what to Milne was critical for a theorist: a "theory of knowledge," which required adherence to self-consistent mathematical law. For Milne, everything had to be linked mathematically, whereas Russell championed anything that worked: he was able to make headway, for example, because he saw nothing wrong with examining the outer layer of a star separately from the inner layer, which allowed parameters in each, like density or composition distributions, to be adjusted separately through the use of ad hoc devices to make them fit at the boundary. This gave Russell flexibility. It was just another form of puzzle solving, guessing at the shapes of the pieces as well as how to fit them together. But Milne strove for "grace and elegance." No arbitrary parameters could be involved.[50] Crying out to Chandrasekhar in late December 1937, Milne lamented that his differences with Russell were as yet "unresolved, because he either considers the equ[libri]um of the outer layers, or else that of the deep interior, but never both together, & he wholly ignores the theory of knowledge."[51]

Russell treated all theory pragmatically. He was never outwardly troubled with the fate of his theories because, unlike Milne and Eddington, who were highly possessive of their models, Russell did not rue the loss of what was, merely, as he described in his 1937 AAS presidential address, "a tissue of approximations." He celebrated the heuristic power of these tenuous methods because they provided a useful framework upon which a stronger fabric might someday be woven; he regarded his achievements, and those of Saha, Payne, and Moore, as "illustrations of the power of this method." Russell believed that "[m]athematical approximations are involved in all our theoretical discussions, since Nature is far too complex to analyze as a whole." In tune with other Americans like P. W. Bridgeman and John Clarke Slater, he preferred numerical methods to probability functions, like Gaussian statistics, because he feared that such practices removed physical insight, or physical control, from the worker. All methods had their limits, however. Russell cautioned against the wholesale use of harmonic analysis, for instance; though powerful, such techniques often yielded predictions for phenomena that were physically impossible. Rather he preferred David Webster's use of approximations in the solution of Emden's polytropic equations—what Webster had called "exact qualitative analysis." Russell always championed their use, as they would "often give very valuable aid when quantitative numerical results are unattainable." Thus Russell was far more willing to explore numerical solutions than was Milne, who remained firm in his conviction that the "computational approach did not lead to real understanding."[52]

In time, numerical techniques and approximations such as those Russell championed were accepted by astrophysical theorists, though Milne's concerns were never forgotten. In the 1930s, Cambridge-trained theorists adhered to mathematical orthodoxy to raise the status of theoretical astrophysics, whereas by the 1950s, the same theorists could admit, as Russell always did, that stars were such complicated beasts that "one is compelled to use numerical methods, whether the atmosphere was built up from physical theory numerically or analytically."[53]

Atkinson's proton synthesis mechanism, Hubble's intermediate timescales based upon his velocity-distance relationship, and Russell's own determination of the high hydrogen abundance in stellar atmospheres led him to abandon the idea that the giant and dwarf branches of the diagram were evolutionary tracks. He continued to entertain his notion that two types of active matter were "turned on" at successively higher temperatures as stars continued to collapse, and to believe that stars were giants before they became dwarfs. He was slow to abandon entirely the possibility that matter annihilation played a role. It would take one more step to make his shift complete.[54]

With improved opacities and composite modeling, Russell soon found that his approximate models worked better if he used a high hydrogen abundance. In February 1932, as he invited Eddington to America to lecture on the expanding universe, his own thoughts were with the stars: "As for hydrogen, I am becoming more and more convinced that it is really altogether the main constituent of stellar atmospheres."[55] Hydrogen made up as much as "99 percent or more of the atmosphere by weight," he argued, but Eddington continued to worry about opacity—where would it come from with a pure hydrogen atmosphere? Eddington wanted to know whether Russell's "99 percent was a calculation or a conjecture," and Russell soon admitted that it was an intuitive guess, even though he did not know how to make a pure hydrogen star fit the mass-luminosity relation or the timescale requirement, "for the transmutation of one percent of the hydrogen won't keep the Sun going for very long."[56]

But in short order, Strömgren and Eddington found that a high hydrogen content worked nicely for the stellar interior.[57] Strömgren used H. A. Kramers's new opacity theory in his numerical study of a set of polytropic models to forcefully show in 1932 that models built out of hydrogen (one-third by weight) did indeed predict the observed luminosities of stars. Further, he demonstrated how stellar models using a high hydrogen abundance could be "correlated with positions in the Russell-diagram."[58] Strömgren's analysis convinced Russell that his old giant-to-dwarf theory was dead. In a major address to the AAAS in December 1932, Russell spoke only of the problem of the "Constitution of the Stars," pointing to three problems to be solved: what was the nature of the equation of state of the gas inside a star; what was the opacity of the material in the star; and by what process (or processes) was the internal supply of heat of the star maintained? These, Russell suggested, were questions for physicists, and at the present they had supplied "a satisfactory answer" to the first question, "a fairly good one" to the second, but hardly a clue to the third, "though there is hope that we may know more soon." Even so, the interplay of physics with astronomical evidence had yielded some remarkable successes that provided clues to the answers for all three questions, and Russell typically outlined them in detail.[59]

The mass-luminosity relation was, for Russell, the "first great triumph," which was now made even better with a high hydrogen abundance. Next came white dwarfs; because of Fowler and Milne, "[white dwarfs] changed their rôle from most perplexing to the best understood class of stars." Third came Atkinson's theory of proton synthesis, which promoted the intermediate timescale. Chadwick's discovery of the neutron dealt a blow to Atkinson's scenario, but it had already proven its value by demonstrating that nuclear processes required temperatures in the millions of degrees, not billions as proposed by Milne or required by matter annihilation. Atkinson's temperature range had been verified by the experiments of Cockroft and Walton and fit beautifully with Russell's own theory of the dwarf sequence, which now represented not an evolutionary path but a locus of equilibrium states for stars whose central temperatures were in the range of tens of millions of degrees.[60]

On the astronomical side, Russell felt that the "Vogt theorem" made the construction of stellar models a practical art. He spent considerable time in his AAAS address delineating how it was done, highlighting the heuristic value of iterative integration schemes to reproduce observable characteristics. Russell painted an optimistic picture, reciting Eddington's hope "that in a not too distant future we shall be competent to understand so simple a thing as a star." But he admitted that the "life-history of a star presents the most difficult problem of all." He knew all too well that "the path to knowledge was "strewn with the corpses of dead theories." Left unsaid was that his own cherished theory was among the dead. Nothing definite would replace it, however, for at least two decades; in the interim, Russell considered stellar evolution to be in such a state of flux that it should not be covered at all in survey courses on astronomy.[61]

Gatekeeper and Commentator

Through the rest of the thirties, the progress of Russell's commentary can best be described as a migration away from Eddington's increasingly inflexible style of stellar structure modeling. Russell discussed Eddington's work no fewer than forty-two times in his *Scientific American* column over the years, giving greatest notice to works about stellar structure. In all of these reports, Eddington was the authority. Russell delighted in repeating his quips and jabs, his homilies and opinions. Thus in an essay entitled "Impossible Planets" written in May 1935, Russell outlined Chandrasekhar's theory of relativistic degeneracy for white dwarfs making the cautionary remark that, although it was a provocative new way to describe these strange objects and was gaining the attention of physical theorists, Eddington had fundamentally opposed it, so the matter had to be left open.[62]

Chandrasekhar's case illuminates the gulf between theoretical physics and mainstream astrophysics, which continued to be defined by Eddington. Chandrasekhar's highly mathematical treatment of relativistic degeneracy had shown that there was a limit beyond which stars could not form degenerate cores and so could not become white dwarfs. By 1934 he had established "an exact differential equation to describe the degenerate state," which he thought would lead to a complete general solution for the structure of white dwarfs.[63] But even Milne knew that astronomers would not accept such a radical concept when couched in highly abstract mathematics. As Cowling later recalled, "Both observing astronomers and physicists tend to wax critical of the mathematician, and sometimes with reason." Appeals to formalism and elegance had their virtue, but "in the theory of stellar structure one must look for other things than elegance in one's models."[64]

Russell best personified this attitude, and so the deeply sensitive Chandrasekhar viewed Russell as Eddington's agent. He was angered and further alienated when Russell blocked him at the Paris meetings of the IAU General Assembly in July 1935. Eddington was president of Commission 35, and Russell was secretary. "With Russell presiding," Chandrasekhar recalled, "Eddington gave an hour's talk, criticizing my work extensively and making it into a joke. I sent a note to Russell, telling that I would wish to reply. Russell sent back a note saying, 'I prefer that you don't.' "[65] It is likely that Russell's refusal to let Chandrasekhar reply, which Chandrasekhar remembered differently on separate occasions, was due in part to his philosophy of airing differences in private. Less concerned with feelings and insensitive to the injustices shown Chandrasekhar by Eddington, Russell felt obliged to keep matters moving along during sessions where time was always tight. But Russell was also very much an Eddington man then. He would not have blocked Chandrasekhar if he thought that a corrective was needed.

Russell soon modified his stance because he listened to physicists. He knew that Chandrasekhar's models made it even more difficult to gauge the fate of massive stars, as he made clear in a *Scientific American* column, but he reviewed his theory in extenso in his 1936 Gibbs Lecture before the Mathematical Society. Always noting Eddington's opposition, Russell made clear that Eddington was now alone in his opinion, and that the consensus view was moving toward Chandrasekhar's "elegant piece of analysis." Thus partly owing to Russell's support, it did not take three decades, contrary to Chandrasekhar's biographer, for astronomers to come round to his way of thinking and accept the "Chandrasekhar limit" as fundamental to stellar structure.[66]

Changing Tracks on the Diagram

With the giant and dwarf sequences no longer evolutionary tracks, the field was wide open for a deep adjustment. By the end of the decade, two episodes forever changed how astronomers thought about the lives of stars. The first was created by a trio of astrophysicists, and the second was due solely to physicists.

For some years Bengt Strömgren had been exploring anew the theoretical distribution of stars on what he called the "Russell Diagram" at first, but, as a loyal Dane, soon changed to the "Hertzsprung-Russell" Diagram. His real goal was to find a way to adjust polytropic models to fit observed distributions of mass, spectra, and composition on the diagram. In 1933 he used Vogt's theorem to show that the position of any star on the diagram was determined solely by its mass and hydrogen abundance, and if hydrogen were transformed into heavier elements in nuclear transmutation processes somehow, possibly along the lines Atkinson suggested, and the by-products were fully mixed throughout the star, then the position of that star on the diagram would change even though its mass remained constant. He constructed theoretical distributions of stars, varying hydrogen content and then varying mass, and found that as hydrogen content diminished through the process of element synthesis, main- (or dwarf-) sequence stars would move up and to the right on the diagram, *toward* the region of the giants. Lines of evolution on the diagram were hence lines of constant mass. Strömgren was the first to suggest that as hydrogen was consumed, "the stars expand."[67]

Visiting Oxford in June 1933 to deliver the Halley Lecture "On the Composition of the Stars," Russell had a chance to talk about Strömgren's work with Eddington and Chandrasekhar. Chandrasekhar was very interested in Strömgren's detection of a "systematic variation of hydrogen in the Russell diagram" and reported to Strömgren that Eddington, however, still preferred "the hydrogen to be constant for all the stars, rather than have another disposable constant in the theory." But, Chandrasekhar speculated, Eddington was "really (secretly!) pleased" because Strömgren's models made centrally condensed radiative solutions more likely: "[T]he perfect gas theory combined with your [Strömgren's] hydrogen-content curves in the R-M diagram really contain all the information summarized in the Russell-diagram." Russell was then exploring centrally condensed models, using Chandrasekhar's work on distorted polytropes, and at an RAS discussion highlighted how they agreed with observations by Dugan and Redman.[68] With Eddington cautiously mollified and Russell using his findings, Strömgren saw his work gain quiet favor among those most influential in the field, though his conclusions about the evolutionary state of giants remained unappreciated.

In late 1935, Gerard Kuiper, then at Harvard, looked for ways to link Strömgren's models to the observations of star clusters by Robert J. Trumpler, whom he had come to know and respect during his years as a Lick fellow. Trumpler originally classified his clusters by relative age, using Russell's theory as a guide. But now Kuiper realized that this was wrong, and wanted to discredit Russell's theory in the process. After meeting Chandrasekhar and listening to his lectures at Harvard in February 1936, Kuiper knew how to reinterpret Trumpler's cluster observations in terms of Strömgren's theory. As he reported to Otto Struve, who had just invited Kuiper to Yerkes:

> In particular we discussed Trumpler's very massive cluster stars (PASP Oct 1935) which at first seem[ed] to present great theoretical difficulties. The way Chandrasekhar cleared these difficulties almost overnight, and showed the importance of convection for the interiors of these massive stars, resulting [in] a small concentration of mass towards the center, has convinced me once more of the genius character of this man.[69]

When Chandrasekhar lectured at Yerkes a few weeks later, Struve was similarly awed: "Chandra discussed the problem of the hydrogen content of clusters in one of his colloquia at the Yerkes Observatory, and I was very much impressed with the brilliant character of his presentation, with the remarkable results that you have obtained in explaining Trumpler's diagrams and with the importance of Strömgren's early work on the subject."[70] Kuiper reworked Trumpler's cluster diagrams, using Strömgren's theoretical work on hydrogen content. Superimposing several clusters on the same diagram, adjusting them vertically using spectroscopic parallaxes, Kuiper found roughly that main sequences were "crowded closely together" in the lower portion of the diagram and diverged for the higher luminosities. With Strömgren's assistance, Kuiper derived rough hydrogen abundances using masses, absolute bolometric magnitudes, and effective temperatures. He found that clusters with differing hydrogen content diverged systematically from the main sequence, and all of the divergences led to the giant region. Based upon the strong assumption that "the typical spectral sequences in galactic clusters found by Trumpler represent lines of constant hydrogen content in the Hertzsprung-Russell Diagram," Kuiper concluded that "[n]o other than 'racial' differences between the clusters can explain such differences."[71] Unlike Strömgren, he did not state that giants were evolved stars, or that there was a "generational" difference, because a detailed analysis of the Hyades yielded equivocal results. He used this observation to argue, however, that whereas the positions of highly luminous stars were influenced by hydrogen content, stars on the lower main sequence were not influenced by composition, thus explaining the convergence of the cluster sequences.

Struve was then intent upon developing a strong theoretical department at Yerkes and hired Kuiper, Strömgren, and Chandrasekhar in 1936 and 1937, creating for a time the most powerful observational and theoretical team in astrophysics in the world (see chapter 21). Strömgren's contribution was to reestablish the diagram as the descriptive framework upon which theories of stellar evolution might be tested and visualized. Kuiper's descriptive analysis had strengthened the empirical basis of Strömgren's post-main-sequence scenario, but, as Chandrasekhar constantly reminded them, they still lacked any rational argument explaining why composition changes produced motion on the diagram. The physics had not been worked out yet, and the observational data Kuiper had applied to derive masses and abundances were far from consistent. Strömgren returned to this question in spring 1938, as did Chandrasekhar, after Carl F. von Weizsäcker found a way to transmute hydrogen into helium nuclei, and, soon after, Hans Bethe clinched the argument.[72]

Transmutation Processes

Russell had been following von Weizsäcker and Strömgren's work closely, as his *Scientific American* columns attest. Thus when Russell was invited to attend the Fourth Annual Conference on Theoretical Physics in Washington, organized by George Gamow, Merle Tuve, and Edward Teller, and learned that it would be devoted to the problem of the energy source of the Sun and stars, it is surprising that he turned the invitation down. He wanted to go, but he knew that by the time of the conference, late March 1938, he would be madly rushing to prepare for a long European tour that would end at the Stockholm meetings of the IAU in August. A deeply personal and emotional milestone during the trip, which would start in early April, was a visit to their daughter Lucy in Greece to see their grandson, who had been born in Princeton in December 1935, but whom the grandparents had not seen for several years.

Lucy's husband, George H. Gardner, whom she had married on 30 May 1935, graduated from Princeton in 1931 and attended the Princeton Theological Seminary for a year. He then studied at New College, Edinburgh, and finally obtained his divinity degree from Union Theological Seminary in New York.[73] He was licensed and ordained by October 1935, and was serving successfully in Columbus, Ohio; but when it became obvious that Lucy was in advanced pregnancy, giving evidence of an indiscretion unbearable to that community, his Columbus Presbytery suggested that he demit himself from the ministry. Russell, who was then a trustee of the Princeton-Yenching Foundation, set up interviews for George for a teaching position in Canton, China. Though resourceful and deeply penitent, George could not raise the necessary funds for travel, nor, evidently, could he rely on Russell, who uncharacteristically groused about not knowing his son-in-law's address during the summer of 1936 when he was teaching at the Columbia Summer School in order to secure enough funds to travel. George finally secured a post at the American Farm School in Salonika: "My call was not to the American pulpit," he noted many years later. Russell only noted dryly at the time that his call to Salonika "has my approval under the circumstances."[74]

Thus the trip in 1938 would be an important reunion, and, it was hoped, a reconciliation for the family. Elizabeth and her caretaker Jean Hetherington were going as well, which made the preparations unusually complex, precluding Russell's attendance at the Washington conference. So as his family made preparations, Russell pushed his projects to completion, including an extended correspondence with Milne on fitting atmospheric envelopes to stellar cores. He also managed to place his very promising graduate student, Lyman Spitzer, at Harvard, continued to refine his eclipsing binary reduction techniques to further explore apsidal motion, and reconciled his dynamical parallaxes with Mount Wilson spectroscopic parallaxes and the mass-luminosity relation, debating the details with Gustav Strömberg but holding back Strömberg's papers from the *ApJ* until he was satisfied that they agreed. With Stewart and graduate students he revised the textbook once again, rewriting sections on stellar evolution, incorporating von Weizsäcker's theory and tentatively mapping out the possibility that a star could indeed expand against gravity. Russell was indubitably very busy yet sorry to miss the conference.[75]

The conference date had slipped into the spring and actually had been set so that Strömgren could attend before he returned to Denmark. As Gamow told Chandrasekhar, Strömgren was the "ace" of the conference, which would bring theoretical physicists and astrophysicists together: "Then we will have a big fight here."[76] But Gamow still tried to convince Russell to attend by sending him a manuscript on nuclear resonance effects and their application to von Weizsäcker's theory for an evolved star. Gamow's ideas intrigued Russell, especially because he tried to preserve the dwarf sequence as an evolutionary track. But though he felt that he could not comment on the physics, he advised Gamow to be more sensitive to the observational evidence; the dwarf sequence Gamow used was far narrower than astrophysical evidence suggested. Russell knew this from his reconciliation of spectroscopic with trigonometric parallaxes, which he felt gave "some opportunity for diversity of mass and of hydrogen content."[77]

Russell missed out on what turned out to be a watershed event in the history of the stellar energy problem. As Theodore Sterne recalled later in the year, the conference was the "big fight" Gamow gleefully hoped for and consciously designed to maximize debate. Gamow had insisted that open disagreement and criticism were to be encouraged. "The meetings on *nuclear astrophysics* in Washington were excellent," Sterne told Shapley, explicitly defining this new specialty, "precisely because there was no public; and no loss of face when Tuve would disagree so vigorously with Bethe, for instance . . ."[78] As a result of this conference, Cornell physicist Hans Bethe took a closer look at Strömgren's evidence and von Weizsäcker's theory and within several months developed his own version, based upon the fusion of carbon and hydrogen nuclei into nitrogen and oxygen—what has come to be known as the CNO cycle. Bethe's full discussion did not appear in print until March 1939. It was delayed because it was being considered for the Morrison Prize and so had to remain unpublished. Thus, as Russell traveled to Europe for the IAU meetings in Stockholm that summer, only vague reports of Bethe's work filtered through during the inevitable junket that accompanied the triennial General Assembly; it received some notice during the meetings of Commission 35, where Strömgren and Atkinson provided summaries of what von Weizsäcker, Bethe, and Gamow were doing and then led a conversation about neutron capture processes and how well these nuclear processes fit observed abundances. Eddington felt that they did not fit at all well for helium, and Russell noted that his recent work on the empirical mass-luminosity relation, using his reconciled spectroscopic parallax technique, would eventually improve knowledge of the celestial abundances of hydrogen and helium.[79] W. H. McCrea reported back to Chandrasekhar that nothing much was decided at Stockholm regarding stellar energy. The really exciting events, McCrea felt, were H. Zanstra's account of novae and Bernard Lyot's new motion picture films of the solar corona and prominences taken out of eclipse with his marvelous coronagraph. The latter was also the highlight for Russell, who wrote a *Scientific American* column about Lyot's "amazing new technique" as he and his wife headed home.[80]

Accepting Bethe's Process

Soon after Russell returned from his trip, he received a breathless letter from Gamow promoting Bethe's process: "[T]he question about nuclear reaction which gives rise to stellar energy seems to be finally settled," Gamow insisted in his trademark broken English, mainly because Bethe could reproduce the Russell mixture for the lighter elements very nicely. Russell's standards for agreement were more stringent than Gamow's, however; the ebullient Russian émigré was well known to quip, "A physicist doing stars feels happy as long as he dose not nead to tuch the astronomical tables."[81] Spectroscopic evidence indicated less nitrogen in stellar atmospheres, for instance, than Bethe predicted.[82] Russell remained equivocal about Bethe's process into the new year, mainly because of abundance anomalies, neutron capture processes, and problems with the mass-luminosity relation. Possibly the heavy elements in nature were older than the stars, formed in an early phase of the expanding universe, Russell mused, and the stars themselves inherited a mixture that was already enriched. This might explain the nitrogen excess that was observed, and so "would also fit in with Bethe's work," which he had finally seen in abstract in the *Physical Review*. Still, Russell was much more excited by the birth of his new granddaughter in Greece, named Lucy Ann, "a fifth Lucy in direct descent."[83]

What Russell hoped for was a decisive astrophysical test, but after decades of searching, astrophysics had failed to supply it. Since his days of high optimism dreaming with David Webster about astronomical solutions to the stellar energy problem, Russell had believed that conditions in stellar interiors might someday be well enough defined to make it possible to pinpoint the physical conditions that powered the stars. But by the late 1930s, even before Bethe's work appeared, Russell was reconciled to the fact that too many astrophysical unknowns remained. Thus it was highly unlikely that astronomers could be of much further help in determining the mechanism of stellar energy generation. It was a problem that was more likely to be solved by an appeal to nuclear physics than through the continued study of the stars. The abundance anomalies did not provide direction.[84]

Until he met Bethe, and read his full discussion, Russell had only secondhand knowledge through Gamow, Atkinson, and Strömgren. In the interim, he agreed to co-organize a "Symposium on Progress in Astrophysics" at the American Philosophical Society, where he would speak on "Stellar Energy and the Evolution of Atoms."[85] He had been enough impressed by von Weizäcker's theory to feel that the time was right for a review, and believed that by the time of the symposium, Bethe's papers would be in print. But as Russell prepared his own review, he learned that Bethe's paper would be delayed because it had to remain unpublished to qualify for the Morrison Prize. This put Russell in a bind; he certainly planned to include Bethe's work. He therefore put off preparing his remarks until early February 1939 and did so only after Bethe visited Princeton, had an extended discussion with Russell about his work, and left his manuscript with him so that he could prepare his talk fully informed. Russell immediately recognized that "I have been very lucky about this."[86]

Meeting Bethe face-to-face made all the difference: "it looks as if we now really understand why the sun and the other main-sequence stars shine," he wrote King. He still worried that the "very success of the theory" did not explain away the abundance anomalies, but now he trusted Bethe's intuition, even his elaboration of von Weizsäcker's speculation that the heavier elements might have been formed in a prestellar early universe when densities and temperatures were extreme, the latter in the range of a billion degrees. "I may add on my own account that I don't see how the hydrogen which is still in the stars could have been part of the same mess," Russell noted, but for the moment he was happy to "leave the origin of the heavy elements 'buried in obliquity' as the little boy said."[87]

Russell was convinced by the breadth of Bethe's argument and, most definitely, the expertise he demonstrated manipulating both nuclear physics and astrophysical evidence. He had dealt with all the astrophysical constraints, including the Russell mixture, better than any other physicist and was able to show that only a small range of hydrogen abundance would work, and it agreed nicely with the now-canonical value of 35 percent by weight that Eddington and Strömgren had found. Bethe also drew upon extensive laboratory data on collisional mechanisms, probabilities of nuclear reaction occurrence, reaction rates, and product lifetimes.[88] Russell acknowledged all this at the Philosophical Society symposium, where he apologized to his audience for not being Hans Bethe, "the man whose recent and brilliant work has inaugurated a new and very promising stage of astrophysical study."[89]

Russell was now a convert and devoted the next three issues of his *Scientific American* column to the subject of what keeps the stars shining. He was most attentive to assigning proper credit to Bethe, because, though he had been clear enough in his Philadelphia talk that he was only commenting from the sidelines, a *New York Times* reporter made Russell the man of the hour. Russell was embarrassed and annoyed, and complained to colleagues as well as to the editor of the *Times*. To make matters worse, he was awarded the Lewis Prize of the Philosophical Society for his talk, probably owing as much to the fact that he had served as the society's vice president and president earlier in the decade.[90] If so, the timing was poor. At least he had the satisfaction of being the judge when Bethe and Marshak's essay on white dwarfs won the $500 Morrison award. And after Bethe's original paper had appeared in the *Physical Review*, Russell proclaimed it "The Most Notable Achievement of Theoretical Astrophysics of the Last Fifteen Years."[91]

Russell was called upon several more times to consecrate Bethe's work. At a conference organized by Shapley at the New York Academy of Sciences, he reviewed observational analyses for determining the internal density distributions within the stars using his apsidal motion test, which had gained new prominence owing to the rapid increase in stellar modeling by theorists like Cowling, Chandrasekhar, Biermann, Öpik, Sterne, and others. Russell responded from his strengths to discuss this yet unfulfilled application. He was still in the game.[92]

With Bethe's success, a major chapter closed in Russell's life. He now knew, with all astrophysicists of his generation, that modern physics once again had conquered a critical astrophysical problem. Russell took the realities, and the limitations, of his profession to heart. In February 1942, at the first Inter-American Congress on Astrophysics, held in Puebla, Mexico, Russell again spoke on the state of the theory of stellar evolution. There he admitted that even though a great mass of data had been collected about the stars over the previous half century, in the hope that these data would inform astronomers about the lives of the stars, "the empirical methods of natural history proved inadequate to suggest a good theory . . . The problem of stellar evolution is not one of natural history, but of physics."[93]

Even though Russell once harbored hopes that astrophysics would help to discover the nature of the stellar energy source, he had also long argued that astronomers had to accept physics as their guide. Natural philosophy, not natural history, was most likely to reveal nature's process. Now he had only admiration for Bethe's process as a new and extremely powerful guide in the exploration of the energy budget of the stars, suggesting how the expenditure of this budget caused stars to change. But even with Bethe's success, the relative ages of stars and how their lives could be interpreted on the Hertzsprung-Russell Diagram remained unsettled throughout the 1940s and well into the 1950s. Russell quietly followed Strömgren's theoretical arguments for post-main-sequence expansion in the late 1930s and took some notice of Gamow's speculative shell-burning models a few years later, both of which implied that some red giants were old, evolved stars. Even though Russell knew well that Gamow's work was severely criticized by Chandrasekhar in the late 1930s and early 1940s, he also knew that Chandrasekhar, with Louis R. Henrich and Mario Schönberg, was developing rigorous solutions for hydrogen shell-burning models which showed for the first time that changes in composition could induce evolution away from the main sequence.[94]

Following these clues, Russell was always ready to speculate when new evidence presented itself. After visiting Menzel and Shapley at Harvard in 1948 and talking about Walter Baade's sensational identification of two populations of stars, Russell commented that the high-luminosity stars constituting Population I found in regions of heavy gas and dust clouds had to be young stars, thus implying that the brightest red giants in Population II were old, evolved stars, and that stars could have more than one luminosity maximum in their lifetime. Although Russell's intuition again turned out to be right, at the time the consensus opinion, uttered by Baade, was that "all the astronomical Darwinism of the last decades had led only into blind alleys."[95] Russell was, of course, keenly aware of this, but as an "astronomical Darwinist" still disposed to a natural history of the heavens, he just couldn't help himself.

• CHAPTER 17 •

Binary Stars and the Formation of the Solar System

On Motive

Of all his areas of interest, binaries held Russell the longest, from his first paper on visual binaries in 1898 until his last on eclipsing binaries in 1956. Starting with his second paper in 1899, Russell maintained a single motive: to search for the fastest and most efficient ways to use binary data to solve astrophysical problems. Russell loved to exercise the equations of celestial mechanics, using the simplest means at hand. He was a deft manipulator of slide rules, including circular and drum devices. The latter were the equivalent of a slide rule some one hundred feet in length, which Russell was known to use for hours on end in his office.[1] He tolerated mechanical calculators like arithmometers, electric multipliers, and adders, but he directed his computers and students to perform linear interpolation mentally and for more complex functions to use freehand curve fitting.[2]

In December 1946, delivering the first Henry Norris Russell Lecture before the American Astronomical Society, Russell spoke on the "The Royal Road of Eclipses." His talk would cap a daylong special session on eclipsing binary techniques, and Russell wanted to provide a larger view. Russell's title reflected his feeling that "after 40 years' work in the field . . . eclipsing variables really do provide the Royal Road of *remarkably easy access* to the solution of other inaccessible problems. More and more of them are coming in sight as we proceed."[3] He never let anyone forget that binaries could answer "questions relating to the statistical equilibrium of systems of great numbers of stars, to the internal constitution, rotation, and vibrations of gaseous masses, and to theories of the evolution of the Universe and its separate portions."[4] As he devised new ways to examine binary systems, he looked for more ways to use them to address astrophysical questions.

DYNAMICAL PARALLAXES

Russell constantly looked for ways to refine his hypothetical parallax technique, finding in his first visits to Mount Wilson copious evidence for what he, Hertzsprung, Adams, Seares, and others had long suspected: that luminosity and spectrum were linked to mass. Based upon the assumption that the masses of the stars in a binary system are equal to the mean value for field stars of the same spectral type and absolute magnitude, he applied the spectrum-luminosity relationship and improved photometry from Mount Wilson to assign masses and use them to determine distance. He called these determinations "dynamical parallaxes" in 1922 to distinguish them from other forms of distance measurement, and, most of all, to give the technique a certainty that "hypothetical" lacked.[5]

Direct contact with Seares and Adams made all the difference. By the time of Russell's second visit in late 1921, Seares had just completed a reconnaissance of the masses and densities of the stars, based upon a statistical study of visual binaries, Cepheids, and isolated field stars using techniques similar to those Russell had developed. This work convinced both Russell and Seares that "the variation of the mean magnitude of binary star components with spectral type is the same as that of single stars." Seares then derived masses for single stars along the dwarf branch of the "Russell Diagram" and confirmed that the "average mass decreases continuously with advancing spectral type."[6] In what was a tour de force for a Mount Wilson observer, Seares drew upon Eddington's theory of Cepheids, and called upon theoretical discussions of ionization equilibrium (aided by J. A. Anderson and Paul Epstein at Caltech), to explore ways to derive densities in stellar atmospheres and interiors, testing theory against Adams and Joy's spectroscopic data. He also confirmed that the dispersion of masses for dwarf stars of any one spectral type was small, again stimulated by Russell's earlier work, as well as by discussions with Russell at Mount Wilson that helped to convince him that there were no systematic effects in spectroscopic data that might conceal a larger mass dispersion.

In all of this Seares worked firmly within the evolutionary framework established by Russell, and used the Russell Diagram to explore the frequency distribution of the masses and densities of stars, showing that lines of constant mass lay orthogonal to the dwarf sequence, and that density diminished along those lines with increasing distance from the dwarf sequence. Russell, of course, was delighted with Seares's analysis, seeing it completed during his second visit. It was an important exploration of the evidence supporting his theory of stellar evolution, just the sort of activity Hale had hoped Russell would stimulate. Accordingly, Russell appended a short note that appeared immediately after Seares's on the calculation of masses from spectroscopic parallaxes—not to push his ideas on evolution, for they seemed quite secure for the moment, but to explore the possibility of systematic errors arising from the use of spectra. Russell concluded that spectroscopic and dynamical parallaxes gave consistently reliable results, but that surer knowledge had to await more and better trigonometric parallaxes.[7]

With Moore keeping a tidy card file for all known systems with spectroscopic or trigonometric parallaxes at hand, Russell amassed information on some 1,600 systems by the end of 1922.[8] He hesitated several years before publishing the chief results from this compilation, waiting until Frank Schlesinger's improved trigonometric parallaxes were available. In the interim, Eddington announced his theoretical mass-luminosity relation, which was powerful confirmation of what Moore's card file revealed. Now at last a rational argument could be made for assuming a mass for the stars in a binary system, deriving a luminosity from that, and then adjusting the assumed mass using Eddington's theory. "Over the whole range covered by observation, this theory represents the observed relation between mass and absolute magnitude almost within the errors of the best observation," Russell concluded in late 1927, after he had had a chance to revisit Moore's card file, dust off their old manuscript, and bring it up to date.[9]

In following years, his goal was to maximize certainty by routing out all statistical error. He found the iterative schemes of a bright Eddington student, R. O. Redman, very useful and adapted them in 1928 to find statistical error and selection effects.[10] In following years Russell and Moore continually expanded and refined what had by now become a major catalog project, using it to recalibrate and test the spectroscopic parallax technique, reducing the limits of error as much as possible. In one flurry of work in 1933 Russell labored to refine knowledge of giant star masses and how they varied with spectral type, but by the time of the 1936 Harvard Tercentenary he decided that there remained too many internal inconsistencies in the various parallax methods to permit an all-out assault on the masses of the stars. Even so, by then, dynamical parallaxes had become an accepted part of practice; observers like Robert Aitken, the heir to Burnham's mantle, concurred that Russell and Hertzsprung had shown how it was "unnecessary to have accurate orbital elements to derive dynamical parallaxes which will be nearly correct, *on the average*, and which can therefore be used safely in statistical investigations."[11]

By the 1930s, Russell was the chief arbiter of practice for all matters dealing with binary systems. Otto Struve, now editor of the *ApJ*, routinely deferred to him, and Russell would engage authors directly if he felt there were issues to raise. When Gustav Strömberg's calibrations using Mount Wilson data did not meet with Russell's approval, Russell held both his own and Strömberg's papers up for a year, debating the details until they could reconcile their methods.[12] Once he and Strömberg agreed upon method and terminology, Russell finally decided that it was time to revise his own catalog of dynamical parallaxes and masses for 2,529 systems, which he published with Moore in 1940.[13]

Russell did not publish until he was sure he had demonstrated that the mass-luminosity relation applied to all stars (except white dwarfs).[14] This allowed him to better define the natural width of the main sequence, which he did at the 1938 Stockholm IAU, and later he commented on how this astrophysical constraint influenced Bethe's process because the width limited how much hydrogen content could vary with mass. Further, it provided a way to determine the relative numbers of stars to be expected in different stages of evolution, critical for assessing star formation rates and the character of the theoretical luminosity function.[15]

Apsidal Motion

Just as dynamical parallaxes led to a better understanding of the generality of the mass-luminosity relation and how stars evolve, Russell's exploitation of apsidal motion in eclipsing binaries led to a reassessment of how solar systems form. The history of the detection of apsidal motion, and its exploitation, requires some backtracking and highlights Russell's dependence upon Dugan's tireless observing routine and the extreme care he brought to the enterprise.

Around 1900, a few observers noticed that the amount of time between primary and secondary eclipses was not constant in the short-period eclipsing system Y Cygni. This suggested that the line of apsides of the orbit, a line drawn in space between the stars when they were at maximum separation, shifted in space. Theorists argued that perturbations caused by the highly distorted ellipsoidal shapes of the stars produced this phenomenon, and that it could be used to deduce the density distributions within the stars.[16]

Russell was drawn to this problem in the late 1920s in order to test Jeans's theory of "liquid stars," which, if they were stable configurations, had to have a specific degree of central condensation. As usual, he thought he could improve on previous methods of reduction; no one had actually done the job. He also knew that Y Cygni had been on Dugan's observing list since 1915; by 1928 Dugan had accumulated over a thousand observations, augmenting them with earlier data from Harvard.[17] In a few weeks of work at Mount Wilson that spring, Russell derived a relationship that equated the rate of the advance of periastron to the internal density distribution of the star, which could be of any configuration ranging from completely homogeneous to having its mass concentrated wholly at the center. He assumed that both stars were rigid bodies and were warped into ellipsoids by rotation as well as by tidal distortion. He calculated the perturbative effects expected for Y Cygni and found a high central condensation, but, to his delight, it was between one and two orders of magnitude lower than Jeans's theory demanded. Originally, Russell wanted to say that his results did not "confirm the very much higher central condensation which [Jeans] advocates." But he thought twice, knowing that he had produced only a rough solution. Dugan advised caution, so Russell excised the sentence.[18]

Once Russell presented their work at the RAS in June 1928 and then at the Leiden IAU in July, he moved off to his many other projects while Dugan remained tied to his telescope and observing lists, unsatisfied with his Y Cygni data. During his sabbatical at the Lowell Observatory in the spring of 1929,

Dugan felt that it would take several more years and thousands of observations to cover the three-day orbit sufficiently. Y Cygni was then the only star to show clear evidence of apsidal motion, whereas other stars on Dugan's list, such as RT Persei, could not be "explained so simply."[19] So Y Cygni was indeed a very important system, and though Russell may have skimmed off the first cream, as was his wont, Dugan knew there was more to be learned, and he collaborated with other observers to fill out the orbit.[20] From time to time Russell examined Dugan's data, looking for new clues to exploit. In 1930 they refined their analysis to account for limb darkening. Apsidal motion was still suspected, but they both agreed that more work was needed because they could not rule out the existence of a third body.[21]

In the 1930s other mathematical theorists began to look more closely at Russell's method. Kurt Walter of Königsberg tried out a rigid body model, and Chandrasekhar investigated the behavior of centrally condensed distorted systems as polytropes. Chandrasekhar's solutions appealed to Zdeněk Kopal, a student at Charles University in Prague, who found them "more accurate than the results of earlier Russell computations."[22] But in 1938, when Kopal, now a postdoctoral fellow at Cambridge with Eddington, tried to generalize his solutions, he found that Russell's idealized treatment was more physical and represented the observed motion of the line of apsides best. Cowling agreed that Russell's method was still the best, even though it did not lead to an exact solution. Cowling studied the influence of tidal distortion when such forces were appreciable, and found that Russell's original formula needed only slight modification.[23] At the end of the decade, Theodore Sterne at Harvard provided an exhaustive summary and evaluation of the state of knowledge, agreeing with Cowling.[24]

Russell watched Sterne's progress closely at Harvard, critiquing his manuscripts as Shapley sent them. Sterne's thorough analysis caused him to rethink a good bit of his own theory and technique.[25] In June 1939, Russell collected together data for some twenty systems that showed evidence of apsidal motion, and found that the shapes of gaseous stars were due to "instantaneous tidal forces." Exact solutions were still far off, but there was little value in pushing a theory beyond the capabilities of observations to test it.[26] The phenomenon could still be due to the presence of a perturbing third body, something Russell had been pondering for over five years. It was by this route that Russell was led to solar-system cosmogony, which he viewed as a direct link to stellar evolution.[27]

On the Origin of the Solar System

"From more than forty years' observations of the ebb and flow of astronomical theory," Russell mused in 1941, "I am convinced that the most difficult of all fields is that of cosmogony. There is hardly any problem more difficult, and in which less progress has been made, than the origin of the solar system."[28] Russell was then the foremost commentator on the state of the field, arguing that none of the major theories, including the old nebular theory and newer encounter theories, explained the present system of planets, satellites, and comets reasonably enough to be acceptable.[29]

At one point Russell tried valiantly to make one particular form of the encounter theory work. He came to it when he and Dugan realized that apsidal motion phenomena could also be due to perturbations of a third body in a close binary system. He therefore asked mathematically inclined students to examine three-body interactions in spectroscopic binary systems; he speculated that a three-body encounter might just account for some of the objections that he and others had raised against two-body encounters.

Russell, of course, had been raised on Laplace's venerable Nebular Hypothesis that envisioned the formation of planets as a consequence of the formation of stars out of diffuse nebular material. Since the turn of the century, however, there had been a powerful alternative created by the geologist Thomas Chrowder Chamberlin of the University of Chicago. Chamberlin, assisted by Forest Ray Moulton, knew that the present distribution of angular momentum in the solar system was centered mainly in Jupiter and the other Jovian planets, a fact which could not be accounted for by Laplace's ring-ejection mechanism. To account for angular momentum, Chamberlin and Moulton argued that solar systems were formed by the tidal encounter of two stars, drawing from one of them two filaments of hot gases that quickly condensed to form small solid bodies which would collide and stick together and grow by accretion into planets.[30]

The Chamberlin-Moulton theory gained favor and was developed and modified through the 1920s by James Jeans and the young Cambridge geophysicist Harold Jeffreys, both finding independently that very close encounters of stellar bodies were not required. Russell liked Jeffreys's exposition, proclaiming the "Fall of the Nebular Hypothesis" in his June 1924 *Scientific American* column. Favoring Jeans's and Jeffreys's accounts, Russell described a solar system "Born by Chance."[31]

But by the end of the decade, Russell had misgivings. How could such very hot filaments condense? Wouldn't their heat make them dissipate quickly? He mentioned the problem in a 1929 review of Chamberlin's book *The Two Solar Families: The Sun's Children* and wrote to Jeffreys and others about it on a number of occasions.[32] By 1931 he was worried about how the view of ordinary stars as having relatively dense cores with temperatures in the millions of degrees could be reconciled with tidal theories. He now labored over the angular momentum problem and the difficulty of envisioning such an extended series of massive planets resulting from any form of collision. "There is a great deal still to be found out before we have a really satisfactory theory," Russell concluded, "but real progress appears to have been made . . ." Russell soon decided, however, that fatal flaws haunted all theories of cosmogony.[33]

The Page-Barbour Lectures

In early 1933 Russell agreed to deliver the Page-Barbour Lectures at the University of Virginia; his topic would be the origin of the solar system, which he thought bold, considering that cosmogonists "tear each other's hair." He wanted to lay out the situation and show that no one had the inside track. "I shall at least be hated impartially," he told S. A. Mitchell, his host. "I have wanted to get my opinions on this subject out of my system for years, and this is a good chance."[34]

For years Russell had been both contributing to and absorbing a wide range of information relating to the origin of the solar system, mainly planetary and geochemical. In 1921 he used uranium isotope abundances in the earth's crust to estimate the planet's age; later in the decade he encouraged Adams and Dunham to study the composition of planetary atmospheres; in the early 1930s he collaborated with Menzel to compare Earth's atmospheric composition with that of nebulae, looking for deviant abundance ratios that would hint at the chemical evolution of the earthly atmosphere. He also studied the impact features around Meteor Crater in Arizona and became interested in the thermal conductivity of the lunar surface during his service on a Carnegie Institution of Washington Committee on the Moon. He continued to push for spectroscopic and radiometric work on the planets by astronomers at Mount Wilson and Lowell where, especially, he guided people like Carl Lampland to develop physically meaningful reduction procedures to determine planetary temperatures.[35]

As he prepared his lectures in the spring of 1934, he consulted with Jeffreys, Georges Lemaître, Willem Luyten, and William Markowitz about what he would say, looking for arguments that might soften or alleviate his criticisms. To Carl Lampland he admitted that "[t]he more I dig into the subject, the less I think that anyone knows about the origin." He did not relish preparing his lectures for publication, however, writing Dugan, who was spending the year in Pasadena, about "those wretched Virginia lectures."[36] Despite his usual level of overwork, Russell remained deeply committed to the project, clearly frustrated with the state of cosmogony, as letters to Jeffreys attest.[37] His lectures would be a trial in more than one way, however.

In late April, the Russells decided to travel to Charlottesville by automobile, a bold move for a couple that still used a fringed surrey and a horse to get around town. They headed south through Trenton and into Pennsylvania. But they got no farther than Lancaster. Feeling the freedom of a newly minted two-lane concrete highway, Mrs. Russell, at the wheel, decided to pass another car and lost control as she accelerated to forty-five miles per hour. The car left the road, rolled over, and providentially pitched the Russells through the cloth roof. Russell ended up with a broken arm and later learned he had a cracked rib. Mrs. Russell was shaken and bruised.[38]

Mitchell was horrified when he got the telegram that day and immediately canceled the lectures and the attendant festivities, though within a day he and campus officials asked Russell to try again that fall. It would be a postponement, not a cancellation, and Russell happily agreed. To Slipher, Russell admitted that their little test had been a failure: "This thoroughly disposes of any plans we had to drive west."[39]

The Russells could not miss their Flagstaff summer, as they were to be Mrs. Lowell's honored guests, staying in the "baronial mansion." Russell took the opportunity that summer to regale the staff with his Page-Barbour Lectures, making them as provocative and visually appealing as possible, collecting new images of the planets at Lowell. With each successive round, as Frank Edmondson recalls, he thoroughly demolished the Chamberlin-Moulton hypothesis and the tidal theories deriving from it.[40]

The lectures were rescheduled for November, and this time they took the train. Much of his published commentary was didactic, and all of it was descriptive. His last lecture concentrated on theories of origin; in it he stated, as if stating a creed, that "[t]he solar system is most evidently not a product of chance." Too much order appeared in the dynamics of the system, and too many regularities were present. The small eccentricities and inclinations of the larger bodies were proof enough for him that some commonly acting process had taken place to bring the solar system into existence.[41] He emphasized how isolated the solar system was, as were all stars from each other, which made any encounter theory problematic if one wished to live in a universe teeming with planets and life, as Russell evidently did. Concentrating on dynamical and physical arguments, he ruefully concluded that all present theories "have ended in an impasse." Yet the greatest piece of evidence was that at least one solar system existed and "must have had an origin of some kind."[42] Every alternative therefore deserved attention.

Russell speculated that some form of encounter might have been at play. The Sun could originally have been the primary component of a binary system with a high mass ratio, where the smaller secondary orbited at the mean distance of the Jovian planets. What would happen, he mused, if this secondary encountered, or even collided with, a passing star? He also wondered about the implications of the expanding universe and the confluence of timescales: the solar system was formed when the universe would have been much smaller than it is now, at a time when collisions were more probable.[43]

Despite this speculation, Russell's published lectures, *The Solar System and Its Origin*, became "one of the most influential books about the solar family written between the world wars" because he presented detailed criticisms of all competing theories and much useful descriptive material. One of his main predictions was that astronomers would need help to make headway. Jeffreys agreed that the problem had to be attacked on an interdisciplinary basis. Russell's prediction fostered "a cooperative style of research" among some scientists, though the field hardly coalesced until the 1960s.[44]

Raymond Lyttleton's Tenure at Princeton

One of Russell's suggestions in his Virginia lectures came directly from studies at Princeton. He had set a Jane Proctor visiting fellow from Cambridge, Michael Heriot Huth Walters, to the task of examining the perturbations of a close binary by a third body in the system, to test alternatives to apsidal motion. Russell wanted to see what would happen if the third body was in a highly eccentric long-period orbit that brought it near enough to one of the stars to cause tidal disruption. Russell paid little attention to Walters's progress until the poor fellow died of a heart attack in late March 1934 and left Russell and Dean Eisenhart to console his family and rescue from his room what they could of the work. Russell was working on Walters's research notes when he had his automobile accident. But after consulting Walters's father and Eisenhart, Russell sent the notes to Cambridge, to William Smart, Walters's tutor and a highly accomplished celestial mechanician, to see whether he had made any real progress.[45]

Smart had another advanced student at Cambridge named Raymond A. Lyttleton who was analyzing perturbations in the famous eclipsing binary system Algol, then suspected from spectroscopic evidence to be a triple system. Lyttleton obtained the Proctor Fellowship that became available upon Walters's death and, as Russell told E. W. Brown, arrived at Princeton in the fall of 1935 to "attack my pet problem . . ." Aware of his own limits and demands upon his time, Russell asked Brown to be Lyttleton's interim adviser when Russell was away in Pasadena: "You know better than anybody how much I don't know about such subjects."[46] Later, upon his return home from Pasadena just before Christmas, barely in time to enjoy the birth of his daughter Lucy's first child in the Princeton hospital, Russell admitted to Smart that Lyttleton's work "would go faster if I knew enough about the subject to give him much help." Russell had found a way to treat perturbations in vectorial form that he wanted Lyttleton to develop, but wanted Brown's assurance that it was physically realistic.[47]

Once at Princeton, Lyttleton took up Walters's work and read *The Solar System and Its Origin*. There he found a solution to the angular momentum problem that incorporated Russell's three-body encounter.[48] Though Russell had rejected the idea, Lyttleton found an elegant mathematical solution and presented it straightaway, to Russell's great astonishment and delight: "[Lyttleton] just had a brilliant inspiration about the origin of the planets," Russell wrote Adams in March 1936. Taking the idea he had rejected, that the Sun was originally in a binary system which experienced the intrusion of a third body, "he has shown that both these stars may have escaped into space, leaving plenty of material captured by the Sun with a distribution of angular momentum quite satisfactory for the planets. His new scheme clears up so many of the minor difficulties that I really think it is a major advance in cosmogony."[49]

As Lyttleton's fellowship drew to an end, Russell applied for an extension from the Commonwealth Fund Fellowship, writing Smart and Eddington that Lyttleton had found a safe passage through what Russell had thought were treacherous waters. He could now account for the distribution of angular momentum in the planetary system and was able to extend his theory to explain the origin of planetary satellites; "with one blow" Lyttleton had removed all the problems inherent in the cosmogonies of Jeans and Jeffreys. To the vice-chancellor of Cambridge Russell pointed out, "It is decidedly unusual that a young man should make so important a contribution to astronomical theory at so early a stage of his academic career." Lyttleton had more than satisfied the requirements for his Cambridge degree.[50]

Smart and Eddington agreed. Lyttleton, in return, gave Russell full credit for the idea, though he did not fail to mention that Russell had originally rejected it. At Russell's suggestion, Lyttleton did not try for an exact solution, arguing like his Princeton mentor that it was both "impracticable" and "unnecessary." His was an "orders of magnitude" discussion, a first step, where the encounter was visualized vectorially in velocity space, as Russell had advised and Brown approved. This scheme neatly defined both the problem and the parameters at hand.[51] Russell especially delighted in the fact that Lyttleton's dynamical argument used a minimum of special conditions and subordinated mathematical formalism to physical intuition and the judicious use of approximations. This was theory Russell-style.

As Lyttleton continued to work toward a more rigorous solution, he ran into difficulties beyond those Russell himself could handle. Russell therefore let his mind wander off to other areas where work could be done. But he also kept an eye on the problems, mainly the nagging difficulty of getting the filaments to condense into planets before their heat caused them to dissipate. Russell reported to Smart that Lyttleton was "pretty well stuck at present . . . I suspect that this is the next big hurdle in the path of cosmogony."[52]

Reaction to Lyttleton's Theory

With Russell's backing, Lyttleton's work was cautiously accepted: most felt that an important step had been taken at last. W. H. Wright at Lick, hardly partial to theory, or the policies of F. D. Roosevelt, quipped to Russell that "[i]f he knows how to make a planet I suggest that he do so, and that we get aboard and leave the New Deal."[53] Smart reported that Eddington was never "very keen on tidal theories—he thinks it is a very 'messy' way of creating the solar system!! He did not say, however, how he would prefer the job to be done."[54] Whatever he felt about the theory, Eddington informed Russell some months later that a fellowship awaited Lyttleton at St. John's, but meanwhile Russell knew that Lyttleton had come up against a real block: the condensation problem. He hoped that modern physics would come to the rescue and managed to get H. P. Robertson to examine conditions promoting condensation. Robertson, however, after being lectured by Russell on Poynting's old work, discovered that the existence of any resisting medium had to be a temporary affair, in the tens of millions of years at best. This put some constraints on Lyttleton's scenario but no more so than it did any other form of the encounter or nebular theories.[55]

The greatest resistance to Lyttleton's theory came from the brilliant but irascible Dutchman Willem J. Luyten, who expressed surprise that Russell could champion a theory which did not explain very much.[56] Luyten retraced Lyttleton's steps using Russell's vector method, but he soon got stuck and with the help of his campus colleague, the Minnesota physicist E. L. Hill, reworked the problem in terms of the energies required to "rub off the planets" from the stars. He soon got bogged down trying to handle the distribution of angular momentum and the preservation of hydrogen in the giant planets. Luyten appealed to Russell for advice, asking at the same time for endorsements for fellowships. Luyten regarded Russell as a critical patron and so was very deferential to Lyttleton's theory at first, doing all he could to make it work. Russell believed that Luyten had raised good issues, but found flaws in his analysis of the perturbations. Luyten agreed to make corrections, thanking Russell for his fellowship endorsements.[57]

Luyten, however, thinking he now had Russell's favor, attacked Lyttleton. They had clashed before, and now, no matter what he tried, Luyten could not reproduce Lyttleton's ability to keep the filament gravitationally bound to the Sun.[58] Luyten also became suspicious, convinced that Lyttleton had slighted the work of others, including his own, and started making claims to Russell, who sternly advised him to concentrate on scientific criticism and stay away from ad hominem remarks. Luyten well knew that Russell's imprimatur on a theory would incline others to accept it; he hoped that if he stayed on Russell's good side, none of the Englishmen would be able to slight his own work. Still, Luyten stated his mind and eventually rejected Russell's and Lyttleton's vector solution.[59]

As Lyttleton completed his stay in Princeton, received his degree, and moved back to Cambridge, where he took the fellowship at St. John's, Russell continued to spar with Luyten. Luyten keenly knew that if he wanted to publish his ideas in the *ApJ*, he had to accommodate Russell, since Struve put all dynamical papers in Russell's hands. Predictably, Struve asked Russell or Lyttleton to append a rebuttal. Instead, Russell kept Luyten's papers at bay until Luyten behaved: "I would very much rather thrash the matter out by letter and avoid even the appearance of public controversy."[60] Luyten, frustrated, sent his paper to the *Observatory*; Lyttleton read it and immediately wrote Russell asking "how to pacify so rabid a critic."[61] Russell assured Lyttleton that "Luyten has bad manners in controversy, but means well, and is entirely honest and straightforward." He had made some good points, which, with a bit of work, could be answered. All Lyttleton needed, Russell felt, was to allow for a closer collision and a larger velocity for the interloper. For Russell, it was "a question of degree and not of principle."[62]

Luyten's criticisms, however, began to worry Russell, who looked for the constructive elements in them while trying to get Luyten to tone down his rhetoric. By the summer of 1937, Russell's patience grew thin. Telling Luyten that by "over-stating your case you injure it," Russell finally got Luyten to back down, and Russell let the paper be published.[63] What really bothered Luyten, beyond the dynamics, was his abhorrence of the deus ex machina character of all collision theories. With many astronomers, Russell included, he wanted to believe that eventually, somehow, an "internal" or uniformitarian model would prevail. There was simply too much order in the solar system to permit theorists to ignore such a mode of origin. There was simply no reason to accept that "the Sun in any way [was] unique or singular."[64]

Others, including Fred Hoyle, Atkinson, and Jeffreys, participated and commented at times. But some of those watching the smoldering debate, like Gerard Kuiper, concluded later that it had all been quite unnecessary. Luyten's doggedness did, however, keep the issue alive.[65] Although he concentrated on the probability factor, lurking beneath the debate was the deeper physical problem that Russell and Luyten had articulated but did not pursue: the question of what would happen to the filament once it formed. Lyttleton tried to extend Jeffreys's ideas to explore dynamical mechanisms of fission and condensation, but he and everyone else avoided the physics of the situation until another Russell student took a closer look.[66]

Lyman Spitzer's Analysis

Lyman Spitzer overlapped with Lyttleton at Princeton and certainly was aware of what was going on. During meetings in the library with Russell, Dugan, and Stewart where they planned their textbook revision, much discussion dealt with planetary physics, including new infrared observations and cosmogony: "It was partly at that time that my interest began in the theories of the origin of the solar system."[67] Spitzer's immediate concern was to complete his thesis on expanding atmospheres in giant stars, but he was also interested in the physics of the interstellar medium and how matter behaved undergoing radiative energy loss. Thus he was well prepared to study Lyttleton's filaments during his postdoctoral year at Harvard.

Spitzer had a solid grounding in physics. Graduating from Yale in physics, he spent a year studying with R. H. Fowler and Eddington at Cambridge before entering Princeton for graduate study. Much of Spitzer's reading and research at Princeton in astrophysics in the years 1936 through 1938 centered on theoretical stellar atmospheres. He studied under E. U. Condon and Eugene Wigner, preferring Condon because he was most concerned with "the physics of what was actually happening."[68] Spitzer completed his general examinations in physics and then wrote his thesis under Russell. As was typical for a Russell student, his data came from Mount Wilson: exceptionally fine spectra that Russell had secured from Adams. Unlike other students, however, Spitzer was able to spend a summer at Mount Wilson working directly with the plates and doing the microdensitometry himself.[69]

Spitzer went on to Harvard with an NRC fellowship: "He is probably the best student whom I have had for a considerable number of years," Russell claimed. Spitzer, like Dunham, Russell judged, was of good character and lineage: "[I]n personality and presence he has exceptional advantages—he obviously comes of very good stock indeed."[70] At Harvard, on Sunday evenings, Spitzer attended informal graduate student gatherings at Bart Bok's home where the "active scientific problems" of the day were thrashed out. Wild speculation was always welcomed in the safety of the exuberant Dutchman's intoxicating intellectual web, and after one such session sometime in the spring of 1939, Spitzer recalls, he returned to his room in Dunster House ready to take on filaments as polytropes.[71]

He suspected that filaments would dissipate rather than coalesce into planets because of their initially high temperatures and pressures. After a few days of work, he was convinced this was so. Radiative cooling was simply much slower and much less efficient than convective expansion. Spitzer sent off his finding to Russell, who immediately grasped its significance.[72] Russell advised Spitzer to take some care in how he presented the problem, mainly because Luyten was involved. Spitzer found that a considerable amount of material ejected in the filament could be retained by one of the colliding members in the encounter, and though it would not condense into planets, it would, once it cooled sufficiently to become optically transparent, become "an enormously extended atmosphere of some sort around one of the two stars . . . reminiscent of the Laplace nebular hypothesis," but with the right amount of angular momentum. Thus Spitzer combined the encounter and nebular theories but left open "whether or not a non-uniformly rotating atmosphere could condense into solid bodies."[73]

Russell was delighted with Spitzer's achievement. Not only was he proud of his own student, but he relished a return to some form of the Nebular Hypothesis. He used his March 1940 *Scientific American* column to talk about the "race between expansion and radiation" in a filament. Spitzer's work left "no room for doubt" that expansion would win.[74] Uppermost in his mind was that Lyttleton needed to face Spitzer's challenge. "Spitzer is obviously right," Russell told Struve; "it is really not worth while to spend much space on any form of the encounter theory" in the *ApJ* "unless some fallacy in his physical argument and conclusions can be shown." At the moment, "the whole thing looks to me like flogging a dead horse."[75] Lyttleton, however, did not agree. He eventually countered, arguing for the importance of convection, liquification, and other ad hoc influences Spitzer had neglected. Luyten declared that Spitzer's work was trivial as he continued to fume at Lyttleton; they all looked to Russell to sort things out. But Russell was now weary of the whole affair, disappointed with the ever-sharper tenor of Lyttleton's rejoinders and Luyten's intransigence.[76] Russell recited an old fable to get Luyten to see the light:

> A crusty old judge once found only eleven jurors in the box on a Monday morning in an important case. He asked the foreman, "Where is Juror number five?" receiving the reply, "Your Honor, he has at least five good reasons for not coming" and thereupon followed a long harangue from the bench on the vital importance of jury duty and the disgrace of neglecting it. Then suddenly interrupting himself, the Judge said, "But the law judges no man unheard. Mr. Foreman state your reasons." "Your Honor, the first is that he is dead."

"Why not let it go at that?" Russell suggested.[77] Luyten got the point but still wanted his say, as did Lyttleton. Both Struve and Russell told all three to work out their difference by correspondence. Publication could wait until all parties agreed.[78] Frustrated with eclipsing binary work at Harvard, worried about the safety of his daughter Lucy's family in Salonika as the war intensified in 1941, occupied with plans to teach navigation again, and anxious to join the scientific advisory board to the Aberdeen Proving Grounds in Maryland, Russell had other things on his mind.

The debate was eventually swallowed up by the war, although Lyttleton managed to publish rejoinders first to Luyten in the *ApJ* in 1941 and later to Spitzer.[79] Lyttleton stuck to his conclusion that his theory formed the most viable basis for further exploration and refinement, and that, despite Spitzer's claims, he had established the basic lines along which future cosmogony would progress. Unlike Luyten, Spitzer listened to Russell and decided that the most prudent course was to concentrate on problems with solutions.[80] Luyten wanted the last word, and Struve gave it to him, embarrassing him when he appended an editorial stating that "[u]ntil fundamentally new information becomes available, this discussion of the origin of the solar system will be closed herewith."[81] By not responding, Spitzer escaped Struve's wrath.

Russell's behavior in this episode reveals once again his continued conviction, championed by Struve, that controversial matters had to be ironed out in private. By 1940 Russell had served as collaborating editor for the *Astrophysical Journal* for two decades, the longest tenure for anyone not on the Yerkes or Mount Wilson staffs. His is a good example of how much of the internal workings of American astronomy depended upon consensual behavior governed by a senior elite. This was a far different policy from that obviously favored by English editors, who flourished on controversy. American astronomy at midcentury was still a rather small, elite community. Although many hundreds of people called themselves astronomers, only a few defined norms of practice.[82] Spitzer understood this fact and would prosper as a result.

At another level, we find Russell readily changing course when new arguments or evidence emerged. He listened first to Lyttleton, then to Luyten, and finally to Spitzer, always following what worked best. He was more pragmatic than Eddington, who in 1938 viewed the course of cosmogony with disdain: "[T]here is no excuse for those who treat cosmogony as a field for unguided speculation of a type which would not be tolerated in any other branch of physical science."[83] In this Russell agreed, but he himself was not reluctant to speculate and was quite aware of what he was doing. He wanted to maintain an open attitude, looking always for fresh, even rather wild, new ideas. In 1945 he was delighted to receive a manuscript from Chandrasekhar on von Weizsäcker's new theory of the origin of the solar system, characteristically remarking that although he had questions about it, he deferred to Chandrasekhar's physical insight. Above all he hoped it was "a chance for real progress."[84]

The Binary Solution: Life in the Universe

If Russell had a private hope or bias, regardless of what science and statistics argued, it was that a cosmogony would be found that would not leave humans alone in the universe. In his Dodge Hall talks on science and religion in 1917, he preached that "[t]he ultimate basis of the Cosmos is adapted to the evolution of life." For Russell, evolution, intelligence, and character were "not 'accidents' but are in the nature of things."[85] He was prudent in his textbook in 1926, helping to establish the prevailing view that since encounters were "extremely rare," planetary systems had to be "infrequent" and the existence of life elsewhere a matter of "pure speculation."[86] In 1930, however, writing a chapter entitled "Life in

Space and Time" for a textbook on human biology and "Racial Welfare," Russell still cited the same dreary statistics that pointed to the rarity of planetary systems. But now he argued that in the vastness of space, there might be millions of them and millions of potentially habitable worlds: "Life appears in the solar system in two out of three of the planets where it has any chance of existing. Hence the number of actual abodes of life in the universe may be great." Deeper into his manuscript text, Russell pondered the lessons of Mars, concluding that "[l]ife, amazing as it is in the complexity and delicate adjustment of its processes, is not confined to our world alone, where we might suppose its origin to be due to some happy, but almost infinitely improbable, combination of favorable conditions."[87] Unlike Eddington, Russell did not believe that humans were intellectually supreme in the universe. There had to be conscious sentient life elsewhere.[88] Yet despite his inner beliefs, Russell still could be fully engaged by Lyttleton's collision theory and followed it to its ultimate conclusions. He admitted to no conflict in this. In his Dodge Hall lectures, he identified himself as a scientist immersed in a "fairly active" scientific life for twenty years, yet he felt that he retained "a belief in every article of the Apostle's Creed." It was quite possible, he claimed, to "retain these two intellectual attitudes simultaneously and concurrently, without the least conscious division of my mind into the 'water-tight compartments' whose contents must on no account be allowed to mix for fear of the ensuing reactions."[89] Nevertheless, Russell held his breath through the 1930s; he was immensely relieved when Spitzer shot Lyttleton's theory down. Even though there was a direct relationship between theories of cosmogony and official statements about the ubiquity of life in the universe during this time, in Russell's case at least his personal convictions—sometimes public, most often not—were fairly constant.[90]

In commentary in *Scientific American*, popular speeches, and radio talks, Russell harbored a deep desire to believe that human beings were not an accident in the universe. For a talk at Bowdoin College in Maine in April 1941, Russell relied on the humor of Gilbert Chesterton, who once defended his belief that he had been born, even though he had no recollection of it and could not test it. Similarly, Russell believed that solar systems had to exist, even without an explanatory theory or observational evidence.[91] When evidence emerged in the 1940s from astrometric studies that hinted at the existence of extremely low-mass invisible companions to some late-type dwarfs like 61 Cygni, Russell was delighted, proclaiming that "among the stars at large, there may be a very large number which are attended by bodies as small as the planets of our own system. This is a radical change—indeed practically a reversal—of the view which was held a decade or two ago."[92] Hardly a month after he suffered his first heart attack in March 1943, during a visit to Elizabeth, who was then living in Manitou Springs with Jean Hetherington, Russell was excited by this new evidence, for it relieved his fear that the old anthropocentric view of human uniqueness would never die:

> Now this last stronghold of the old way of thinking has fallen, and there is no longer a basis for supposing that either this world or its inhabitants are unique, or in any way the "first, last and best of things." The realization of this should be good for us.[93]

Life in the universe was an issue that touched closely Russell's sense of being. And since he felt that the subject was most powerfully addressed within binary star astronomy, it was binary work that remained closest to Russell's heart; his abiding interest in binaries was nothing less than a deep personal need to find himself in God's universe. This fact will help us appreciate why he reacted so forcibly when his dominance in the field was tested.

• CHAPTER 18 •

The Royal Road

"A TISSUE OF APPROXIMATIONS"

In December 1937, taking the podium at the Bloomington, Indiana, meeting of the American Astronomical Society, Russell gave his retirement speech as president of the society. He was dressed as always, in his high-top therapeutic shoes, high starched collar, and expensive specially tailored suit. White hair neatly combed, his sixty-year-old face still lean but deeply lined, Russell was definitely "the old Princetonian," even though he habitually wore his King's College pin announcing his Anglophilic bond.[1] His sartorial precision, and his bearing as he addressed his audience, stood in contrast to his topic, for he spoke on "the place, utility, and limitations of approximate methods in astronomical work."[2]

If anything characterized Russell's mode of operation at the interface of theory and observation, it was his penchant for approximate methods, his delight in crafting what he now described as "a tissue of approximations," and then fighting for it with all the powers he could marshal.[3] Just as he had rebuked Doolittle's resistance to his hypothetical parallaxes, just as he had resorted to classical models of the atom to develop his theory of two-electron interactions, and just as he had marshaled arguments based upon the classical atom and evidence from astrophysical data to argue for hydrogen's abundance in stellar atmospheres, Russell's taste for approximate methods in astronomy would place him in sharp contrast to more orthodox European theorists.

Russell employed his "tissue" metaphor to make a political point. In his address, which he couched in the form of a lecture on "astronomical economics," he celebrated the heuristic power of his "tissue of approximations" because the tissue provided a useful framework upon which a stronger fabric might someday be woven. Recalling the achievements of Saha, Payne, Moore, and his own, as "illustrations of the power of this method," Russell reminded his audience that he had scored "a remarkable success" in his abundance work, performed before transition probabilities were available, using "a doubly and trebly approximate method." His analysis of line intensities in multiplets gave a "first theoretical approximation" which had led others like Milne and Minnaert to apply the "Russell mixture" to create new model stellar atmospheres that worked "very much better than might have been expected."[4]

As the most senior astrophysical theorist in a largely observational American landscape, Russell engaged in a style of theoretical inquiry that was both practical and pragmatic, characteristic of American physical theorists who were more interested than their European counterparts in how theory could be applied to solve real problems.[5] This characteristic of Russell's scientific style was evident in his interactions with mathematical theorists like E. A. Milne, whom he debated over fitting procedures for composite stellar models. Russell's writings, scientific and religious, and his correspondence with Milne, reveal how he tried to balance physical insight with sense data, resorting whenever necessary to "exact qualitative approximations" and rough numerical solutions to approximate reality.[6] His constant testing of insight against observational evidence, searching for the best fit, rhetorical or scientific, was the puzzle he loved to play. He knew that the pieces would never be known perfectly, so he constructed models that were just good enough to hint at how the pieces went together, hoping they would eventually lead to iterative schemes to complete the puzzle. He held to his view tenaciously, never more so than in his disputes with Zdeněk Kopal over methods of reducing eclipsing binary data.

Rules of the Road

Binary stars were Russell's "Royal Road," as he said in 1946, the most favorable path to new knowledge. But the rules of the road were very different from those familiar to traditional celestial mechanicians. Unlike solar system dynamics, Russell pointed out many times, where observations were both plentiful and precise, the determination of orbits for visual, spectroscopic, or eclipsing binaries required very different standards since observations were crude. Thus it was rare that fully analytical methods were warranted. Trust no single observation, he would say time and again, but make the best fit of the scatter plot of observations that exist, and from them compute a preliminary orbit to see how well it fits the data. Russell used simple iterative schemes where each new set of elements described the data better than the last. Above all, he lectured students at Princeton and colleagues everywhere, "[l]east-square solutions should never be attempted until the resources of graphical adjustments have been fully exploited" (see fig. 8.1). Prudence avoids waste: "The computer thus escapes the heartbreaking experience of finding, after the heavy labor of least-squares, that the second-order terms were sensible, and the whole job must be done over."[7]

Fig. 18.1. Russell and Shapley confer aboard a steamer during the 1938 International Astronomical Union General Assembly in Stockholm. Photograph by Dorothy Davis Locanthi, courtesy E. Segrè Visual Archives, American Institute of Physics.

In a day when computers were people, Russell abhorred mindless repetitive work, at any level. Astronomical computing was "not the mere mechanical application of standard formulae." The analysis of orbits required intuition and insight, experience and care, and, above all, "intelligent appreciation of the accuracy available in the data . . ." As he well knew from his first decade of exploring hypothetical parallaxes and the analysis of orbital systems, "[t]o get the best results out of data of low precision is as much of an art as a science."[8]

Russell practiced this philosophy, not only because he believed in it, but because he knew it was accessible to observing astronomers who were more attached to their telescopes than to mathematical analysis. By the 1920s, his reduction procedures were widely accepted; as one recent commentator concluded, for the first half of the century "Russell governed the whole evolution of the theory of [binary] orbits."[9] Robert Aitken, Lick Observatory's consummate double star observer, quickly warmed to the "Russell Method" and looked to Russell as his guide, basing his own treatment of eclipsing systems "entirely upon [Russell's] investigation" in his 1918 text *Binary Stars*. The "Russell Method" was taught widely and accepted almost universally; it became his "Royal Road" to uncovering the origins of the earth, planets, and stars, and held out the potential for revealing humanity's place in the universe. Such power, authority, and conviction were hard to deny and, surely, even harder to relinquish.

Harlow Shapley awoke one spring morning in 1945 with a most felicitous thought. His old teacher was nearing retirement, and something had to be done to mark the event. Shapley was then president of the AAS and, after deliberating over various options, decided that the society should establish an endowed lectureship in Russell's name. The "Henry Norris Russell Lectureship" became a campaign: Shapley enlisted Russell's colleagues in physics and astronomy and, with the approval of the AAS Council, had the society secretary organize the rank and file as he went after the deep pockets—including the Mexican ambassador to the United States, because Russell had been the leading light at the dedication of the Mexican National Observatory at Tonantzintla in 1942. He told Harold Dodds, president of Princeton, "I am, of course, very much pleased that the one who taught me the tricks of the stars will be through this lectureship indefinitely honored as the trickiest star of them all." He petitioned Arthur W. Butler to support the lectureship, in the name of "America's leading astronomical genius." Joseph Boyce and William Meggers pitched in, writing to physicists.[10]

By February 1946, $8,500 was in the coffers from several dozen major donors. To raise more, Shapley sent out a blunt demand to the rank and file ("According to our records you have not yet joined your colleagues in subscribing . . ."), and 250 members signed up. Shapley's techniques were hardly subtle, and a few balked, like W. H. Wright at Lick, who wrote directly to Russell and Shapley that honors like these should be posthumous, and, in any event, he really resented the implication that those who did not contribute were not Russell's friends. There is no evidence that Russell resisted this effort, but he may well have been embarrassed a bit by Shapley's tactics; he did admit to Berkeley's R. T. Crawford in November 1945 that "[w]hen I first heard of it, I asked some of my friends if they wanted to label me as a 'dead 'un,' but it didn't stop them." The endowment reached its $10,000 goal in time for the December 1946 meetings of the AAS, where Shapley could present Russell as the first lecturer. By then, however, his scheme involved far more than a tribute to his old teacher.[11]

Shapley was delighted when Russell chose to speak on "The Royal Road of Eclipses" as the first Russell lecturer, because he had already been planning a Symposium on Eclipsing Binaries for the December AAS meeting. This was not the first time Russell spoke of a Royal Road. He used the phrase to describe spectroscopy in 1932 and radioactivity in 1934. He liked the well-worn metaphor, and he used it most frequently to describe binary stars. Russell believed that they provided insights into the way stars behaved as living, aging bodies that was available in no other way.[12] He may well have chosen the title, however, to counter allegations by Kopal that there was no royal road in binary star work. His lecture was a distillation, in fact, of his position in a continuing debate with Kopal over the most effective way to analyze binary systems that had been going on for about seven years. The debate generated both political and intellectual ramifications because what started as a happy thought in Shapley's mind, to honor his old professor, ended up as an instrument to separate Russell from the

Fig. 18.2. Detail from group photograph at AAS meeting, December 1938, Columbia University. Russell's students and associates among the attendees. Most of the members of the Harvard-Princeton binary star team are present, save for Dugan. *Left to right, front to back*: Shapley, Mrs. Kopal, Kopal, Hemmendinger, Stewart, Russell, Payne-Gaposchkin, Menzel, Edmondson, Gaposchkin, Sitterly, Moore-Sitterly, Schwarzschild, Pierce, Goldberg, Aller, Spitzer, van de Kamp, Green, Merrill. Photograph courtesy Lick Observatory, Mary Lea Shane Archives, University of California, Santa Cruz.

forefront of his field. Although Russell would chair the symposium held earlier that day, he was not one of the speakers. Kopal, with Shapley's backing, had orchestrated the event to discuss "Some Unsolved Problems in the Theory of Eclipsing Variables." Russell's long and stormy relationship with Kopal provides further insight into Russell's style as theorist.

Eclipsing Binaries at Harvard

By the time Kopal arrived in Cambridge, Massachusetts, in October 1938 on a Czechoslovak traveling fellowship and International Education Board grant, eclipsing binary work at Harvard bore Russell's stamp. In the early 1930s Russell, Payne, and Dugan planned out a photometric program and hired Frances Woodworth Wright as a computer to collect and combine all available observations on eclipsing binaries. Despite her obvious promise as a spectroscopic astrophysicist, Shapley had turned Payne to photometric work, which was languishing at Harvard, and so made Payne the organizer subject to Dugan and Russell's direction. She wrote voluminous letters to astronomers asking for data, sources of error, differences between magnitude systems, always holding Russell's name up as her source of inspiration for the intrusion. The basic plan was that Wright would provide photographic data for the eclipsing binary systems Dugan was interested in as he and Russell continued to search for evidence of apsidal motion. They needed good light curves and accurate timings for secondary minima. Payne, as head of the Milton Variable Star Bureau, supervised Wright's work.[13]

Although the Milton Bureau, one of several topical work groups Shapley maintained in his updating of Pickering's factory system, was responsible for all types of variable stars, eclipsing binaries grew in importance throughout the decade, especially after Payne married Sergei Gaposchkin in March 1934. Gaposchkin was a flamboyant and voluble Russian émigré whose passion for eclipsing binaries matched Russell's.[14] When Kopal arrived, the Milton Bureau was a beehive of activity. Theodore Sterne was very interested in the theory of apsidal motion and was making headway reviewing the state of the field. The Gaposchkins and Wright were producing and analyzing data, largely under Russell's watchful eye, and the bureau also benefited from Henrietta Swope and from Shapley's mathematically adept wife Martha, who was equally able to host seventy people at a picnic feast and to compute orbits. The Milton Bureau was constantly being reinvented as the nature of the problems changed from Cepheid monitoring to eclipsing binary analysis, and as the staff changed in consequence of graduation, retirement, or recruitment from Nazi-dominated Europe.

Kopal had worked for a time with Eddington and had come to America with enormous hopes and expansive dreams about ways to accelerate his career and propel himself back to a professorship in Prague.[15] Kopal chafed when Shapley assigned him the usual observatory chores; nothing would deter him from his destiny: to overhaul completely the theory of eclipsing binary systems, to construct new tables for their solution, to write a seminal monograph on the theory, and to publish a comprehensive catalog of orbits for all systems that had been well observed. He told Shapley that he was "at a most critical time when a young astronomer is expected to show his best and build up some credit on which his future can be based."[16]

Shapley was deeply impressed with Kopal, whom he described as the only "Harvard exile who is not Jewish." Kopal's theoretical expertise would be critical to the Milton Bureau eclipsing binary campaign, Shapley told potential sponsors in his successful effort to maintain support for Kopal after his fellowship ran out. Shapley specifically cited Kopal's goal of creating entirely new methods of binary analysis, "replacing the methods of Russell and Shapley of twenty-five years ago." Shapley's sales pitch kept Kopal at Harvard through 1941, when in effect he did become a refugee.[17]

Kopal soon learned, however, that Shapley did not have the final word on how the Milton Bureau operated. Russell was Shapley's appellate judge for all binary work, holding court on all major assignments and manuscripts for publication. The idea was, as usual, to smooth out differences and to suggest corrections before anything went public. When Kopal submitted his first manuscript within months of his arrival, Russell was impressed by his ability to handle problems like phase distortions due to reflection effects. After checking Kopal's work with his own, Russell was delighted to find that the residual errors were very small. He gave his stamp of approval: "A beautiful bit of work! H.N.R."[18]

By the end of 1938, keenly sensitive to the fact that Dugan's health was failing, Russell believed that the future of his own eclipsing binary work depended upon Harvard expertise, so he and Shapley planned out a strategy for continued cooperation, with Russell calling the shots. By February 1939, however, with Kopal and Sterne working madly on the theory of close binary systems, and very much out of harmony with one another, and with Sergei Gaposchkin cutting a rather broad swathe through the Harvard plate vaults and making some wild claims about what he found, Shapley complained to Russell, "There are already two or three generals in the Bureau and only three or four privates." Both Shapley and Russell worried that good workers like Wright and Martha Shapley might become demoralized by the lack of direction too many demands could bring, or by the growing attacks that both Gaposchkin and Kopal were making on the traditional methods of data reduction. Only one émigré astronomer was proving to be manageable: in March 1939, with Richard Prager safe from the Nazis, Shapley reported, "I hear a computing machine buzzing away down the hall—freely and merrily. The operator is R. Prager."[19]

Shapley had to find a way to make his "generals" cooperate. Martin Schwarzschild was then at Harvard and followed the debates between Kopal and Sterne. Apsidal motion was the great problem of the day, but, he wrote Chandrasekhar, the struggle between Sterne and Kopal "has left me in a great confusion about the density distributions in early type stars." They differed over everything: how to use polytropic models, the influence of gravitation on surface brightness, and how tidal forces might induce convective motion in the stellar envelopes.[20] Although Kopal's mathematical techniques were very rich, they were highly complex and hard to verify. Sterne also resented Kopal's encroachments.[21]

Russell's first impressions of Kopal were positive, especially concerning his application of spherical harmonics, which reminded him of some of his own old work. He was not able to devote as much time as he wanted to the subject at the moment, however, partly because he was overcommitted and was also worrying himself over the threatening war clouds, which propelled him into various humanitarian activities. When asked by Struve to present a paper at the dedication of the new McDonald Observatory in Texas, Russell felt he was once again "riding furiously in all directions," but he agreed to the obligation. His distractions had consequences, however. In January 1939, Russell had eased into a mellow mood playing with Margaret's children. When they were about to leave on a chilly Sunday morning for the trip home, Russell knelt down to ask little Peggy, "What shall I do when you are gone?" Peggy pertly replied, just as her mother might have done years earlier, "You'd better go to your 'serbatory and do some work."[22] And that he did. As he caught up on the reports from Harvard, his mellow mood rapidly darkened.

What lay before him were manuscripts by Kopal and Sterne on theory and technique, the beginnings of Prager's study of cluster-type variables with changing periods, and Sergei Gaposchkin's published orbital analyses. Russell did not find Gaposchkin's work at all to his liking; it showed poor judgment in the treatment of observational error, conclusions drawn from inadequate data, and gross sloppiness in even the most simple calculations. Russell was red-faced when he scrawled long letters to Shapley about Gaposchkin's antics. If any grandchildren had been around, they would have been whisked from his presence. Harvard needed a real "watch-dog" like Charlotte Moore Sitterly, Russell demanded; possibly Miss Wright, who by then was an expert in the details of computation. He did not know what to do about Sergei, "without worrying Cecilia half to death." Shapley promised to keep him in check, but to little avail.[23]

Turning to Kopal's papers, Russell remained comfortably positive. His mathematics seemed fine, but his analysis required a broader understanding of the physics of the situation, which was already available in the literature.[24] As Russell continued to read paper after paper through March, he found problems with all three "generals": Sterne, Gaposchkin, and Kopal in their competition all exhibited various degrees of sloppiness and high-handedness. Of course, all three had something to offer, even Gaposchkin, who loved to ferret out utterly fascinating pathological systems. But, Russell fretted, he was making a mockery of himself and of Harvard. Russell knew that critics like Gerard Kuiper damned the "light-hearted Harvard spirit" that bred sloppiness, and Russell shared some of this concern.[25] He worried that his own venerable reduction techniques would be given a bad name. He had to take control personally. No one would be spared.

Russell sarcastically proposed to Shapley that they form a new society, the "S.D.N.M." or the Society for the Damnation of Numerical Mistakes. Sergei was his primary target, but Kopal came into his sights as well, because with increasing frequency Russell was finding simple numerical blunders and inconsistencies that a mathematician of his quality should not have been making. Obviously, Kopal was in a mad rush to publish, and Russell felt the time had come to take him down a notch or two. After a local Neighbor's Meeting, Russell reported that "Kopal is certainly a bright chap," but he was "not quite so sure of his judgment."[26]

"Iteration" vs. "Regula Falsi"

Through spring 1940, in the heat of competition, Kopal became more aggressive about his analytical methods, all the while making simple blunders in his calculations. Kuiper caught them, as did Struve, and Russell knew he had another problem on his hands.[27] Russell became particularly annoyed when Kopal continued to claim that the field of eclipsing binary analysis was moribund, beyond redemption, and required nothing less than a wholly new mathematical framework, fully fit for the future needs of modern astrophysics. Kopal, Russell realized, was not interested in fitting in; he was intent upon taking over.

Indeed, from Kopal's perspective, Russell was a fossil, "living off his past; and concentrating on efforts to leave a record of [his] work for posterity in an organized form." Lacking the "ancestor worship" he found among his Harvard colleagues, especially what he called in several places "Russell-worship," Kopal recalled feeling that Russell had to be benched; he was not about to adapt to Russell's mode of practice and was bound and determined to replace it with his own.[28] What Kopal attacked were Russell's approximate techniques. He scoffed at his graphical solutions, nomographic techniques, and slide rules. The reams of correspondence Russell, Kopal, and Shapley issued over the following months and years, well through the war and beyond, boiled down to several issues surrounding how to treat observational data and how detailed the reduction procedures needed to be, given the quality of these data.

In Russell's idealized method, one drew freehand curves through scatter plots of observations. Next the computer would mark specific points on the curve as "false data" points and use them to determine the elements of an idealized orbit using nomographs and tables that Russell and Shapley had devised, and which had been refined over the years and were now undergoing extensive elaboration by John Merrill, a former Russell student and now sometime-associate at Princeton. Beyond that, depending upon how well the idealized solution fit the original curve, corrections could be made from sets of orbital phase tables (called "associated alpha-functions") to account for the various anomalous effects that were important in the study of apsidal motion. The effects included, among others, partial eclipses, variations in color between the two stars, variable limb darkening, variable ellipticity, reflection effects, and the effect of the varying distance between the components, which could influence all the others. The tables Merrill was calculating based upon the most recent versions of Russell's theory could account for these effects, but to a lesser degree than could Kopal's more rigorous treatment. Kopal's associated alpha-functions promised to be able to manage the most complex, and usually the most interesting, systems, but only when the data warranted the effort. They were overkill for the general analyses Russell desired.

Kopal thought they were always needed. No smoothing of the curve by eye would satisfy his sense of mathematical rigor. Every data point came into his solution, expressed in higher-order spherical harmonics, and the ultimate light curve would emerge by iterative least squares fitting procedures. He rejected outright Russell's "regula falsi" method in favor of successive iterations by least squares fitting. Kopal admitted that Russell's method was sufficient when "the second order variations are negligible." But this was the "point at issue" since any system that was the least bit interesting (i.e., unusual or anomalous) required higher-order analysis.[29]

From 1941 through 1943, Kopal found that it was becoming harder to get his papers approved by Russell, even though Russell still considered him part of the "cooperative plan" between Princeton and Harvard.[30] He was more concerned, however, when Russell began to block progress of his intended monograph, which he saw as an essential stepping-stone in his advancement. Nevertheless, Kopal could appreciate why Russell resisted. When he carried his own harmonic analyses to second-, third-, and fourth-order terms, he created mathematical expressions that were very hard to handle, and more than once Kopal abandoned his elaborations to return to Russell's "classical" solution.[31] Yet when Russell argued that astronomers wanted efficient means of analysis, and that Kopal's were needlessly complex, Kopal retorted that "whether a given method of solution is quicker than another is largely a matter of practice and experience."

He was hardly happy when Russell charged that his methods were "merely a waste of time."[32]

Russell took out his frustrations on Kopal when he was most fearful about his family's safety as the European war spread rapidly. Lucy and her family were still in Salonika, a region being bombed regularly by the Italians. The Gardners barely escaped the German Second Panzer Division tanks in April 1941, and Russell heard they were safe only in June, after Robert Atkinson located them through British Intelligence.[33] None of this was evident to Kopal, of course. Russell let up on the fellow briefly only when he suffered his first heart attack in the spring of 1943. As Russell recuperated at home through the summer and fall, his movements were limited, but not his mind nor his hard lead pencils. Taking the time to think things through, shuffling between bed rest and study, Russell decided that the real difference between Kopal's method of iteration and his of "regula falsi" boiled down to "a purely mathematical argument," which took Russell four closely packed pages of algebra to explain, with the aid of his secretary Henrietta (Young) Boughton. "From the practical standpoint," Russell concluded, "I am as strongly as ever convinced that it is an unsound policy to make solutions by least-squares, until a light curve which satisfactorily represents the whole course of the observations has been obtained by the most rapid available methods, making free use of "trial and error." Kopal shot back that times were changing. Accurate photoelectric techniques and "the facilities offered by modern computing machines and other means of mechanical calculations" were altering the landscape. The time was now right for "precise orbital work." Kopal was convinced that the results were well worth the effort, and that no method should be condemned "only because it is laborious." This was where, he felt, he and Russell differed the most.[34]

Russell's criticisms did force Kopal to refine his techniques. Kopal admitted in his autobiography that his original procedures in 1941 left "much to be desired"—the least squares solutions might well produce the "most probable values of the desired elements" but could not provide information on how well the successive iterations converged. Kopal neglected to mention that Russell had pointed this out to him, though he did so in a manner Kopal was unable to accept. Only when S. L. Piotrowski came from Poland for a year's stay at Harvard after the war to collaborate on the problem did Kopal reach the first "complete solution," which Russell, after the usual critical round of letters and discussions, eventually approved.[35] Indeed, as the various iterations of his papers, and his monograph, moved between Harvard and Princeton, many kinks and vagaries were worked out, and both Kopal and Shapley thanked Russell, the former for his insight, the latter for his perseverance. Privately, however, Kopal continued to seek ways to placate the old man without giving any more of his methods away. In April 1945, contrary to an autobiographical recollection, he suggested that dedicating his monograph to Russell might do the trick. "I have every hope to believe that Russell is now so completely appeased that such a dedication might possibly please him, and help to remove too the last traces of bitterness at my darned innovations—if any is still left."[36]

Kopal's *An Introduction to the Study of Eclipsing Variables* was primarily a theoretical treatment. Russell had blocked its publication in the prestigious *ApJ* monograph series, or even as a publication of the American Philosophical Society. "He seems blithely indifferent to the enormously greater labor in his methods," Russell told Shapley, "and presents them as the last word." In October 1943, Russell claimed no interest in preserving "the methods which you and I developed when Kopal was a small child," though his actions spoke otherwise. He simply could not condone "a slower and clumsy method presented as an improvement. This is all a great pity because Kopal has in his hands a real and very important addition to the theory of eclipsing variables."[37] Four months later he admitted to Shapley that he was indeed prejudiced on the matter. It had been his way of life since he created the "Russell Method," and he was not about to let it go. Shapley, whose loyalties were most definitely torn in the matter, published Kopal's book as a Harvard monograph, but he still wanted Russell to write the introduction.[38]

New Tools and Organizations

Goals can dictate methodologies. Except for apsidal motion, Russell always wanted to collect as much information about as many normal binary systems as possible to allow him to come to general conclusions about the nature of stars. He therefore created accessible reduction techniques that any astronomer or computer could manage. Kopal always wanted to examine single systems, preferably exotic ones, in exhaustive detail to showcase his mathematical skills; he designed his techniques to account for subtle effects at a time when apsidal motion was the most pressing issue at hand. Whereas Kopal wanted to establish a new technique, knowing full well that there is always a "strong emphasis on originality" as a "fundamental requirement of the researcher's role" in "making a contribution," Russell wanted to show that his old methods, suitably refined, worked just fine. The resulting competition between the two thus became generational: the young attempting to displace the old with the senior person using his influence to avoid displacement.[39]

Goals can also dictate the choice of instrumentation.[40] Russell always claimed that the observations were not precise enough to warrant the labor of Kopal's techniques. Kopal countered in two ways: first, by trying to accelerate the introduction of photoelectric techniques into binary star work, and second, by trying to convince astronomers that high-speed electromechanical computers would relieve the drudgery of computation. Russell agreed that photoelectric techniques were a great improvement over visual and photographic methods. He had nothing but praise for Kopal's industry in bringing photoelectric photometrists into binary work during the 1940s. When C. D. Shane, the new director of Lick Observatory, invited Russell to be one of two Morrison fellows for the summer of 1946, Russell at first hesitated, worried about his health. But after John Irwin visited Princeton and had shown him "the finest light curve that I ever saw," Russell was anxious to visit the mountain to see what stars Gerry and Katherine

Gordon Kron were working on. There was nothing like a good photoelectric light curve to get Russell excited. For a new generation of young astronomers who had spent the war working in military laboratories getting familiar with electronics, photoelectric photometry with the RCA 931 photomultiplier and its improved version, the 1P21, became astronomy's newest tool.[41] But it was not a technology that could be rapidly adapted and used by any observer. Few astronomers had experience with electronics, and when it came to acquiring the expertise, some observatory directors decided that "we could probably train a physicist in astronomy better than we could train an astronomer in electronics."[42] Russell believed that it would take time for photoelectric techniques to displace the older methods, and he was right.

High-speed computers, however, were not so openly welcomed. Russell and Kopal held very different views about computers, and these differences came out in the ways they designed their methods for data analysis. Whereas Russell insisted that his computers were people who had to think, Kopal held out little trust in the intuitive powers of the average binary star worker. Kopal believed that human computers rarely had the qualities Russell demanded of them.[43] Kopal gained this perspective from his war work at MIT's Center of Analysis, working with its differential analyzers to compute firing tables and other military data. He led a team of some fifty people in three shifts, working around the clock for the navy, and, like many who used these new machines, he saw them as symbols of the way science would be done in the future.[44]

In late 1943, Kopal warned Russell that "[t]he literature of the past thirty years bears ample record" of how an inept trial-and-error analysis may lead the computer astray. His own method of least squares, however, was "safer in so far as it replaces to a large extent personal judgment by impersonal criteria." Kopal predicted that "with the aid of modern computing machines," his methods would be as quick as Russell's slide-rule computations. He envisioned an "unparalleled invasion of mechanical methods in astronomical computations" after the war, "which should render the solutions of many problems, shirked so far because of their laboriousness, both practicable and easy. Therefore the sooner we shall reform our lines of approach to make full advantage of this forthcoming trend, the better." These were certainly prophetic words, bluntly stated by one who had been in the heat of battle with Russell now for almost five years. Computers and the independence from Harvard and Russell that his war work provided only strengthened Kopal's convictions.[45]

Russell was no stranger to automatic computing since he had been on the Board of Managers of the Astronomical Computing Bureau at Columbia University.[46] He sensed through his wartime contacts at Aberdeen, however, that even though such computing machinery as Bush's differential analyzers showed great promise in the hands of Harrison, Rosseland, Marshak, or even Kopal, the potential for wide and practical use of these machines in astronomy remained largely unrealized. Russell was more like those who populated the MIT Department of Electrical Engineering in the 1920s, which introduced the calculus to freshman "in a manner which emphasized graphical presentation and intuition over abstract rigor." His world remained the slide rule, the planimeter, and the nomographic chart. In like manner he was well enough versed in least squares methods to know that they should never replace physical intuition. His curve fitting and parameter variation techniques were squarely in line with engineering practice, especially in American aeronautics, where rigorous mathematical theory was secondary to empirical testing of aerodynamical shapes.[47]

Russell feared that Kopal was more interested in mathematical technique than in what was really happening in binary star systems.[48] He believed in the integrity of the human computer for science. Kopal's reliance on unskilled operators of computers was dangerous: "I object to this on principle. In such methods the computer proceeds by [rote], not seeing what he is doing." There was no place in astronomy for a computer who could not think his way through a situation, or for a machine that was improperly handled. For Russell, astronomy could never be forced into an industrial template.[49]

Kopal and Russell stuck to their positions throughout the war, which made Shapley nervous. Shapley quietly sympathized with what Kopal was doing, but openly tried to appease Russell. Photoelectric techniques were slowly being adopted, he knew, but high-speed computers would remain inaccessible in the foreseeable future, though he found them exciting. One way to break the impasse was to create a new form of organizational structure. This was something Shapley knew how to do.

As AAS president, Shapley was also the head of the American delegation to the IAU Executive Committee, which convened in Copenhagen in March 1946 to chart its postwar course. Among the many issues on the table was how to reorganize variable star research, which had been the responsibility of the venerable Astronomische Gesellschaft and the Berlin-Babelsberg Observatory. Shapley, with A. Danjon, G. A. Shajn, and Hertzsprung formed an ad hoc committee to deliberate over the issue. The matter of a new headquarters was, in Struve's view, "[o]ne of the burning questions" of the conference, and eventually the Russians and Americans cut up the pie. Their solution was to form a Panel on the Orbits of Eclipsing Stars, which would consist largely of activist members in eclipsing binary research in the United States. It was mainly an expansion of the original Harvard-Princeton collaboration, but now Shapley made sure that photoelectric observers were represented, because that was where the action was. Shapley lobbied effectively and gained IAU approval. Struve cheered, "American astronomers have good reason to be grateful to Dr. Shapley!"[50]

Russell was designated chairman of the panel and was happy to accept as long as Kopal, designated executive secretary, did all the work. In fact, Shapley had conferred with both Russell and Kopal before his trip, and so the outcome was no surprise. The charge given to the twelve-person panel was to oversee the construction of a catalog of "Orbits of Eclipsing Binaries."[51] As the panel took shape in April 1946, Russell added John Merrill and Newton Lacy Pierce from Princeton, and both Russell and Kopal added Gerald Kron, the congenial and agreeable Lick astronomer who was a Joel Stebbins protégé and master of photoelectric technique. They also added "several people interested in the physical side of the thing—Joy, Struve and Chandrasekhar." Kopal would keep lines of communication open by issuing a periodic bulletin.[52]

Russell left the catalog to Kopal, but he would "have a finger" in the work, to be sure, mainly to keep Kopal from "over-elaborating" the material and Sergei "from treating it too slightly."[53] He was more interested in retirement but was still working with Merrill to develop improved nomographic solutions for limb darkening and thought they were making good progress.[54] Still, he was feeling his age. He had suffered another heart attack en route to Chicago the previous September, which upset him greatly and sent him to Stanley Cobb again to consult "upon the psychological question" of his health. He feared evermore that his days were numbered, and there were other worries piling up too: his successor at Princeton had still not been named, his son Henry was ill, he had lingering NRC duties to attend to, Ginn wanted another revision of "RDS," and he was the acting president of Commission 35 (Constitution of the Stars), an office that he had assumed at Eddington's death in 1944. Russell needed Kopal to manage the details.[55]

Kopal's Fate

Kopal did manage the panel and its bulletin, using them both, with Shapley's blessing, as a platform to push for better observations to counter Russell's criticisms of his reduction methods.[56] By now, his monograph had been published and was receiving mixed reviews. Kron felt that Kopal had neglected practical matters, especially how to apply his methods to observations. Others were more critical: "The greater the mathematical requirements of a subject, the more careful an author should be in its presentation," Henk van de Hulst argued as a preamble to an extended discussion of mathematical errors, inconsistent notation, and lack of any correspondence with observational tests or numerical tables that would allow others to make those tests.[57] Arthur Beer wondered rhetorically why Kopal spent so much time describing Russell's geometrical methods. It was because, Beer surmised, most detached systems were described just fine by Russell's methods. Only the most distorted systems required Kopal's dynamical analysis, and even then Beer felt that the most intractable systems lay beyond any solution. Russell's ideal spherical case remained "fundamental to the analysis of any light-curve." John Merrill, Russell's closest colleague in these matters, agreed, with Beer adding that Kopal was unable to prove that his methods were better. Merrill's partisan views were echoed by most reviewers who worked with limited computational expertise. Observers wanted direct aid in computational technique, with only enough theory to give them a physical feeling for what they were doing.[58]

Kopal's monograph symbolized how distant he was from the problems of the observer. Photoelectric astronomers warned him that his claims for the new technique were seriously overblown and would only mislead neophytes.[59] And Russell never ceased in his own criticisms, which he carried to Lick Observatory in 1947, where he lectured on binaries, telling anyone in reach why Kopal's methods were overly cumbersome and still unnecessary. He applauded the high accuracies of photoelectric observations but cautioned that even probable errors of observation which were "rarely much better than 1% of the range" required "numerous observations closely defining the light curve." Russell predicted that Merrill's tables would greatly facilitate the reductions, with accuracies comparable to any analytical method then available.[60] Lick staff and Berkeley graduate students keenly recall Russell's lectures. "I did all of my work based strictly upon Russell's [methods]," Kron later recalled. "I used his tabulations . . . and things like that . . . methods that he had devised. They were very ingenious." Gibson Reaves, then a student on the Berkeley campus, recalls that "nearly everyone at the Berkeley department who could get up to Mount Hamilton was present, not so much to learn about binaries as to hear the great man speak." Even though the sixty-nine-year-old Russell was frail and "appeared quite doddering," Reaves remembers him only as a large figure, intellectually, "a brain on stilts." Russell's apparent openness, sharing insights and practical hints, was his chief attraction.[61] Russell openly encouraged photoelectric observations but did not feel that the time was right for a wholesale revolution in reduction technique.

Kopal drew criticism from many quarters; in fact, Russell was the least of his critics, though the most powerful among them. After the war, Kopal saw little future in MIT's Center of Analysis since he was still a contractor, and Shapley could find nothing acceptable for him at Harvard. There had been a nibble from the Caroline University of Prague (Charles University), and, unknown to Kopal, Russell recommended him strongly, noting his industry and facility in applied mathematics and his contributions to eclipsing binary theory. Otto Struve, however, was less kind: he severely criticized Kopal's abilities in astrophysics and deplored his carelessness. Russell's recommendation weighed more, so Kopal was offered the job, but when Czechoslovakia went Communist in early 1948, Kopal and his wife applied for American citizenship, and Kopal was advanced to a temporary post as associate professor of numerical analysis at MIT.[62]

Kopal continued his organizational work for the panel, and his theoretical papers continued to flow. He hoped to establish an astrophysics theory group at the Center of Analysis and enjoyed his lectures in numerical analysis, later publishing a seminal text in the field. No matter what Kopal thought of Russell, at the time, Russell was essential to his own future. As much as they bickered, Russell always supported him, seeing to it that his *Theory and Tables of Associated Alpha-Functions* was published with a subvention from the National Academy of Sciences. And when Russell finally allowed his name to be associated with Kopal's original monograph, it was a better book than it had been at the start. Russell's criticism played a role in the ultimate improvement of the book, which, though it still drew criticism for the very reasons Russell had raised, sold out in two years.[63]

Predictably, there was much in the monograph that annoyed Russell, especially Kopal's assertion that there was "no royal road to the direct determination of elements of distorted eclipsing systems."[64] Russell could not let such a statement stand. His December 1946 "Royal Road" lecture was once again a powerful statement in favor of intuitive, iterative methods. His message remained that one could gain a "good first approximation to the actual character of the system," which was sufficient for statistical purposes when determining the general properties of stars was a goal.[65] Given the quality and quantity of data at hand, Russell believed that one could best travel the royal road at high speed, straightening out the twists and turns. There were too many stars in the sky to labor over every one. Obtaining general characteristics led to broader conclusions and to the solution of larger problems. If he said it once, he said it a hundred times: binaries provided "important tests for *astrophysical theory*," and their general study formed the basis for improved analyses of the general population. Although Kopal could have argued rightly that overly rough approximations could produce misleading information, Russell remained confident that his intuition and physical insight dictated where balance was to be found.[66]

Russell continued to support Kopal's programs, because he regarded them as collective property, or in the collective interest of either the Panel on the Orbits of Eclipsing Stars or the new Commission on Eclipsing Binaries that had been established at the August 1948 Zurich IAU meeting. At that meeting, Kopal was made president of the new commission, which replaced the panel. Kopal always believed that Russell was terribly upset by his ascension, but when Russell heard about the new commission, he told Shapley in September: "I'm not at all sorry to be free from the responsibility of the chairmanship. We'll probably have some lively discussions in private about technical points; but 'its differences of opinion that makes horse-races.' " Russell had just suffered a mild stroke and was taking his mortality seriously once again, trying to relieve himself of all responsibilities.[67] Yet his support, and criticism, continued undiminished through the rest of the 1940s, though his energy was waning, along with his patience.

By 1950, after endless battles, Russell finally wrote off Kopal as a lost cause. When Shapley tried to defend Kopal in the face of a fierce critique by Fred Whipple, Russell would hear none of it: "[W]hatever Whipple's personal relations with him may be . . . I must own that I am convinced that Kopal has gone off the deep end in his love for extensive computations by least-square." Russell had just visited Harvard and had again been put off by Kopal: "His energy is admirable—but he is the hardest man to convince, or even to debate in matters of purely technical disagreement that I know."[68] When Kopal's new manual, *The Computation of Elements of Eclipsing Binary Systems*, appeared later that year, Russell felt that Merrill's nomograms were still a better means to determine the basic nature of eclipsing systems.[69] And when Kopal's contract work at the Center of Analysis began to intensify even as funding for the center was being cut, and as his dreams for an academic "inter-departmental Laboratory for Theoretical Astrophysics" were dashed, Russell decided that between his government contracts and his binaries, Kopal should finish the former and drop the latter. This "might be a very good occasion for him to ease off gracefully and retire from the work which he has been doing on overtime," Russell told Shapley. Knowing only too well how Shapley could mishandle such a suggestion, Russell was prepared to tell this to Kopal himself.[70]

As we will see (chapter 20), Russell always fought to save a good researcher for research, but in Kopal's case, as in a few others during the war, he found his limit.[71] Most certainly Russell was threatened, and was weary of the bickering, but equally likely Russell was totally convinced that Kopal's work, though brilliant, was not going to be adopted by astronomers in the foreseeable future. If this was his view, he was partially correct: both methods had their adherents, as we will see in the next section.

Kopal was saved when his old teacher E. Finlay Freundlich, then at St. Andrews, Scotland, wrote him about a new professorship in astronomy at the University of Manchester, which was at the time just developing a major computational center. With recommendations from Chandrasekhar and an endorsement from Shapley, by late August Kopal knew that he had the job.[72] Manchester was almost too good to be true; since 1946 it had been one of the few places in the world that, like MIT, was building large-scale digital computers. Russell, of course, did not appreciate the value of such machines; he only snorted, "How he loves them!"[73] But over the years Kopal built a distinguished career there, in both astronomy and numerical analysis.

From his bitter commentary in his 1986 autobiography, it is clear that Kopal never reconciled with Russell and believed that Russell was equally troubled over their impasse until his death. When he first encountered the man, Russell was over sixty years old and, in Kopal's mind, as we noted, "living off his past. . . ." Kopal complained about Russell's perceived frailties from the start, but Shapley would always remind him, "You should have known him twenty years ago." To Kopal, Russell's charm was the "transparent sincerity" with which he "accepted all accolades as his rightful due." Russell's self-esteem was thus a "bit inflated . . ." because he had always been the focus of admiration and attention, "thoroughly spoiled by a doting family as well as friends and pupils." For Kopal, "Russell was never a hypocrite or an operator scheming to take advantage of others for selfish purposes." But he was the de facto Harvard director, acting as if Shapley was his second in command.[74]

There is much that rings true in Kopal's unhappy recollections, which do offer valuable insight into how Russell was most likely perceived by those most in competition with him. Generational differences, the clash of strong personalities, and conflicting scientific styles all were manifest in the debate over reduction procedures, which were never resolved by either side. Russell never relented, thinking Kopal's methods, and evidently Kopal himself, impractical and burdensome. Kopal dismissed anyone who followed Russell's methods, which he felt were "kept alive by a dwindling band of epigones, whose devotion to what they had learned when they were young helped to prolong the lifetime of obsolescent ideas well beyond their natural term." Kopal's idealism was more important to him than any need to make his methods accessible to mainstream observing astronomers. He never appreciated Russell's practical goals and, like Milne, was annoyed with his casual pragmatism. Deeper still, Kopal felt that Russell and his followers held back the whole field of double star astronomy, impeding "the natural advance of scientific progress."[75] Kopal's contention, although not strictly testable here, can still be examined through a look at how observers did respond to the computational choices before them.

Observers' Choice

In 1950 Kopal predicted that the "observer of the future" would be "primarily an expert in intricate photoelectric techniques of light measurements" who would depend heavily upon better, faster, more reliable, and more accessible computational techniques and the machinery to handle them.[76] Nevertheless, as Russell predicted and as both Shapley and Kopal feared, observers were on the whole reluctant to accept Kopal's challenge, even those with considerable computational skills. Most of the reviewers of his monograph and his 1950 manual shared this concern. John Irwin, then at the Flower Astronomical Observatory at the University of Pennsylvania, was particularly annoyed: "I am sure that the average photoelectric observer will look with horror at the complexities of the equations of condition . . . and will reach for the nearest Merrill Table or Nomogram."[77]

Russell's and Kopal's comparative impact can be explored through citation counts to their papers. Let us look just at citations to Kopal's 1946 monograph and to those citing Russell's primary publications on binaries in the mid-1940s: through 1961 some thirty different workers continued to cite Russell's methods, many referring to his 1912 paper as the work defining the basic method of analysis.[78] Some observers also cited Kopal before 1959, but in most cases, Kopal's methods were used by observers only when they had someone available to handle the computations, or when Kopal agreed to help them out himself. Wherever possible, observers either tested the results of one method against the other or sought a useful accommodation for the particular binary they were analyzing.[79]

Kopal soon came to realize that few astronomers, even those mathematically inclined, were ready to adopt his methods. Observers like C. M. Huffer depended upon his help; otherwise they stayed away.[80] In 1951, even mathematically adept astronomers like Dirk Brouwer of Yale gave up trying to figure out Beta Lyrae using rigorous mathematics: "It appears to be so complicated that one can hardly expect to hit upon the correct solution at once; perhaps a number of fruitless attempts have to be made first."[81] Kopal also found Katherine and Gerry Kron less than happy with his methods. "We liked the Russell/Merrill nomograph approach because we could visualize what was happening as we changed values," she recalled recently. Only much later, when digital computers were accessible, did the Krons use more complex analytical techniques to get beyond the limitations of the Merrill tables.[82] Leendert Binnendijk published a textbook for double star observers in 1960 which made clear that the method of choice among U.S. workers was Russell's, with "only a few later refinements" added, which were facilitated by the tables and nomographs by Merrill.[83]

Through the 1950s, then, observers kept Russell's methods alive, although astronomers tended to adopt the most useful aspects of both Russell's and Kopal's methods in a complementary manner, and their choices depended as much upon their goals as upon their willingness to compute. As Russell well knew, there could be no end to the methods available to the astronomer to pursue binary stars. He resisted Kopal's new methods when he thought they were overly complex or led the common astronomer away from physical insight. Russell always knew that the "Russell Method" was a compromise between theory and observation, an accommodation to practical limitations. His intent was always the same, to search out a framework within which astronomical observations of all types could be planned that would efficiently and effectively help to solve fundamental questions about the nature of the stellar universe. Russell's "Royal Road" was a means to approach useful solutions and deeper insights about the universe. It was never an end in itself.

The feud between Kopal and Russell might be interpreted as the result of conflicting disciplinary frameworks, but it was more likely a clash in choice of appropriate mathematical tools within a single framework. Kopal failed to create a school of thought among astronomers to replace the one Russell represented.[84] Thus our excursion down Russell's royal road demonstrates not only Russell's style as a practical theorist but his dominance as gatekeeper and stakeholder for a major specialty in astronomy. His manner of enforcing this practice was well illustrated in his dealings with Shapley, Kopal, and Harvard's binary programs, as it was in his quieter debates with Milne. The influence he exerted at Harvard was, indeed, enormous, just as it had been at Mount Wilson in the 1920s. During this entire period, from the 1920s through his retirement, Russell also maintained an especially close but very different relationship with V. M. Slipher and the Lowell Observatory. As he did with all observatories, Russell started by trying to influence the course of research at Lowell, bringing the observatory's denizens out of their severe isolationist mode and into the mainstream. He faced a very different challenge, however, with the men of Mars Hill.

• CHAPTER 19 •

A Summer Place: The Lowell Observatory

ON HIS TRAVELS by rail to and from the West Coast, Russell, like many astronomers of his day, often stopped at the Lowell Observatory just outside Flagstaff, Arizona, where the staff were happy to guide friendly astronomers to the best scenic sites. Unlike most, however, Russell returned frequently, brought his family and servants, and became part of the summer community at that isolated, beautiful spot.[1] On his first trip west to Mount Wilson as a research associate in 1921, Russell stopped to visit the high desert and Indian country, camped in Walnut Canyon, viewed the prehistoric Indian ruins there, and saw the Grand Canyon.[2] As he did at Mount Wilson, Russell lectured and offered advice copiously and freely; he hardly had to be asked for his opinion once he was shown a gem from the plate vault on Mars Hill. But Russell knew that the Lowell and Mount Wilson observatories were very different places, about as different as two astrophysical institutions could be, in history, governance, stature, and outlook. Even though by the 1920s some of their research lines intersected, the two observatories remained very far apart.

In his *Scientific American* column, Russell had always been kind to the Lowell Observatory, even at the height of the controversy over life on Mars, when Percival Lowell claimed that his observatory site was superior to those elsewhere, allowing him to perceive the canals and water vapor when others using larger telescopes failed. Lowell's assertions, and the fact that he sought an independent public forum for his views, threatened elite observatory directors including Hale, Campbell, Pickering, and Frost. Russell, less threatened, assured his readers that such controversies were positive indicators of healthy debate within a self-regulating community of astronomers.[3]

Early in his student days, Russell had heard Charles Young say that Mars might be the abode of life, since the spectrum of its atmosphere yielded evidence for oxygen and water vapor. He blithely wrote in his 1895 "Astronomy" notebook: "Mars is the only body visible with the telescope on which human beings could possibly live . . . Inhabitants might be gigantic."[4] In 1899, like many students of his day, Russell made drawings of Mars at the Students' Observatory at 14 Prospect Street and, like most youngsters, was smitten by the visions Lowell's theories presented. Even though Young's textbooks sternly cautioned that, until the question of Mars's surface temperature was firmly settled, Lowell's theories "appear rather premature . . . ,"[5] he also freely acknowledged the "extreme popular interest" caused by the "rather sensational speculations and deliverances of Flammarion, Lowell and others." In 1901, recuperating at home and writing his *Scientific American* column, Russell chose the Lowellian vision for one of his earliest commentaries: "Perhaps the best of the existing theories, and certainly the most stimulating to the imagination, is that proposed by Mr. Lowell and his fellow workers at his observatory in Arizona."[6] As he always did in his column, he wanted to excite his readers to the scent of the hunt, and thereby, above all, to promote astronomy, so Russell never publicly criticized Lowell.

Russell, however, was no Lowellian. In private he despised Lowell's patrician style, bristling as much as anyone when he was subjected to it.[7] He felt that the Martian polar snows were insufficient to irrigate any substantial area. He also was an agnostic about the canals, feeling that Mars was more rugged than Lowell's vision demanded. Although he scoffed at Lowell's claims, he told Edwin Frost that all he would say in public was that "the differences in the reports of the canals as seen by different observers must be due to a sort of subliminal personal equation." He said just this in delicately phrased commentary for the *Outlook* upon Lowell's death in 1916.[8] He assured E. E. Barnard that he did not "for a moment subscribe to the idea that the only good seeing in the world is to be found at Flagstaff"; and, as he repeatedly lectured his students, if Lowell's geometrical canals were in fact real, others like Barnard would surely have seen them too.[9]

By the 1920s, Lowell Observatory senior staff included V. M. and E. C. Slipher and Carl Otto Lampland, all farm boys who had trained at Indiana under John A. Miller and Wilbur Cogshall, a former Lowell staff member. They had learned their craft under Lowell and were now limping through a decade of Lowell family bickering and fear of professional ostracism. They had few friends but thought of Russell as an ally because of his visits and his friendly monthly column. When Roger Lowell Putnam became the observatory's trustee in 1927, he wanted to put the place right. His proximity to Boston led him to Shapley for advice, and Shapley gave him a copy of Russell, Dugan, and Stewart's new two-volume text, *Astronomy*, to read. Putnam was pleased that much of it was comprehensible, and that it contained photographs from Lowell. Russell, he decided, was just the man to help him guide his observatory.[10] Approached by Putnam, Russell immediately killed a Shapley idea that the Lowell staff publish a popular book on Mars. Instead, he argued, the staff had to publish their work in reputable journals. Only in that way could they build the prestige of the observatory. This

19.1. Russell with Roger Lowell Putnam, trustee of the Lowell Observatory. Photograph by Clyde Fisher at an IAU meeting, probably September 1932, held at Cambridge, Massachusetts. Neg. No. 280362. Courtesy Department of Library Services, American Museum of Natural History.

was a plan the Lowell staff liked. If they wanted publicity, Russell promised Putnam, he would give them all they needed through the *Scientific American*.[11]

Russell became more than welcome at Flagstaff. He was royalty. His 1927 visit was a major orchestration including children and servants. Celebrating his newly acquired C. A. Young Research Professorship, Russell took a grand tour of American observatories and physical laboratories, meeting his family in West Yellowstone and taking them first to Mount Wilson, where they stayed in the Kapteyn Cottage. Elizabeth, with Jean Hetherington and Miss Eden, the cook, were already at Flagstaff when the rest of the Russell family arrived on 7 July, making the party eight in all, converging for the first of many "Russell invasions."[12]

When Russell was in residence, all attention was on him. The staff would meet the Russells at the station and then listen to the news. He would talk nonstop about visits to Mount Wilson, recent meetings with Putnam, and his own work.[13] He would constantly walk the halls, inspect research in progress, and deliver lectures (some planned, most impromptu), mainly trying to introduce the Lowell staff to modern atomic structure as he did at Mount Wilson. In between, the staff arranged fishing trips, camping in the Grand Canyon, shopping, and driving lessons from Cogshall (Russell's first solo in their Ford was the highlight of early August 1927, "an event," recorded Lampland). Russell constantly pried into the work of the staff, looking for publishable morsels, trying to get the Sliphers and Lampland to bite.[14]

Over the next several years Russell acted as Putnam's agent. In Russell, Putnam found both champion and mentor and made him his "unofficial adviser." Not only would Russell constantly nag the Lowell staff to publish, but he did what he could to convince Putnam to smooth the way—for example, persuading Putnam to set up a publishing fund so that the staff could have access to the *ApJ*.[15] It was a good try, but it made no sensible difference in the output of the staff. There were deeper problems lurking on Mars Hill.

Russell's contact with the observatory continued for years. At first he tried to introduce changes, to reform the staff and get them thinking along physical lines. Next, he defended members of the staff whenever they crossed paths with influential astronomers, such as W. H. Wright at Lick, who tended to ignore E. C. Slipher's photographic studies of planetary detail. Third, he took the side of the observatory itself when celestial mechanicians doubted that the planet Lowell had predicted was actually the one Clyde Tombaugh discovered. And finally, the observatory was the site where his youngest daughter, Margaret, met the astronomer Frank Edmondson, a meeting that created a new focus for Russell's manipulative energies.

Staff Training

Russell tried hard to reorient the observers of Mars Hill. After the dispute with Wright in 1927 and 1928, which centered on priority for using certain combinations of photographic emulsions and filters to enhance planetary surface detail, Russell realized that the Sliphers and Lampland were tireless observers but had huge gaps in their training. His lectures during the summer were not enough. He saw his chance to do more by arranging for Lowell staff to come to Princeton for a sustained dose of astrophysics, Russell style.

The first to come was A. L. Bennett, a junior member of the Lowell staff who showed great promise. The twenty-two-year-old graduate of Union College arrived in Princeton for advanced training in 1928 after a long delay partly due to V. M. Slipher's reluctance to let him go. Slipher, who believed in observatory-based apprenticeship training, was finally compelled by Putnam and Russell, but as he feared, Bennett's horizons widened sufficiently that he never returned to Flagstaff.[16] With Bennett in hand, Russell's next step was to invite Lampland to spend a semester at Princeton in trade for Dugan, who would observe at Lowell on the plan of faculty trades that Russell had created.[17] Lampland, four years older than Russell, was a deadly serious, indefatigable worker who at various times specialized in photography and radiometric measurements of planetary heat. He was totally focused on instrumentation and observational technique, not interpretation or publishing. Unfortunately, although Lampland was unable to do much with his data himself, he was loath to see it exploited by others.[18]

Russell hoped that a semester at Princeton would help Lampland learn how to turn his excellent radiometric observations into Martian surface temperatures, a major element in the Lowellian defense. Lampland was highly competitive, fearful of competitors like Edison Pettit at Mount Wilson. But, lacking knowledge of the physics involved, he had little confidence in what he was doing, and little understanding of why his results differed from those obtained at Mount Wilson. In consequence, Lampland doggedly continued to observe. Russell, never one to worry about the last decimal place, assured Lampland that his observations had value; all he needed was some training to set it all right. Lampland believed that "attacking new lines is inviting grief in abundance," but he (and Slipher) agreed to the swap with Dugan, again pressured by Putnam. Slipher admitted that closer ties with Russell were essential: "[H]is high standing today makes the interchange with Princeton University . . . the best we could possibly choose."[19]

Princeton, to Lampland, was like a surreal dream, not a pleasant one for the Mars Hill astronomer. It was filled with exciting distractions and potential opportunities for growth, a smorgasbord both tempting and forbidding. Lampland opted for the familiar whenever he could. The only fellow he felt comfortable with was Bennett.[20] During his six months in residence in 1929, Lampland was officially a visiting faculty member, which meant he had to prepare lectures on his research, a ploy Russell had planned out to get him writing. Russell was now the taskmaster, and his word was law. For the first time Lampland saw Russell at his full powers: "He certainly does dispose of things with some dispatch," he told Slipher in March, "whether letters, lectures, conferences or anything else." Lampland also attended Russell's lectures on stellar atmospheres and took in a few physics lectures by Henry Smyth on band spectra. He even listened at the doorway to a Hermann Weyl mathematics lecture but got nothing out of it.[21]

Russell astounded Lampland with his ability to keep track of a bewildering number of activities. He seemed to know at every point just how far Lampland had gotten with his radiometric reductions, and always demanded more, all the while complimenting his work and never failing to remind him of the value he saw in his and in Slipher's planetary spectra.[22] Lampland did try to focus on his radiometric work, but Russell constantly led him to meetings and conferences, and these distractions became his excuses not to finish his own work. But as he saw the time slipping away and Russell's expectations for his output undiminished, Lampland began to fret: "If one could have a year in a place like Princeton it would help out immensely."[23]

Lampland was beginning to see the light, but it was too little too late. By June Lampland was packing for his return, and no manuscript had been started, much less completed. He therefore reverted to his primary defense, that he needed more observations for the reduction of the radiometric work. Despite the opportunity he knew he had at the moment, his deepest desire was to get back to the security of the routine on Mars Hill.[24] Russell, bound for Europe with his family by this time, was out of the picture.

Lampland's experience at Princeton was far from an exercise in futility. Although there was no tangible product, he was exposed to physics and did change his attitude somewhat. Lampland also saw Bennett being trained and knew he would be a wonderful asset back home. But the cards were stacked against him, as Slipher always feared, since Bennett already had offers from larger observatories and moved on, eventually ending up at Yale under Schlesinger.[25]

In the spring of 1930, knowing that they had lost Bennett, Slipher started looking for another "promising young man" as a new staff assistant.[26] Slipher wanted someone like their "man from Kansas" Clyde Tombaugh, who came in 1928. Tombaugh, "a young man of the self-made variety," suited their needs exactly. Slipher held the Kansas farm boy out as an example of Lowell's real needs. He wanted someone as an assistant, not someone of "the highly trained variety" who "care[s] only to take up new pieces of work for themselves rather than to help us with lines the Observatory has been doing."[27]

In the early 1930s, however, Putnam was not ready to give up. Russell's counsel, Putnam's continuing exposure to Harvard's growing army of well-trained talent, and his visits to Mount Wilson told him that he should not be satisfied with the status quo at Lowell. He and Russell conspired to interest Tombaugh in modern physics, which was a total failure, though they managed to force Tombaugh to obtain college training.[28] Russell's unsuccessful efforts to improve present staff convinced Putnam that astrophysics had to be imported. In the summer of 1933, he saw a chance to make his move as he watched V. M. Slipher get bogged down in his spectroscopic studies of planetary atmospheres. Slipher could not compete with Mount Wilson spectroscopists like Theodore Dunham, but if he had a physicist as collaborator, maybe he could. Dunham and Adams's work on the infrared spectra of planetary atmospheres attracted the attention of physicists at the University of Michigan, who worked in one of the nation's most powerful centers for infrared molecular spectroscopy, built by Harrison M. Randall. Eventually, through a long and complex series of maneuvers in which both an enlightened Lampland and Dunham played significant intermediary roles, Putnam hired Arthur Adel, a very promising Michigan student who had been working on the theoretical structure of the carbon dioxide molecule and was anxious to verify his theory with laboratory experiments. From the moment Adel arrived at Lowell in the summer of 1936, however, the young man from a Detroit working-class émigré Jewish family clashed with Lowellian culture. Though he remained at Lowell until he was drawn into war work, Adel was far from a happy man, and there is no evidence that Russell fought on his behalf, aside from his general support of Putnam's plan.[29] Russell's failure to do so marks another transition in his career.

From Adviser to Patron: Defending the Observatory

Sometime after Russell returned from Europe in 1931, his role at Lowell shifted. No longer was he Putnam's eyes and ears, the resident critic or academic dean. Lowell Observatory had become his playground, and this arrangement paid off handsomely for the staff. As chair of the Draper Committee of the National Academy, Russell saw to it that V. M. Slipher received the Draper Medal for his spectroscopic work.[30] Now he worried less about the lack of theoretical expertise at Lowell and more about how to defend it against attacks. Russell's transition from adviser to patron became most visible when Russell was called upon to defend the Lowell Observatory's discovery of Pluto.

The notoriety attending the discovery of the new planet in February 1930 created an immediate problem for the Lowell staff that they could not handle: calculating the orbit of the object and proving that it had been predicted by Lowell in 1915.[31] Celestial mechanics was not a strength of the Mars Hill staff, and so, with Russell away in Europe, they could not deal with questions being raised by leading experts like E. W. Brown and A. O. Leuschner, who concluded that the discovery was "purely accidental." Brown applauded Lowell for the discovery but knew it did not come from the tiny residuals in the Uranian orbit. Certainly the faint object Tombaugh found could not have sufficient mass to be Lowell's predicted perturber. Shapley only wanted to know, "Is the name to be Osiris, or Bacchus?"[32]

Slipher stubbornly resisted repeated calls to share their data while he frantically called upon J. A. Miller, his former teacher, to perform the orbit calculations. Miller, however, was a bit rusty and came up with a nearly parabolic solution with a period greater than three thousand years. Not realizing the problem, Putnam sent the report to Shapley for distribution. The seasoned celestial mechanicians chuckled. Slipher was mortified.[33]

In April 1930, returning to Athens after wintering in Egypt, exhilarated by a Nile cruise and a visit to Breasted's excavations in the Valley of the Kings, and after a long circuit through the Holy Lands and Asia Minor via the "Baghdad Railway" to Constantinople, Russell found an urgent letter from Slipher and marshaled what evidence he had to make a case for his friends.[34] He prepared a series of *Scientific American* columns proclaiming, from the data Slipher sent, "There is no doubt that it is actually a new major planet." He admitted that the planet's brightness and distance indicated a smaller diameter and probably lower mass than predicted. But its general motion, inclined orbit, and distance agreed closely enough to Lowell's mark to let Russell conclude, not knowing of Brown's opinion, that Lowell's prediction "is near the truth." It was "most gratifying that this long and devoted search has been crowned with full success."[35] In a second article written in the summer from Cortina d'Ampezzo in the Italian Dolomites, Russell explained that the Lowell staff delayed announcing their orbital elements because the analysis was so complex.[36] Defend his Lowell friends as he might, editors at *Scientific American* ran a sidebar on the Berkeley and Mount Wilson results, which highlighted a serious mass deficiency. Now Russell began to look rather foolish, or at least out of touch.

Only in Paris, delayed because of his son's illness, did Russell gain access to a full astronomical library and learn the whole story. Russell spent his mornings with Henry and afternoons at the Paris Observatory library to read and prepare his third defense of the Lowellian prediction.[37] Russell explained away Miller's erroneous orbit by bad data and once again proclaimed that the planet Tombaugh discovered had been mathematically predicted by Lowell. "[I]t is quite incredible that the agreement can be due to accident," Russell argued. "[T]he actual accordance is all that could be demanded by a severe critic." If Brown's conclusion were justified, Russell felt, it would be even harder to explain how errors in calculating the orbital elements canceled each other out precisely to explain the agreement that was found.[38]

Back home in the fall, Russell repacked his bags for the West, stopped in at Lowell to reconnoiter, and then moved on to Mount Wilson to examine the most precise observations of the planet's position. At Mount Wilson, Seth B. Nicholson and N. U. Mayall found that if they included a 1795 observation from Lalande to compute the planet's mass, it came out similar to the earth's. Russell rejoiced at this apparent vindication of the Lowellian prediction and, without checking, wrote it up for *Scientific American*, well before Nicholson felt ready to publish. Russell heralded the denouement of doubt: "So all contradictions vanish and we may welcome Pluto to a fully accredited place among the major planets of our system."[39]

Nicholson, however, was worried. If Lalande's observation was spurious, Pluto's mass would be drastically reduced. He wrote Brown for advice, and Brown immediately saw flaws in his analysis. And when Brown read Russell's *Scientific American* article, he could not let it stand. Brown confronted Russell as only Brown could, calling for more than a retraction: "[W]ith your very extensive experience of least squares and your knowledge of the residuals, it was somewhat astonishing that you did not notice the discrepancy right away." The damage, thought Brown, was not so much to Russell's reputation as to the integrity of mathematical astronomy. "The trouble is that journals like 'Nature' quote you from the 'Scientific American,' and when Nicholson's paper comes out, there will be general confusion as to what is the fact."[40] Brown, of course, knew well who he was talking to, and so to drive home his point, and how deeply he felt about it, he described how his world of exactitude differed from Russell's world of approximation:

> The trouble is, however, that there have been some bad howlers from the gravitation point of view. If it were a question of physical astronomy where a good deal of latitude may be permitted with doubtful data, it would not matter much. Where it is a question that can be settled by recognized methods or where it is a mere question of mathematics, I think it is not fair on the coming generation to leave the matter uncorrected.[41]

Brown could not have better described the differences between traditional astronomy and the new quantitative astrophysics as Russell practiced it. It wasn't only a matter of making a mathematical blunder; it was a difference in modes of practice. Russell admitted to the error privately and appended a note to his next column that made the error appear as a typo in the carbon he had received from Nicholson. But Russell was chastened.[42] His blunder came as something of a shock, and he did all he could to minimize the damage. What he thought of his objectivity is unknown. But he was determined to uphold the Lowell discovery;

though not a prediction, it was still a major event, one the Lowell family should well be proud of.

The Pluto affair shows not only how far Russell had moved away from the world of mathematical rigor but how far Russell would go to favor friends. This loyalty extended to aiding A. Lawrence Lowell's preparation of a biography of his brother Percival. In 1935, Russell prepared a set of essays on Lowell's contributions to astronomy, and Lawrence Lowell published them as appendixes to his biography. He also quoted extensively from Russell's earlier *Scientific American* columns supporting the Lowellian prediction of Pluto, stressing that the similarity of the observed orbit to the one predicted could not be due to accident. Neither Brown nor Russell felt that it would be politic to object; it was their responsibility to the discipline to assure the Lowell family that no matter what the efficacy of Lowell's prediction, Lowell deserved the credit for the discovery. Above all, the family had to be encouraged to continue supporting the observatory.[43]

A Summer Place

"The van guard of the Russell invasion is nearly ready to start . . . ," Russell warned Slipher in his usual fashion in May 1931, as the family plotted their separate itineraries to the Southwest.[44] Life was now comparatively quiet and dull around the house, Russell reported; he and Mrs. Russell felt "rather lonely with three young folks away at college at once."[45] It may have appeared that way to Russell because he took little part in the details of the household. During the year, however, Lucy Russell kept busy enough, running the house and managing the family expenses (except for the taxes and large-ticket items, which Russell took as his responsibility). She was an active member of the Present Day Club of Princeton, following in Aunt Ada's footsteps, loved to play bridge, and attended to the endless social obligations of a professor's wife and active church member, which meant that she constantly hosted afternoon teas.[46]

In concert with Henrietta Young, his secretary, and Charlotte Moore, Lucy May Russell's mission was to keep her husband's life in order. She also managed her considerable inheritance, visiting her New York broker routinely. After the 1929 crash, she suffered little in the way of real losses of holdings; "her stockbroker said rather ruefully that he wished he had done as well!"[47] Russell's substantial salary combined with his wife's considerable investment income allowed for a secure life of comfort and gentility, although in the dark days of 1933, with bank closings and economic skepticism rampant, Russell admitted to Adams that the "insane economic outlook has wrecked our plans for a summer abroad with the family." He was still planning to travel to Oxford with his wife for his Halley Lecture, but they had to leave the family behind. "Nothing has hit us personally so far, but I don't like the outlook, on account of an inveterate and profound distrust of idealists in politics." Russell was not above inflating the economic crisis to avoid responsibilities. In May 1933, he informed the British Association that he could not remain in England after his Halley Lecture to attend their meetings in Leicester. He was more direct with Harry Plaskett, admitting that he really wanted to return to Clark's Island, to "live cheaply" and get a rest, and then journey to Flagstaff.[48]

Fiscally conservative, the Russells shunned any form of material ostentation, though they lived very comfortably at the peak of the elite intellectual aristocracy. What money they spent was devoted to their children's welfare, to travel, and to maintaining their sizable household staff of two maids, a laundress, and a cook, in addition to Jean Hetherington. Russell's academic year was filled with travel and work, and his life was taken care of at home. The family would gather routinely at dinner, where often they would wait while he finished a thought or a calculation in his study:

> All the time that he was awake he was working. That is to say he was at his desk under bright lights with a yellow pad and again with the hard pencil writing on it—he was at this continuously. If mother called him to come to the table, "yes my dear, yes my dear, coming in a minute my dear" and he was at it for another ten minutes before he came to the table.[49]

And even at the table, talk would be of science, at least at Russell's end of the table. His wife deftly manipulated his world: she placed her husband at the opposite end of the great mahogany table her father had given them, which, when extended, could seat fourteen people with ease. Henry jr. was on his left side, and Margaret on his right, "because we would talk nothing but science loudly—and mother had Jean Hetherington on her right and my sister Lucy on her left. And they talked ladies' talk." Elizabeth, who rarely ate with the family, had already been tended to by Hetherington or the upstairs maid. With the cook commanding the kitchen, the downstairs maid brought in the food, and Mrs. Russell served. "And when dad wanted something, he stuck his plate out in front of him and continued talking. And mother was way down there, and I took the plate and passed it down."[50]

Family dinners were rarely quiet, since all the Russells talked. One might get a word in sometimes when Russell took one of his frequent long drinks of water. His son speculates that he may have had a form of "water diabetes" because he habitually drank about a gallon of tap water at a sitting, always made ready for him in a proper silver pitcher sitting on a large silver tray. "And he just poured himself glass after glass after glass of water . . ." at every meal. Russell liked simple foods, few garnishes, and tolerated his wife's insistence that everyone had to eat vegetables. He would have been perfectly happy having oatmeal "every morning of his life," his son recalls, but for dinner it was usually roast beef or fish. Cheeses, dried fruits, bread, pilot crackers, and canned salmon made for a fine picnic or, in the case of the first two, desserts, as did apples from their backyard orchard. Overall the Russells lived the virtues of moderation and temperance. "He was very very righteous and he was going to do things exactly according to the law. So during the period of prohibition we didn't have anything alcoholic at all in our house. And when we went to Europe then he took this opportunity to introduce us to the fact that the French had wines and the Italians had wines and we got the opportunity to sample some of these"[51]

In the 1930s summers became the family's time together, and the Lowell Observatory became one of their favorite converging points, especially for Elizabeth and Jean Hetherington because of the recuperative powers of the climate, rent-free accommodations on Mars Hill, and very low rental rates in Flagstaff when Russell himself was not in residence. Complicating her palsy, a serious illness had struck Elizabeth in her late adolescence; she had remained bedridden for several years and suffered congestive attacks and sinus problems that would last the rest of her life.[52] Her condition compelled Russell to seek out milder and more healthy climates for her starting in the late 1920s. They tried summers in northern Arizona or the White Mountains, and then winters in Tucson or Pasadena. In September 1933, Russell and his wife were deeply torn over the question of whether they should send Elizabeth away permanently to live in a warmer, dryer climate, asking their New York doctors to help them balance the pros and cons. Russell considered three criteria: medical, psychological, and economic. Winter in a warm climate was "doubtless desirable," but was it necessary? In her state of health, susceptible to infection and dependent upon others, she saw very few people other than Hetherington and the family. But if she were sent away to live alone with Hetherington, she would miss the stimulation of her active family, which might further limit her social and intellectual growth, a factor that cerebral palsy specialists were just beginning to appreciate in the still very few centers devoted to its study and treatment. The Russells feared this option but seemed resolute to place her physical well-being uppermost. Never mentioned was the possibility of institutionalized physical therapy using a team approach, which was then being suggested and tested by practitioners in southern New Jersey and reported before the New York Academy of Medicine. Naturally Elizabeth would have her own home, with Hetherington at her side. Despite their fears, Russell claimed that Elizabeth's future would be made on the basis of medical and psychological advice alone, since, fortunately, the family could afford either option.[53]

What advice they received caused them ultimately to decide to send Elizabeth to Pasadena for the winter, because of Russell's considerable contacts there. Flagstaff also became one of their favorite sites for the summer, alternating with Tucson and Pasadena in the winters. Finally, by 1939, she and Jean Hetherington had moved permanently to Manitou Springs, Colorado, at the foot of Pike's Peak, a legendary site for tubercular sufferers in the late nineteenth century and a place that had attracted many figures known to Russell, including Teddy Roosevelt, an asthmatic.[54] Some summers were still spent in Flagstaff when the family was there, and winters in Tucson. On his later travels west, Russell would always stop in Manitou Springs.

Summers at Lowell Observatory for the Russells were therapeutic in more ways than one. Life was a concoction of delights, diversions, and surprises. New faces started showing up on Mars Hill in the 1930s, made possible by the revitalized Lawrence Fellowship. Though Slipher had succeeded in gaining Putnam's approval to expand the scope of the fellowship to other universities beyond Indiana, one of their most promising young men, Frank Kelley Edmondson, had been a Hoosier since the age of five.

Just as he had with the Sliphers and Lampland three decades earlier, Cogshall groomed Edmondson for Lowell.[55] Russell's *Astronomy* was one of the first of his freshman purchases; "that was the Bible . . . in those days." In his junior year, he used it to calculate the solar motion based upon the recently published *Yale Bright Star Catalogue*.[56] In the summer of 1934, as he continued to work on his master's thesis on the solar motion, using a set of unpublished radial velocity spectra of globular clusters Slipher had taken, he met Russell and his daughters. By the time of their third invasion in 1934 the children were adults.

This was the summer when Russell lectured the Lowell staff on the origin and nature of the solar system, preparing for his delayed Virginia lectures.[57] Captivated though they all were by Russell, Edmondson and the other young men of Mars Hill could not help but notice that there were young Russell women about too. Lucy had just graduated from Smith, and Margaret had just finished her freshman year there.[58] Frank and Henry Giclas doubled-dated Lucy and Margaret, and soon Frank and Margaret were together. Margaret was just twenty years old then and had decided to leave Smith for Bryn Mawr to pursue classical archaeology. The last thing on her mind was marriage when she arrived in Flagstaff; so when she announced their intentions that fall, Frank recalled, it created a bit of a shock. "Propinquity propinqued," Russell was later fond of saying, but at the time, he reported to more than one astronomer that "Margaret has given us a surprise."[59] Margaret, the youngest, was the first to leave the family orbit.

Russell now paid more attention to Frank than he did to the future of the Lowell staff. Frank's future became Russell's mission, and he decided that the place for Frank after Indiana was at Harvard under Bok, an excellent galactic structure specialist. Frank, however, had been thinking about Michigan. But Russell, with Shapley's and Bok's eager assistance, kept pushing. Russell made the situation very clear for them both in a remarkably heavy-handed way, even for a father-in-law: Margaret had an inheritance coming eventually; it would be available immediately if they moved to Cambridge.[60]

The marriage took place in the big living room at 79 Alexander on Saturday, 24 November 1934, with close family and a few friends attending, "as is customary for quiet weddings in this country," he told Giorgio Abetti.[61] Russell noted to Putnam that they were married "in the same room where my mother and father were married fifty-eight years before." After the ceremony, while Frank and Margaret packed their car to return to Flagstaff, the rest of the family went to the Dartmouth-Princeton football game. "He is a thoroughly nice boy" with a good academic record, Russell told Putnam. Above all Russell admired the "courageous way in which he has worked through college and made his position."[62]

Outwardly, Russell was proud of Frank's accomplishments and promise, but he would work very hard in the following five years on "[t]he problem of Frank's future."[63] For the present, Frank had to finish the second year of his fellowship, not only working for Tombaugh while he and Margaret set up housekeeping in the Tombaugh house, but also finishing his globular cluster work and readying it for publication. Russell was delighted that the newlyweds were to start out their life at Flagstaff, but privately he was not sanguine about Frank's plans. Frank would enter Harvard and do well, but ultimately he doggedly resisted his father-in-law's wishes. Their differences illuminate Russell's view of the state of the astronomical profession and reveal the lengths to which Russell would go to micromanage his son-in-law's career.

After his master's degree from Indiana, completed at Lowell, Edmondson wanted to return to Indiana, where a job was waiting for him. Russell would not hear of it and interceded with his professors and the president of Indiana University. Frank had to get a firm grounding in astrophysics at Harvard, he told K. P. Williams, from people like Bok, Menzel, and Payne-Gaposchkin, "who had an intimate knowledge of the best methods." In short order Russell convinced Williams that Harvard had to come first for the good of astronomy, Indiana, and Frank Edmondson. Russell planned out every step, telling Frank which seminars to take, which physicists to consult (like John Slater), and even where to live. Meanwhile, Shapley saw to it that Frank won the prestigious Agassiz Research Fellowship. Whether he liked it or not, Frank was now on a fast track. He and Margaret moved to Cambridge in the summer of 1935, all the while hoping to keep the door open at Indiana.[64]

That summer was filled with family business. Lucy and George Gardner were married in May (see chapter 16), set off for a Florida honeymoon in a small car the Russells gave them, and later ended up in Columbus, Ohio, where Gardner had been called to minister to students at the university. Frank and Margaret were just setting up housekeeping in Cambridge when they announced to everyone's considerable surprise that the Russells' first grandchild was on the way. But Frank and Margaret, living in the world of astronomy, especially at Harvard, escaped the scrutiny Gordon and Lucy's indiscretion would be subjected to in the world of the ministry. The Cole family summer home, now at Jamestown, near Newport, Rhode Island, became the center of activity, as Margaret moved in with her parents to bring Margaret Jean into the world in early July. Frank drove madly between Cambridge and Jamestown that summer, for he was bound to attend the inaugural session of Shapley's graduate Summer School in Astronomy, which brought dozens of the brightest astronomers together for several weeks of intensive work and play in an atmosphere of heady excitement only Shapley and his Harvard staff could generate.[65]

Frank barely had a chance to take a breath before he plunged into his first semester at Harvard. He managed to keep up correspondence with Lampland, whom he adopted as his Dutch uncle, and who applauded Russell's direction for Frank: Harvard would lead him to "certainly one of the important branches of research at the present time. More and more we must substitute quantitative for the qualitative . . ."[66] By the end of his first semester Frank knew that Russell had been very right. Though Flagstaff and Indiana remained close to his heart, Harvard was where he learned his profession. But what came next?

Indiana was still Frank's choice, but once again Russell advised against the move to a place where administration and teaching would no doubt dominate. He had to take a proper research position at a good observatory whose facilities matched his specialty in galactic astronomy. Russell knew that a position would open at Virginia because he had arranged to have Peter van de Kamp leave to take the directorship at Sproul Observatory. S. A. Mitchell responded by offering Edmondson a research instructorship, knowing that this would please Russell. To Russell, it was a growth position, and it was East Coast.[67] "This is better luck than you deserve," Frank recalls Shapley saying. Two job offers when jobs were really scarce. But Shapley kept his own mind this time, favoring Indiana because he did not trust Mitchell and liked the idea of the young fellow's striking out on his own. Bok, who evidently had a hand in the initial offer from Virginia, too, strongly sided with Russell. The Indiana offer's greatest advantage was the opportunity to leverage more from Virginia, Russell argued.[68]

What Russell feared about Indiana was that Frank was definitely in line for the chair of the department, since Cogshall dearly wanted to retire. As he had lectured Menzel and others, he told Frank that administration was anathema to good research:

> I suppose that I am as independent theoretically as anybody well could be, with my research professorship and the directorship of a small observatory. But for the last couple of months I have had to devote most of my time and energy to administrative problems . . . It is the worst of all enemies of real active research . . . I should be seriously concerned to see you or any other man of your age, take a position which involved any considerable amount of administrative duty or responsibility.[69]

Russell was by then also president of the American Astronomical Society, a duty he had long avoided but could put off no longer. Dugan had been in Pasadena for a term in 1934, mainly to recuperate from his arthritic condition, and he was now back in Princeton but in poor shape again. There were several graduate students around who had to be worried over, and there was little help or relief, even though Stewart managed most of the teaching.

Russell pressured, but Frank quietly and firmly resisted. After a visit to Virginia, Indiana looked better than ever to Frank. Mitchell would not make a clear commitment of resources, the cottage he offered them was very small, and they had to share part of it with Mitchell's servants. In the end, Russell reported wearily to Mitchell, after an exhausting effort, the decision was for Indiana. Russell was shaken: "[O]f course he knows and likes the place," Russell rationalized, but he regretted that "they will lose the professional opportunities and the pleasant life which they could have had in Virginia." V. M. Slipher was delighted when he heard the news from Russell. Slipher immediately wrote Cogshall congratulating him, for this meant that Cogshall would be relieved of his administrative load, and gave him more time for his own work.[70]

Although he certainly tried to micromanage Frank's career, Russell acted consistently, having given others he favored much the same advice (see chapter 20).

Russell always maintained one message: get as close as possible to the best facilities for research.[71] What Russell evidently never considered, outwardly at least, was that Edmondson had very different priorities: he relished a healthy research life but was equally excited to be involved in administration, teaching, and tending to the profession, responsibilities Russell consistently neglected.

In fact, Russell generally ignored such factors such as another person's priorities, but he could also let his own priorities drive his actions. After he inspected the facilities at Indiana, Russell wrote to Frank at Cambridge, criticizing their new photographic telescope, cutting it, and with it the whole place, to shreds.[72] Frank looked carefully at Russell's criticisms, which he knew his father-in-law had also expressed to Lampland and maybe others. In fact Russell had told Frank that he discussed the matter with Lampland, asking him to help fix the telescope, which was most irritating. Frank knew he had to explain things to Lampland. First, Russell had been "visibly disappointed when I decided to accept the offer from Indiana," Frank reported, and pointed out that this was why Russell was so negative when he inspected the Indiana telescope. "I am happy to report," he added, "that not a single one of his 9 objections carries any weight." To explain Russell's behavior, Frank stated what must have been obvious to anyone who knew him: "You perhaps realize that Dr. Russell's very active interest in my future has been somewhat embarrassing at times. Dr. Shapley's diplomatic cooperation was a great help to Margaret and me in the Indiana-Virginia decision."[73] "I admit that Dr. Russell's letter annoyed me more than it worried me," he added, but he also felt trapped. Even though his inspection fully vindicated his faith, he still had to answer his father-in-law. There was no way he could ask Lampland to get involved without embarrassing himself and Cogshall. Yet he knew that Russell had taken it upon himself to made this suggestion to Lampland and felt obliged to set things straight with his Mars Hill mentor. He did so by distancing himself: "The defect in Dr. Russell's advice is that it neglects the fact that Prof. Cogshall is my friend, and I want him to continue to be my friend. There are alot of things that need to be done in the department, and I hope to bring about changes and improvements by making use of this friendship."[74]

Indiana was still a poor place. It had little funds to bring Lampland or anyone else in. And further, Frank wanted to do the work himself, for better or worse. "It may take longer to get some things done, but I prefer to do it that way than to create an unfriendly atmosphere." Frank had finally gotten hold of something that was his, and he wanted to keep it that way.[75] Frank looked forward to starting his career at Indiana in the fall of 1937, fully on his own and among friends. He and Margaret also looked forward to summers in Flagstaff, not Harvard, as Russell wanted. Whatever Lampland and Slipher thought, it is clear that for a while they feared getting drawn into any tiff where Russell was involved. Slipher well knew that the supplemental funding they had just received from the American Philosophical Society to continue their trans-Neptunian planet search (still searching for the one Lowell had predicted) had Russell's fingerprints all over it, and that he was the observatory's most formidable ally.[76]

Although Russell began as a reformer at Lowell, he ended up an advocate and ally. After he tasted the hospitality of Mars Hill, and found that summers in Flagstaff did improve Elizabeth's health, the Lowell Observatory became a high-priority summer haven, a respite from worry. Even at the beginning of his formal visits, planning his first grand tour in 1927, Russell turned down tempting invitations from other summer venues. Ralph Curtiss had asked him to lecture at the Michigan Summer School, an emerging forum for modern theoretical physics. This would give Russell a chance to parley with physicists, and Curtiss had hinted that Milne might show up, but Russell was clear in his priorities: "[W]e are going [to Flagstaff] for the climate and a vacation. I expect to talk shop a little incidentally with my good friends there, but they all understand that this is only an incidental feature. The main point is the summer in that fascinating place, and the good society of those very delightful people. I need a rest anyhow, and I could not consider coming east so great a distance."[77] Russell had nothing to hide. He could never be accused of being idle, especially at an observatory. Over the years, he would consistently prefer Flagstaff to any other offers that came his way, although he would adjust his visits when major invitations, such as the Darwin or Halley lectureships, beckoned. Of course, Russell never ceased work: Slipher typically fed Russell recently acquired spectra asking for his advice and counsel, and Russell was always helpful.[78]

But as contact between Russell and the Lowell staff continued over the years, there was less talk of science and more of politics, family, and plans for the summer. In the beginning the Lowell staff hunkered down to avoid controversy by focusing on purely observational work to legitimate themselves in the wake of the criticism left by their founder, and they needed Russell's sympathies and support. But by the 1930s, E. C. Slipher was firmly entrenched in local politics, Lampland was equally confirmed in his ways, whereas V. M. Slipher had gained all the national and international recognition he or Putnam could want for Lowell. The observatory could now claim the discovery of a major planet. They no longer needed Russell as they had, but still wanted him around, and for good reason, for it was through Russell that many of these honors and awards flowed.

Apart from temporarily enlightening Lampland and helping Slipher interpret his planetary spectra, Russell had little long-term influence on the observatory itself. The Lowell staff remained one of the most conservative of major American observatories up to midcentury; even though it engaged in some innovative observing projects, it remained isolated from the expanding forefront of modern physics and astrophysics. Even when Russell secured the Draper Medal for Slipher in 1933, he knew that Slipher would probably not show up for the ceremony in Washington. He had already been away from Flagstaff once that year: "Their devotion to observation is almost unbelievable," Russell told Campbell.[79]

Transforming Lowell had been a mission for the 1920s that had less urgency in the 1930s. Astronomy in America had changed in profound ways in the decades during which Russell had visited Lowell. In 1921, American observatories were still largely untapped storehouses of spectroscopic and photometric data; studying the physics of the stars was limited largely to compilations with a bit of empirical deduction. Fifteen years later, however, there were many institutions moving in Russell's direction. Harvard under Shapley was, by then, as Russell told Edmondson, the best place to learn modern astrophysics, with Payne-Gaposchkin, Menzel, Bok, Whipple, and Sterne. By then the Yerkes Observatory under Otto Struve was becoming a formidable astrophysical engine, with Struve's phenomenally gifted team of Kuiper, Strömgren, and Chandrasekhar complementing solid observers like W. W. Morgan and C. Elvey. Impressive new observatories were being built in Canada, where at least one of them, at Victoria, was an important astrophysical center.[80] In 1921, there were few places beyond Mount Wilson and Harvard to which Russell could turn to find data and expertise capable of exploiting Saha's ionization theory. By 1935, there were over a dozen laboratories and observatories loosely linked by Russell's NRC committee, cranking out the spectral data he felt were essential to a full understanding of the atomic structure of the elements.

In the 1930s, the flood of brilliant refugee talent from Europe accelerated the change. By the end of the decade there was enough theoretical expertise around to assure Russell that there was a robust cadre of theorists capable of carrying on in the directions he had established. Thus his campaign to reform the Lowell Observatory was no longer necessary. It was a summer place: a nice place to stop on his way west to Mount Wilson, an ideal place for his family to be together to regain their health, and a place where, above all, he was treated like royalty. Indeed, he was the closest thing to academic royalty, since by then he was known as "Dean of American Astronomers."

• CHAPTER 20 •

Influencing Institutions and the Profession

"Dean of American Astronomers"

In the 1920s and well into the 1930s, rank-and-file astronomers in the eastern part of the country looked to three men—Russell, Schlesinger, and Shapley—as leaders; Peter van de Kamp called them "the Generals." They decided appointments, advised on research programs, and controlled much of the largesse the National Academy of Sciences earmarked for astronomy. Their influence was the greatest within an informal but influential eastern network, known as "The Neighborhood Club," that Schlesinger created when he moved to Yale. Among them, Russell's imprint was deepest.[1]

In 1937, when Russell was elected a foreign member of the Royal Society, *Nature* cited not only his contributions to science but his influence "among the American observatories" where he was long regarded as an "unofficial ambassador at large, co-ordinating work of various types and often taking an active part in the solution of the problems encountered." Editors at *Scientific American* added that "many speak of him as the dean of the profession." In March 1939, the *Sky* proclaimed Russell "Dean of American Astronomers" outright. Shapley, in his eulogies and encomia for Russell, reminded the world that the Dean of American Astronomers "grew up with astrophysics" and was its chief arbiter. Van de Kamp put it another way: "Russell was *the* referee, if ever there was a referee."[2]

"General" and "Dean" convey images of power and authority. If power is understood to be the ability to mobilize resources or to influence the course of careers, Russell indeed had power.[3] We have already seen how Russell was able to mobilize physicists and astronomers for term analysis, and Harvard staff for binary reductions, and, in both cases, to influence careers. Van de Kamp and others owed their livelihoods to Russell, and Cecilia Payne remained vigilant never to cross the man. Otto Struve advised his students that Russell's favor could insure a career: "Nothing could help you more in the application of a fellowship," Struve told William Markowitz in 1932, "than the support of Dr. Russell who, I believe, is on the board of the [National Research] Council."[4] Here we examine additional examples of how Russell gained, and then wielded, influence in the community.

To appreciate Russell's influence, we must recognize that in the 1920s and 1930s, the "Generals" commanded a rather small army. The professional astronomical community in the United States between the wars included a few hundred active members, led by about a dozen observatory directors, tied together by common practice, three journals, one regional society, and one national organization. It was a closed universe; appointments both high and low, from instructors to observatory directors, were typically determined in private. Only the leaders of the community were consulted. Rarely were the rank and file involved at all.[5] On a national scale, however, American astronomy was not monolithic. Since the turn of the century competition for resources pitted director against director. Pickering competed with Boss, Campbell, and Hale, and in the next generation Schlesinger and Shapley vied with Adams, Aitken, and W. H. Wright. There was considerable dissension at times between observatories of the East and West Coasts. A smaller power-bloc, led by Struve, also formed in the Midwest by the 1930s. No one, therefore, held the community wholly in check, though Russell came closest.

Unlike most observatory directors, who commanded large staffs, built great observatories, or, as in the case of Campbell, went on to higher academic office, Russell focused only on doing science, using ad hoc networks or collaborations to pursue it. His liaisons with Pickering, then Hale and Adams, then Meggers and Shapley, provided Russell with both enormous resources and compelling authority, which he used to great advantage as successive editors of the *ApJ* sought his advice, and as his *Scientific American* column became, in the words of a young Jesse Greenstein, "a kind of monthly classification of the state of astronomy." First Frost and then Struve long enjoyed Russell's eager assistance reviewing papers for the *ApJ*. In 1941, when the journal was taken over by the American Astronomical Society, Struve noted that among his collaborating editors, Russell was "perhaps the only one who has invariably been willing to review papers, and during the past ten years he has spent an appreciable amount of time in helping me decide whether a paper should be published."[6]

Among them all, it was probably Hale who did the most to make Russell a power center on a national scale. After his NRC 1919 essay, and continued organizing efforts on Hale's behalf, Russell was drawn more and more into the political infrastructure of the National Academy. His first major contribution was to assess the needs of astronomy for the planned National Research Fund in the late 1920s. By calling upon Russell to perform this function, Hale made him a focal point for American astronomy.[7]

The NRC's role in marshaling scientific manpower in the winning of the war convinced Herbert Hoover, secretary of commerce in the mid-1920s, to lead a campaign to raise $20 million for science from corporate America, to match what it had poured into industrial research.[8] Hale helped map out a ten-year campaign organized into discipline-based committees to determine the needs of science; he asked Russell to be an adviser to a newly formed National Board of Trustees for what was called the National Research Fund.[9] To Hale the campaign meant more than money; it would be a wake-up call that science was central to national health. The fund would also act as a stimulus to universities and other institutions to do their part.[10]

At first, Russell thought he was acting ad interim for Hale, but in early 1928 found himself appointed formally to the Scientific Committee and asked to create and to chair its subcommittee on astronomy. By then some six million dollars had been pledged, and the future looked bright.[11] After a bit of procrastination, on 30 April, knowing that he would be leaving for the IAU in Leiden in less than two weeks, which also meant that he would miss the meetings of the fund's Scientific Committee, Russell hastily wrote the directors of fourteen major observatories asking them to state their priorities and needs. He noted that the board wished to support individuals, not institutions, that institutions had to provide in-kind support, and that proposals should be for the completion of research underway "in which results will be attained more quickly than in new undertakings."[12]

Under such constraints, the responses Russell collected from Shapley, Aitken (Lick), S. A. Mitchell (Virginia), Schlesinger (Yale), and the other directors were hardly groundbreaking or innovative. In fact they hearkened back to Pickering's old plans for more routine assistants to get more of the same work done. What was remarkable was that neither Adams at Mount Wilson nor Russell himself asked for anything. Russell made it clear that he urgently felt "the need for *time*, rather than money, in his personal research and would be very glad to be relieved of his administrative duties in connection with this Committee."[13]

Clearly Russell wished to be free of this chore and dispatched it as efficiently as possible. Countering the goals of the Pickering era, the board had encouraged problem-based proposals that promised quick return, and also wanted to support individuals more than institutions. But Russell in his haste did little to carry this out. He queried only "salaried directors" because that was the way astronomy was structured.[14] Russell in fact ignored suggestions by Stebbins and Adams to push for innovative technologies, like photoelectric systems. Nor did he consider helping younger astronomers, as Leuschner and R. H. Curtiss suggested, or doing anything to promote theoretical astrophysics.[15]

Russell was in great haste to be done with the report so he could leave for Europe. When he returned in October, however, the job wasn't over. Pledges to the fund were running much lower than expected, so projected funds had been cut. Nothing could be spent until the entire amount had been collected. Russell created an ad hoc panel of observatory directors (the Smithsonian's Abbot, Hale, Campbell, Shapley, and Schlesinger) to evaluate a second round of requests to deal with the cuts.[16] They quietly fielded new requests and entertained second thoughts and opinions. But smoldering underneath all this was a growing rift between East and West Coast astronomers that centered on Hale's independent efforts to secure funding from the Rockefeller Foundation for a 200-inch telescope.

In this second round, Russell started to question his original priorities, made, he knew, in haste. Now he felt that established institutions did not deserve preferential treatment at the expense of support for junior researchers. He felt that such a plan would be demoralizing and would stifle creativity. Russell also did not feel that grants should be standardized, and argued against establishing what today would be called overhead: it would be "disastrous to make any hard and fast rule concerning additional financial contributions by the favored institution." And finally, in seeking to reduce regimentation, Russell argued that "[t]he astronomers, biologists, and others, are convinced that much better results can be obtained by considering individual programs and adjusting the sums awarded in proportion to their needs."[17] Russell pushed for freedom of action by individuals, not institutions. He had broken ranks with the more conservative elite among his peer observatory directors, but he brought them all along in the process.

These deliberations took another year, and Russell left them behind as he and his family went on their fifteen-month European tour in June 1929. He was therefore spared subsequent meetings of the committee, and by the time he returned, the fund was stillborn, killed by the Depression and by indifference among corporate managers to the idea that "technological progress in their industries required industrial investment in pure science." None of his cohorts expressed any deep concern to Russell about the demise of the fund; indeed most seemed content to work within preexisting limits and thanked Russell for his attention to their needs.[18]

For Russell, the National Research Fund had been an administrative burden. In fact it demonstrated that he could be partial to his friends. In the cutting process, Shapley's proposals remained on top even though Adams protested. Aitken and Schlesinger's proposals were reduced less than most, and V. M. Slipher's was supported beyond its merit. Those more distant from Russell, like A. O. Leuschner, H. D. Curtis, and C. G. Abbot, were favored the least. Yet the failure of the fund did nothing to lessen Russell's status. In fact his continuing service on the Executive Board of the NRC and his memberships on the Gould and Draper Fund Committees as well as many other ad hoc assignments only enhanced his position. He gained additional ground in 1932 when he helped to quell a motion to combine astronomy with geophysics into one section of the National Academy of Sciences because he did not want to see any "dilution of our Section by members who have no serious astronomical interests."[19]

Much of Russell's power flowed initially from Hale through Mount Wilson and the National Academy. But it did not come without a price. Hale's success in securing six million dollars from the Rockefeller Foundation to build another huge telescope came just as the NRF was struggling to meet its goals, which made Shapley and Schlesinger seethe. In late July 1928 Shapley described the channeling of so much money into the "small autocracy in Pasadena" as "criminal." A few weeks later, growing even hotter, he told Schlesinger that Hale's success was "very little short of embezzlement." Both Shapley and Schlesinger had been championing a competing proposal to the Rockefeller Foundation and now feared that all was lost.[20] They both spoke out, "vehemently damning the new project," Roger Lowell Putnam reported to Slipher. "It is apparently going through, and Russell will be somewhat identified with it. They feel it is a terrible waste of six million dollars, and so do I."[21]

Russell was indeed one of the few outsiders Hale trusted with his open secret in the spring and summer of 1928 as he maneuvered between the Carnegie Institution and the Rockefeller Foundation to make his dream come true. Russell had long been willing to lend a rhetorical hand, telling the *New York Times* in 1926 that a great new telescope such as what Hale envisioned was "not an idle dream, but a practical project; yet without the necessary pecuniary support, it can come to no more than a dream."[22] But Russell was not good at keeping secrets when deals were being made, especially when he thought he was among friends. There were many troublesome wrinkles between the Carnegie Institution and the Rockefeller International Education Board (IEB) that Hale had to iron out, including the governance of the new facility and where it would be built. Russell had opinions about both and injudiciously let Shapley know about them during travels to the IAU meetings in Leiden, knowing full well that Shapley and Schlesinger had asked the International Education Board to support a South African site for a large installation, partly in support of cooperative programs interconnecting Harvard, Yale, and the Dutch. Russell admitted to Shapley in July that he had serious qualms about Hale's plan because the Southern Hemisphere was largely untapped. Shapley eagerly agreed, writing Russell, "I fear that your fears are well founded."[23] What Russell could not imagine was that Shapley would use his words in his campaign to stop Hale; he managed to get the IEB to reopen the site issue. Hale wanted the telescope within a few hours' driving distance of Pasadena. Shapley wanted it in South Africa and enjoyed sufficient influence with the Rockefeller Foundation, as an adviser and petitioner, to make a frontal attack.[24]

Despite his reservations, Russell wrote supporting letters for Hale earlier in the spring that buttressed his arguments for governance and site selection.[25] But when the IEB asked Hale to justify why the telescope should be in southern California and not South Africa, Russell realized what Shapley was up to, and took decisive action, unequivocally supporting Hale, which only made Shapley more strident in his criticism that "Russell has sold out."[26] Both Russell and Hale argued that South Africa was too distant from established astronomical centers to attract good staff or to support programs where "prime advances in principles or methods of observation are likely to proceed." Hale also argued that the Northern Hemisphere, being better mapped than the southern, was ready for the next deeper plunge.[27] Russell argued strongly that "the great telescope must be the nucleus of an *observatory* and not a mere *observing station*." Based upon the evidence that Hale's staff had accumulated, Russell added, the best observing sites in the north were in southern California.[28]

It was quite true that the Southern Hemisphere was hardly as well mapped, as Hale stated, but the north needed more consistent and deeper mapping as well. Their arguments about the problems of isolation were more to the point, however, considering the enormous lifeline that would be required to manage such a large and complex instrument. But right or wrong, Russell had to undo the damage he had helped to create. He had to muzzle both Shapley and Schlesinger. "I have been hammering at Shapley on this matter for a long time," Russell told Hale in September 1928, "and do not know how much of a result I have produced (not much!!!)" Indeed, earlier in August Russell warned Shapley and Schlesinger that though he had had some minor qualms at first about Hale's plan, now he was fully convinced that it was sound.[29] Schlesinger soon backed off, which convinced J. A. Anderson, Hale's project manager for the new telescope, that Shapley would soon retreat too, as he reported to Hale: "I feel that Russell has done much to remove opposition. Am sure that after your next trip to New York all will be smooth."[30] Shapley, however, remained bitter and pressed Russell to endorse his Southern Hemisphere interests to the Rockefeller Foundation. Russell was willing to do so, but his support came with a price: he sent a very clear message that Shapley had better stop this divisive action. If he didn't, Russell was ready to state to the IEB that Shapley had grossly overstated what had been a casual conversation in Leiden. Above all, he warned, astronomers had to appear united; it would "be a real calamity if anything should diminish the completeness of this team-work in the future." Just as he had sermonized students before the war in Dodge Hall, where he called for "devotation and self-sacrifice in the full" over "satisfaction of the individual," Russell gave Shapley a dose of Kipling, one of his favorite poets and one in whom he found much inspiration. "A great deal depends upon the conviction that 'The game is more than the player of the game, and the ship is more than the crew.' On this matter I am really very much in earnest."[31]

Shapley soon backed down, and Hale thanked Russell for his intervention. With Russell's endorsement, the Rockefeller Board gave Hale no more serious trouble and assumed he would conduct a thorough site survey in the Southwest. Trying to make amends, Shapley wrote Hale that "Russell has unalloyed enthusiasm for all phases of the project. He is certainly a valuable asset wherever ideas are needed."[32] Russell went further, using his *Scientific American* column to signal the end to the site controversy. In late November Russell reported to Hale that everything was under control; he harbored some guilt for his inopportune remarks but now felt he had done his duty to kill the site controversy: it was "dead and it doesn't even deserve Christian burial."[33]

Russell's presence at Mount Wilson was now even more welcomed by Hale and Adams; this episode demonstrated to them that Russell controlled the eastern establishment. They even toyed with the idea of inviting Russell to build an astrophysical department at Caltech to solidify their position, but they knew they could never convince him to leave Princeton permanently.[34] For Russell, this was an important life lesson in managing conflicting loyalties but a lesson he learned slowly. As his power peaked in the late 1920s and 1930s, Russell would more than once let himself get caught between a sense of impartiality and duty to friends. Many of the choices he made were unpopular ones, which ultimately eroded the confidence others had in his opinions. These dealt mainly with staffing American observatories and membership on the Council of the American Astronomical Society. In both areas, Russell struggled with an entrenched American xenophobia.

Staffing American Observatories

Lick Observatory

Lick Observatory's Robert Aitken, who succeeded Campbell as Lick director in the 1920s, held to a firm view of proper practice in astronomy: "Observatory work, in the main, is necessarily program work. We must secure data of the same kind for a large number of similar objects before we can generalize safely, and that means laying out extended programs requiring years or even decades for their completion."[35] Places like Lick, owing to the tenacity of directors and staff, possessed masses of largely unexploited data ripe for analysis. The problem, as Russell saw it, was how to gain access to those data. What was needed was fresh blood on the staff able not only to exploit what had been carefully collected but to design new and more efficient ways to continue collecting, adopting methods informed by specific problems and guided by physical theory. In the 1920s, European-trained astronomers, especially from Holland, tended to know how to do this, and so did some of Russell's better students, like Donald Menzel and Theodore Dunham. Dunham had found Mount Wilson sluggish but capable of change; Menzel found Lick totally unresponsive and hostile.

"Professor Russell had spoken highly of my abilities to analyze spectra," Menzel recalled, so Lick hired Menzel in 1926. Through Russell's urging, he was given access to their incomparable storehouse of solar eclipse plates, accumulated from many expeditions to the far ends of the earth. As Russell hoped, Menzel devoured these materials and, after several quick discoveries about how the spectrum changed during the flash event, started applying his training to study the physics of the chromosphere in great detail, using chromospheric spectra taken by Campbell at eclipses from 1898 to 1908.[36]

In his first year, Menzel was deep into his work, analyzing data. It helped that Mrs. Menzel was very sociable in the isolated mountain community. But they never quite fit in. Menzel continually encountered resistance as he fought to make a niche for astrophysics.[37] Berkeley staff like C. D. Shane were far more interested in his solar chromosphere work, which was applauded everywhere but Lick. Menzel's penchant for physics marked him as a maverick on Mount Hamilton, a place that resisted a visit from Arnold Sommerfeld in 1926 and hated anything Germanic. Menzel, in character and name, looked German to the Lick staff, although he was a third-generation American. Even in 1939, Shane testified that "Menzel denies that he is a Jew, but everyone else assumed."[38]

Menzel found Lick "most uninspiring." Other than Robert J. Trumpler, Swiss-born and -educated but naturalized American, there was no one on the mountain who "seemed ever to crack a book."[39] He was deeply disheartened after he submitted his monumental study of the solar chromosphere to Aitken in 1930, for the *Lick Observatory Publications*, and found that Aitken threatened to delete his quantum mechanical discussion of absorption coefficients. Aitken, Menzel recalled, was adamant: " 'After all,' he would say, 'this is an *Observatory*! Your responsibility is to make the observations and record them. Leave the theory to the poor, underprivileged British astronomers, such as Milne and Eddington, who don't have an observatory.' " Menzel fought to retain large sections on the calculation of the absorption coefficients in his Lick memoir, based upon the wave mechanics of J. Robert Oppenheimer, Sugiura, and Gaunt.[40]

Menzel grew very unhappy at Lick and spent more and more time at Berkeley, where he learned much of the theory he needed from Oppenheimer and spectroscopic technique from R. T. Birge. He had a growing family and wanted a normal life. Even though Leuschner and Birge urged Aitken to assign Menzel to a joint appointment, Aitken was reluctant to accommodate, since he felt that Menzel was not the sort of man for his mountain. Aitken was willing to approve a faculty trade for a semester at Berkeley, but this only helped convince Menzel to look around for another position. He had nibbles from Columbia and the University of Minnesota and turned to Russell for advice.[41]

Russell did not care for the options. Menzel was too valuable at Lick and would be distracted by teaching at Minnesota or Columbia. He urged him to stay put. "It seems to me that you are disregarding the fact that your research work goes permanently to the credit of your professional career, whereas your teaching work, however successful, would be of very little value to you elsewhere." As he had advised Edmondson, research potential was everything; it was the only thing that would be considered some fifteen years thence when, as Russell envisioned, the senior men in the best positions started dropping off. Menzel followed Russell's advice and stayed another year at Lick, with a slightly improved salary made possible by the Minnesota offer. He was welcomed at Berkeley, but the mountain remained oppressive. In 1932, Russell advised that good jobs were "appallingly scarce at present," and urged Menzel to "stick to the devil you know." Lick had great professional advantages until something equal or better came along. Meanwhile, Russell managed to secure the Minnesota position for Willem Luyten, then at Harvard, who he felt worked best away from others. After Harry Plaskett also left Harvard for Oxford in June, Shapley had at least two slots to fill. Russell made sure Menzel got one of them. Whatever the western astronomers thought of Harvard, Russell advised Menzel, "[i]t is one of the most attractive positions in the country."[42] Indeed, Menzel found that theoretical astrophysics was most welcome at Harvard; the physicists on campus helped him make "it respectable for astronomers at the Harvard Observatory to engage in such calculations themselves and to count them as astronomical research."[43]

Lick had lost Menzel, but Aitken was not so entrenched that he was blind to new talent, especially if it was a double star worker. When he started corresponding with a young Hertzsprung student in the late 1920s, Gerrit Pieter Kuiper, he had a plan. Aitken was impressed with Kuiper's industry and creative ideas for a systematic program to improve statistical knowledge of the number and types of double stars in space. At Aitken's urging, Kuiper sought and won a Kellogg Fellowship to work at Lick in 1933, and his tenure was extended by a Morrison Fellowship through 1934.[44] Unlike Menzel, Kuiper thrived on observing, searching out faint visual doubles, determining their colors and spectra, and extending his observational program to white dwarfs. Not content to merely catalog his observations, Kuiper refined the observational mass-luminosity law and luminosity function, both important astrophysical quantities critical for stellar structure and evolution.[45]

As Kuiper completed his second year, Aitken was planning to retire, and there was talk that the young Dutchman might succeed him. His research work was very fresh and exciting, and he very much wanted to stay. Aitken liked Kuiper, even though, like Willem J. Luyten, he appeared brash and impetuous, and some felt that he was a brazen self-promoter. Although Aitken hoped he could stay, there was considerable pressure on him from more nativist Lick colleagues like Wright, Kuiper told Bart Bok, "only to appoint young Americans."[46]

Kuiper came to Russell's notice in a number of ways; in late 1934 Russell congratulated him for his careful and systematic searches for low-mass nearby dwarfs. They met during Russell's visits to Lick as well, but Russell also knew of Kuiper from Shapley and Bok, a former Leiden classmate now at Harvard and a most powerful and persuasive ally.[47] One of Bok's passions, in concert with Hertzsprung, was the health and welfare of Dutch astronomers in America, Kapteyn's legacy, and he fully expected Kuiper to become Aitken's successor. When he heard in the fall of 1934 that Kuiper might leave Lick, he admonished his friend to fight it out: "Lick need[s] fresh blood more than any other American observatory." In January 1935, Bok cornered Russell at Harvard, anxious to get some real heavyweights involved.[48]

In February, W. H. Wright was named Lick director by Robert Gordon Sproul, UC president. Wright had campaigned for the position, but a faculty committee from Berkeley made the recommendation upon the condition, in light of the Menzel fiasco, that Lick strengthen its staff by hiring "competent young observers and men trained in the modern aspects of astrophysics in relation to atomic theory." Wright and Aitken proposed that Arthur B. Wyse, Nicholas U. Mayall, and Kuiper all be hired. Aitken was thankful for Sproul's support but knew that the budget was still tied up in the state legislature, and if it were not approved, they would have to worry more about staff reductions than about hirings: "[W]hat can he do?"[49] He soon learned that Lick could hire only one new astronomer, and Wright's first choice was Wyse, his own student, although money was also found for Mayall from operating funds. Aitken agreed with the decision, for both men placed Wyse and Mayall above Kuiper, a decision that angered Trumpler. Part of Aitken's reasoning, as he told Sproul, defending himself against Trumpler, was that Kuiper had announced that he had to make a quick decision about his future because he had received a call to a Dutch colonial observatory—the Bosscha Sterrnwacht in Lembang, Java—which he accepted, thinking he was doomed at Lick. "In view of all the circumstances, we therefore withdraw the recommendation made that Dr. Kuiper be appointed as a member of the staff of the Lick Observatory for the year 1935–36."[50]

But Kuiper dearly wanted to stay at Lick: "Here *everything* is ready; there is nice material here, dating from 1900, for the bright stars, which only needs some completion to form an excellent basis for some very important statistical work."[51] He appealed to Russell, warning that his departure meant the death of double star work at Lick. Kuiper needed the "voice of a leading American astronomer" to change the Lick state of mind and asked Russell to intervene on his behalf with Sproul and Leuschner, who had been on the search committee for Wright.[52] Russell sprang into action, writing letters to Aitken and Leuschner, and an "official" letter to Sproul, sent through Aitken. Russell's official letter charged that losing Kuiper "would be a disaster to double-star astronomy, and to the Lick Observatory." He was more personal with Aitken, asking whether the decision had been based upon Kuiper's nationality. If so, as Bok had alleged, then Russell was ready to do battle: "I wish to register a most emphatic protest against the policy of exclusive nationalism in pure science." Russell then cut deeply into Aitken's hide: "The days of routine discovery of pairs by the thousand are over, but [Kuiper] has shown that there are *better* fish in the sea than ever were caught, and has actually caught them. If the Lick Observatory cannot rise to an opportunity like this, I am very greatly concerned for its future." This offended Aitken, who flatly denied that nationality was a factor, and told Russell not to worry about Lick's future.[53]

Russell's attack did stimulate Aitken to search for alternative sources of funding for Kuiper, and in doing so, Aitken demonstrated that, indeed, there were sensitivities about Dutchmen at Lick. Writing to the consul general of the Netherlands in San Francisco, he suggested that the Dutch government pay its fair share for Dutch astronomers. Simply put, in an era of rampant nativism and tightening budgets (see chapter 21), Aitken felt that Dutch astronomers were overrepresented on American soil, which carried "certain embarrassing aspects."[54] Nothing came of this, but in the meantime Bok managed to secure a temporary appointment for Kuiper at Harvard. Above all, as Hertzsprung had advised, Kuiper wanted to stay in the United States and leave Lick on good terms because he knew he would need its data. Russell was delighted that Kuiper would be at Harvard, though he worried about Kuiper's command of the language, since he would be teaching.[55]

Russell's opposition to nationalism and his annoyance with Lick overrode what political acumen he may have possessed. He vowed to Shapley that he would fight the "policy of exclusive nationalism applied to pure science." Because of its insularity, Russell feared, Lick Observatory was on the wane. Losing Kuiper accelerated the process.[56] Therefore Russell continued the attack, this time in a *Scientific American* column where he wondered aloud how Kuiper could discover "one of the most remarkable double stars on record" when two generations of Lick staff had searched the area thoroughly. The answer was that Kuiper was looking for what he found. Rather than employing the Lick "drag-net" style, Kuiper searched for companions to faint red high proper motion stars because he knew that they would inform statistics on binary mass ratios. The existence of very low mass companions implied the existence of planets. Russell made a big point of this, implying that Kuiper's brief tenure at Lick had yielded information of greater significance to astrophysics than the oceans of data Lick astronomers had accumulated.[57]

At a time when directors like V. M. Slipher, Aitken, Wright, and even Adams resisted the steadily increasing influx of highly qualified foreigners, whether they were political refugees from the growing horrors in Europe or postdoctoral scholars in search of clear skies, large telescopes, and rich plate vaults, Russell remained outspoken that only scientific worth should be considered in hiring decisions. Russell put science first and could appear to be insensitive to social circumstance, human sensitivities, and political and economic reality. He was intent upon opening Lick up to modern astrophysics, using either Menzel or Kuiper, but he could react differently under different circumstances. Why, for instance, did he not invite Kuiper to Princeton? Dugan was then suffering from his increasingly painful arthritic condition, and there was funding available for someone to assist him in double star work, a position soon filled by Newton Lacy Pierce.[58] Russell might well have hired Kuiper, but he had just invited Rupert Wildt to come to Princeton before he learned of Kuiper's fate, and neither Kuiper nor Wildt was the type of scientist one would expect to engage in Dugan's brand of work. In fact, in hiring Wildt, he anticipated Shapley by only a few weeks, and apparently the funding Shapley had scrounged to grab Wildt was immediately turned over by Bok to pay for Kuiper.[59]

Russell's efforts to influence staff hiring at Lick were not immediately successful, though he did have some impact later. As we shall see, Russell could also set priorities based upon personality or nationality, but he did so only when other overriding professional factors were at play. This trait will become stronger as we follow his continued efforts to influence hirings elsewhere, as well as, ultimately, at Princeton.

Yerkes Observatory

In June 1931, Walter Sydney Adams, knowing that Edwin Frost was due to retire in a year, told University of Chicago dean Henry Gale that Frost's successor had to be an American. "I shall be interested to hear what Russell says on the subject" because, he warned Gale, "Russell's opinion, especially on the personal and administrative sides of individuals ought to be taken with considerable reserve." Although Russell continued to count Adams as one of his closest colleagues, Adams had been annoyed by Russell's promotion of Luyten for the Minnesota directorship, which he thought was a great mistake. Of course, Adams added, this did not apply to a man like Otto Struve, the leading candidate to replace Frost, "but what I wish to emphasize is that Russell's appreciation of ability carries him away from any other possible considerations."[60]

Adams knew that Russell would weigh in heavily on the matter of Frost's successor, so did what he could to inflame Gale's well-known dislike of foreigners. Both worried that Struve was too much of an individualist, which was something that would not concern Russell as long as he was not personally threatened.[61] Struve, in fact, did not have an easy time succeeding Frost. The remaining Yerkes staff wanted Joel Stebbins. Adams expressed the general feeling about Struve when he told Gale that "it would be difficult for him as a foreigner to represent the observatory adequately." V. M. Slipher urged Philip Fox of Northwestern to make a run for the Yerkes job: "Surely American astronomy is not at such a low ebb that it must go to Europe for the director in this case. Power to you!"[62]

Gale had to ask Russell's opinion, and predictably Russell hoped that astrophysics would be strengthened by "a man of standing" who carried weight on campus and in the community. It would take a strong person to break Yerkes out of its rut of "limited means and hampered activity." He had to be a fighter, but most of all, "He must be willing to come." Russell knew that the best astronomer on the present staff was Struve, but did not think he was the right man for the job. Nationality played no part, which would not have surprised Adams or Gale. But he felt strongly that Struve's talents and drive were better spent doing research than being worn out by administration. Reviving Yerkes would be a chancy and difficult fight, and, frankly, Russell thought the place was not worth Struve's energy. He said as much to Shapley because he thought Struve would do better at Harvard, where Shapley's plans for two 60-inch telescopes would give Struve the light-gathering power he needed, and where a vibrant astrophysical staff would propel his ambitions in high-dispersion stellar spectroscopy.[63] Yerkes needed a director who was an administrator, and who knew how to exploit a large refractor, Russell argued. In all this Russell claimed to be an impersonal adviser, but he was hardly so. Though he told Shapley that he had written frankly and openly both to Gale and to Struve, which "ought to make me as objective as possible," he added, "I hope that you get him."[64]

When, through Frost's and Struve's efforts, the McDonald bequest was secured to build a new very large telescope in Texas to be run by Yerkes, Russell was doubly convinced that managing two observatories would ruin anyone for research. He had thought this in the case of Stebbins, who was better off at Wisconsin where he was free to work at Mount Wilson for three months out of every year, like himself. Struve should take the same tactic, Russell told Gale and Shapley. "I am advising Struve that the main danger consists of overwork for himself." Russell could not understand why Struve would want to direct two observatories. At the least, he might consider managing Yerkes alone, and being designated research associate in Texas, "where he would have the run of the place with none of the responsibilities."[65] Throughout all of this, Russell never mentioned the question of a foreign-born director. That was not a factor for Russell any more than it was for the progressive University of Chicago president Robert Hutchins, who finally appointed Struve and kept xenophobic deans like Gale in check.[66]

Just as Struve frequently consulted Russell on spectroscopic matters, he also sought his advice about Yerkes's future, expressing gratitude that Russell had "taken an exceedingly kind interest in the affairs of the observatory and in my personal plans, and I owe a great deal to his encouragement and advice."[67] Since his sabbatical with Eddington in 1928–1929, Struve had resolved to develop a wide range of quantitative techniques to better understand line broadening mechanisms in stellar spectra, a specialty that kept him at the border of modern physics and astronomy. It became his mission as director to build up institutional resources at Yerkes and McDonald that could first collect the spectral features of the stars with as much fidelity as possible and then "find their explanation in simple physical laws."[68]

Struve's overall strategy was to staff Yerkes with a balance of theorists and observers who could work closely within an observatory context. He knew what was needed, as he later told President Hutchins: "Throughout the entire history of astrophysics, the educational institutions of the United States have depended upon the European countries to provide nearly all the theoretical work in astrophysics, with the sole exception being Professor Russell's department at Princeton University."[69] In late 1934, Struve saw his way clear to establish a new and more powerful center. The infusion from the McDonald bequest increased campus support too. His goal was to import the best theoretical talent and mix it with domestic observing expertise to create a training ground to produce fully competitive astrophysicists. He searched for a theorist, failing to secure Albrecht Unsöld or Svein Rosseland by the end of the year.[70] After another year of searching for an observer adept in theory, he found Kuiper.

Struve wrote to Russell before he discussed matters with Kuiper, who, though hired, was still at Harvard in late 1935. Struve was considering young theorists like Strömgren and W. H. McCrea, as well as Chandrasekhar, though he seemed to be "altogether too much of a mathematician to be of value." Struve asked about Americans, too, because he was very aware of contemporary feelings. Menzel, O. C. Wilson, and even Leo Goldberg, who was still a graduate student at Harvard, came to mind. But Struve dismissed the American names; either they were not available, or they needed seasoning. Russell was sorry to hear that Rosseland had turned down Struve's offer, but Marcel Minnaert, Strömgren, or Wildt would be fine. Chandrasekhar was the "most brilliant of the group" Struve had named, and, Russell predicted, "bids fair to be one of the best men of the coming generation in theoretical astrophysics." He was now visiting Harvard and was about to start lecturing, and already Shapley was sending back glowing reports about him. Russell, however, thought that "his political views are pretty radical, but I don't imagine that would prejudice him with President Hutchins." Even so, Russell added that "I should want to know a little more about Chandrasekhar's personality before tying up with an Oriental," though he agreed that everything he had heard about him to date was favorable.[71]

Russell well knew that Shapley had been trying to snag Chandrasekhar as a Harvard fellow for over a year. Failing more than once with the Harvard authorities, he complained to Russell, "I can not see who would deserve it more . . . perhaps it is best to be an Englishman rather than an Indian."[72] He partially succeeded late in 1935, and so Chandrasekhar packed his bags and left Cambridge for three months of lecturing on "Cosmic Physics" at Harvard so he could be scrutinized. He was then delicately balancing his family's wish that he return to India against the possibilities and problems attendant on staying, somehow, in the West.[73]

Knowing Shapley's wishes, Russell hatched a plan. There was another theorist Struve should consider. Russell now felt that Robert d'Escourt Atkinson of Rutgers was the best young theorist available. Atkinson had deeply impressed Russell with his abilities in theoretical physics, especially his contribution to the stellar energy problem. Although Russell did not feel that Atkinson was in Rosseland's class, he was the equal of all the others. "I don't see why I didn't think of him before," Russell admitted to Struve. "He is very much the kind of man whom you might want."[74] In February 1936, Russell learned that Strömgren had accepted Struve's call, but that a second position was also open, and so redoubled his campaign for Atkinson, touting his capabilities rather than criticizing Chandrasekhar. But when Struve raised the specter of Chandrasekhar's "nationality and political views," which, he told Russell, were "such as to make it difficult for me to defend him properly on the campus," Russell sympathized: "I also see very clearly the other difficulties which you mention . . . It seems to me very hard for westerners and asiatics to become really acquainted."[75]

Shapley pulled out all the stops. Since Chandrasekhar was sympathetic to the national movement in India, he alleged to Struve, he had to be a communist and so would be safer at Harvard. Struve, Shapley well knew, was very sensitive to the fact that archconservatives like Gale fiercely opposed Chandrasekhar, and that "only the President and some of the trustees are full-heartedly in favor of his candidacy." Struve worried that Chandrasekhar would suffer the fate of the "American Negroe" on the Chicago campus, far from Yerkes and Struve's protection.[76] When Struve also expressed concerns about Atkinson's productivity, which were in fact quite justified, Russell rationalized that Atkinson had been burdened with a heavy teaching load at Rutgers. At Chicago, he would bloom. Only when Hutchins weighed in, demanded to meet Chandrasekhar, and then pressed Struve to hire him, was the deal done. Hutchins knew that they were in a race with Harvard, and did not want to lose the strongest theorist he had met to date.[77] Struve was probably relieved. Certainly Kuiper was. As noted in chapter 16, both keenly appreciated how Chandrasekhar had helped Kuiper interpret Trumpler's cluster diagrams in terms of Strömgren's reinterpretation of the Hertzsprung-Russell Diagram. They would make a terrific team.[78]

Although Struve and Russell agreed over the race issue, and let Kuiper know it, Kuiper felt that there were deeper motives at play. Kuiper puzzled over Atkinson—"I do not know why Russell recommends him," he told Struve—but as he continued to observe Russell in action at Harvard and at Neighbors' meetings, where Russell was always the center of action, he sensed what he thought was a scheme: "The only interpretation I can give to these points is the (unconscious) desire of the great American astronomer to keep a certain 'balance' between the various observatories; that one institution having *two* good theorists would be relatively too strong."[79] More likely, Russell believed that having Atkinson at Yerkes would strengthen Russell's position there, whereas having Chandrasekhar at Harvard made him more than accessible to Russell because of Shapley. This mix could have been very important to Russell, intent as he was on building circles of trust and confidence at various institutions. As Russell claimed to William Smart of Cambridge, "Chandrasekhar's renewed interests in astrophysics [are] to be credited mainly to the very stimulating environment at Harvard." He was ready to do all he could "to keep it going."[80] If Russell was gerrymandering the astronomical community, as Kuiper alleged, it was for a complex of reasons, both professional and personal.

On his way back to Harvard after meeting Hutchins and lecturing at Yerkes, Chandrasekhar stopped for two days to confer with Russell and recalled Russell's concern about racial issues. He was moved, but not convinced, putting off his final decision until he had a chance to talk with Eddington and Milne. Both urged him to work with Struve and Strömgren. In the end, Chandrasekhar made his own choice. He felt far more compatible with Kuiper and Strömgren than with anyone at Harvard, especially Theodore Sterne, who criticized Chandrasekhar incessantly during his lectures.[81]

If Russell harbored any disappointment, he kept it to himself. As much as he wanted Atkinson at Yerkes, by the New Year Russell applauded Struve's staff building. As he told Adams: "Struve has made a magnificent start. He has an extraordinarily fine group of young men—Morgan, Chandrasekhar, Strömgren and Kuiper—and they ought to do a lot."[82]

A Referendum on Astronomy

As Russell had predicted to Menzel, by the 1940s many of the directors of the largest American observatories were at, or well beyond, retirement. This was true at Mount Wilson, Princeton, Lick, and Lowell. Russell advised on all these positions, and his advice, as well as that of others, not only highlights again the xenophobia of the time but offers a glimpse into how astronomy was resistant to change. When Wright was appointed, it was well known that he would soon retire, and so the process had to be repeated in 1940 and offers a particularly revealing case of Russell's still running against the tide, but, more experienced now, trying for a practical compromise.

The search was conducted by R. T. Birge, since 1933 the chair of the Berkeley physics department and a longtime Russell correspondent in spectroscopic matters. Birge had also been the chair of the search committee that chose Wright in 1934–1935, and knew that Russell was the only person not to approve of the choice.[83] Now, in 1940, Russell was not alone in casting his negative vote for Wright's heir apparent, J. H. Moore. Both Adams and Russell agreed that there was a serious lack of good men among Americans of the right age, but, in tune with Birge, Adams wanted the new Lick director to be an American, "or at any rate an Anglo-Saxon (Canadian or English)." This brought Adams around once again to worrying about what Russell had said to Birge.[84]

Adams knew Russell's power and worked hard to counter it: "The man of the greatest ability might be impossible as a director, a point on which I have always felt that Henry Norris Russell has a blind spot." Adams agreed with Russell on many points: A director should have superior scientific standing and be active in research organizations around the world, he felt, but he also must be willing to "accept responsibility and make decisions (a quality rarer than I used to believe) and preferably he should have administrative experience." For Adams, Russell's decisions were based upon overly narrow criteria: scientific value, not citizenship nor personality.[85]

The narrow search process in 1934–1935 had raised resentment at Lick, so Birge now queried many more people and formed a committee of Lick and Berkeley staff. He questioned directors and rank and file; in many ways, his twenty-four pages of closely typed text in a series of three reports to Sproul in February 1941, based upon twenty-one letters from eleven people and numerous interviews, constitutes a referendum on modern observational astronomy. As Adams feared, Birge paid close attention to Russell's suggestion, ignoring rumors flying about the Berkeley campus about Russell's opinions. The top rumor, promoted by Leuschner, was that Russell no longer looked upon Kuiper with favor.[86]

Contrary to Adams's opinion, for reasons I examine later, Russell had grown very sensitive to political issues in the late 1930s. First, Russell described Shapley and Struve's directorships as good examples of enlightened leadership and hoped Birge could find someone equally adept at administration and modern research techniques. Next, he told Birge about the Kuiper episode at Lick in great detail, which caused Birge to alter the questions he asked people. And indeed, even though Russell favored Kuiper, he did not recommend him for the job. It was a difficult decision, he admitted, because he believed that institutions and instruments defined research traditions in observational astronomy, and that Kuiper fit Lick's potential perfectly. But Russell well knew that the Lick staff would revolt, as would most American observatory directors other than Struve and Shapley. Russell thought that Trumpler, who had befriended Kuiper, would suffer the same fate, since his years at Lick had also been less than comfortable and had propelled him eventually to the Berkeley campus. Trumpler was the most qualified scientifically, Russell believed, based upon his exemplary performances in astrometry, statistical astronomy, cluster studies, and galactic structure, but Russell wanted to save Trumpler for Yale, when Schlesinger retired, where the instrumentation better suited Trumpler's talents.[87] As usual, Russell thought broadly about the community.

Paul Merrill of Mount Wilson was Russell's first choice, and eventually Birge's choice, partly because his research strengths were in line with Lick's spectroscopic capabilities, but also because he had proven himself to be an intellectual leader. Most of Struve's people were too young, Russell felt, knowing full well how foreigners would be received at Lick. Menzel, he also realized, was out of the question too: he possessed "marked ability, but I would not quite recommend him," for reasons similar to those that prevented him from endorsing Kuiper. He did not think the Lick staff should move to Berkeley, which was also a factor in Birge's deliberations, so Russell felt that Shane was a poor choice, because he advocated the move. As Russell had argued on Hale's behalf, an observatory devoid of its staff was a mere observing station, which was not what Russell had in mind for Lick.[88]

In his analysis, Birge was able to discern a definite east-west split on many issues. Most easterners agreed with Russell that the scientific standing of the director was of utmost importance, whereas a significant number of the westerners thought of the institution first and the need to upgrade instrumentation, as long as the director was otherwise acceptable in mind, body, and national origin. The general feeling was that the future lay in astrophysical practice, defined by the spectrograph. A very few thought that if the university decided not to invest any new energy in Lick, it might best be run by a traditional person, possibly a statistical specialist with theoretical inclinations. A theoretical astrophysicist was never considered.

Only a very few, like Russell and Struve, worried about the importance of maintaining a theoretical infrastructure anywhere. Looking for a way to train students in theory more effectively in 1940, Struve tried to establish a theoretical group on the Chicago campus, centered on Chandrasekhar:

> The rapid decline of astronomy in Europe and the threat of losing contact with theoretical astrophysics is so great that one of the principal tasks in the United States is to establish a center for theoretical research which would become the successor to the great theoretical departments in England and Germany. With Russell's retirement there will be no department, or even an individual astronomer, left in the United States who is competent to continue the work of Eddington, Milne, Unsöld, and others, except Chandrasekhar who has all the qualifications necessary for the task which I have in mind.[89]

He did not suggest this for Lick, however. Nor did Russell. Neither did Russell seem to worry about the need to upgrade instruments at Lick, for what was, clearly, a once-dominant but now definitely outclassed observatory. His ambivalence reveals his distance from those like Struve, Zdeněk Kopal, and especially Ira S. Bowen at Caltech, who believed that the future lay in new instrumentation.

In any event, Birge's recommendations were eventually put on hold, overtaken by the war, and Moore was installed as interim director. Until the war was over, astronomy at Lick was in a holding pattern.

• ☆ •

The Dean of American Astronomers did have blind spots caused by the often conflicting demands created by personal loyalties, scientific standards, and political necessities. Except for the 200-inch telescope episode, he typically played politics on Shapley's behalf, and, from the examples covered here, had a mixed record of accomplishment in influencing the course of institutions. There were, indeed, consequences to the choices Russell made. By the end of the decade, Adams let their political differences influence his opinion of Russell's continuing scientific worth to Mount Wilson. More faithful to his longtime colleague F. H. Seares, who was about to retire, and annoyed with Russell's continuing promotion of foreigners, in late 1939 Adams preferred Seares to Russell as a research associate after his retirement, if a choice had to be made. Budgets were very tight, and Adams had his own loyalties. Although Russell remained a Carnegie research associate until his own retirement in 1947, there were other consequences of his blind spots and of his actions on behalf of foreign-born astronomers, as the next chapter reveals.[90]

• CHAPTER 21 •

Astronomical Isolationism

As American observatories became attractive sites for foreign-born astronomers in the twentieth century, especially after World War I and into the 1930s, these astronomers found that they were less than universally welcomed. Even though he was a strong advocate of hiring foreigners, Russell became the target of foreign-born astronomers who believed that he and Shapley personified American arrogance, and by implication they defined the nature of the discipline and its practice. The best example of the sensitivities of foreign-born astronomers to what they perceived was American indifference to their achievements is the story of how the "Russell Diagram" became the "Hertzsprung-Russell" Diagram.

Russell's advocacy of foreign talent also got him into hot water among American colleagues. Adams and Slipher, among others, felt that he held unrealistic ideals and was insensitive to the need to support native astronomers over foreigners. As we saw in the last chapter, Russell encountered the growing nativism of the 1920s and 1930s when he pushed for Kuiper at Lick and Luyten for Minnesota. This pattern was reinforced to some extent by his advocacy of Peter van de Kamp for Swarthmore, though this appointment raised fewer hackles than the first two.[1] At the end of the decade, Russell faced nativism once again in a dispute over the composition of the Council of the American Astronomical Society.

These episodes help us appreciate at a deeper level why foreign-born astronomers posed such a threat to isolationist American astronomers and themselves felt threatened as a result. They also reveal one way Russell's influence in the American community was compromised by his ideals.

The Renaming of the Diagram: Contrasting Styles

Of all the astrophysical relationships linking Russell's name artificially with others, such as the Vogt-Russell theorem, the most contentious was how the spectrum-luminosity diagram, called by American and English astronomers the "Russell Diagram," became the "Hertzsprung-Russell Diagram." Danish and Dutch astronomers were never happy with the "Russell Diagram," although some, like Bengt Strömgren, used both terms in the early 1930s.[2] What really annoyed people like Kuiper was that starting in 1926 at least, people like Eddington and Milne ironically tended to call Russell's troubled theory of stellar evolution the "giant and dwarf theory of Hertzsprung and Russell."[3]

Personality and national pride were ingredients in the long-smoldering dispute, which revealed fault lines created by different modes of practice. But the fact remained that it was Russell's prominence, due largely to his forceful demonstration that giants existed, and his dogged campaign to establish his theory of stellar evolution, that indelibly linked the diagram to his name. By the 1930s, however, as Russell's theory faded into oblivion, and as Kuiper sought to replace it with Strömgren's new tracks, the ardent Dutchman also campaigned to alter the very symbol of Russell's authority and reestablish his old teacher as the creator of the diagram.

Kuiper's passion was born not only of loyalty. He was provoked. After leaving a party at the Gaposchkins' home during a Neighbors' meeting in March 1936, where Russell presided, Kuiper was livid. He had come hoping to discuss the evolutionary meaning of Trumpler's star clusters and wound up in a skirmish with Shapley and Russell over the name of the diagram. Kuiper objected to the "Russell Diagram," insisting, as he claimed Luyten and Strömgren did, that it was more properly the "Hertzsprung-Russell Diagram." Kuiper later told Hertzsprung and Chandrasekhar that his suggestion annoyed Russell considerably. Kuiper had demanded top billing for Hertzsprung and was astonished when, he reported, Russell barked back, "Hertzsprung did not make it!" This "terribly unfair statement" propelled him from the Gaposchkins' home to the Harvard Observatory library to prove Hertzsprung's priority.[4]

This exchange came just as Chandrasekhar was being courted by both Harvard and Yerkes, and Kuiper tried to use it as proof that Harvard was no place for a foreigner. The name of the diagram revealed what he believed was a deep bias among Americans. Russell slighted all earlier efforts and took the credit all for himself, Kuiper alleged; but only his now-discredited theory of evolution was new. Everything else belonged to Hertzsprung. To Kuiper, everything that was wrong with American astronomy—its insular and derivative nature, its lack of standards—was due to Russell: "It is incredible how the astronomers have been hypnotized by the name Russell diagram." Kuiper had even heard Bok claim that Russell called Hertzsprung "a perfect failure as an astronomer, although he has had brilliant ideas." Even Kuiper could not believe this rumor; he must have meant that Hertzsprung had good ideas, "but did not carry them through." Even so, Russell was "definitely wrong." Kuiper also reported that when he confronted Shapley with the evidence, the man only sneered, "The Germans claim now everything for themselves," and then sauntered off shouting " 'Deutsche Physik.' " If Shapley had indeed made such a thoughtless and ill-tempered remark, Kuiper's already extreme sensitivity was more than inflamed; he vowed thenceforth to use only the title "Hertzsprung-Russell Diagram."[5]

Whatever the truth of the matter, Shapley could be terribly sarcastic and Russell blithely insensitive over what was, to be sure, a very serious matter for proud Europeans. In private, Russell did call it the "Russell Diagram" but did not push the appellation in his publications.[6] Underneath the tension was the fact that Hertzsprung's style of research was quite different from Russell's, and they had stepped on each other's toes many times in the first decades of the century. Kuiper well remembered Hertzsprung's own anger during his lectures in Leiden, his claims of how Russell had stolen the show. But now, Hertzsprung's goal was to preserve the foothold his students had gained in America. It was all-important to understand the Americans and learn to live with them. He wanted Kuiper to focus his energies on the opportunities, such as the new Texas telescope, not on his animosity for Russell and Shapley. "I've always thought it was my first duty to use the available instruments to their full extent," he added. "In doing so, I changed from theoretician into practician." Moreover, defensive tactics were wasteful: "If the criticism is justified, it is good that one gets to hear it, if it isn't, it's worse for the one who passes it. Self-criticism is best—it is the source of all progress."[7]

Hertzsprung's advice reveals his priorities. In September 1937, awarded the Bruce Medal of the Astronomical Society of the Pacific, Hertzsprung spoke "On Collaboration in Astronomy," hailing the wealth of double star data that the Lick staff had amassed. Knowing full well that with the retirement of Aitken ("the Nestor of double-star astronomy") and Kuiper's departure, Lick no longer was active in double star research, he hoped that Aitken's successors would freely share their hard-won data with their less well-equipped European colleagues.[8] In sharp distinction to Russell, he was a passionate advocate of "sacrifice . . . to do work of which others later will earn the fruits, [for] this is the only manner in which we can pay the debts we owe to our ancestors." Playing to his Lick audience by taking a dig at Russell, he decried the growing trend in America of downplaying routine work: "As everyone familiar with the matter knows, good observations require a kind of skill which is not given to all and seldom to the theoretician." Denying much of his own early history, Hertzsprung suggested that most of the important discoveries were "made by the unconscious but eager observer rather than as the result of theoretical considerations." "I should be the last to deny the importance of theory," he admitted, "but let us be modest and not underrate what we owe to the ardent eye of the observer."[9]

The issue of the naming of the diagram smoldered for years. Finally, in the late 1940s, when W. W. Morgan and Chandrasekhar were shepherding the *Astrophysical Journal*, Chandrasekhar, tiring of Kuiper's continued rants, settled the matter by declaring "H-R Diagram" standard nomenclature for the journal. Morgan at first objected to what he called "unscientific science," but he admitted that "Kuiper's treatment of Russell and Shapley has always gotten in my hair a little and I probably lean too far the other way in consequence." So he deferred to Chandrasekhar: "I think you are the best informed person about the merits of the diagram label and if you intend to use H-R and feel it is the proper course I shall be glad to follow suit."[10]

Beyond the dispute, the diagram remains today the central framework for stellar astronomy defining the interface between theory and observation. The diagram reflects the sharply contrasting styles of its two creators. Hertzsprung certainly created the diagram, but Russell turned it into an interpretive astrophysical tool and fought for its acceptance using arguments that overcame strong resistance. The polarization over the naming of the diagram mirrored national boundaries and reveals deep fault lines between Hertzsprung's followers and Russell's. It is not inconsequential, therefore, that Kuiper's emotional outburst came at a time when many Americans were less than happy with the presence of an increasing number of foreigners commanding astronomical resources on American soil. And it is not without irony that the staunchest defenders of the rights of foreigners to work at American observatories were the same two astronomers Kuiper despised. However annoying and arrogant Shapley and Russell could be, they had done more for Kuiper than any other Americans had.

Foreigners Become Refugees

Kuiper had a serious attitude problem, and Hertzsprung was partly responsible. Responding to Kuiper's claim that the Dutch were being singled out, Hertzsprung agreed that "many Dutch astronomers come to America, is certainly right, but that's because the Dutch astronomers are very good at their work and one can't blame them for that."[11] Hertzsprung hoped that Dutchmen would be welcome if they behaved and paid their way. But they both believed that American data and telescopes were better off in Dutch hands. The Dutch, after all, had spent centuries colonizing the world, and the world, in consequence, had become a better place.[12]

It wasn't only the Dutch. After Robert Hutchins declared in April 1936 that the Yerkes astronomy faculty was Chicago's "League of Nations," W. H. Wright tore the page out of *Science* and sent it to Sproul to illustrate his fears: "Under these circumstances I can see no justification for loading up American observatories with foreigners, to the discouragement of our own young men."[13] By then, however, the "foreigner" problem had deepened into the refugee problem, with more and more people fleeing the Nazi threat. Ever more restrictive immigration laws made their plight even more dire as the "national origins clause" became more strictly enforced in what was a "triumph of nativism" in the 1930s.[14] Astronomical "refugee scholars" encountered the same resistance those in the larger world experienced, since opportunities were severely limited. To make matters worse, by the mid-1930s, Ph.D. production at American observatories had doubled (from five to ten per year), and hirings flattened out or decreased owing to the Depression. For the first time, observatories were saturated and newly minted astronomers were not finding jobs, as Adams reminded Russell constantly. Thus in a climate ever more clouded by insecurity and fear, xenophobia was rampant. The nativism Kuiper encountered at Lick, discussed in the previous chapter, was widespread.[15]

Complicating the situation was the fact that many of the refugees were elite practitioners who expected both professional and cultural privileges unknown in even the largest American institutions.[16] Their expectations, coupled with the already substantial inroads Kapteyn and Hertzsprung's descendants had been making in American astronomy, fueled the fears of observatory directors. In the course of the decade there were no fewer than nine foreign-born astronomers and physicists at Yerkes; comparable numbers moved in and out of Harvard and Mount Wilson. The prominence of many of them only increased their impact on what was still a small community.[17]

Russell was not immune. He initially felt that choosing a foreigner as his successor at Princeton "would be risky at best," and his professed "New England caution" made him uneasy about offering positions to people who were temporary, with the prospect of having to worry about them later. Russell took some time to decide to hire Wildt, and then he failed to promote the man to a permanent faculty position. He also was not above using Chandrasekhar's skin color and supposed radicalism to further Shapley's ends and certainly harbored some of the typical antisemitic phobias of his day, although he was willing to entertain Jews as graduate students.[18] Even so, Russell's attitude sharply clashed with those of some of his closest colleagues. The worst was V. M. Slipher, who felt that Struve had stolen his work and so all foreigners were suspect: "I have seen enough of this through the years," he told Russell in 1939, "to feel we must realize we cannot treat them as we do our American colleagues or expect them to conduct themselves as such." Above all, in tune with Wright, Adams, and many others, Slipher warned that it was irresponsible "to go on turning over American observatories to Europeans, if we are fair to young Americans."[19]

Russell became involved in the refugee problem through campus colleagues Oswald Veblen and Rudolf Ladenburg, usually following their lead.[20] He had already teamed up with Schlesinger and Campbell in 1934 to lobby the House of Representatives' Foreign Relations Committee to support the International Council of Scientific Unions, mainly on behalf of the IAU, which was under pressure from Nazi astronomers to oust Jews. And he worked with Shapley, Schlesinger, Bok, Dugan, and others to lead astronomers into the fight to save their European brethren.[21]

In addition to lobbying Congress, writing letters, and seeking out positions for scholars, Russell also spoke from pulpits and conference platforms trying to shake his listeners awake from their isolationist slumber and prod them to take an active role in what he believed was a growing crisis. Speaking before the National Conference of Jews and Christians at the Hotel Astor in November 1938, in the wake of the infamous Kristallnacht, Russell mocked the idea of an Aryan or a Jewish physics and warned his audience that "[w]herever autocracy reigns we are in peril. Perhaps our heads may be among the stars; but our feet are on earth, and the currents of tyranny will catch them and drag us under with the rest." At every opportunity he campaigned to admit refugees from all nations and to hire them equally.[22] Throughout the rest of the decade and until he became consumed by the widening war and by his own declining health, Russell advocated universalism. In 1938, Russell rejoiced that the International Astronomical Union was still active, preserving "our own fraternity of astronomers," even in Germany.[23]

Foreigners in the AAS

The American fraternity was, however, severely tested in these times, especially when, in late 1938, its president, secretary, and some members demanded that nominations to office thenceforth be restricted to native-born Americans. Russell found this especially painful since he was the chair of the Nominating Committee, a standard obligation of the past president of the society. Russell naturally gave the job low priority and created a slate in a rush only when the meetings of the Council loomed in December. The slate he created with Struve and Jason Nassau included Europeans George van Biesbroeck, Luyten, van de Kamp, and Jan Schilt as candidates for councillors, along with S. W. McCuskey and J. A. Pearce. Bok and Cecilia Payne-Gaposchkin were also suggested for various posts.[24]

Aitken, president of the society, was shocked by Russell's slate. There was only one native-born American for the Council, he observed, because Pearce was Canadian, which meant that the Council would be taken over by foreigners. Aitken demanded "*at least*" three astronomers born in the United States. John C. Duncan, the society secretary, agreed with Aitken, as did Adams.[25] Russell objected: everyone on the slate was a seasoned senior professional, devoted to American astronomy, and all were recognized as American contributors. Making any distinction between native-born and naturalized American citizens was "inconsistent with the spirit of our National constitution and institutions, and with the spirit of scientific inquiry." Russell believed that this issue was "of vital importance to the future of American astronomical research." "We older men have certain responsibilities as directors," he argued, which included securing the means to perform "the greatest practicable amount of valuable astronomical research." Vacancies on the staff, therefore, should go to the "best man, quite irrespective of his place of birth." He knew well that factors like character and the ability to work with others were important, "especially in a small or isolated community." But they were "second only to professional capacity." He therefore once again "strongly opposed . . . the introduction of a nationalistic spirit into American science."[26] To illustrate his point, Russell took another hit at Lick:

> [T]he history of American astronomy in the last few years seems to me to be equally convincing. I think no one can question that in the last decade two observatories east of the Mississippi [River] have advanced greatly in relative as well as absolute distinction, while the great observatories of the Coast have made relatively much less progress.[27]

The ascendancy of Yerkes and Harvard were "due mainly to the presence [upon their staffs] of investigators of distinction; and many, though by no means all of these, are 'foreigners.' " No American of greater or equal ability, to his knowledge, had been held back, Russell alleged; in fact, there were few if any who could compete. There was still a major need for competent midcareer astronomers, but "first-rate men who are American born can be counted on one hand." In a few years, many of the directors of the largest American observatories were due to retire. Appointing second-rate Americans instead of first-rate foreigners was "much the greater evil."[28]

Russell had been oblivious to the issue of nationality when he prepared the slate, and was mildly amused when he realized that they had named "three Dutchmen." He did not fear the Dutch pipeline as did Donald Shane, who believed that "[a]ny Dutchman put in would try to bring in other Dutchmen."[29] As much as Russell protested, however, he agreed to keep the matter quiet and to reach some accommodation. Russell also knew that he was outnumbered if the issue was taken to the full membership: "This nationalist feeling is an evil and I think it is very important to prevent its getting more headway." Aitken knew he had the numbers, and warned Russell that the majority of the membership were "disposed to take a nationalistic view." He demanded that Russell submit a new slate, or he would take it to the Council.[30]

Despite his demand, Aitken felt that he was a moderate compared with his Lick colleagues, who chastised him severely when he brought Nicholas Bobrovnikoff and then Kuiper to Lick. He reminded Russell that though he was warned, he still accepted Jewish students like Louis Berman. These actions were proof, Aitken hoped Russell would agree, that he too regarded "fitness for the job" as the primary consideration for appointments. But in a professional society, Aitken felt, standards were different: people have different beliefs, and this fact had to be taken into account to avoid controversy that might break the society apart.[31]

Russell knew that open debate was futile. Checking the rules, he dropped Schilt's name when he realized that he had been proposed at the last election and had lost. Conferring with Struve and Nassau, who sided with Russell, he reluctantly added two Americans so that it would be possible to elect an all-American Council.[32] This satisfied Aitken, though Duncan still wanted more Americans on the list. "What you say about the dearth of American-born astronomers who are qualified to be directors of observatories may be true," Duncan admitted, but the professor from Wellesley College wanted nothing less than an all-American slate for an American society.[33]

The wrangling continued quietly, and though the positions taken were distinct, they were interspersed with personal gossip, warm news of family and friends, and, as was always the case, requests for Russell's advice or thanks for his aid. Aitken and Duncan may have taken a stand on nationalism, but they were not about to lose Russell's patronage. Duncan in particular was always careful to follow Russell's advice when accepting or rejecting papers for AAS meetings.[34]

When the slate was brought to the Council, someone demanded that van de Kamp be dropped because there was already a member from Swarthmore. Eventually, the slate included four native-born American and two foreign names. Clearly soured by the experience, Russell reported back to Aitken that he had gone to the limit his conscience allowed in " 'appeasing' a nationalistic feeling which seems to me hardly more excusable than the one which we associate with this unlovely word." Any more pushing in that direction and, he implied, he would quit in public protest.[35]

There were a few Council members who sympathized with Russell's position, but none were as distressed as Struve, who claimed that no foreign-born astronomer would now "have any personal ambition of any sort in connection with the society." Struve, however, knew how sensitive the political landscape could become after Henry Gale asked him to poll his staff before the 1936 presidential elections, to be sure they could prove that Chicago astronomers, Hutchins's "League of Nations," were not Communists. So he took no action.[36]

Aitken, whose parents were European, felt deeply torn. All the names on the original slate he considered to be his personal friends, except for Luyten. "Luyten and I have never hit it off well, to use the language of the street." Not because he was born in Holland, but because "he was *Luyten!*"[37] After the election, van Biesbroeck became the only foreign-born astronomer on the Council. American-born C. T. Elvey (McDonald Observatory) and G. M. Smiley (director of the Ladd Observatory, Brown University) were also elected that year.

Battling Isolationism

Amid all this bickering, Russell delivered the 1939 Lincoln's Birthday address in Alexander Hall at Princeton on behalf of the American Committee for Democracy and Intellectual Freedom. Under the rubric "A Natural Scientist Evaluates Intellectual Freedom," with a text that would eventually be toned down considerably for publication as "Science and Freedom," Russell confronted his audience with the challenge that "[t]his [was] no time for philosophical disquisitions . . ." but was a time for action, a time to take a stand. Inspired by contact with the noted ethnologist and social activist Franz Boas, who as national chair of the committee had asked him to be a signatory to his "Manifesto on Freedom of Science" that spring, Russell filled his original draft with heat, anger, and concern for those he knew of in Europe who were suffering at the hands of Fascism. Evidently he was also thinking about the fate of his slate: "If some vagueries of politics and economics at home look like Bedlam, half the world looks like Hell. And Hell has one great moral advantage: *there*, there is no shedding of innocent blood."[38]

For Russell, freedom, tyranny, and injustice knew no national boundaries: there were good Germans, to be sure, who were "deeply shocked" by Goebbels, "the devil incarnate," and by the horror of the pogroms. Jews and Catholics have been persecuted everywhere, Russell added, "and who knows who will be next." On Lincoln's Birthday, of all days, Russell felt that there was a lesson to be learned:

> We have all become half-blind and callous—perhaps by an inevitable defense-reaction, lest we go mad with seeing too clearly. It is hideous to think of what the world has seen in the twenty years of "peace" since the Armistice—Smyrna, the Siberian mines, Ethiopia, Spain, Nanking.[39]

Now it all came bursting out. Drawing upon letters from Robert Atkinson, who had returned to England, Russell ripped into Chamberlain's policy of appeasement. He called for the cessation of commerce with despots and railed against any accommodation in return for the release of refugees. And his most passionate concern was for preparedness; everyone had to become actively involved preparing for war. Above all, America could not cease to be humane; the nation's doors had to remain open for refugees: "[T]o close our gates tighter now would be dishonor."[40]

Loosing frustrations he must have felt after his doomed slate, he called for open resistance and democratic action. "There may be some cases, indeed, in which a patriotic executive may go a good way to avoid an explosion; but it is a better service to democracy to find some plausible reason than to admit yielding to a threat. For this invites a show of violence in future, and that way madness lies." As he said to Aitken, he had been pushed to "the extreme limit" in appeasing the nationalistic appetites of his brother astronomers.[41]

Russell reworked his Lincoln's Birthday speech for other venues over the next few years and was well known for his views. After his address he was invited by the local chair of the NAACP to attend Princeton's first "inter-racial gathering" at the First Baptist Church, which he thought was an "excellent plan" with which he was "heartily in sympathy." He only asked whether "ladies" were invited, because Mrs. Russell "hopes to be present."[42] In early 1940 he spoke in Pasadena on "The Price of Intellectual Freedom," railing against Nazi and "Marxian" thought and again lashing out at Aryan physics and proletarian cosmogony. These issues were "[n]o joke," he warned, lamenting the sad fates of Russian astronomers like Numerov and Gerasimovich, which, for Russell, only heightened the refugee problem once again and the resulting job crunch. In Pasadena, Russell had to face the ire of Adams, who was constantly worried that there were "young, and in many cases able, astronomers who are badly in need of positions." This was a fact of life in the larger world too.[43] Russell advised youngsters aspiring to be astronomers that the competition for good jobs was "at present stiff, on account of the considerable numbers of men of conspicuous ability who are refugees from the appalling conditions on the continent of Europe." For Russell, however, along with Struve and a few others, this fact of life had a "silver lining."[44]

Russell was one of the few who did not fear what the refugees really represented: superior training and a combination of theoretical and practical talent that was rarely seen among American graduate students. In fact Russell had much in common with Europeans who were eager to exploit the data stores at American observatories. Many of them were, after all, like Struve, Kuiper, and even Kopal, modern extensions of his own scientific philosophy. In October 1941, dealing with a particularly vitriolic letter V. M. Slipher had sent to Cattell for *Science*, Russell wrote that Slipher had missed the real point: the suffering in Europe was a golden opportunity for American science. It was a chance "for this country to enrich itself by securing the services of men of notable ability and distinction, most of them it would have been very difficult to 'pry loose' except under the present tragic circumstances."[45] As we shall see in the next chapter, Russell had just managed to secure Svein Rosseland for Princeton, which would have been impossible in normal times. Men like Rosseland, Russell hoped, would revolutionize American astronomy just as European émigrés had done for American theoretical physics, which had done very well for itself by accepting them early and often. But unlike American physics, which felt it had to catch up to Europe, American astronomy's world dominance in observational astrophysics did not provide much incentive to rise to the challenge. Other than Struve and Shapley, there were few observatory directors in America willing to mix theorists with observationalists.[46]

As he worked to accommodate refugees, secure in the conviction that this strategy would eventually improve American astronomy, Russell also fought isolationism on every front. In May 1940, after Germany's invasion of Belgium, the Netherlands, Luxembourg, and France, Russell was alarmed by the American Association of Scientific Workers' isolationist manifesto of neutrality. Russell and Veblen worked feverishly to gather signatures and mount a protest to the *New York Times*. Willem Luyten helped out on his Minnesota campus, collecting signatures denouncing the "peace" resolution of the association and fearing that the holocaust would reach American shores and end all science.[47] Russell even joined the leftist "Fight for Freedom Committee" headed by Bishop Hobson, telling Adams, "This is not a holy war, but it is a religious war." Russell knew he was siding with liberals but blamed his decision to do so on the state of the world: "They are much more radical than I economically," Russell admitted, "but I don't think that is the main issue."[48] He had also been writing to Sol Bloom, of the U.S. House of Representatives Foreign Affairs Committee, as well as other representatives within his reach, lobbying strenuously for the lend-lease bill.[49] Russell had his limits, however. As a middle-of-the-road conservative who still quoted Kipling and Teddy Roosevelt, he warned Franz Boas that he would lend his name only to groups that opposed "all tyrannical governments, whether of the Right or the Left." He would support Spanish Intellectual Aid but would not favor either of the warring factions. With no sympathy for Spanish Republican forces, he broke with Boas over support for the "popular front." He also objected to Boas's efforts to perpetuate the New Deal by electing Franklin D. Roosevelt for a third term and became especially disheartened when Boas's committee involved itself with the choice of textbooks in public schools, which Russell regarded as an ephemeral issue.[50]

Close to home Russell reorganized the department to anticipate the demands of wartime training in navigation and meteorology. He was also delighted to become a member of the Scientific Advisory Committee of the Ballistics Research Laboratories at the Aberdeen Proving Ground in Maryland. "At last," Russell rejoiced to Webster in October 1941, there was "something resembling defense work on which I can put in part of my time."[51] After December 7th, Russell devoted his passions to ballistics. He relished the meetings and the debates with staff and committee members. Ladenburg was there, along with Schwarzschild, Atkinson, and, part time, Chandrasekhar, all foreigners, as well as a host of domestic talent like Theodore Sterne and Dorrit Hoffleit from Harvard and Edwin Hubble from Mount Wilson. Although Chandrasekhar grimaced at the military character of the place, Russell loved it. Ballistics was a great application of orbit theory, and, as in the first war, he sensed his mission: evaluate and secure the fastest and most efficient means for computing ballistics trajectories.[52]

During the war, Russell continued to spar with Kopal about binaries and attended to ceremonial and political duties. At Shapley and Bok's behest, he led a high-profile delegation to Mexico in 1942 to celebrate the opening of the National Astrophysical Observatory at Tonanzintla. He also became collaborating editor of the *ApJ* in that year, at Robert Hutchins's request. He watched his son go off to war in New Guinea in September 1944 as a medical officer, and, to the extent his health would allow, he continued to lecture from campus platforms and pulpits, expounding on "Paramount and Absolute Obligations" and "Determinism and Responsibility" in 1942, and "Science and Intercultural Understanding" in 1945, as a continuing participant in Louis Finkelstein's Conference on Science, Philosophy and Religion under the auspices of the New York Institute for Religious Studies. He also said goodbye to many old colleagues and friends, writing three obituaries for Dugan, two for Schlesinger, and one each for Alfred Fowler, J. S. Plaskett, Hinks, and Eddington.

The isolationism and xenophobia Russell fought against before the war festered in his mind at the end of the war. Although he still believed in the sanctity of science, he was less sanguine about his fellow scientists. He became deeply troubled by the existence and application of nuclear weapons, which led him to speak on "Ethical Aspects of Atomic Energy" in 1946. Later that year he cooperated with Hugh Taylor, Eisenhart, Henry DeWolf Smyth, John Wheeler, and Eugene P. Wigner to use Princeton's bicentennial as a venue for a September conference on the "Future of Nuclear Science."[53] Russell gave the public lecture, on the responsibility of the scientist in the nuclear age. In "The Ivory Tower and the Ivory Gate," he argued that scientists could "play no favorites" in their profession. Characteristically (as if to shame his colleagues), he held up astronomy as an exemplar of international cooperation: "A star-catalogue or orbit calculation which rejected any data except those provided by some national, racial, or ideological group would be ridiculous." Russell, possibly still ruminating on the abominable behavior of some of his closest colleagues, lamented that "the ivory walls were too thin to keep out all sound of the approaching storm." In consequence, as if in retribution, he reaffirmed that "[p]ersonal relations between individuals will depend upon past records and present convictions; but this has always been true, and ought to be."[54]

Although he was speaking globally, of the failure of the ivory tower to speak out effectively during the Nazi terror to "expose and refute the twisted theories of race and class and meet the specious claims of a degenerate science," he was thinking locally, of the divisiveness of his xenophobic colleagues, and of the lingering effect he feared it would have on his profession. "I believe that science and the body politic may both be injured by listening to the enthusiasts who strive for the participation of science, as such, on either side of the divisions raised by political or social ideologies."[55] Of all the sciences, Russell believed, astronomy had the best chance to remain truly universal. Sharing his thoughts with Wigner before the conference, Russell recalled a "spirit of unspoiled and sincere friendship" at the 1913 Bonn meetings. He lamented the break caused by the "First War" and, contrary to fact, celebrated the "steady healing" process in its wake. Russell wanted to remember it this way because he shared Kipling's belief in the "white man's burden." Only for Russell, the burden was on science. Though he was "very far from a pacifist or an appeaser," he felt moved to make a "real plea for cooperation now[,] and science can go far toward leading the way."[56]

Russell hoped that the wounds of the war could be healed once again, and that science could rebuild itself as it had before. Astronomy seemed to be making some progress, since the Executive Committee of the IAU had met in February 1946 to plot its future course. But closer to home, and deeply coloring his thoughts and fears at the Bicentennial Conference, was the fact that the future of astronomy at Princeton was then very much in doubt. He had been trying for over a decade to appoint and groom his successor and, thus far, had met only with disappointment.

• CHAPTER 22 •

Searching for a Replacement

A Proposal to the President

February 1934. The Halsted was closed for renovation, and Dugan was on leave in Pasadena to ease his worsening arthritis. Russell wrote frequently to keep him updated on the telescope, and just to chat. There was a touch of melancholy in his words, for life had limits. It had been one year since his brother Gordon had died, and the loss still weighed heavily.

Gordon Russell had stayed close to the family, ministering in Merchantsville, just outside Philadelphia, and spending his vacations in Princeton. He was devoted to Russell's children, and there was talk that he and Jean Hetherington might be wed, and that they would care for Elizabeth together.[1] But in January 1933, Gordon contracted pneumonia and was terribly sick for about two weeks, though he seemed to be on the mend toward the end of the month. On Saturday the 28th, an urgent phone call to Princeton propelled Russell to his fifty-two-year-old brother's side. But all he could do was say goodbye. Gordon's death shook his elder brother, who dropped elaborate travel plans he had made to sail through the Panama Canal en route to the West Coast, and seemed unable to cope with his profound loss and a deepening sense of mortality. While attending to Gordon's burial the next week in the family plot in Princeton, and to assessing his estate, Russell resumed some scientific correspondence, though it was perfunctory. He had just been elected president of the American Association for the Advancement of Science, but his preoccupation was with closing out his brother's life.[2] He was not himself for quite some time.

His brother's affairs were not settled for over a year; although Alex Russell agreed to act as trustee to the estate, using his expertise in business and financial matters, as general manager of the Rochester and Ontario Lakes Water Co.,

Henry was still very much involved in the transactions with their brokers and lawyer, and his mortality was very much on his mind. His wife, in concert with their doctor, was again worried about his health.[3] Russell was now fifty-seven years old, and after a short visit to Harvard in February 1934, where he evidently unburdened himself on Shapley, he wrote darkly to Dugan about "our own possible successors." Russell was definitely thinking more of Dugan than of himself, worrying how much longer his colleague could hold on in the face of crippling disease. Russell knew he had to find someone to continue either his or Dugan's projects, preferably both. "If the present third member of our staff cared for either . . . ," Russell told Dugan, they would not have to think from scratch.[4]

Russell had already given up on Stewart, who continued to teach but had dropped out of astrophysics in favor of his deep interest, gained from his days at AT&T as an electrical engineer, in what he later called "Social Physics," the application of physics to demographic and sociometric problems. His "work diary" in the mid-1920s was filled with ways to evaluate and improve urban conditions, national income distribution, and mail-routing systems.[5] By the 1930s Stewart became interested in weather prediction, especially hurricanes, which gave him some notoriety. He was also an enthusiastic solar eclipse chaser, enjoying himself immensely. He conducted routine photographic investigations looking for the Einstein shift, which yielded in the late 1930s "what Russell calls 'sound but modest' results."[6] He certainly started out with great promise in astrophysics but found the competition a bit too rough, and Russell unable to support him adequately. As long as K. T. Compton was around, Stewart held out hope that "borderline researches" would have a place in the physics department, or even that a collaborative astrophysical unit might be formed there. But when Compton left for MIT, Stewart lost his most sympathetic campus ally.[7] Russell's inability to keep Stewart under control was due both to his lack of interest in administration and to the fact that the Princeton University Observatory existed more as a campus department than as an autonomous observatory. Although Russell felt that Stewart devoted too much time to demographic studies, which to him were merely "an avocation in space-time," still he honored Stewart's independence of choice: "[T]his division of his time is his own responsibility."[8] Still, Stewart was active in the department, directing the laboratory researches of a number of graduate students in areas of direct interest to Russell and Shenstone.[9]

Thinking from scratch was not Russell's strength when he was faced with administrative matters like that of his or Dugan's successor. He was very proud of the fact that Princeton's graduates were now staff members or directors at Mount Wilson, Harvard, Yale, Princeton, Dartmouth, Wesleyan, Swarthmore, Allegheny, and Dominion Observatories. Russell had no intention to alter the focus of the department away from research and graduate instruction, nor did he want to change his connections with other observatories. As he mused to Adams in April, "The privilege of work elsewhere, which fortunate circumstances have attached to the Research professorship which I hold, is an enormously valuable one, not simply for the holder, but, I think, for astronomical science."[10] What Russell finally asked for in 1934, writing the new Princeton president, Harold

W. Dodds, was to supplement his adequate but aging staff with a younger person, someone to groom as director when he and Dugan would be required to retire in 1943. He hoped to find a "skilled and enthusiastic observer" who was also a fully competent theorist.[11]

The best candidate in America, Russell lamented, "alas, is a woman!,—not at present on our staff." He was thinking of Payne-Gaposchkin, but neither Princeton, nor Harvard, nor Russell, of course, would dare consider a woman for a faculty position in that day. Just as he promoted women's suffrage in the abstract but denied it in practice in the 1890s, Russell remained socially a middle-of-the-road conservative, unwilling to fight for radical change. Less radical was the foreigner option, but even this, Russell then thought, would not be a popular alternative at Princeton. Russell expressed no interest in expanding the support staff; if it ever was to be enlarged, it would be by the "*new*" director.[12] He named no one in his memorandum to Dodds but clearly had someone in mind.

First Attempts

Russell made two attempts in the 1930s to hire someone he could groom as his successor. Both choices reveal Russell's priorities and his limitations. His first choice, a man who came closest in overall capability to Payne-Gaposchkin in Russell's mind, was Theodore Dunham. Russell opened negotiations with Adams in April 1934, arguing that Dunham might be shared on a joint appointment between Mount Wilson and Princeton. "The continued liaison" would be maintained, he told Dugan, but to ease the shock on Adams, who was also interested in Dunham as his own successor, he suggested that Dunham would not be burdened with observatory duties, elementary courses, "or wasting time in correcting papers." These tasks belonged to Dugan and, especially, Stewart, who, Russell claimed, insensitive to what his staff were really doing and what their contributions meant to Princeton, "has a light schedule and is doing little or nothing else."[13] Funding not spent on the Halsted renovation was enough to bring Dunham for at least the first semester.

Dunham was Russell's kind of fellow. He came from a fine professional family, had deep New England roots, and had distinguished himself at Mount Wilson working with Adams on high dispersion planetary and stellar spectroscopy, usually providing deft theoretical analyses for Adams's work. Dunham was congenial and a hard worker, and more than anyone represented Russell's legacy at Mount Wilson. By the time Russell wrote Adams, he had a complete strategy. Russell had heard that the theorist Rupert Wildt of Göttingen, who was being noticed for his planetary atmosphere research, wanted to work in the United States and had already applied for a Rockefeller Fellowship. Russell thought he would be a good replacement for Dunham. Wildt would probably want to stay in the United States because he was engaged to a woman of Jewish ancestry: "If he marries her he will be sacked." Finally, the Carnegie money that would be saved by sharing Dunham, Russell added, could be used to bring Robert King back to Mount Wilson.[14]

Russell developed his scenario firmly convinced that it was the best he could do for himself, for Princeton, and for astronomy. He was careful to weigh all factors he could grasp, and knew that he had to think strategically lest his legacy and destiny be somehow denied: "All this is for the next few years. What happens later must be left for Providence, and to the various politicians, boards of trustees, and other persons who keep putting their fingers into the designs of that august Power. I am beginning to realize the obligation that rests on me at least to look a few years ahead. . . ."[15] Clearly Russell thought he knew best what had to be done, but characteristically he failed to take into account one critical factor. Indeed, Russell thought he had it all figured out, except for the fact that Dunham was none too enthusiastic about the idea. He loved telescopes more than teaching, and only his loyalty to Russell compelled him to break away to become a part-time associate professor on a three-year appointment.[16] "Our loss will be your gain," Adams wrote in March 1935. By then Wildt was happily working away at Mount Wilson, and Robert King was back as well. Russell's plan seemed to be working.[17]

Dunham did lecture in 1935, and again in 1936, but spent most of his time reducing his data, talking with Russell, planning for future improvements to Mount Wilson spectrographs, and pining for clear skies.[18] When at Princeton, he longed to return to Mount Wilson. Russell always wanted him to stay longer, and Adams was resolved not to let him go permanently. Dunham and Walter Baade were "the most able" of our young men, Adams told the Carnegie's John C. Merriam, and of the two, Dunham was wholly attached to the mountain. This was true. Dunham, in fact, had been looking for a way to get out of his Princeton commitment for 1936 and tried to send Rupert Wildt instead, whose fellowship and visa were both running out. Russell agreed to hire Wildt as a research assistant but still insisted that Dunham show up on time.[19]

Dunham felt trapped and, in his frustration, let Adams know his true fears. He sensed that Adams and Russell were competing for his time, and he also resented what little lecturing he had to do. Though he did not come right out and say so, it was clear that Dunham found Russell to be a very demanding boss. After several more cycles, he feared, he would lose his own professional identity and become Russell's satellite.[20] Dunham also had a serious nibble from Harry Plaskett, who had left Harvard for Oxford and was looking for a spectroscopic specialist who could build up the instrumentation at the Radcliffe Observatory at Pretoria.[21] Dunham was intrigued because, above all else, he relished freedom to build instruments. There was not enough freedom even at Mount Wilson, and none at Princeton. "I have talked with Russell about the possibility of establishing a Princeton observing station for stellar spectroscopy in a favorable site," Dunham told Plaskett, "but he is not anxious to push any such new project at Princeton while he is Director."[22]

Russell's resistance to growth proved to be Dunham's last straw. Russell did not deny that Dunham's plans were admirable, but he would not raise the "very substantial endowment and an increase in permanent staff" that would be required. Russell asked for time to ponder the situation, to take in Dugan's and Shapley's advice, but in the end rejected any move toward expansion. "Russell agreed that I ought to continue to do observational work, rather than devote myself primarily to theory, and that decision seemed quite definitely to eliminate Princeton."[23]

Russell's wish to keep things small at Princeton overrode his desire to keep Dunham; he balked at Dunham's experimental needs, which, to be sure, were insatiable. As Adams later told Henry Gale, who with Struve was interested in hiring Dunham too: "[H]is main interest in life is the design of spectroscopes for astrophysical work. As you probably know, Russell wanted to get him at Princeton, and groom him for his successor . . . Dunham is very able, but usually has too many irons in the fire."[24]

Despite his failure to keep Dunham, Russell always maintained a strong affection for his devoted student. Indeed, Dunham, who held a Harvard medical degree, was a great consolation when, in October 1936, Lucy was diagnosed at Princeton Hospital with a malignant growth in her cervix. Russell had just returned home from Yale after delivering the Gibbs Memorial Lecture on "Model Stars," and the news put him into a panic. He immediately wired Dunham: their local Princeton physician suggested radical treatment in Philadelphia, followed by radium and X-ray therapy. Russell hoped that Dunham could advise.[25] Dunham demanded a second opinion. Within days, Lucy May was in New York and, to their collective relief, found that the initial diagnosis was wrong. Over the following months Russell often relived this episode in letters to colleagues; it clearly left him in a state of dread, keenly sensitive to his, and his wife's, mortality. As if to acknowledge Dunham's devotion to his family, Russell let him know in November that whatever decisions he made about his life and career, there would always be a warm spot for him in Princeton, where "the latch string is always out . . ."[26]

He said much the same, however, when Robert Atkinson turned him down in 1937. He saw a good bit of Atkinson, whom he had tried to place at Yerkes but enjoyed having close by in New Brunswick. Although Atkinson was not a complete theorist, he was more adept at mathematical theory than Dunham, which Russell told Dodds was a plus: "While competent astronomical observers are not very difficult to find, good investigators in theoretical astrophysics are rare . . ."[27] Russell wanted Atkinson to take over Dunham's lectures for spring 1937. Smitten more by Atkinson's charm and Anglophilic roots than by his actual accomplishments, and seeing more in his potential than in past performance, Russell admired Atkinson's grasp of mathematical and physical theory and liked his style. They were in frequent contact, and Russell had every reason to think that Atkinson would prefer to work at Princeton, where the teaching load was far lighter than at the state university. On several occasions Russell wrote glowing letters of reference for Atkinson for other positions, always cautioning him to "look twice, and then twice again" before accepting a job calling for a teacher: "Almost all the colleges are strong on teaching and give very little opportunity for research, and I am afraid you might bury yourself."[28] He was confident that Atkinson would answer his call to Princeton.

Atkinson, however, parlayed Russell's offer into a raise at Rutgers, where faculty salaries were then tax-free, but he eventually answered a call to be chief assistant at the Royal Observatory, Greenwich.[29] He deeply regretted turning down Princeton, "where I could have played indefinitely with the insides and outsides of stars, with a friendly guide such as few ever have the luck to meet; Greenwich seems a little like a factory where one gets down to business and forgets all one's iridescent fancies."[30] Russell accepted Atkinson's decision as part of the game, and the two remained in warm contact over the years, talking science and politics as Europe plunged toward war. Although he hardly shared Atkinson's radical political views, he found them refreshing against the American backdrop of an "unthinking pacifism that glorifies non-resistance as an absolute good, and damns the use of force as an absolute evil, whatever the circumstance. I think this has been the most narcotic ingredient in the mixture that has drugged us—for we are drugged—rather than, I think, yellow." Russell lamented the pacifism he found both in church and on campus. As he became more and more distracted by the growing war, he heard that Atkinson might be asked to take a post in Washington with the British Embassy, which made him think of old ties: in July 1940 he reassured Atkinson, as he had Dunham, that "a certain latchstring is still out, and a warm personal welcome awaits you both."[31]

Dunham and Atkinson had been two of Russell's favorites, and his failure to bring either of them to Princeton as his successor bears witness to his own limitations, at least to his unwillingness to alter the course he had set for Princeton astronomy, or to fight harder to win his chosen successor. He wanted to find a person he could groom on intimate terms, who would accept his direction and gain his complete trust. Although he wanted a theorist, he had to admit that there was an observatory to run.

Years later, looking back during a commemorative session honoring Russell, his son remarked that Atkinson was, for his father, something of a "spiritual or at least intellectual son." Atkinson fulfilled something in Russell that ran very deep; Russell saw the younger man not only as an intellectual successor but as someone within whom he could find a bit of himself, an extension of his spirit and life.[32] Dunham and Atkinson and their wives had proven themselves highly compatible socially as well as scientifically. Twice disappointed, for the moment Russell decided not to look again. When he again took up the search, he cast a much wider net.

Interim Arrangements

Russell now knew not to look for an ambitious instrument person, or for someone with deeply conflicted loyalties. Russell converted the funds allocated for Atkinson to extend Wildt's temporary appointment in "astronomical physics" and to hire as a temporary research assistant John E. Merrill, who had recently graduated but was without a position. There was also enough money to support Newton Lacy Pierce, who had come as an advanced graduate student from the University of Michigan in 1935. Dugan's arthritis grew so severe in a few years that he had to curtail observing and so came to rely on Pierce, in whom he found a strong kindred spirit. Pierce quickly mastered Dugan's visual polarizing photometer, "in the best Dugan tradition of thorough accuracy."[33] Pierce in every way became Dugan's apprentice, and Dugan, without formality, groomed him to be his own successor. He was not brilliant, Dugan admitted, but neither was Dugan.

After Merrill resigned in August 1937 to take a teaching job, Pierce stayed on, slowly but surely assuming more and more responsibility for running the daily routine of the observatory.[34] Russell, in his early sixties, was still traveling constantly with his wife: in 1939 he journeyed to Texas in May for the dedication of the McDonald Observatory; then, as a newly elected correspondent of the French Academy of Sciences, he presided at the Paris Colloquium on Astrophysics in mid-July, taking his wife, Elizabeth, and Jean Hetherington so that they could visit Elizabeth's twin sister Lucy in Salonika in August to meet the newest Russell granddaughter. They were back in France (Bordeaux) after war was declared in September and did not return to Princeton until the latter part of the month. All the while, someone had to mind the store, which became all-important upon Dugan's death on the last day of August 1940. Dugan's death was hardly a surprise and to those closest to him came as a sad relief. He had been in terrible pain and was virtually immobilized by his arthritis. Those who continued to harbor hope were either too close to him, or too distant, to realize how gravely ill he had been for some time.[35]

With Dugan's death Pierce became indispensable. Moore-Sitterly was working part-time, as was Merrill, and Wildt and Stewart looked after the teaching, what little of it there really was. The observatory had to be managed, however, and Russell appreciated Pierce's practical expertise in the area. He never seemed to consider Wildt appropriate as his own successor, though Wildt had demonstrated by then exceptional talent as a theorist and was held in high esteem by his colleagues. Wildt had been advanced to research associate and instructor in 1939 in the wake of his spectacular identification of the importance of the negative hydrogen ion as the source of the continuum opacity in the solar atmosphere. Russell was, of course, very interested in Wildt's research, especially in his studies of partial degeneracy in white dwarfs and planetary interiors, and always encouraged him to publish in the *Astrophysical Journal*. Although he always praised Wildt's work highly, he seemed not to be comfortable with Wildt personally, or the comparatively leisurely pace Wildt favored. By this time Pierce was drawing a higher salary than Wildt, even though he had been advanced to instructor a year after Wildt. Pierce also had use of the apartment in the new observatory on Fitz-Randolph Road.[36]

Pierce offered Russell the assurance that Dugan's legacy would continue and that the observatory would be managed. But Pierce could not assume Dugan's professorship. No one at hand fit Russell's long-term needs. The type of men he would turn to in the next six years were very different from Dunham, Atkinson, or Pierce.

The Silver Lining in a War-Torn World

In 1935 Russell told Luther Eisenhart, dean of the Graduate School, that the Norwegian theoretician Svein Rosseland was "one of the most distinguished of living astrophysicists."[37] Ever since he had encountered Rosseland at Mount Wilson in the 1920s and found that his suggestions about metastability paid off handsomely in Bowen's identification of the chief nebular lines, Russell always regarded him as top-notch. He had built the first institute devoted to theoretical astrophysics, attracted excellent students, wrote seminal review papers on quantum theory, and prepared monographs on astrophysics and atomic theory, all of which were highly regarded.[38] By 1940 Rosseland had visited the United States a number of times, most recently for the 1939 Harvard Summer School in Astronomy, where three of his disciples were in attendance. One of them, Martin Schwarzschild, secured a fellowship at Harvard after a postdoctoral position in Oslo; his continued brilliant research into the theory of Cepheid-type pulsation models was a constant reminder of Rosseland's intellectual strength.

Rosseland had already turned down offers from Harvard as well as from the Yerkes Observatory, preferring to build up his institute and theoretical programs in Oslo. But his fate turned drastically for the worse in April 1940 after Germany's invasion of Norway.[39] Rosseland's situation was dire when Dugan died. Still very much shaken in late September, Russell thought only of Dugan, "whom we all miss acutely." But in November, with a professorship to fill, Russell wrote Shapley, Struve, and Adams asking what they thought of Rosseland, as well as Jan Oort from Leiden, also in German occupied territory. The silver lining in the horrors of war might be turned to benefit local needs, Russell realized; the question was, which of the two would fit in best on the Princeton campus? Adams still wanted an American to take Dugan's place but thought Oort would be "more at home" with Americans than Rosseland, if a foreigner had to be chosen. Everyone Russell consulted on campus favored Rosseland, except Dodds, who liked Bart Bok. Taking Bok away from Shapley was unthinkable, however, so by December, Russell convinced Dodds to call Rosseland: "[I]t is desirable that Princeton should continue work in the same general field of astronomy, rather than shift to another." Russell saw Rosseland not as Dugan's successor but as his own.[40]

Shapley learned through a relative of Rosseland in Pittsburgh, who had called Oslo by shortwave radio, that "he would be glad to come!" Rosseland confirmed by telegram and hastily sought out a passport, asking Shapley for help securing visas and bookings, knowing that the process required assistance from America.[41] University secretary Alexander Leitch contacted Breckinridge Long, a Princeton trustee and assistant secretary of state, to cut red tape. As Leitch described the situation, "Professor Russell will retire in five years, so that the future of Princeton's work in astronomy will depend in large measure on our ability to get Professor Rosseland."[42]

Rosseland's wife had given birth to their son Halvard during their first visit to Mount Wilson in the 1920s, so the U.S. consul in Norway quickly granted the visa, but the real hurdle was an exit permit from the German occupation authorities.[43] The Germans first refused, but more appeals from the State Department yielded a year's leave for Rosseland and his family. Now assured that he had Rosseland, Russell tried to console Annette Dugan, who he knew felt strongly that "Raymond preferred a good American, if there is one to fill the place, to a foreigner." Rosseland, he reported, was everyone's choice: a "distinguished man, and we could not get him here were it not for the war."[44]

Rosseland did not trust the Germans or the Quislingites, so "he and his wife and son *walked* through the forest across the Swedish frontiers, meeting no guards at all and carrying only what they could take in knapsacks." In March 1941, Russell expected them to arrive no later than July.[45] The Rosselands feared that a western route would expose themselves to the Germans, so they traveled across Russia, with the help of Peter Kapitsa in Moscow, and then took a steamer to California, after more than their fair share of mishaps, jailings, appeals for their release from the State Department, and, finally, a harrowing passage aboard a Japanese ship during the Indochina crisis. The ship turned back once but providentially reversed course again and left its American and European passengers in Honolulu. "In one week they had two Wednesdays, no Sunday and two Mondays," Russell exclaimed. "Rather an epitome of the present time!"[46]

They finally arrived in Pasadena in August and traveled east to attend the AAS meetings at Yerkes in September, where Russell was anxiously awaiting them. Shapley greeted them with open arms but with bad news from home. Through his extensive intelligence network managed by Bok and Harvard's burgeoning entourage of refugees, they learned that Rosseland's student Jaakko Tuominen had been seriously wounded in a battle a month earlier. This news made Rosseland's first days and weeks in Princeton hardly carefree, though, he reported to Shapley, his wife and son were delighted by the great rambling twenty-odd-room observatory residence: "[T]he Russells have parted with so much of their furniture that the house looks quite well."[47] Russell got Halvard into a private school at Lawrenceville and saw to it that the observatory residence was as comfortable as the Princeton University carpenters, painters, and plumbers could make it. In several month's time Russell and Ladenburg managed to have Rosseland elected into the local chapter of Sigma Xi.[48]

Rosseland found that he was in powerful company in the fall of 1941. Russell had followed a suggestion by John von Neumann to invite Chandrasekhar to spend a term at Princeton, as a guest of the Institute for Advanced Study and the Princeton physics and astronomy departments. The prospect of having both Rosseland and Chandrasekhar at hand was so exciting for Russell that he did not travel west that fall, and he created a new colloquium series to showcase theorists.[49] Martin Schwarzschild, by then on the faculty at Columbia and a constant visitor to Princeton, gave a memorable talk on pulsating stars in November that

Russell had long anticipated. Russell reported contentedly to Atkinson that he felt "more and more that Princeton is very fortunate."[50]

Chandrasekhar's impressions of Princeton were somewhat different from Russell's; although he agreed that the colloquia were refreshing, he chafed at the gentility of the place and found Rosseland aloof. He and his wife Lalitha lived comfortably enough in the attic of a large rooming house on Chambers Terrace and were neighbors of the mathematician Kurt Gödel, but his office was on the third floor of Fuld Hall at the Institute, in a warren above the palatial quarters of Marsden Morse, Einstein, Veblen, von Neumann, and Weyl: "It is really incredible," he wrote Kuiper, "the luxury in which these men live and work—it even makes a past fellow of Trinity faintly jealous!" The Institute was a real "Ivory Tower" even after Cambridge. As for the observatory, "What a contrast to the Institute!" The Observatory of Instruction's drab rooms and squeaky floors were more reminiscent of old outbuildings at Yerkes. "But then there is Russell inside it. And of course Rosseland."[51]

Chandrasekhar found Rosseland a puzzle: "He is peculiarly reticent with me regarding scientific matters. I said peculiarly, because I know he has discussed extensively with Gunnar [Randers] and Martin [Schwarzschild]. May be Rosseland does not want to let me in on his trade secrets."[52] In contrast, Chandrasekhar found Russell very approachable and deeply interested in working out details of the theory of eclipsing binaries with him, as well as discussing stellar structure, since the two were combined in Russell's mind. Chandrasekhar, however, sensed that "in those years Russell was rather lonely." Dugan was gone, and Rosseland turned out to be a brooding, quiet person. "He rarely talk[ed] to people, at least not during the time I knew him."[53] Russell needed someone compliant and accessible to talk to and missed Dugan more than ever.

Chandrasekhar always accepted Russell's frequent invitations to talk in his office, or dine with him at 79 Alexander Street, or walk with him on the sheltered and tree-lined paths of Princeton. Their conversations ranged far beyond astronomical matters as Russell opened up his life to the young theorist, recalling humorous stories about how his mother once intimidated a would-be burglar by confronting him in her home with the demand "Where did you come from?" Chandrasekhar also saw much of Schwarzschild, who was only too happy to engage in deep and detailed discussion, and they all took in lectures at the Institute and in the physics department.[54]

What Chandrasekhar may not have appreciated was how deeply distressed Rosseland was over the war and over the loss of his country, institute, family, and friends. Ladenburg and Russell encouraged Rosseland to speak out publicly about Nazis in Norway, and he did, giving what Russell described as "a masterpiece—a quiet effective statement, quite unemotional and very impressive." Even so, it was not a very satisfying way to fight back. Chandrasekhar sensed only competition, an impression Kuiper abetted.[55]

Chandrasekhar did come away with many warm memories of Princeton. He was delighted to be working with von Neumann, and to be in Russell's company. He also watched Russell in action, finding him the dominating force. Even among the physicists on campus, "Russell was the great figure."[56] One day, Chandrasekhar strolled behind Russell and von Neumann as the two headed to a meeting. Chandrasekhar listened while von Neumann chided Russell for managing to avoid responsibilities like committee work and asked him what his secret was. Russell revealed a strategy that he also shared at various times with others, like Shenstone and Atkinson: "There is one principle by which you can get off committees. When you are on a committee, no matter what the subject is, talk endlessly, prevent other people from talking. They won't put [you] on another committee after that."[57] Von Neumann confirmed that Russell indeed did just that. But Russell was also good at putting others on committees in his place. Stewart bore the brunt of the duty, and Russell asked Rosseland to get involved, as his heir apparent.

With Rosseland, Chandrasekhar, Wildt, and Russell in residence, Princeton became a powerful magnet for anyone seeking astrophysical theory. Astronomers and physicists alike started visiting more frequently. Franklin Roach applied for a Guggenheim Fellowship to spend a term working with Wildt and Rosseland on the spectroscopic data he had secured at Mount Wilson.[58] Robert Marshak, Bethe's student, came to confer with Chandrasekhar, Rosseland, and Russell, looking for new approaches to astrophysics.[59] Russell revised the curriculum for the 1942–1943 year to give Rosseland a chance to create new courses in astrophysics and theoretical and practical meteorology. Rosseland and Wildt would team-teach "Elementary Astrophysics," which replaced the old undergraduate course in celestial mechanics. This, Russell argued, would give "competent undergraduates . . . an opportunity to work with Professor Rosseland in the field in which he has achieved distinction." They would be exposed to the newest insights into atomic theory, the theory of spectra, and applications to astrophysical problems. Rosseland, Russell told Shapley, also knew "all about the new meteorology from its start under Bjerknes," which was a real asset for the war effort.[60]

A Depleted Department

The department's new lease on life was exhilarating but brief. In fall 1942, Wildt left for a real faculty position at the University of Virginia, and the British Admiralty asked for Rosseland's assistance, at the suggestion of the Norwegian government in exile. Russell could not object; the department had largely closed down anyway since the sole computer, Dorothy Davis, had left for war work, and those who were left, Pierce and Stewart, were teaching navigation.[61] As observatories closed or were turned to war work across the land, Russell shared the common concern among directors that programs would be forgotten. He hoped that his spectroscopic reduction and analysis programs, like Charlotte Moore's *Multiplet Table*, would not die. Even in the present emergency, Russell argued, these programs had to continue, to preserve staff expertise.[62]

During Russell's long convalescence after his first heart attack in March 1943, Stewart managed as acting director. Shapley took over Russell's National Academy responsibilities, which were not great at the time, but no one was really looking after the longer-term needs of the department. Confined to his bedroom and study that summer, Russell was depressed and frustrated. He could work a bit on spectra and multiplet structure but felt at times "at the end of my rope." He was happy at least that Rosseland was prospering in England; he had been named George Darwin Lecturer that year and had Eddington close at hand.[63]

By the end of June Russell felt a bit better. He finished a paper on white dwarfs and small companions, but, he told Shapley, it did not add much to what the younger people could do. He thought of all the talent now in war work: Chandrasekhar, Rosseland, Dirk Reuyl, Kaj Strand: "Chandrasekhar would beat anybody else on theory." On the bright side, Russell's physical health was returning: he had attended church the past three weeks, could now walk a mile, and felt "pretty well out of jail." He dearly wanted to escape to the Cole summer home at Jamestown, Rhode Island, to be with grandchildren and his work on gadolinium, to "keep me from brooding."[64]

Throughout the 1944–1945 academic year life at the observatory did not change. Rosseland's accumulated salary had made it possible to invite John Merrill back to augment the navigation courses and continue binary work with Pierce and Russell. Russell promoted Pierce to assistant professor and extended Merrill's appointment through the summer, hoping to regain their prewar momentum. Charlotte Moore Sitterly remained a part-time research associate though she was living in Middletown with her husband.[65]

In February 1945 Rosseland returned, not to Princeton, but to a new war project in New York City. Russell hoped that Rosseland would soon return to campus, and fought off campus efforts to claim his office and phone, as well as the observatory residence, which had been taken over temporarily by the university organist but was still under the control of the astronomy department.[66] When he did return to campus in January 1946 Russell pushed him into university life, making him representative to the Committee on the Graduate School. Typically, Russell failed to inform Princeton promptly that Rosseland was to be put back on the payroll, a fact that caught the attention of Hugh S. Taylor, the dynamic new dean of the Graduate School.[67]

Taylor knew that the department was in a holding pattern. Russell released accumulated departmental funds to hire former students returning from war, like Frank Bradshaw Wood, as well as a few others who showed special interest in binaries, like Mrs. Felix Recillas (Paris Pishmish Recillas) from Tonantzintla. Rosseland was given a raise to $6,500 plus the observatory residence, but Taylor was most concerned that Rosseland confirm his commitment to Princeton.[68]

Rosseland and Chandrasekhar

Throughout the war and in its wake Russell worried that Rosseland would not stay. His wife had turned seriously ill in 1945 and had to be hospitalized and then placed in a sanitarium while Rosseland remained in New York. Though she seemed to be the picture of health as they settled back into campus life in 1946, she remained unhappy living in Princeton, and, more to the point, Rosseland even seemed uninterested in developing the curriculum. Their son had graduated and had returned to Norway to work as a meteorologist.[69]

The shakiness of Rosseland's commitment also concerned Princeton's new Advisory Council for Astronomy appointed by Dodds with Russell's advice. Adams, Struve, and Shapley worried that with Russell's delayed but pending retirement, and with Rosseland hardly secure, Stewart was next in line. "There is no astronomer in the world, not even Rosseland, who can be a worthy successor to you," Struve warned, "and that consideration is certain to be quite troublesome no matter what the final choice may be." Struve strongly advised against promoting Stewart to full professor because such a promotion "might result in a weakening of the astronomy department at Princeton." At the least it would increase the "difficulties of the director." Russell agreed with Struve, asked that he advise Dean Taylor directly about Stewart, and, holding out hope that Rosseland would stay, still agreed that a wider search for his successor was warranted.[70]

Russell and Rosseland agreed on one thing: with Russell's pending retirement in spring 1946—delayed owing to the war from the original mandatory date of 1943, when Russell was sixty-five–the department needed to fill the second full professorship. They and the Advisory Council had no trouble convincing Taylor to invite Lyman Spitzer to a three-year appointment as associate professor, since they all agreed that "the greater part of [the department's] energy" in the future would be on the theoretical side of astrophysics. Spitzer was one of those "rare types," Russell had argued in his passionate brief to campus officials, who could manage both theory and a robust observational program. But Yale countered the offer with a "substantial promotion and aid in research," mainly by letting Spitzer set up a joint venture, an "astrophysical unit" linking astronomy and physics, which provided the autonomy he sought. Yale also gave him an assistant professor slot to fill and an annual operating budget well beyond what Princeton provided its entire department.[71]

Russell deeply regretted losing Spitzer but knew that Princeton could not match Yale's counteroffer. "I don't wonder that it tempted you," he wrote Spitzer, adding: "My sympathy for organizational and promotional work is from a somewhat respectful distance, as I hate the thing myself and have had good sense enough to leave it to others who can do it better. If you can do it, more power to you, but think where Shapley is and tremble."[72] Spitzer's decision did worry Shapley, Struve, and Taylor, however, who called a meeting to deliberate over the situation. Taylor, a chemist, was skeptical of the proclivity of observatory directors to become permanent fixtures and thought that Princeton astronomy had become moribund as a result. Russell defended tradition, arguing that it was a practical custom because of the long-term nature of astronomical programs. Rosseland, however, was a pure theorist and seemed even less interested in the observatory than Russell had been, at least concerning the details of its maintenance and upgrade, and this bothered Taylor. To muddy the waters, sensing that he was not in line to assume the directorship, Pierce started to hint that he was leaving for a better offer, but Shapley checked and decided it was all a bluff. What was revealed, however, was that Russell had apparently led Pierce to believe that he was indispensable and deserved better than he was getting.[73]

Spring 1946 was a time of worry and deepening depression for Russell. When he was asked to speak at the Newton Tercentenary, he needed Stanley Cobb's advice "upon the psychological question." Problems were "arising in the family": his son Henry had just gone through a divorce. His wife thought it would do him good to get away, "and help to keep me young." But a trip to London, notwithstanding how much he looked forward to it, was too strenuous a prospect, and so he passed up the Royal Society offer, which only deepened his distress.[74]

Russell now depended more and more upon Shapley to deal with Taylor: "I feel that there is a real danger here that if we try for too much we may get very little," he worried to Shapley.[75] Neither Shapley nor Taylor, however, shared Russell's fears. They had lost Spitzer because they were too cautious, and would not make that mistake again. Taylor thus set out to find and secure the strongest astrophysical intellect on the planet. On a trip west, he stopped in Chicago to have a long talk with Chandrasekhar, who he felt had worked well with Rosseland during their semester together in 1941. Taylor found Chandrasekhar willing to consider a research professorship at Princeton. He was not interested in being director, but, Russell reported to Shapley, he was also "definitely unreceptive if Rosseland was to be the Director." Russell surmised that "there seems to have been some sort of past 'wrangle,' " but he did not know its source or nature, clearly unaware of how Chandrasekhar had felt jilted by Rosseland during their tenure together in Princeton. He asked Shapley to intercede, pointing out that they had several choices: make Rosseland director and research professor; split the positions between Rosseland and Chandrasekhar and hope for the best; or appoint Rosseland director and let him find his own research professor. There were problems with all three options, but above all, he feared, any additional delay would jeopardize the endowment given over to the research professorship.[76]

Shapley's solution was to make Rosseland director and Chandrasekhar an independent research professor not in any way subject to the whims of the director. This plan, he hinted, might be "reasonable and possibly acceptable" to Chandrasekhar. The problem, they both knew, was that "Chandra is both very sensitive and highly critical of other people's work. Rather naturally he considers himself head and shoulders above his Yerkes colleagues—at least head above." Shapley had to be very tactful with Struve, too, as he pursued Chandrasekhar once again. Shapley also grew more skeptical about Rosseland after talking with one of his students, Gunnar Randers. Rosseland was an enigma: he had shown "so much promise and inherent ability, such a sane attitude toward the hydrodynamics of the astrophysical problems, and so persistent an evasion of astronomy in the raw."[77] Shapley further worried that Chandrasekhar might be called to take Eddington's post in England, or that if Milne were, Chandrasekhar would get Milne's chair. Ever faithful, Shapley convinced himself that Chandrasekhar's best prospects were in Princeton rather than Oxford or Cambridge. If Rosseland

and Chandrasekhar could just cooperate, they could make Princeton the strongest theoretical astrophysics center on the planet.

Shapley was also convinced that Mrs. Rosseland's health was linked to her expressed desire to return to Oslo, which greatly surprised Russell even after Rosseland confirmed it. In any event, Russell brooded to Shapley, he did not want to see the cherished research professorship vacant for any length of time. Shapley knew what was at stake. Princeton needed a director but had to be constantly reminded that it also needed the professorship, which was far more precious. Not even Taylor's repeated assurances gave Russell comfort that his endowed research professorship would live beyond his retirement. After all, in failing to build a strong infrastructure for astronomy at Princeton, as he implied to Spitzer, he knew he had done less than he should have to preserve it.[78]

When Rosseland finally announced in May 1946 that he was leaving, Russell thought he might still stay if Princeton would let him build a "center for physical and astrophysical theory." If not, he told Shapley, "[t]hey *may* ask me to carry on . . ." to avoid the Stewart issue. Russell was now more than upset that had they let Spitzer slip through their fingers five months earlier; had they known what was to happen, Russell added, they would have been well advised to have offered him the directorship outright. Coming as close to an expletive as he dared, Russell fumed: "I am tempted to say WORD!"[79]

Dodds did ask Russell to delay his retirement again. To sweeten the deal, he gave him carte blanche to designate a visiting professor for the next year. It didn't help; Russell's frustration only deepened as he continued to fight Kopal. "I regret this very much," Russell told Shapley. He had to cancel a trip west to Pasadena, as well as postpone visits to Lick and Harvard; both had invited him to be a senior research associate. Meanwhile, Shapley continued to work away at Chandrasekhar; with Rosseland out of the picture he found Chandrasekhar willing to visit Princeton to take another hard look.[80]

All attention now focused on Chandrasekhar, who visited the Russells in Jamestown and talked to people on campus in July. After his visit, Russell believed that Chandrasekhar was interested in the research professorship, not the directorship. He assured him that the research professor would have complete autonomy, and that Chandrasekhar did not have to "fit in" with any department policy. Chandrasekhar would also get the observatory residence. Russell told Taylor that Chandrasekhar and his wife would fit in beautifully; they had been "exceptionally attractive as guests. Mrs. Chandrasekhar begged to help Mrs. Russell and Margaret in the kitchen, and did so as if she was one of the family. They are exceptionally fine people, and will be a great addition to Princeton."[81]

At the end of August, still in Jamestown, Russell heard that Chandrasekhar had accepted Taylor's offer. Russell was elated, convincing himself once again, after his disappointments with Dunham, Atkinson, and Rosseland, that he had found just the man he wanted to succeed him. He looked forward to seeing Chandrasekhar at the upcoming conference on nuclear energy honoring Princeton's bicentennial and settled down for what little remained of his tortured summer at Jamestown.[82]

During this time, Shapley was enormously busy with his own life, which by then he had filled with a wide range of humanitarian and social activities that included testifying before Congress in favor of lifting "the secret of the atomic bomb," fostering however he could "a marked sympathy for Soviet Russia," and urging the formation of a National Science Foundation, or "national scientist foundation," which for J. Edgar Hoover was evidence of subversive Communist activity. He was therefore drawing the rapt attention of FBI informants, who were watching his ever-growing observatory empire from the inside, and who reported on his myriad of activities.[83] Preoccupied and consumed as he was by public and university affairs, nevertheless Shapley met his duties to Princeton astronomy. His devotion is a mark of his loyalty and respect for the legacy of his mentor, but it also suggests that he believed Russell to be incapable of managing his own succession. Indeed, the two differed radically in their view of what was needed.

It only made sense to give Chandrasekhar the professorship and the directorship as they were now, combined in one position: "This is in some ways a trivial matter," Shapley argued. Russell, however, saw things in a different light. Thinking that Chandrasekhar harbored values close to his own, Russell gave Shapley "five powerful reasons" why it made sense not to combine the research professorship with the department head's position, including his innate dislike for executive work and his need to escape from it; the fact that members of the department "(even J.Q.S.)" approved of Chandrasekhar as research professor but not as director; and the importance of the preservation of the observatory residence. Dodds had told Russell that there was tremendous pressure to claim the house for other campus uses, and he could not guarantee it permanently for Chandrasekhar, who did not need such a large house and in some quarters was not thought to deserve it. Russell gave Chandrasekhar all the time he wanted to make up his mind, quoting extensively from Scripture to illustrate his feelings about the matter.[84]

In the postwar world, just as Russell had predicted long ago to Menzel, there were many good positions open and too few good candidates. Bowen had replaced Adams at Mount Wilson, Michigan was courting Bart Bok, Lick now was headed by C. D. Shane, and upheavals were brewing at Harvard and Yerkes as Shapley and Struve were distracted by politics and the increased demands of their large staffs. There were now at least fifteen openings at places with telescopes and no good takers, Shapley estimated, and all around there were personality problems. Princeton was a particular challenge: its instrumentation was so specialized that those who would make good use of the antiquated 23-inch were not modern astrophysicists. What would Chandrasekhar do with a refracting telescope?[85]

Worse, what would Chicago do? Not surprisingly, forces in Chicago shifted into high gear. When Robert Hutchins found out what was afoot, he marched Chandrasekhar into his office and offered him a coveted Distinguished Service Professorship that more than matched Princeton's offer. The only thing Hutchins could not offer, Chandrasekhar reflected, was the "honor of succeeding Russell because we don't have a Russell." Hutchins knew how to play on Chandrasekhar's sensitivities. He told him that he had to make up his own mind about the difficulties of being someone's successor. Quite often it was not to the advantage of the replacement. "Who succeeded Lord Kelvin[?]" Hutchins asked, and thereby made his point, winning the round and the battle. Chandrasekhar stayed put.[86]

Before he received the bad news, Russell realized that he was not totally satisfied with Chandrasekhar. He missed Dugan more than ever and had second thoughts about the future of binary work with the old Halsted. Dugan had been an admirable administrator. Only after his death did Russell realize the burden he had carried, "and how much better he could do it than I. We were a good team for many years, I hope, certainly we were very fond friends." Could Chandrasekhar be the faithful colleague Dugan had been? Russell worried: "What I want to leave at Princeton is another good team."[87]

It was in this frame of mind that Russell participated in Princeton's Bicentennial Conference on the Future of Nuclear Science in September. During a general discussion over the future complexion of science, his comments reveal that he was brooding over his successor. Seconding remarks by C.E.K. Mees and M. S. Vallarta, who had emphasized that it was less the size of the laboratory that was important than who was running it, Russell observed:

> In forty years experience in astronomy I have watched many groups, watched them grow, watched them decline, and watched them come back; and in every instance, and without mentioning any names, I think I can say that my colleagues in my own profession and I all agree that these advances and these retrogressions had to do more than anything else with the man who directed the group.[88]

But in his depressed state, taking rejection personally, possibly because he knew he had not been the most effective director, Russell now felt duty bound to support those who had proven faithful to him, and seems almost to have given up hope. John Merrill deserved better since he was carrying on the legacy of the "Russell Method." Pierce knew the field of eclipsing binaries thoroughly and was a competent administrator. "He is the heir to Dugan's tradition," Russell cried, and even Stewart approved of Pierce as head of department. If he were to leave because he was not given the directorship, Russell brooded, "Princeton's whole photometer tradition would be lost."[89]

At the end of October, Hutchins wrote Russell explaining that "[w]e put terrific pressure on Chandrasekhar." Above all, Hutchins wanted Chandrasekhar to hear from Russell that "you do not regard him as an ingrate." If Russell would do this, Hutchins felt, "I am sure you will be performing an important service to one of your devoted admirers."[90] Russell at first was silent but soon found he could happily grant Hutchins's request because by then Shapley had found another candidate.

Shapley never stopped talking to Spitzer, letting him know what was happening at Princeton. Spitzer made it clear that he and his wife Doreen were still interested in Princeton if it meant the directorship. "[W]e hadn't papered our dining room [in New Haven] until the situation became more definite."[91] "A solution may well be at hand," Shapley reported to Russell in October 1946. Bringing Spitzer would solve many problems: Rupert Wildt, whom Spitzer had just hired at Yale when Virginia could not continue his appointment, could now look forward to a real job at Yale. Russell contacted Spitzer immediately, wild with enthusiasm: no solution to the problem would please him as much "professionally or as much personally."[92]

It seemed as though everyone would emerge a winner. Chandrasekhar's position was strengthened at Chicago, and Russell found that in a pinch, Shapley really could deliver the goods. Russell was right to leave the negotiations with Spitzer to Shapley: "in the hands of some one who has the 'know-how'!" He was beside himself with joy at the outcome, as was his much relieved wife, who, Russell told Shapley, fretted helplessly as she watched her fragile husband worry himself to distress: "[S]he has been worrying about *my* getting worried!"[93]

Russell had thought he had lost Spitzer to Yale permanently because Yale supported Spitzer's plans to develop a hybrid astrophysical unit funded by the navy's new Office of Research and Invention.[94] But, unknown to Russell, Spitzer's Yale Observatory director Dirk Brouwer, a mathematical astrometer, was concerned lest his astrophysical unit get out of hand. Spitzer was also having a difficult time recruiting staff. Leo Goldberg turned him down, and when Spitzer tried to convince Martin Schwarzschild to leave Columbia and take a position with him at Yale, Schwarzschild decided that the unit Spitzer was building was something of an "administrative oddity." He preferred the solidity of a traditional department and so elected to stay at Columbia for the time. Brouwer did not impede Spitzer's progress, although he did ask some rather pointed questions, which Spitzer was mulling over when he responded positively to Shapley's inquiries. Thus Shapley found that the door was still open.[95]

Schwarzschild had offers from Mount Wilson and Caltech, where they wanted him to head up a new theoretical program to complement the 200-inch telescope. But Chicago, Lick Observatory, and, evidently, other places were inquiring too, some asking him to head their programs. Schwarzschild was, however, too canny to be swayed by a directorship or chairship. He felt that he was still too European to understand American politics and knew too where his priorities lay: his decision was based less upon the prestige of the institution than upon whom he would be working for. "You don't want to be yourself the head in the future. You'd better go to the place which has the best head." Since his days at Harvard, the most promising fellow he had met was Lyman Spitzer. They both were theorists at heart, but both had an equally healthy respect for the importance of observational data. If Spitzer were to build a real program at Princeton, that was where Martin Schwarzschild wanted to be.[96]

But would Princeton play ball? There were many questions still unanswered. Although he had grown very rapidly into an astronomer of distinction and promise, Spitzer at thirty-two was still quite young, his list of publications moderate though impressive, and few at Princeton appreciated his organizational abilities, which came out during his National Defense Research Committee work on undersea warfare at Columbia and later in his leadership of a sonar analysis group.[97] Struve plotted with Russell and Shapley to find rhetoric that would convince Dodds and Taylor. He promoted Spitzer as a "first class observational astronomer" who could plan "for some kind of active cooperation with Mount Wilson of the same character [Russell had] maintained with such great success."[98] But Dodds and Taylor were convinced only after they had interviewed Spitzer and absorbed his "Proposed Long-Range Plan for the Princeton University Observatory." At Shapley's suggestion, Spitzer mapped out a detailed plan that called for the clarification of university policy regarding astronomy, offered clear guidelines for administration and operating levels (including the retention of unspent funds), and stipulated that the house and all apartments now under the department's control remain so. At a time when such activities were not yet commonly practiced, Spitzer wanted to be absolutely clear about his ambitions: Princeton would have to allow Spitzer to continue to consult for the government and industry, and must support his use of outside funding, including governmental and military grants and contracts such as those he already enjoyed. Spitzer would chair the department and direct the observatory, at the rank of full professor, but he shied away from Russell's coveted mantle of research professor, though he wanted assurances that it would not be taken away; if he worked out, he would assume it, "say in seven years."[99]

Spitzer also wanted to hire Schwarzschild as associate professor, to assure that Princeton's primary strength would be in theoretical astrophysics. Equally important, he wanted "to keep theory in touch with current observational problems" and so suggested that the two new members of the department would "each spend one academic term out of every four in a major observational center such as the Mount Wilson Observatory," supported entirely by Princeton. In this manner, following Russell's lead, Spitzer hoped to ensure that the fruits of the world's greatest telescopes would continue to flow to Princeton. To keep the place alive, and to make Princeton a leader once again in theoretical astrophysics, Spitzer called for support to bring an outstanding theorist to campus for a semester each year.[100]

Spitzer did not ignore tradition. He wanted someone to carry on Dugan's work, using photoelectric techniques, and also asked for two additional slots at the instructor level, nontenured, mainly for theorists but some observationalists too, picked from the ranks of the best graduates in the country. Some of these positions might be supported by funding from a governmental "Science Foundation," Spitzer felt, if Princeton positioned itself correctly. He included an itemized departmental budget, describing clearly and succinctly what could be done at what he felt was the minimal level of funding needed to attract the type of people who would make Princeton a worthwhile place to pursue theoretical astrophysics.[101]

Spitzer's proposal convinced Taylor that he could both build and manage a competitive department that would exploit new opportunities in the post–World War II world, such as government support of science. Advised by Taylor, Dodds also agreed to the periodic leaves of absence to work at Mount Wilson. "I was so thrilled with the happy ending of a year's worry," Russell told Spitzer at mid-month, after they had an audience with the president and knew that many major hurdles had been cleared. By then Spitzer had asked Martin Schwarzschild to join him if all went according to plan. Russell's only wish was that the two would come as soon as possible.[102]

Russell saw much of himself in Spitzer's plan, updated, to be sure. This only made him more worried that, somehow, Dodds would not accept it. Both he and Shapley agreed that if Dodds bought the proposal, "Princeton would be assured of thirty years of distinction for the Observatory and for astronomy."[103] Throughout the negotiations, Spitzer began to call Russell his "Honorary Uncle" or "Uncle Henry," and Russell now addressed Spitzer as "My Dear Lyman" or simply "Lyman"—a familiarity Russell reserved for very few people beyond family. Indeed, all along, since he had set his eyes upon Theodore Dunham, Russell had searched for an extension to his family, to himself.[104]

Princeton did meet Spitzer's demands, which, as it turned out, required no major increase in support. Owing to Russell's frugality, the department retained a large reserve that could cover Spitzer's requirements for the next five years. Recognizing this and anxious to end Russell's regime of fiscal restraint, Taylor recommended that Spitzer's program be put into effect immediately since it promised a greater "breadth and scope than has heretofore been attained." Spitzer would strengthen ties with physics, with other university groups, and, most important, with new sources of funding. Even at present funding levels Spitzer would be able to "create a vital center of astronomical research at Princeton." Dodds shared everyone's pleasure that they were calling back a Princeton man. On 17 January 1947 Spitzer formally accepted Princeton's conditions and announced that he would be coming in July.[105]

The Board of Trustees approved the appointment in April, and Russell expressed true joy; at last he could retire "with a feeling of great confidence as to the future of the Department of Astronomy." Russell could now accept renewed invitations from Shapley and from Shane, to be the Morrison fellow at Lick that summer. "It all seems too good to be true!" he cried to Spitzer.[106] In retirement he could now concentrate on completing Merrill's *Tables* and Pierce's *Finding List*: "They are vastly more convenient and useful than Kopal's clumsy methods." Spitzer was more than willing to oblige, suggesting in June that Merrill take a leave of absence from Hunter College to finish the *Tables* at Princeton. Sensing that Spitzer's kindness and support might have limits, since he clearly had other priorities, Russell kept up the pressure not only to complete these projects but to have the observatory ready for his arrival, now less than a month away. He badgered campus crews to fix leaks and replace rotted timbers, and acquired two new Marchant 10 × 10 calculating machines with automatic multiplication and division for Schwarzschild and Spitzer, a major purchase for Russell.[107]

Russell's cathartic housecleaning reached fever pitch as he sorted out years of accumulation to ready his office for Schwarzschild. Now, after years of neglect, he struggled to separate official from personal or scientific correspondence. He also regretted that he had failed to keep minutes at department meetings, and wished, too, that he had maintained clear guidelines for professorial rank. Moreover, he wished that he had created a confidential correspondence file.[108] All of these failures now haunted him as Pierce made claims that could not be verified. Furthermore, Pierce was acting strangely, blocking the private entrance to Russell's retirement office, a small room next to the classroom. When he confronted Pierce, he was met with hostility: "I don't know what is wrong," Russell wrote in July to Spitzer, who was then enjoying his first tour at Mount Wilson. "[H]e showed no signs of this disposition here a few months ago."[109] Russell could not comprehend the possibility that Pierce was challenging the man who had failed as chair and director, who ignored those who had served him for years, and to whom he may well have promised advancement in the past. Pierce's anger and frustration would have to be dealt with, but, Russell likely sighed with relief, it was Spitzer's problem now.

Russell was dropping responsibilities right and left now: he retired from the presidency of IAU Commission 35 on stellar structure and gave over his chairship of the NRC Committee on Line Spectra to Meggers after he suffered his stroke in 1948. He called it a "cerebro-vascular spasm," which laid him up for a few weeks of enforced "conscientious laziness" that was "rather hard for a man of my Puritan ancestry to learn." At age seventy-two, he had no intention of being "put up on the shelf yet," he told Shapley and others. "If I play safe, and I'm a-gointer." He hoped for years of productive work with Merrill and Moore, as long as his health permitted.[110]

The Department under Spitzer

Spitzer found no surprises at the observatory. He knew its quaint quirkiness and its limitations. It was a relic. Even though it had been the bastion of America's premier theoretical astrophysicist and its greatest pioneer astronomical spectroscopist, he was amused to find that the library had been maintained with two separate indexes, "Astronomy" and "Astrophysics"—"and they were quite separate. The astronomers had been unwilling to concede that astrophysics was a legitimate discipline, so it was brought in as a wholly separate topic." Although Russell did what he could to prepare the place for their arrival, both Spitzer and Schwarzschild found that much housecleaning still needed to be done. They started on a regular schedule, every Saturday morning picking a room, evaluating its contents, and discarding everything that was unrecognizable. As their new department grew, they started sacrificing the sacrosanct spaces that held the clocks and transits, sealing off the observing slits to make more livable offices for students and visiting colleagues.[111]

Princeton astronomy had been Russell, and now that Russell was gone, Spitzer and Schwarzschild had a nearly clean slate with very little overhead or entitlements to worry about. Most definitely Dean Taylor felt this way, and so in the first years Spitzer's and Schwarzschild's major efforts focused on building up a new infrastructure that would attract not only the best students but senior theorists and observationalists too. They were very successful in doing both. For Spitzer, it was certainly an honor to return as Russell's successor, and it was also a chance to seek out truly interdisciplinary areas between astrophysics and plasma physics. He eventually built both up to unprecedented levels at Princeton.

Stewart seemed happy enough. Spitzer gave him free reign over his interests and sympathetically reviewed his efforts to apply physical principles to sociological problems, seeking out the advice of demographers at the Institute for Advanced Study and at the Woodrow Wilson Institute for his yearly evaluations. Spitzer gave him higher marks than had Russell, and even tried to promote him. As for Pierce, Spitzer felt that "his indispensability was no longer a factor," but Pierce was encouraged to turn the Halsted to photoelectric photometry, which he tried to do until his untimely death from a cerebral hemorrhage in early August 1950. Spitzer then asked Frank Bradshaw Wood to gather up all accumulated data for eclipsing systems from Dugan's and Pierce's long collaboration and put them into shape for a Princeton *Contribution* in 1951. His was a labor of love, Wood felt, expressed as a "deep debt of gratitude" to the two men who most influenced and aided his life.[112]

Russell did not vacate his old office immediately. Schwarzschild allowed him to leave the majority of the floor-to-ceiling bookshelves filled with his records, which would take much more time to sort through and remove. Russell continued to frequent the observatory and in his correspondence with his colleagues continually expressed delight and satisfaction with the way Spitzer was managing the department and providing support for binary work. Most of all, he was delighted that Bowen allowed Spitzer and Schwarzschild continued access to Mount Wilson.[113] The arrangement continued a long tradition, but now it became a real trade: Mount Wilson and Caltech students and staff started to visit Princeton. Out of this mixture Schwarzschild established collaborations with Walter Baade and William Fowler and their students, as well as Fred Hoyle from Cambridge, that revolutionized stellar structure studies. The mix brought together the best observational evidence with the most insightful theories as to how stars of many disparate types behaved on the Hertzsprung-Russell Diagram. Only then, in the mid-1950s, did the decades-old problem of the evolutionary status of the giants find a solution identifying them as evolved main sequence stars.[114] Russell lived just long enough to see this happen. It must have made him very proud, indeed.

• CHAPTER 23 •

Russell's Universe

IN MARCH 1942, after David Webster had lost a prolonged struggle over control of the Stanford physics department and became despondent watching it move in the direction of military and industrial research, he felt beaten. Russell paid the Websters a visit that month, during his yearly stay in Pasadena, and tried to console his friend. Webster had just received a commission in the navy to work for the National Defense Research Committee (NDRC) and seemed almost relieved to get away, feeling that his career in physics was over. Russell would hear none of it: "[I]t is neither good psychology nor good sense to feel that you are through."[1]

On the return train to Pasadena, Russell wrote Webster with affection and encouragement. Sixty-four years old now, eleven years older than Webster, Russell felt "much nearer the shelf than you are," within one year of his original retirement date. He pleaded with Webster not to give up hope, for there was every reason for him to continue in research. Even though at his age Russell knew he could no longer hope to do the "big constructive things" as the "younger folks will probably do," still "there is alot of filling in—and mopping up that needs to be *completely* done, and the number of people who can do accurate, conscientious, competent work is always limited. When we have to take in sail, we start the *Second Voyage*." Thus Russell described the last two decades of his career as he advised Webster to "[p]ick up your Kipling."[2]

Research was Russell's salvation, the breath of life. The investment Webster made in devoting himself to building the Stanford department was not one Russell had made at Princeton. Naturally, he gave such investments lower weight and advised accordingly. It only slowly dawned on him, therefore, that his friend was in real pain, and so he reached out. They had spoken about Thornton Wilder's *The Bridge of San Luis Rey*, which he thought poignantly framed Webster's despair: "I didn't know . . . how much pounding that Fate (another name for the Disposer of Destiny) had given you. We all get it, one way or another, and we are lucky if we get it young, as May and I did."[3] David and Anna Webster most certainly knew what Russell was referring to: the challenge of giving Elizabeth a good life.

In these few words to Webster, Russell summed up much of his own life. The insights one draws from them stand in sharp contrast to the public Russell, the image he constructed for inspection. Thus in agreeing to prepare his "spiritual autobiography" for Louis Finkelstein in 1949, Russell claimed that he had "no spiritual Odyssey to recount" because his life was "a good case for predestination." He took pleasure in this ironic bit of wordplay and wanted the world to think that the adjustments he had made during his life "involved a minimum of inner conflict."[4] Outwardly, Russell's life does appear to have been remarkably seamless, a steady progression to high attainment, personal fulfillment, and all the rewards astronomy could offer. One family, one home, one institution, one life. Outwardly too, Russell exuded surety. He was sure of his methods, his sense of proper practice, and his value to his science. Though he was sure of his intellectual superiority, which gave a certain strength of conviction, he was far less sure of his ability to manage life. Although he championed approximate methods throughout his life, he always believed that an ultimate truth existed, beyond mortal grasp. Even though he harbored a deep aversion to rigid boundaries, he was never comfortable with uncertainty, either on the atomic level or on the human level. While he relished scientific challenge, he reacted badly to personal challenge. His wife's eclampsia, her later cancer scare, his calls to Yale and Harvard, and, finally, his search for his own replacement all reveal a complex man who constantly questioned his own mental fitness.

His sense of obligation to Elizabeth was unending. As he told Webster, research was his top priority "unless some obligation like the care of Elizabeth's health plus financial stringency" forced him to take on more lucrative tasks.[5] Russell certainly carried that obligation, as a duty, all his adult life. But his substantial salary as well as Lucy's inheritance were more than sufficient to handle almost any circumstance they would be likely to encounter, and so he managed to avoid ever noticeably compromising his research; in fact he rarely sacrificed his priorities. Despite his claim of no "inner conflict," however, he was hardly a blithe spirit accepting his fate but a complex and sometimes tortured soul who worried constantly that he would lose the freedom he so cherished. Research may have been his preoccupation, but security was paramount, as was peace of mind. He never achieved the latter, for he was obsessively manic about his work and life. No wonder that he thought of himself as Stephen Leacock's defiant yet frantic hero, who "rode madly off in all directions."[6]

Structurally, Russell's career and life appear linear and progressive: his elite upbringing, training, postdoctoral years, faculty appointments, subsequent advancement, influence in the American community of astronomers, awards, and honors—the reward system enjoyed by a typical highly successful academic. But once we factor in the science, the kinds of problems he attacked and how and when he chose to attack them, the appearance of linearity vanishes. His career was punctuated (binaries), truncated (stellar evolution and energy), and serendipitous (spectroscopy, atomic physics). Russell was not thinking of extending Saha's work when he was made Carnegie research associate. He was thinking of stellar energy and evolution. But when Saha's papers appeared, and Russell realized that the evidence to test his theory was in the plate stacks at Mount Wilson, he seized the opportunity. He had been preparing himself generally in quantum theory, and this provided a fortuitious payoff that changed his scientific life, opening a path from the atoms to the stars. If a calculus was involved, it was a general philosophy of preparedness and opportunism. Russell did not change his methodology—he remained strongly intuitive, always preferring approximate methods and iterative schemes—but he did change his tools and problem areas to suit his talents and the problems that arose, attacking what was ripe for exploitation whenever and wherever he thought he could make a difference.

Being a student of C. A. Young had certain advantages but certainly did not predestine one for greatness, since it is fair to say that of Young's formal students only Russell is widely remembered. Being identified with Young no doubt eased Russell's entrance to Cambridge and then to the Princeton faculty. But it was his industry, accomplishments, and promise that attracted successive patrons, from the Carnegie Institution's advisers to Pickering and then Hale. Young's legacy to Russell was, moreover, the absence of an overarching program. Russell was never forced into an apprenticeship, nor was he a junior member of an observatory staff devoted to specific goals. The freedom of his training under Young, infused with the intellectual heritage of Princeton; his mentorship under George Darwin; and his success at directing data-gathering programs at other observatories following Kapteyn's example—all these are factors that created America's first astrophysical theorist.

This last factor leads to the question: Where would Russell be without Shapley, or Shapley without Russell? One could ask the same about Russell's interdependence with Pickering, Hale, Moore, Dugan, Payne, or Stewart, or, to be sure, Lucy May Russell. Without Pickering there might have been fewer astrophysical data, or no patronage. Without Hale the infrastructure for astronomy in America might have looked very different, lacking his great solar observatory and incomparable collection of spectroscopic data. Without Moore, Russell might well have lost his mind; without Saha or Payne, his inspiration to link the atoms to the stars; without Dugan, his observatory; without Lucy May Russell, his family. Such counterfactuals highlight one of the most poignant lessons of Russell's life: how the many relationships he established and maintained made it possible for him to shape his life and career in the way he did. He exploited the work of others yet offered a wealth of insight and opportunity in return. Russell and Shapley were exact opposites; but both being opportunists, each saw in the other a liaison of great value. Russell provided intellectual power and authority, and Shapley responded with resources, social contacts, and, most definitely, a ready and willing audience. It was the same with all Russell's professional relationships, more or less.

Russell's universe was made up of stars, not galaxies. As Shapley recalled, his old teacher "never seemed to interest himself in Milky Way structure, or in the many problems of the star clusters and galaxies."[7] This fact is not surprising, considering that galactic research was an area where he had misjudged the evidence rather badly during the crucial years before Edwin Hubble confirmed the extragalactic status of the spirals. In September 1916 and then in March 1918 Russell had hailed V. M. Slipher's observations of the great velocities of the spiral nebulae, which, along with the work of the other "Pacific Coast Observatories . . . has so greatly widened, at a single bound, the limits of distance at which our investigations may operate." But, as we have noted, he was soon swayed by Shapley's dramatic announcement that the Milky Way was enormous, and by van Maanen's detection of proper motions in spirals. Russell therefore argued that the spiral nebulae could not be of the same order of magnitude as the Milky Way, a conclusion he reversed immediately when faced with Hubble's observations.[8] Quite likely, then, extragalactic research was not an area Russell wished to tackle, considering too that it would constantly pit him against his closest ally, confidant, and friend.

But Russell's grasp was still unusually broad, even for his day. He wrote popular essays on Einstein's theories of relativity and toyed with the visible consequences of intense gravitational fields for his *Scientific American* audience. And the only time he left the galaxy was to show that Shapley's speculation on why spirals were in recession was wrong. Even though Russell never tried to contribute to general relativity or theories of the expanding universe, he was fully aware of the impact both had on modern astrophysics, writing about them with clarity and style, making them accessible to the general reader and rank-and-file astronomer alike, and employing the constraints they placed upon the age of the sidereal universe.

Shapley's comment implies that Russell was of a generation when it was conceivable, though still remarkable, that one person could embrace the entire discipline, and Russell came close. He lived through a period when the parochialisms produced by observatory-based apprenticeship training and optical observational techniques lent an odd unity or sameness to the discipline. Though few astronomers breached their particular specialties as much as Russell did, specialties in astronomy then were not formal separations nor isolated separate communities. There was a substantial degree of permeability, and Russell, in fact, was an extreme case of someone who did not recognize boundaries.[9]

The radical shift Russell made after 1921, which surprised even physicists like Meggers, foreshadowed how astronomy itself would change in time, led in large part by the example Russell set and the goals he elucidated in his NRC essay "On Some Problems in Sidereal Astronomy." It was not only a conceptual change but a behavioral change, one making approximation acceptable. Indeed, this shift, epitomized by his alliance with Hale and his rejection of Pickering's mode of investigation, marked a shift from the empiricism of astronomical spectroscopy (as a "natural history of the heavens") to the analytic, causal, and interpretive framework of modern astrophysics (in the sense of "natural philosophy"). Although his conception of what constituted theoretical astrophysics changed over time, from being a branch of mathematical astronomy to being an indistinguishable part of modern physics, Russell had always practiced it by working from mental images to solve specific problems. His shift from Pickering to Hale was therefore a political statement, as embodied in his NRC essay, and in his actions, showing how modern physics could be applied to the stars. At the midpoint of his career, he set out to rationalize astrophysics, raising it out of its still largely empirical framework. He worked at the interface between observation and theory, making theory more accessible to observing astronomers by constructing approximate methods and creating visualizable conceptual tools to attack problems in sidereal astronomy. As a result, he helped transform American astrophysics into a wholly physical discipline. Establishing this framework, more than any one discovery or application that bears his name, is Russell's greatest legacy to astronomy.

Yet the relationships and methods that bear Russell's name have nevertheless become his legacy as constructed by the community of American astronomers. This fact is all the more remarkable because so much of his work was not really his own, not new. To say that his many methods and discoveries were "anticipated" by Hertzsprung, Payne, and others, would be silly and Whiggish. What we have seen in Russell's life and style, rather, is that he typically seized on the incomplete work of others and tried to confirm or refute it. Often he was in a conscious race to develop something that he knew other well-equipped opportunists were actively pursuing. His constant overlap with Hertzsprung before World War I, most poignantly represented by giants and the diagram, and then the Vogt theorem, two-electron interactions, and finally the hydrogen abundance problem, all speak to his aggressive nature. As Russell told Doolittle in 1914, he wanted to have his fun and see the fruits of his labors during his lifetime, jumping from problem to problem as they came ripe for exploitation. He was not surprised to see his work superseded, though he certainly resisted whenever he thought the replacement was not worthy.

His opportunism raises once again the issue of how credit is assigned by the community of astronomers. John Herschel's ancient dictum, "He who proves, discovers," begs the question: What constitutes proof?[10] For Russell, critical factors included an overwhelming mass of evidence, carefully crafted and presented in terms comprehensible to the community and the establishment of commonly held standards among peers in the community. Proof also required mutual trust, which Russell established in several ways: through compelling intellectual power, through alliances, and through a successful track record. Trust also depended upon familiarity, that unique social circumstance to which Russell seemed especially attuned. For clearly he favored those he knew, those who, one way or another, met his needs. There are many paths to power, through merit or political patronage, past track record, powers of persuasion, ability to marshal evidence, successful recruitment of allies, and institutional affiliation. Russell's power was a product of all of these, but it rested most firmly upon merit. He certainly was no builder of institutions and was rather clumsy in politics, but he knew how to forge intellectual alliances, gain the trust even of his competitors, and take advantage of ripe opportunities. His powers of persuasion were mainly intellectual, as in his advice to Payne, or in the way he crafted his 1929 paper on the composition of the solar atmosphere. But as we have seen in his dealings with Shapley, Russell was fully capable of calling upon political and moral standards of righteousness and self-sacrifice when larger issues, like Hale's telescope, were at stake. Further, as in his dealings with Kopal, he was sensitive to the political necessity of acceptable practice.

The perception of merit is also dependent upon norms of community consensus and acceptance, as well as upon the status (or trust) of the person whose work is under inspection. What was difficult for him early in his career, convincing others of the veracity of his hypothetical parallaxes or of the course of stellar evolution, became progressively easier as he gained status and seniority in the community. Burnham had easily blocked Russell's hypothetical parallaxes in 1907, when Russell was young, untested, and vulnerable. Becoming a gatekeeper himself by the 1920s gave him greater access to the pursuit of his whims but did not relieve Russell of the need to make a convincing argument. He still had difficulty getting others to agree with his findings when his admitted "tissue of approximations" wore too thin, which would lead him to adopt new strategies to keep his tissue intact. The fate of his theory of stellar evolution illustrates this nicely, but more poignant is the hydrogen abundance episode, which reveals a process whereby Russell first had to become convinced and then chose a particular strategy to convince others. Only when Russell was faced with every bit of evidence from astrophysics, after being confronted by it in the work of Payne, Stewart, Moore, Unsöld, and Menzel, did he reverse course, but he was far from comfortable relying upon astrophysical evidence alone. Thus he chose to present his findings to his peers in a way that made physics central to the process and astrophysics the beneficiary. He knew that Payne had argued from a narrow line of astrophysical evidence, and he quite properly tried to persuade her against it. What physics she used was weakly constructed. She certainly did not convince Russell, and if Russell knew anything, he knew that her argument would not convince others. After four more years of work, trying to make the problem go away, Russell believed that the only way he could convince others of his new conviction was to adopt a different strategy, one that seemed to depend not solely upon the accumulated astrophysical "tissue" but upon a framework that was undeniable: the structure of the hydrogen atom and the legitimacy of physics. It was therefore for political, not intellectual reasons, that he could not acknowledge Payne's prescient role, or Stewart's, more than noting their work as confirmatory at the end of his paper. If he had, his rhetorical argument, and strategy, might have collapsed, or so he may have thought. Much the same can be said to explain why Russell resisted Kopal so strongly, or ignored Hertzsprung's work at first. In both cases, he was convinced that he knew what the community would accept, how far he could stretch the tissue. And in both cases, he was right. Payne probably knew this. Kopal evidently did not.

Just as with Saha's revolutionary work—which pointed to incontrovertible evidence for the relationship between the structure of the atom and the role of temperature in altering spectra but did so using a technique that was not acceptable—Russell managed finally to convince himself and everybody else that hydrogen dominated stellar atmospheres, but knew just as clearly that his was a stopgap measure soon to be superseded by the sophisticated physics available to Unsöld and his generation. One can thus readily appreciate why Russell dug in his heels over his reduction methods for eclipsing binaries. Not only did his technique stand at the very core of his being and promise to continue to yield useful information, but he remained convinced that his methods were still the best way to keep astronomers willing to compute. He knew that Kopal's methods were potentially more powerful, but unlike Unsöld's or Minnaert's methods, which evolved into the "curve-of-growth" technique, they were not reduced to a form accessible to the common astronomer. All-important for a pragmatic theorist like Russell was knowing how to craft technique and evidence so that they would be acceptable to the community. To be acceptable something had to be useful. Utility was as important as validity as long as it promised to lead to new knowledge in an accessible manner. This is where the essence of Russell's contribution lies.

Russell's material universe was confined by the Bohr atom at one end, and the galaxy on the other, which marks him as a transitional figure in the emergence of modern theoretical astrophysics. His physics was limited to "exact qualitative analyses," as he put it, not full-blown quantum mechanical treatments that one could apply to the stellar interior and stellar atmosphere, to the interstellar medium, or to intergalactic space. His gravitational dynamics was confined to double star systems, not n-body calculations assessing the behavior of vast assemblages of stars. Even though the vanguard of modern physics moved beyond Russell's grasp by the end of the 1920s, Russell found ways to remain very competitive in what was then still a slippery borderland between physics and astronomy, one that Russell felt was defined as much by funding sources as by anything else. As such he symbolized the centrality of physics to astrophysical practice and served as an example for others that independent problem-oriented methodologies, balancing observation with theory by relying equally on physical intuition and evidence, represented the course astronomy had to take in the future.

The problem-oriented approach was a course not without costs. It certainly made the field more competitive, especially when, in more recent times, government funding patterns and cost-accountability have driven goal-oriented research to new heights of specificity and novelty. As Payne-Gaposchkin mused at the end of her own life: "The young are in too great a hurry, too eager to be the first to get the credit for a new discovery, a new idea. I have come to feel a nostalgia for those early days when an astronomer was content to accumulate the basic facts, and knew (as the observer of double stars still knows) that it is not for him to reach the final interpretation." E. W. Brown expressed the costs in another way, when he scolded Russell for his sloppiness in his advocacy of the Lowellian prediction of Pluto's position. But it was the course astronomy chose to take, given the power of the new physics and the persuasiveness of adherents like Russell.[11]

The rise of the new physics did not create an identity crisis for astrophysicists like Russell as it seemed to for chemists, who had developed their own language for dealing with the nature of matter in its various forms and struggled to resist the infiltration of physicists by retaining a vestige of their own style and "thought forms" in quantum chemistry.[12] Since astrophysicists never developed an indigenous set of conceptual tools for dealing with the interaction of matter and light, they could be reasonably open to physical theory, or at the least be indifferent, since they had nothing to lose in the bargain—so they thought. Of course Russell lived and worked very far from the turbulent centers of theoretical physics in Europe, where much if not all of theoretical astrophysics was emerging. Nor was he a professional physicist subject to the peer pressures of the world of physics. Safe in cloistered and stable Princeton, he enjoyed ample opportunity to stake out his territory and pursue his muse with enormous enthusiasm and single-mindedness. He could pick and choose the kinds of physics he found most applicable to astronomical problems.

The new physics was useful, to be sure, but Russell was never comfortable with its deeper implications. For example, in 1928, when a reader of his Terry Lectures on "Fate and Freedom" asked whether mankind's imperfections might be a manifestation of God's ongoing process of "becoming," Russell answered delicately that he did not agree with such an idea, since his mechanistic world consisted of order and purpose worked out from the beginning of time by God's Design: "I suppose it is because I am a physicist that I find it hard to think of God as "becoming."[13] Believing that humanity could never know the nature of the "Power" driving the universe, Russell nevertheless wanted to believe that there was such a power, operating with exactness. On the eve of Hubble's revelations, he expressed the common belief in what was, overall, a static universe subject to exact physical law, even though there was "plenty of development and evolution of individual parts of the universe, from stars to souls; but these are all particular cases of the operation of invariable laws of nature."[14] Uncertainty, probabilistic wave mechanics, and other trappings of the newest physics left him cold, but their practical application was most exciting.

During Russell's lifetime, advanced training in astronomy became university-based, and hence physics-based, as modern physics demonstrated time and again its interpretive and analytical power in solving fundamental astrophysical questions. Thus a greater fraction of astronomers could handle quantitative methods in astrophysics. Astronomy had moved in Russell's direction, and Russell could take considerable pride in the fact. Others certainly did on his behalf.[15] As a transitional figure, Russell trained or inspired a small but significant group of students over some thirty years of graduate instruction. Shapley was the only student of Russell's prior to the First World War to rise to world prominence and become a central player in the discipline. He also stands in distinction to those who came later, such as Menzel, Dunham, and Spitzer, all of whom embraced modern physics and excelled at the interface of observation and theory. Menzel in particular carried theory to the Harvard College Observatory and became a bridge to the Harvard and MIT physics departments. He enjoyed an independence denied to Payne and was able to create a new group of workers adept in astrophysical theory, with early products being Lawrence Aller and Leo Goldberg. As Russell's successor, Spitzer established a robust graduate and postgraduate program infused with physics. And, unlike his mentor, Spitzer harbored few prohibitions against institutional growth.

The astronomical community regarded Russell not as administrator, builder, or institutional director but as an intellectual leader, as guide and consultant, as "Dean." Russell exercised enormous influence through his many venues, from his *Scientific American* column and encyclopedia and textbook writings, to his visits to observatories, to his advisory roles to successive editors of the *ApJ*, and to his membership on prestigious award and funding panels for the National Academy. By the late 1920s and through the 1930s, gaining Russell's approval could be a critical ingredient in a successful career. Russell, of course, did not judge purely on merit and could be swayed by practical utility and personal favoritism in pursuit of his own interests. His early search for someone to groom as his successor points to a certain naïveté or, at worst, ineptitude. However, none of those he favored were weak or unqualified. All had their strong points. Merit was always a high priority for Russell, one that definitely overrode the petty nativist fears of many of his colleagues. American astronomy was hardly a pure meritocracy, but with people like Russell at its center, research competence was a necessary qualifying condition.[16]

Russell lived in a spiritual and moral universe, as well as a material universe. Heaven for Russell was "not a place but a state," his ultimate reward. He often alluded to it as a place, but a place beyond the stars, a place beyond astronomical knowledge. When he took his children outside to view the stars, it was clear to them that for their father the firmament was "as much a part of a spiritual world as a material one and [they were] sure he was also interested in how much we could see of it." When he first saw an ultraviolet spectrum of the Sun taken from beyond the atmosphere in October 1946, Russell told Leo Goldberg that "[t]hese rocket spectra are certainly fascinating. My first look at one gives me a sense that I was seeing something that no astronomer could expect to see unless he was good and went to heaven!"[17]

Living long enough to see a spectrum of the Sun taken from space, as a man who started his career six years before the Wright brothers' first powered flight, Russell experienced profound change not only in his science but in the fabric of society. Enjoying the privilege of gender, race, creed, and class himself, Russell shared many of the stereotypical prejudices of his contemporaries. He possessed progressive, even liberal, sensibilities at times, though they rarely broke through in action. Women deserved suffrage, Russell argued in the abstract, just as he recognized that Payne was his most worthy successor, or that Moore was sadly undercompensated. But practically, Russell knew that the political system he lived in and was a part of would not allow equity, in New Jersey in 1897, or Princeton University in 1934. In the same fashion, his deference to intellectual worth in hirings and promotions was more in evidence in the advice he gave other institutions than in his actions at his own. What he did not share with many of his contemporaries, especially on the West Coast, however, was an abiding fear of foreigners. He was free of the doctrinaire fundamentalism that was fighting to retain dogma, creed, and tradition on campus, although he never sacrificed himself to break down discriminatory barriers. He could argue against compulsory prayer while sponsoring voluntary "Sunday Seminars" on science and religion because his religion was ecumenical and universalist: he led a family where both parents prayed each night with each child, individually, and felt that church attendance per se was more important than which church one attended.[18]

Russell also tried to live in a responsible universe. After World War II, he worried about the future of science and the uses to which it might be put. During Princeton's bicentennial in September 1946, as we have seen, he warned that scientists had to act responsibly lest they lose control of their discipline. The "Ivory Tower" needed to be preserved as a place of freedom to work, but it was neither a place to harbor political or social ideologies nor a place to hide. "This is no academic matter," he warned, since science was likely to be supported by governments in the future, which would put control in the hands of nonscientists unless scientists came out of their ivory towers, fought collectively and honestly, and acted responsibly. Thinking darkly of the bomb, he argued that "[w]e need a new Hippocratic code today, for the physicist, and for all others who deal with nuclear energy." Science was "no place for the negligent, the neurotic, the seeker after personal glory and profit." Supporting the recent observation by Robert Merton that "the activities of scientists are subject to rigorous policing, to a degree perhaps unparalleled in any other field of activity," Russell called for increased self-control, beyond the purpose of establishing mere academic truths, by the institutions of science. "But," he argued, "if we stand intransigently for the dream of an absolute and irresponsible freedom of the individual scientific worker, we may see some catastrophe which will be followed by a stern Draconian control, in which we have little part."[19] By then, he had experienced Webster's bitterness, and the bomb, and feared for the future.

In the 1950s, Russell remained as active as his health permitted, visiting Harvard frequently, acting as "Special Consultant" to Lowell Observatory, and traveling to Aberdeen and Washington, D.C., as long as the advisory committee was in existence. He continued to shepherd the department's waning interests in binary stars and term analysis. His Morrison Fellowship at Lick had been a rejuvenation. In February 1952 Russell felt fit enough to lead his family once again to Egypt and the Mideast, mainly to visit Lucy and George, who were then stationed in Cairo. It was a harrowing trip. They encountered gunfire, rioting, and bombings of British offices and were saved only after taking sanctuary at the American University. Nevertheless, Russell, at seventy-four years of age, relished the excitement and managed to get in a few lectures he had prepared on spectra in the laboratory and in the stars. But as the years moved on, the coronary he suffered in 1943 continued to haunt him: he encountered ever more frequent dizzy spells brought about by advancing arteriosclerosis.

As Russell neared eighty, he never stopped work entirely, save for the times when he surrounded himself with his grandchildren, whom he loved to watch grow and gambol on the shore at Jamestown. He delighted in crafting little origami figures for them, as he did for any youngsters who came within his reach.

He was a constant font of limericks, rhymes, and aphorisms, spontaneously erupting at gatherings where he and Mrs. Russell would polarize their captive audiences with parallel monologues on every conceivable subject. Though his children remained dispersed, they made sure that summers in Jamestown with the grandparents were filled with life and laughter. But Russell's energy was largely gone now. There was a growing recurrent tremor to his hands and body, and his memory was weakening. Outwardly he remained, as Stewart always thought of him, the "old Princetonian." Inwardly, without evidence, we can only speculate. As he told Webster, he seemed content to fill in details, which he did in the last decade of his life.[20]

At Russell's funeral on 21 February 1957, four days after his heart finally gave out, Shapley sat on the sidelines watching as Russell's body was carried from the First Presbyterian Church down Nassau Street and into his study at 79 Alexander, escorted by Dodds, Eisenhart, classmates, and colleagues, including Merrill, Bancroft Sitterly, Schwarzschild, Shenstone, and Stewart. Spitzer would have been there too, but he had to be in Berkeley working for the U.S. Controlled Fusion Program. Russell's family was by his side, though their grief was doubled by the unexpected death of Jean Hetherington within twelve hours of Russell's in the Princeton hospital. Watching the procession, Shapley could think only of the past, of the countless hours he had spent with his mentor in the ancient book-lined study surrounded by antique grandfather clocks and portraits of the Norris family. Russell would chatter incessantly, of course. Ideas came thick and fast. But what Shapley wanted to remember that day was how easily Russell could lose himself in his own thoughts, his mind leaping to conclusions as he gazed beyond the walls of the room, far out into space, thinly connected to Earth by that symbol of efficiency and approximation, a "10-inch slide rule in his hands."[21]

• ABBREVIATIONS AND ARCHIVAL SOURCES •

Abbreviations for Journals, Societies, Frequently Cited Sources

Abridged Record: "Abridged Record of Family Traits," CIW Eugenics Records Office and the Eugenics Society of the United States of America, attached to Paul Brockett to Members of the National Academy, 22 May 1935. Reel 11:9543–9550, PUL/HNR.
AJ: *Astronomical Journal*
AN: *Astronomische Nachrichten*
ApJ: *Astrophysical Journal*
ARAA: *Annual Reviews of Astronomy and Astrophysics*
BAN: *Bulletin of the Astronomical Institutes of the Netherlands*
BMNAS: *Biographical Memoirs of the National Academy of Sciences*
DSB: *Dictionary of Scientific Biography* (Scribner's)
GHA: *General History of Astronomy*
HNRjr: Henry Norris Russell, jr. oral history, 2 November 1977. AIP
HSPS: *Historical Studies in the Physical Sciences* (renamed *Historical Studies in the Physical and Biological Sciences*)
IAU: International Astronomical Union
JBAA: *Journal of the British Astronomical Association*
JHA: *Journal for the History of Astronomy*
JRASC: *Journal of the Royal Astronomical Society of Canada*
MN: *Monthly Notices of the Royal Astronomical Society*
MRE #1: Margaret Russell Edmondson oral history #1, 21 April 1977. AIP
MRE #2: Margaret Russell Edmondson oral history #2, 25 June 1993. NASM
NASM: National Air and Space Museum
Obs: *Observatory*
OHI: Oral history
PAAAS: *Proceedings of the American Association for the Advancement of Science*
PAAS: *Publications of the American Astronomical Society*
PAASA: *Publications of the Astronomical and Astrophysical Society of America*
PASP: *Publications of the Astronomical Society of the Pacific*
PhysRev: *Physical Review*
PhysTod: *Physics Today*
PIEEE: *Proceedings of the Institute of Electrical and Electronics Engineers*
PNAS: *Proceedings of the National Academy of Sciences*
PopAst: *Popular Astronomy*
Proc. Am. Phil. Soc.: *Proceedings of the American Philosophical Society*
QJRAS: *Quarterly Journal of the Royal Astronomical Society*
RDS 1: Russell, Dugan, and Stewart, 1926
RDS 2: Russell, Dugan, and Stewart, 1927

SciAm: *Scientific American*
Spiritual Autobiography: "Henry Norris Russell," in Louis Finkelstein, ed. (1953)
ZAstrop: *Zeitschrift für Astrophysik*
ZPhys: *Zeitschrift für Physik*

Archival Sources and Their Abbreviations

AIP: American Institute of Physics Center for History of Physics: Theodore Dunham Jr. Papers, microfilm edition (AIP/THD); Karl Schwarzschild Papers, microfilm edition (AIP/KSM); Frank Schlesinger Papers, microfilm edition (UP/FS; Yale/FS); William F. Meggers Papers (AIP/WFM); Otto Struve Papers, microfilm edition (AIP/OS/YOL); Robert d'E. Atkinson Papers, microfilm edition (AIP/RA). C. A. Young Papers, microfilm edition (AIP/DCA/CAY).

AJM: Papers of Norman Lockyer, in the hands of A. J. Meadows, subsequently deposited at the University of Sussex Library (AJM/NJL).

BANC: Bancroft Library, University of California, Berkeley: R. T. Birge Papers (BANC/RTB); A. O. Leuschner Papers (BANC/AOL); Department of Astronomy Records CU-25 (BANC/DOA); President's Files CU-5 (BANC/UCP).

BL: Bodleian Library, Oxford University. E. A. Milne Papers 1896–1950. MSS. Eng. misc. b.423–9, C.870–2,d.1277. (BL/EAM).

CIT: California Institute of Technology Archives: George Ellery Hale Papers, microfilm edition (CIT/GEH).

CIWA: Carnegie Institution of Washington Archives. G. H. Darwin file (CIWA/GHD); H. N. Russell file (CIWA/HNR); Forest Ray Moulton file (CIWA/FRM).

CL: Francis A. Countway Library of Medicine, Harvard Medical School, Papers of Stanley Cobb. HMS c53 (CL/SC).

CUOL: Cambridge University Observatory Library. A. R. Hinks Letterpress books; R. S. Ball scrapbooks.

DCA: Dartmouth College Archives: C. A. Young Papers, Microfilm Edition, American Institute of Physics (AIP/DCA/CAY).

DO: Dudley Observatory Archives: Sebastian Albrecht Papers (DO/SA).

FPCA: First Presbyterian Church of Oyster Bay Archives.

HL: Huntington Library, Manuscripts Division: Walter Sydney Adams Papers (HL/WSA); George Ellery Hale Papers (HL/GEH); F. H. Seares Papers (HL/FHS); Charles E. St. John Papers (HL/CESJ); Paul Willard Merrill Papers (HL/PWM); A. H. Joy Papers (HL/AHJ).

HUA: Harvard University Archives: Harlow Shapley Director's Correspondence, UA V.630.xx (HUA/HSDC); Personal correspondence HUG 42 4773.xx (HUG/HSPC); E. C. Pickering Director's Correspondence, UA V.630.17.7 (HUA/ECP); A. Lawrence Lowell Papers, UA I 5.160 (HUA/ALL).

LAUC: Mary Lea Shane Archives of the Lick Observatory, McHenry Library, Santa Cruz: W. W. Campbell Papers (LAUC/WWC), R. G. Aitken Papers (LAUC/RGA), W. H. Wright Papers (LAUC/WHW); C. D. Shane Papers (LAUC/CDS); Gerald E. Kron Papers (LAUC/GEK); R. T. Crawford Papers (LAUC/RTC).

LC: Library of Congress Manuscripts Division: Simon Newcomb Papers (LC/SN).

LowA: Lowell Observatory Archives. Lowell Papers microfilm edition (LowA/LP), V. M. Slipher Papers (LowA/VMS), C. O. Lampland Papers (LowA/COL), Roger Lowell Putnam Papers (LowA/RLP).

NBS: National Bureau of Standards Archives. Charlotte Moore Sitterly Papers (NBS/CMS).
NYAS: New York Academy of Sciences, A. Cressy Morrison Prize files (NYAS/ACM).
PTS: Princeton Theological Seminary. Archives (PTS/A); Alumni Records (PTS/R).
PUA: Princeton University Archives, Mudd Library. Records of astronomy department (PUA/A).
PUL: Princeton University Library Manuscripts Division. Henry Norris Russell Papers (PUL/HNR); Raymond Smith Dugan Papers (PUL/RSD); Lyman Spitzer, Jr., Papers (PUL/LS); John Q. Stewart Papers (PUL/JQS); Charles A. Young Papers (PUL/CAY).
RUC: Regenstein Library, University of Chicago. S. Chandrasekhar Papers (RUC/SC); Robert Maynard Hutchins Papers (RUC/RMH).
RUM: John Rylands Library, University of Manchester, Zden;akek Kopal Papers (RUM/ZK).
SLRC: Schlesinger Library, Radcliffe College. Margaret Harwood papers (SLRC/MH).
SUL: Stanford University Libraries, Deptartment of Special Collections. David Locke Webster Papers (SUL/DLW).
UA: University of Arizona Archives. Gerard P. Kuiper Papers (UA/GK).
UA/HER: Hertzsprung Papers, Microfiche edition. University of Aarhus.
UCSD: University of California, San Diego. Harold Urey Papers (UCSD/HL).
UMBL: University of Michigan Bentley Historical Library. Horace H. Rackham School of Graduate Studies (UMBL/HHR); R. H. Curtiss Papers (UMBL/RHC); H. D. Curtis Papers (UMBL/HDC); Department of Physics Papers (UMBL/PP).
UP: University of Pittsburgh. Frank Schlesinger Papers (UP/FS).
UT: University of Toronto Archives.
Yale: Yale University Library, Manuscripts Division. Frank Schlesinger Papers (Yale/FS).
YOL: Yerkes Observatory Library. Edwin Brant Frost Papers (YOL/EBF); Otto Struve Papers (YOL/OS); Struve Papers, AIP Microfilm Edition (AIP/OS/YOL).

Oral Histories

Adel, Arthur (with Mrs. Adel). 13 August 1987. (Robert W. Smith) AIP.
Aller, Lawrence H. 18 August 1979. AIP.
Atkinson, Robert d'Esourt. 22 April 1977. AIP.
Bok, Bart J. 17 May 1978; 14 June 1978. AIP.
Bowen, Ira S. 9 August 1968; 26 August 1969. (Charles Weiner) AIP.
Chandrasekhar, Subrahmanyan. 17 May 1977. (Spencer Weart) AIP.
Cowling, Thomas G. 22 March 1978. AIP.
Dunham, Theodore, Jr. 30 April 1977. AIP.
Edmondson, Frank Kelley. 21 April 1977. AIP.
Edmondson, Margaret Russell. 21 April 1977. AIP. [MRE #1].
Edmondson, Margaret Russell. 25 June 1993. NASM. [MRE #2].
Green, Louis C. 14 September; 18 October 1991. NASM.
Greenstein, Jesse. 7 April 1977. (Spencer Weart) AIP.
Herget, Paul. 19–20 April 1977. AIP.
King, Robert B. 8 March 1992. AIP/NASM.
Kron, Gerald E. 20 May 1978. AIP.
Locanthi, Dorothy Davis. 3 August 1977. AIP.
Marshak, Robert. 1975 (Charles Weiner). AIP.
Menzel, Donald H., Manuscript Autobiography, September 1974. AIP.

Russell, Henry Norris. 1951 (James Cuffey), courtesy Frank Edmondson.
Russell, Henry Norris, jr. 2 November 1977. AIP. [HNRjr].
Schwarzschild, Martin, session 1. 30 July 1975 (Spencer Weart). AIP.
Schwarzschild, Martin, session 2. 3 June 1977 (Spencer Weart). AIP.
Schwarzschild, Martin, session 3. 16 December 1977. (Spencer Weart and DeVorkin) AIP.
Schwarzschild, Martin. 18 June 1982. NASM.
Schwarzschild, Martin. 26 August 1991. NASM/AIP.
Shane, Charles Donald. 14 June 1978. AIP.
Shapley, Harlow. 1966. (Charles Weiner and Helen Wright) AIP.
Shenstone, Alan. Manuscript autobiography. AIP.
Shenstone, Alan. 20 March 1979. AIP.
Sitterly, Charlotte E. Moore. 15 June 1978. AIP.
Spitzer, Lyman, jr. 8 April 1977; 10 May 1978. AIP.
Spitzer, Lyman, jr. 27 November 1991. NASM.
Unsöld, Albrecht. 6 June 1978 (Owen Gingerich). AIP.
Van de Kamp, Peter, 9 April 1977; 18 March 1979. AIP.
Walker, Richard Lee. 13 January 1997. AIP/NASM.
Webster, David L. 21 May 1964 (W. J. King) AIP.

• NOTES •

Preface

1. On the use of "fine structure" in assessing scientific change, see Holmes, 1981, and Shortland and Yeo, 1996, 5. These remarks on the importance of examining the individual life to reveal larger issues in history have been inspired by Söderqvist, 1996, 47, and Carson and Schweber, 1994, 286.
2. Herschel, 1879, 77; 79.
3. Gill, 1900, 477.
4. Brock, 1969; Meadows, 1972; McGucken, 1969.
5. Servos, 1990; Gavroglu and Simoes, 1994.
6. Smyth, 1880; Young, 1884, 199–202.
7. There is a growing literature on the study of the observatory as a hierarchical factory. See Smith, 1991; Lankford and Slavings, 1996; Lankford, 1997. A sense of the nature of the community and its sources of support at the end of the nineteenth century can be gathered from numerous sources, including Miller, 1970; Lankford, 1981; 1997; Lankford and Slavings, 1996; Rothenberg, 1981; Moyer, 1992; Doel, 1996; Hufbauer, 1991; Osterbrock, 1984; Plotkin, 1978; Butler, 1993; Wright, 1966; and DeVorkin, 1999.
8. Quotes from Boltzmann, 1992; Struve, 1943, 477. See also Merrill, 1923.
9. Hankins, 1979, 5; 14; and Kragh, 1987, 171. Scientific "culture" embodies a framework made possible by the existence or creation of an agreed-upon set of tools. Pickering, 1992, 2.
10. "tissue," Russell, 1938. "Address of Retiring President," 112–113. On late-nineteenth-century opinions regarding the importance of positional astronomy, the "astronomy of precision," see: Gill, 1900, 475; Eastman, 1892, 29–31. Söderqvist, 1996, 66–67, argues for understanding knowledge production in terms of the scientist's struggle for self-assertion. My sense of the importance of trust in consensus formation comes from Shapin, 1994.
11. Struve to Compton, 22 November 1937. AIP/OS/YOL. Crawford, Heilbron, and Ullrich, 1987. Citation analyses from Small, 1981.
12. On Pickering: Jones and Boyd, 1971; Plotkin, 1978; 1990. On Hale: Wright, 1966; Osterbrock, 1993; 1997a.
13. Browne, 1995, xiii; Dupree, 1959, 42–43; 391–392.
14. Russell to Shapley, 31 January 1920. PUL/HNR. The role of personality is noted by Camerini, 1997, 311.
15. MRE #2, 9. Margaret Edmondson Olson to the author.

Acknowledgments

1. Philip and DeVorkin, 1977.
2. "There were bushels, literally. We kept finding more and more and more." MRE #2, 10. See also MRE #1, 25; 51; 54.

Chapter 1
Religious Heritage

1. Russell to John D. Fitzgerald, 19 November 1934. PUL/HNR.

2. Russell to Vernon R. Loucks, 1 May 1941. PUL/HNR.

3. On spiritual development and extracurricular schemes, see Reuben, 1990, 183; 249–250.

4. Quoted from Schweber, 1993, 39. Reuben, 1990, 320–322; Moyer, 1992. Browne, 1995, 51–52.

5. Russell, "Comments for Conference on Science, Philosophy and Religion, August 1942," 6. Box 111.8, PUL/HNR

6. Kargon, 1982; Christianson, 1995, 41.

7. Visher, 1947, 533–535; Holton, 1978, 30.

8. Hoeveler, 1981, 50; Veysey, 1965, 42. See also Bozeman, 1977, chap. 1; Hovencamp, 1978, chap. 1; Reuben, 1990, 18–21; Moyer, 1992, 135–137; 1997, 49–51; 141; 183–186.

9. "James McCosh," in Leitch, 1978, 301; White, 1896, noted McCosh's secularization of the college. Hoeveler, 1981, 100; Livingstone, 1992, 423.

10. Numbers, 1977, 86–87; Moyer, 1992, 29; Bozeman, 1977, 43; 161. Weinstein, 1981, 3–4.

11. Numbers, 1977, 37. Hodge's mission was to show how Darwinism was antithetical to theological doctrine. Livingstone, 1992, 410.

12. McCosh, 1890, 69.

13. Moyer, 1992, 29; 129–130, quoting Frank M. Turner, "Public Science in Britain, 1880–1919," 591.

14. Quote from Bruce, 1987, 67–69; 67. On Henry's rejection of pure empiricism, see Reingold, 1972; Moyer, 1992, 29. On McCosh's views, see Bozeman, 1977, 102; and, among others he cites, Dupree, 1957, 82; and Rossiter, 1971, 624.

15. Bragdon, 1967, 19.

16. Russell, 1953 (hereafter Spiritual Autobiography), 31. "Russell, A.G." Miscellaneous Family File Information, Colchester Historical Society Records, Truro, Nova Scotia.

17. "Alexander Gatherer Russell," *Necrological Report, presented to the Alumni Association of Princeton Theological Seminary*, 6 May 1912, 176–177. "Alumni of Princeton Theological Seminary," 1877 entry worksheets for A. G. Russell, PTS Library.

18. He was one of the first to introduce the process of rubber vulcanization. "Family Group Record-12 Documentation," 3–5, courtesy Margaret Edmondson Olson. Norris's first wife died in childbirth. MRE #1, 5–6. On the family's Puritan background, see MRE #1, 43–44.

19. H. N. Russell, "Biographical Data" file, statement ca. 1945. National Academy of Sciences archives; "Family Group Record-12 Documentation," 5. "Family Information" courtesy Margaret Edmondson Olson.

20. Russell, 1935a (hereafter Abridged Record); Spiritual Autobiography, 31; MRE #1, 5–8; Perry, 1944, 84; Loetscher, 1967, 39–44; conversations with William O. Harris.

21. Congregational Minutes. Trustees Minutes of the First Presbyterian Church of Oyster Bay, New York, June 1, 1876. Entry for 6 October 1876, Register of Communicants, Oyster Bay, NY, First Presbyterian Church Records, #1845. Presbyterian Historical Society, MF POS. 327.

22. John Cox, Jr., ed., comp., Frederick E. Willits, Daniel Underhill, Edward T. Payne, eds., 1940. *Oyster Bay Town Records volume 8* (1795–1878) (New York: Country Life Press Corp., 273), 373. Johnston and Reifsnyder, 1990, 6–13; 65–66; McCullough, 1981, 109; 141–142; *Encyclopedia Britannica*, 11th ed., s.v. "Oyster Bay."

23. Johnston and Reifsnyder, 1990, 55–57.

24. Financial Records "First Presbyterian Church Oyster Bay N.Y." entries for 16 and 22 May 1882. FPCA.

25. Sulloway, 1996, has continued the long-standing debate over the influence of birth order. "Oyster Bay," *East Norwich Enterprise*, 20 August 1881. Reel #1, 11 September 1880–20 December 1884. W. L. Swan, "Organ and Choir" May 1882–January 1884," records, entry for 28 January 1883. FPCA.

26. William L. Swan, "Address," *In Memoriam, Reverend Alexander Gatherer Russell.* 19 November 1911, Sunday evening service pamphlet. FPCA. F. Irvin, n.d., *Oyster Bay in History*, manuscript draft, 135. Oyster Bay Library. "Oyster Bay," *East Norwich Enterprise*, 27 April 1889. Microfilm Edition Roll #2 1884–1890. Nassau County Office of Administrative Services Record Services Division.

27. G. R. Van der Water, "Address," *In Memoriam, Reverend Alexander Gatherer Russell.* 24 November 1911, Friday evening memorial service pamphlet. FPCA.

28. Johnston and Reifsnyder, 1990, 74. Selden, 1992, 7.

29. "Infants may be Doomed, So Argues one Presbyter of Nassau," fragment, probably from the *Long Island Farmer*, 12 February 1890. A. G. Russell scrapbook. The Presbytery of Nassau, Synod of New York. Quoted in Johnston and Reifsnyder, 1990, 75.

30. Hoeveler, 1981, 330–331.

31. Johnston and Reifsnyder, 1990, 63–67.

32. Link, 1971, 6; Nordholt, 1991, 10. Sulloway, 1996, xiv.

33. "Mrs. Eliza H. Russell," *East Norwich Enterprise*, 3 October 1903. The Women's Home and Foreign Missionary Society of the Presbytery of Nassau was probably her largest official responsibility. By 1885 she was a member of the Executive Committee of the Board, and also secretary of the affiliate from Oyster Bay. "First Annual Report of the Women's Home and Foreign Missionary Society of the Presbytery of Nassau, N.Y.," Jamaica, LI: 1885. A. G. Russell, "Scrapbook," The Presbytery of Nassau, Synod of New York.

34. Spiritual Autobiography, 32.

35. "Saving Souls," 1886 fragment from the *Long Island Farmer*, in A. G. Russell, "Scrapbook," The Presbytery of Nassau, Synod of New York.

36. McCullough, 1981, 162.

37. See chap. 2, below. MRE #1, 4; 19; #2, 14. Anglim to Russell, n.d. ca. February 1921. Incoming file, Russell family, PUL/HNR. The Friends Academy is noted in Johnston and Reifsnyder, 1990, 59. Financial Records, Friends' Academy, courtesy Mary Anne Rearden, academy archivist. See also *Fourteenth Annual Catalogue of Friends' Academy* (New York: 1890).

38. The phonograph demonstration was noted in "Oyster Bay," *East Norwich Enterprise*, 7 July 1889.

39. Russell to Harry H. Henderson, 1 November 1921. PUL/HNR.

40. Excerpts from "Abridged Record of Family Traits." Johnston and Reifsnyder, 1990, 56.

41. Russell, "Biographical Data." National Academy of Sciences, Russell file, 1945.

42. The *East Norwich Enterprise* reported that the locals were all in the streets with shards of smoked glass. DeVorkin, 1982, 524.

43. W. L. Swan, "Organ and Choir," May 1882, 63, entry for 3 December. FPCA.

44. W. L. Swan, "Organ and Church Records, January 1889–March 1892," entries for 15 November 1891; 9 February 1890. FPCA.

45. W. L. Swan, "Organ and Church Records, January 1889–March 1892," entries for 9 February 1890; 4 January; 1 February 1891. FPCA.

46. Spiritual Autobiography, 32. A. G. Russell, "Diary of a Trip to Europe," entry for 26 June, 1892. Courtesy Margaret Edmondson Olson.

47. Russell, 1927a, 12. On Channing, see Dillenberger, 1988, 224–225.

48. Spiritual Autobiography, 32–33; MacDonald, 1989, 17.

49. Sulloway, 1996, 21; 69, argues that firstborn males identify most with their fathers.

50. A. G. Russell, Diary, 25th June 1892 entry.

51. Diary, 29th June 1892.

52. Diary, 2 July entry, 30; 39. From his description, it is not clear whether he is writing for himself or is representing the family group. Most of his references are first-person singular, but now and then he uses first-person plural. Quote from Diary entry, 41. Spiritual Autobiography, 32.

53. Diary, 74.

54. Diary, 12 July.

55. Diary, 21 July entry, 131–132; 22 July entry, 137.

56. Diary, 22 July.

57. Diary, 24 July, 149.

58. Diary, 5 August, 204.

59. Diary, 18 August, 276–277. Again, the penchant of firstborn males to be concerned with authority and custom has been noted by Sulloway, 1996, 21; 69.

60. Diary, 19–20 August, 285; 287.

61. Diary, 23 August, 302. Pastor Russell thus dismissed the destitute pilgrims headed for the agricultural colonies in Argentina and in the United States created by Baron Maurice de Hirsch.

62. Diary, 28 August, 310; 1 September, 316.

63. See Browne, 1995, 23–27, for an appreciation of the detachment created by removal to a boarding school.

64. Comparisons from Weinstein, 1981, 14.

65. Lears, 1981, 5.

66. Hevly, 1996, 82; 85, following Lears, 1981.

67. Quoted in Nordholt, 1991, 10.

68. Sulloway, 1996, 70, notes these traits as typical of firstborn males.

69. Weinstein, 1981, 8.

70. MRE #1; HNRjr, 8–9; 11–12; esp. 12, noted that living at 79 Alexander placed him within his mother's world. Theodore Cuyler, a contemporary to Alexander Russell, described the overpowering obligations of a pastor that took him beyond his family. Cuyler, n.d, 9. See also Nordholt, 1991, 10–11; Walworth, 1965, 6; Sulloway, 1996.

71. Entry for 1 October 1893, Register of Communicants, Oyster Bay, NY, First Presbyterian Church Records, #1845. Presbyterian Historical Society, MF POS. 327.

72. Spiritual Autobiography, 32; Reuben, 1990, 57.

73. Russell, 1927a, 11. Sulloway, 1996, 69.

74. Russell, 1927a, 13.

75. Ibid., vii–viii; 9–10; 23. In the 1890s, as in the 1920s, the tensions between science and religion were only part of a larger cultural debate over theological viewpoint, church organization, and policy. Longfield, 1991, 4–5; 8–9.

76. Eddington, 1932, 141.

77. Russell, 1943b, 250. Russell to Finkelstein, 13 November 1942. PUL/HNR.

78. Russell to James B. Pratt, 9 October 1941. PUL/HNR.

79. On Russell's arguments to abolish compulsory chapel, see Russell to J. G. Hibben, 27 March 1927; and Russell to the Reverend Dr. H. W. Jacobus, 27 September 1927, Reel 8, frame 6696, PUL/HNR. On his becoming church elder, see Spiritual Autobiography, 37. On abolishing compulsory chapel, see Reuben, 1990, 262–263.

80. Russell to Finkelstein, October 1943. PUL/HNR.

81. Russell, 1927a, 17.

82. Russell to Finkelstein, 10 November 1945. PUL/HNR.

83. Minutes of the University Faculty, 3 January 1889–6 February 1895, 483. PUA.

84. Minutes of the Trustees of Princeton University, Volume 8, entry date 14 June 1897. PUA. Hoeveler, 1981, 332.

85. Under Patton, enrollment more than doubled in the 1890s, to eight hundred undergraduates in 1898. Bragdon, 1967, 202; Veysey, 1965, 41. On Patton's administration, see also Weinstein, 1981, 108–109.

86. Livingstone, 1992, 423–424; Numbers, 1977, 106.

87. Young to E. C. Pickering, 21 July 1885. HUA/ECP. Egbert, 1947, 121; Numbers, 1977, 106.

88. Murray, 1935, 345.

89. Young, "God's Glory in the Heavens," holograph copy, 3; 6. Young Papers, microfilm edition, R3. AIP/DCA/CAY. The lecture was held before August 1895, since it was recalled in Fred L. Kingsbury to C. A. Young, 6 August 1895. AIP/DCA/CAY.

90. Young, "God's Glory in the Heavens," 30–34. Young's evening talk was remembered thirty years later by a Princeton alumnus. Russell to William M. Carle, 7 December 1925. PUL/HNR.

91. Bozeman, 1977, 161.

92. Young, "God's Glory in the Heavens," 45.

93. Bragdon, 1967, 20.

94. Young, 1890, 3–4. He repeated these sentiments almost verbatim in his 1902 *Manual*, 5.

95. Young, "God's Glory in the Heavens," 44.

96. Moyer, 1992; Young, 1891, 313. See especially Watson, 1892, 8; 557.

97. Link, 1971, chap. 1; Walworth, 1965, 6; Weinstein, 1981, 5–6. On pluralism, see Hollinger, 1996.

98. "Scientific Approach to Christianity I." Box 72.1 (a), PUL/HNR; North, 1995, 378–379.

99. Spiritual Autobiography, 31; "Scientific Approach to Christianity I." folder dated 1916–1917. Box 72.1 (a) PUL/HNR; Russell, 1927a; Reuben, 1990, 181–183; 249–253.

100. Russell, typescript review of A. S. Eddington, "The Nature of the Physical World." Russell to Editor, *Yale Review*, 29 March 1929. Box 106.11, PUL/HNR. See also Russell, "Determinism and Responsibility," typescript, June 23, 1942, and galley. "A Paper presented at the Third Conference on Science, Philosophy and Religion, New York, August 28, 1942." Box 111.7, PUL/HNR.

101. Russell, 1927a, 89.

102. Ibid., 107–108.
103. Dillenberger and Welch, 1954, 213.
104. Spiritual Autobiography, 41.
105. Russell, 1927a, 7.

Chapter 2
Russell at Princeton

1. There was substantial training in mathematics and in various scientific subjects. A successful holder of its certificate typically entered Cornell, Swarthmore, or Vassar. Melnick, 1988, 87. On the curriculum, see *Fourteenth Annual Catalogue of Friends' Academy* (New York, 1890).

2. Melnick, 1988, 86.

3. Spiritual Autobiography, 33; MRE #1, 20; #2; Johnston and Reifsnyder, 1990, 59; *Fourteenth Annual Catalogue of Friends' Academy* (New York, 1890), 16.

4. Bragdon, 1967, 219.

5. *Polk's Princeton Directory 1887–1896*, 1896 edition. Princeton Historical Society Records, microfilm. "Venerable Stargazer . . . ," *Princetonian*, 19 March 1937. See also M. Halsey Thomas, "Princeton in 1874: A Bird's Eye View," *Princeton History* 1, no. 1 (1971): 72–78.

6. Bowman, 1981, 7. Ada and one of her brothers were living in Princeton. Henry Lee Norris had died in 1881 from Bright's disease, and Maria Norris died in November 1889 from injuries resulting from a fall down the steps of an elevated railway in New York City. H. N. Russell, "Biographical Data" file, statement ca. 1945. National Academy of Sciences archives; "Family Group Record-12 Documentation," 5, courtesy Margaret Edmondson Olson, "Family Information." Ada Norris's piano playing is mentioned in MRE #2, 29; her society memberships, MRE #1, 43, verified by the Daughters of the American Revolution.

7. (H. F. Lee, et al., Bond and Building Chairman). "Princeton Preparatory School, Princeton," 1923. 372.2 PRI, Princeton Historical Society Records, PUA.

8. MRE #1, 24.

9. "Catalogue of the College of New Jersey." Academic year 1893–1894, 22–26.

10. "Entrance to Freshman Class—Academic" and "College of New Jersey, Preliminary Examination." Russell student records, PUA.

11. Woodrow Wilson's tuition was waived in similar fashion. Walworth, 1965, 17. On college costs, see "Catalogue of the College of New Jersey." Academic year 1893–1894, 152; 154–155. PUA.

12. Hoeveler, 1981, x; quote from 343; Bragdon, 1967, chap. 11.

13. Johnston to Young, 12 July 1887. AIP/DCA/CAY. Bragdon also makes this observation.

14. Bragdon, 1967, 15–18; 30–31; 203–204. Hoeveler, 1981, 322–332.

15. Spiritual Autobiography, 33; *Catalogue of the College of New Jersey, 1893–1894*, 147–148. PUA. Hoeveler, 1981, 322, and Bragdon, 1967, 16–17, note student unrest in the 1870s, and the Trustees Minutes reveal that it remained a problem through the 1890s.

16. Minutes of the Trustees of Princeton University, Volume 8, June 1894–October 1898, "President's Report," 11 June 1894, 3; 8 November 1894, 60. On the lack of standards under Patton, see Bragdon, 1967, 203; 225; and on weaknesses in the School of Science, 292. On the rise of scientific schools, see Bruce, 1987, 328.

17. "Grade Books, Undergraduate, 1894–1897." Minutes of the University Faculty, 3 January 1889–6 February 1895, 531; 537. PUA.

18. Minutes of the Trustees of Princeton University, Volume 8, June 1894–October 1898: "President's Report," 11 June 1894. PUA.

19. Shenstone OHI, 1979, 10.

20. H. N. Russell, "Henry Burchard Fine," *Princeton Alumni Weekly*, 21 January 1926. Typescript. File 104.14, PUL/HNR. Roberts, 1998.

21. "Integral Calculus 1895 /Sp./Fall." Box 69.2, PUL/HNR.

22. H. B. Fine faculty card, PUA. Aspray, 1988, 346–366; 347. H. N. Russell, "Henry Burchard Fine," *Princeton Alumni Weekly*, 21 January 1926. Typescript, B104.14, PUL/HNR. McCosh hoped that Fine would lead Princeton into the realm of the "new mathematics." James McCosh to H. B. Fine, 16 February 1885. Fine faculty folder, PUA.

23. Allan Shenstone, "Physics," in Leitch, 1978, 363. Fine adminstered the exam in algebra, and Thompson those in arithemtic and geometry. "Entrance to Freshman Class—Academic" and "College of New Jersey, Preliminary Examination." Russell student records, PUA. Quote from Luther P. Eisenhart, "The Man on the Cover," *Princeton Alumni Weekly* 29 1. PUL.

24. "Catalogue of the College of New Jersey at Princeton, 1893–1894," 1893, 50. "Sophomore Disputation," Box 69.1, PUL/HNR.

25. "Questions for the Nassau Herald Class of '97." Class of 1897 Class Boxes, Box II. PUA.

26. "Astronomy, Prof Young, H. N. Russell, '97." Student notes, fall 1895. Box 69.3, PUL/HNR

27. Young, 1890, 3–4.

28. "Astronomy, Prof Young, H. N. Russell, '97." Student notes, fall 1895. Box 69.3, PUL/HNR.

29. Warner, 1994, 11–12. See also Holton, in MacKay, 1965, 53; and Numbers, 1977, 77–80.

30. Russell lecture notes; quote from Murray, 1935, 345. "Class of 1897" Class Box II, PUA. The term "Twinkle" came into general use. See V. Lansing Collins to Russell, 8 February 1932, HNR faculty folder, PUA; Russell to Miss Young, 24 April 1927. PUL/HNR. Wertenbaker, 1946, 366; Norris, 1917, 206; Egbert, 1947, 126.

31. Russell, 1954, 100.

32. Ibid., 101. Minutes of the Trustees of Princeton University, Volume 8, 14 June 1897. PUA.

33. "Astronomy, Prof Young, H. N. Russell, '97." Student notes, fall 1895. Box 69.3, PUL/HNR.

34. Young, 1884, 26; 292.

35. Young Faculty Folder, PUA.

36. Meadows, 1972, 60; 68; Hufbauer, 1991, 62–65. "Young, Charles Augustus," *DSB 19*, 1971, 557–558.

37. "Minutes of the College of New Jersey [Trustees' Minutes Volume 5, 1868–1878]," Report of the Committee on the Curriculum, February 8, 1877, 557. PUA.

38. Minutes of the College of New Jersey [Trustees' Minutes Volume 5, 1868–1878], 18–19 June 1877, 590; 593; 627. PUA. Wertenbaker, 1946, 282–283. Williams, 1905, 49, notes that the Halsted building was completed in 1872 but lay empty and unused for over a decade. Egbert, 1947, 212–213, adds that after Young was hired, Halsted mounted a general subscription campaign that yielded $32,000.

39. On Young as a "founder of astrophysics in the United States," see Lankford, 1997, 50; and as the "leading solar observer in the United States" in the 1870s, see Meadows, 1972, 60.

40. Clerke, 1903, 9–10.

41. Clerke, 1887, 198; 339–340; Young, 1885; 1886; 1893.

42. On the opportunism of Young's contemporaries, see Becker, 1993; and Meadows, 1972. Young's health may have been a limiting factor. See Frost, 1910, 92. Russell, 1954, 100.

43. Hentschel, 1993; Hufbauer, 1991, 67; Warner, 1986.

44. Box 2, Data Books, Department of Astronomy Records, PUA. On Young's respect for authority, see Crew to Young, 10 December 1891; 8 April; 8 June 1892; Young to Crew: 5 July 1892. AIP/DCA/CAY; Henry Crew student folder, PUA. On Crew's fate, see Osterbrock, 1984, 106.

45. Young, 1891, 320.

46. Frost, 1910, 101. Young, 1892, 296; Warner, 1994.

47. What impressed Hall was the spectroscopic equipment, laboratory exercises based on it, and access to it by undergraduates. Hall, 1900, 18; "Western University of Pennsylvania," *PopAst* 2 (June 1895), advertisement section. Holden, 1898, 880. Osterbrock, 1984, 141.

48. Hall admired the German system, which incorporated formal study in the context of observatory work. Hall, 1900, 18. Struve, 1943, made much the same observations.

49. But even as a teaching observatory, Princeton was far less productive than Hall's Detroit Observatory, which by the mid-1890s had graduated some two dozen astronomers who went on to establish similarly active programs in Pennsylvania, Berkeley, and Wisconsin. Holden, 1898, 881. Hussey, 1912; Osterbrock, 1990, 95–115. Through 1943, Berkeley topped the list, followed by Michigan and Harvard. Visher, 1947, 171.

50. Holden, 1898, 882–888; Howe, 1901, 173. Among the fifteen colleges Holden canvassed, all of them used *General Astronomy* or *Elements*. Some 130,000 copies of Young's texts were printed. For a breakdown, see Young Faculty Folder, General Catalogue entry notes, January 1908. PUA.

51. Young to F. A. Hill, 29 January 1889. Princeton University Observatory historical files.

52. Young, 1884, 1–2; 10.

53. Ibid., 20. Young was not concerned with "wastefulness against nature" and regarded such worries as manifestations of "ignorance." Ibid., 21.

54. Ibid., 15.

55. Ibid., 15; 19. On the early use of radiation laws to determine the temperature of the Sun, see Kidwell, 1981; and Clerke, 1887, 263–266.

56. Frost, 1910, 108–114.

57. "Questions for the Nassau Herald Class of '97," Class notes, Box II. PUA.

58. Lockyer called the line D_3 following Fraunhofer's system. See Meadows, 1972, 58–61.

59. Young, 1890, 3–4. On Newcomb and scientism, and the context surrounding his own proselytizing, see Moyer, 1992.

60. Russell, "Astronomy" 1895. Box 69.3, PUL/HNR.

61. Ibid.

62. Todd, 1897.

63. Reuben, 1990, 26–27.

64. Fragment, Class scribe, 26 January 1940 [date uncertain]. Russell faculty folder, press clipping file. PUA.

65. Russell, "Student Notes 1896–1897—Theory of Functions—lectures of H. B. Fine," Box 69.7; Computation book, Box 69.9, PUL/HNR. On the tradition of dictation

at Princeton, see Bragdon, 1967, 206. E. L. Crain and S. J. Shirk, "Venerable Star-Gazer Peers at Nassau's Past; Believes Present-Day Changes 'Not for Worse,'" *Daily Princetonian*, 19 March 1937. Fragment, Russell Faculty Folder, PUA; Russell, "Biographical Data." National Academy of Sciences, Russell file, 1945. "Class of 1897" Class Boxes, Box II, "Questions for the Nassau Herald Class of '97." PUA.

66. See "Longitude, Washington and Princeton. Observations at Washington," 23 August 1880. Princeton University Observatory historial files.

67. "Book II." Box 69.4; 69.5, PUL/HNR.

68. Minutes of the University Faculty February 1895–June 1905, entry for 21 September, 1898. PUA. Reed left Princeton in 1901 to work as an electrical engineer at General Electric in Schenectady. He ultimately became president of the Milroy Banking Company of Pennsylvania. Taylor Reed faculty folder, and faculty card. PUA.

69. Russell, Student Notes, "Theoretical Astronomy—lectures of C. A. Young, 1896–1897." January 19, 1897 entry, 69.6, PUL/HNR. These elements of Darwinism have been discussed by Bowler, 1985, 643.

70. Russell student lecture notes for Young's "Theoretical Astronomy," 10. Box 69.6, PUL/HNR. See also Murray, 1935, 345; Hoeveler, 1981, 332.

71. Ibid., Russell Student Notes, "—lectures of C. A. Young, 1896–1897." January 19, 1897 entry. Box 69.6, PUL/HNR.

72. H. N. Russell, "Astronomy, Prof Young." Student notes, fall 1895. Box 69.3, PUL/HNR. See also Young, 1888, 224–225.

73. Computation book. Box 69.8, PUL/HNR.

74. Russell, "Visual Observations of Star Spectra, Thesis in Practical Astronomy," Princeton University, June 10, 1897. Box 25, PUL/HNR. In 1954 Russell reminisced that his senior thesis was his first practical experience in stellar spectroscopy. Russell, 1954, 99.

75. The Faculty Minutes (9 June 1897) contend that Russell was the only one to receive this designation. There is much correspondence over this issue with the Office of the Secretary of the University. Russell faculty folder, PUA.

76. See "Honors for General Excellence," Russell faculty folder, 19 November 1936; Minutes of the University Faculty, February 1895–June 1905, entry for 9 June 1897, 243–245. PUA.

77. Russell, "Salutatio Latina," 16 June 1897. Faculty Folder, PUA. Translation courtesy Noel Swerdlow.

78. *New York Trubune Ilustrated Supplement*, n.d., 14. "Memorabilia." Box 126.6, PUL/HNR.

79. Ibid.

80. Ibid.

81. Fragment, "Memorabilia." Box 126.6, PUL/HNR.

82. *Record of the Class of 1897 of Princeton University Number 1*, 12 November 1898, 46. "Princeton '97: 'Looking over the Editor's Shoulders,'" (May 1951), cites "Pop" Keener to Bob Garrett, 1 November 1898. On Russell's dyslexia, see MRE #2, 40.

83. Cooley et al., 1897.

84. *Nassau Herald, Class of 1897 Sesquicentennial Number* 33 "Statistics" appendix. For the slang, see Bragdon, 1967, 446 n. 4.

85. Russell response form, "Questions for the Nassau Herald Class of '97," Box II. PUA. "Conditions" were notices of deficiency in performace or attendance. See *College of New Jersey Catalogue 1893–94*, 29–30.

86. MRE #1, 16–17; 22–23; #2, 8–9. John E. Hammond to the author, 14 October 1995.

87. *Nassau Herald, Class of 1897 Sesquicentennial Number* 33, "Class Prophecy," 61.

88. Kennedy, 1947, 238.

89. Ibid., 238–239.

90. MRE #1, 20; 31.

91. Bragdon, 1967, 223.

92. Ellen Axson Wilson to Woodrow Wilson, 22 June 1896; 10 August 1896. Link et al., 1971–1976, 9:524–525; 563–565.

93. E. L. Crain '39 and S. J. Shirk, '39, "Venerable Star-Gazer Peers at Nassau's Past; Believes Present-Day Changes 'Not for Worse,' " *Daily Princetonian*, 19 March 1937. Fragment, Russell Faculty Folder, PUA.

94. Minutes of the Trustees of Princeton University, Volume 8, June 1894–October 1898, 8:518. College of New Jersey "Course Catalog," 1893–1894, 122. PUA.

Chapter 3
Graduate Years

1. "Congressional Record—Senate," April 11, 1900, 4019. Bates, 1965, chap. 3. Ph.D. production statistics in astronomy can be found in Lankford, 1997, chap. 4, table 4.1.

2. On Hale, see Wright, 1966; Kevles, 1968; 1978; Osterbrock, 1993; 1997.

3. Fine, Young, Brackett, and Wilson all petitioned for it. Thorp, Myers, and Finch, 1978, 54–61; Bragdon, 1967.

4. Lefschetz, 1969, 73. Hoeveler, 1981, 284–285.

5. Frost had been a Young student at Dartmouth. Frost, 1933, 6–3.

6. "Faculty Committee on Graduate School, Minutes" for the 1890s, 1909–1922 Graduate School Records, Box 16. PUA.

7. The award had been created when Mrs. William Thaw of Pittsburgh donated $10,000 to support astronomy as part of the university's sesquicentennial fund-raising drive. The endowment provided $400 to $500 per year in the first years of its use. Minutes of the Trustees of Princeton University, Volume 8, June 1894–October 1898, entry for 10 December 1896. PUA.

8. The amount of Russell's fellowship has not been determined, but that awarded to Oliver D. Kellogg, '99, in 1900 was $250. O. D. Kellogg student records, PUA.

9. Watson became a standard for professional training. Russell probably used the 1892 edition. Klinkerfues's text included binary orbit methods, and Oppolzer dealt with advanced orbit theory, especially the methods of Encke and Gauss.

10. Young to W. F. Magie, 10 November 1902. A. F. West correspondence files, Graduate School. PUA.

11. In 1898 Lovett was given an unprecedented $1,000 raise. Minutes of the Trustees of Princeton University Volume 8, June 1894–October 1898, entry for 14 June 1897. Link et al., 1971–1976, *10*:552–553.

12. Russell submitted a draft to the *Annals of Mathematics*, then edited at the University of Virginia, "On the Values of Lim $\log^n x$ considered geometrically." Manuscript, Box 100.12. Echols to Russell, 25 November 1897; 20 January 1899; Ormond Stone to Russell 30 November 1897. PUL/HNR. The journal was purchased by Harvard University that year. *Annals of Mathematics* publication history, Library of Congress on-line search service. "Theory of Functions," 2 February 1897; "Higher Plane Curves," 5 February 1897. Box 100.11, PUL/HNR.

13. Young, 1888, 496–497; 499. On the statistics, see Hussey, 1900, 91–103; and Aitken, 1918, chaps. 1; 4.

14. Russell, 1898a, 10; *PopAst* 1898, 149. "A New Graphical Method for Determining the Elements of a Double Star Orbit," n.d. ca. spring 1898. Box 100.13, PUL/HNR.

15. "Our Star," *Record of the Class of 1897 of Princeton University*, Number 1, 12 November 1898 (Baltimore: Williams and Wilkins, 1898), 46.

16. Zwiers, 1896. The editor appended a note attesting to Russell's independence. Russell, 1898a, 9n.. The editorial correspondence does not exist in Russell's papers at Princeton.

17. The volume containing Zweirs's paper was probably received at Princeton in March 1896, from contemporary accession stamps. Aitken, 1918, 80; Smart, 1931, 351. Luyten, 1987, 29, recounts Russell's assertion that the *AN* was at the bindery.

18. *Record of the Class of 1897 of Princeton University*, Number 1, opposite 46. Young to Russell, 11 May 1898. PUL/HNR.

19. Fine to Russell, 2 August 1898. PUL/HNR.

20. Young to Russell, 2 August 1898. AIP/DCA/CAY.

21. Ibid.

22. MRE #1, 20.

23. On Cobb, see White, 1984. On the contemporary importance of foreign training in astronomy, see Hall, 1900, 18. Struve, 1943, 474; Struve, 1949, 382–384. On the value to Princeton, see Hoeveler, 1981, 284–286.

24. "Graduate Club" records, entry for meeting of 20 Jan 1896. Graduate School Records, PUA. Lankford, 1997, 84–85, argues that a European Ph.D. did not provide clear advantages, but his statistical sample was very small and his criteria are open to question.

25. Minutes of the Trustees of Princeton University, Volume 8, June 1894–October 1898, entry for 12 October 1898, 728–729. Oliver Dimon Kellogg faculty file, PUA.

26. Meadows, 1966.

27. Russell, 1899e.

28. Young already had such a screen for his solar studies. Ibid., 285.

29. Ibid., 298–299.

30. Sagan, 1974, 27.

31. Russell to Hale, 10 April 1899; Hale to Russell 20 April 1899. Box 15:9, *ApJ* files, YOL. Quote from [Russell] "Biographical Data," 2. National Academy of Sciences biographical files.

32. Russell to Keener, 31 January 1901, quoted in Keener, 1903, 187.

33. Russell paid his first fee for entrance to the Ph.D. preliminary examination (forty dollars) on 5 May 1899 and passed exams in French (Harper), German (Humphreys), math (Fine), astronomy (Young), and physics (Magie). Russell Faculty folder. PUA. Russell sought physiological factors that influenced observations. "Observing Logs," twenty-three-inch, May and June. Russell, 1899c. Astronomy Department records, PUA.

34. "Minutes of the Committee on the Graduate Department 1895–1901," 49. PUA.

35. Hoskin, 1979; Pierce, 1951, 479–494. Vogel, 1890. Young, 1888, 478–484.

36. Clerke, 1903, 274; Campbell, 1906, 459; Meadows, 1972; DeVorkin, 1984a.

37. See *AJ* 19:169; 1899. *PopAst* 7:129; Young, 1899.

38. On Lane's Law, see Powell, 1988.

39. Young to Russell, 6 September 1899. PUL/HNR. Young, 1888, 174–176.

40. Russell, 1899d, 318.

41. Roberts, 1899, 314.

42. Seeliger, 1900, 247.

43. Russell faculty folder, PUA. Keener, 1900, 187.

44. [A.C.D.C.], 1899. "Discovery of Minor Planets in 1898 (including that of Eros)," *MN* 59:272–274. On nomenclature, see "Discovery of Minor Planets in 1897," *MN* 58:197.

45. Russell, 1898b, 147; Hussey, 1898, 120.

46. He had available continually improved observations, refined normal places, and better elements. Russell, 1899a, 31.

47. Russell, "The General Perturbations of the Major Axis of Eros by the Action of Mars, with the corresponding terms in the Mean Longitude," 27 April 1897 [*sic*], 1. Box 100.14 (a), PUL/HNR.

48. Russell, 1900a, 25.

49. "The General Perturbations of the Major Axis of Eros by the Action of Mars," 2–6. Box 100.14 (a), PUL/HNR.

50. Russell, 1900a, 26.

51. See material in Box 100.14, folders 14a, 14a′, 14a″, 14a‴, a⁗, PUL/HNR.

52. "Minutes of the Committee on the Graduate Department 1895–1901," 9 May 1900 entry. Box 16: "Faculty Committee on Graduate School, Minutes," Graduate School Records. PUA.

53. "Seventeen Little Deans," *Princeton Alumni Weekly*, 3 March 1933, no. 2. "Puritan conscience" from: Russell to Howard C. Watson, 16 March 1939. PUL/HNR. MRE #2, 46.

54. This observation runs counter to Lankford, 1997, 59, who suggests that by the end of the century "[p]hysics was seen as an active, creative, high-status field" by astronomers. Some astronomers, like James Keeler and Hale, did take this view, but they were in the minority, and as Karl Hufbauer has argued ("Crucibles" paper, History of Science Society meeting, Minneapolis 1997), their efforts were not heeded by the majority of astronomers or physicists.

55. Jones and Boyd, 1971, 333–334; Conversations with S. Dick.

56. Clerke, 1903, 9.

57. Russell, 1900a; 1900b.

58. Young, 1900.

59. MRE #1, 36. A. G. Russell Diary, entry for 22 July, 137. Johnston and Reifsnyder, 1990, 74–75.

60. The "senior" first appeared on the cover of the first issue of volume 3 of the *Princeton Tiger*, 20 October 1892. The character driving the quill, "The Prince of Orange of the Tiger," was "a jolly fellow, who will not lose his good-nature, and means his fun in all gentleness" (3). PUA.

61. Russell to Keener, 31 January 1901, in Keener, 1903, 186–187, mentions Naples and Rome. MRE #1, 25, adds Capri.

Chapter 4
Postdoctoral Years at Cambridge

1. Frost to Young, 11 March 1901; 11 October 1901. AIP/DCA/CAY. Reese soon left Yerkes to take up a career in engineering. Reese Alumni folder, PUA.

2. Keener, 1903, 56. MRE #2, 4.

3. O. D. Munn to E. B. Wilson, 29 January 1926, clipped to Munn to Russell, 29 January 1926; to Russell, 1 June 1943. Scientific American Folder, PUL/HNR. See also

C. A. and O. D. Munn student records, PUA. Serviss (1851–1929) was a lecturer and popular writer on travel, history, and astronomy, generally known for his "Astronomy with an Opera Glass," published in 1888. Why Serviss was replaced by Russell is not known. "G. P. Serviss," *Who Was Who in America 1897–1942* (Marquis, 1943).

4. [Russell], "Discussion of Cooling of Perfectly Gaseous Star, Unfinished," November 1901. Box 100.15, PUL/HNR.

5. Powell, 1988, 194, notes that "Lane's Law" was first explicitly stated by Kelvin in 1887.

6. DeVorkin, 1978, 57–62. Hufbauer, 1981, 302, similarly argues that astronomers felt unable to attack the stellar energy problem ca. 1900.

7. Clerke, 1903, 4–5.

8. Russell, 1902, 259. For a review of technique, see Henroteau, *Handbuch der astrophysik vi*, pt. 2 (Berlin, 1928), chap. 4.

9. Russell, "On the Period of Delta Orionis" [holograph, ca. April 1902]. Box 100.16, PUL/HNR.

10. Hale to Russell, 22 April 1902; Russell to Hale, 15 April 1902. *ApJ* correspondence 1:3, YOL. DeVorkin, 1977.

11. Campbell to Russell, 19 May 1902. PUL/HNR.

12. Russell to Campbell, 26 May 1902. LAUC/WWC.

13. Russell to Campbell, 26 May; Campbell to Russell, 3 June 1902. LAUC/WWC.

14. Russell, 1954, 103; Spiritual Autobiography, 33. On his walking habits, see MRE #1, 30–31.

15. Russell's old classmate Nicholas Stahl recalled that Russell needed extra endorsements to enroll at Cambridge because he had applied so late. Stahl to Russell, 29 July 1938. Russell to "The Registry of the University," 15 October 1902. PUL/HNR. Wilson to Russell, 27 August 1902, with copy of Wilson to "Registrar of Cambridge University," 27 August 1902. Link et al., 1971-1976, *14*:112. F. L. Patton to Russell, 7 August 1902. Patton to "Vice Chancellor, Cambridge University," 30 July; to Ball, 7 August 1902. PUL/HNR. Darwin to C. A. Young, 28 September 1898; 17 July 1902. PUL/CAY. H. B. Fine to W. Wilson, 29 July 1905. Link et al., 1971-1976, *14*:163.

16. Compiled from fragmentary records, and noted in passing in Spiritual Autobiography, 33–34. Commencing in fall (Michaelmas Term) 1902, his tutorial and matriculation fees totalled £8. Arthur E. B. Owen (King's College) to David Dewhirst, 14 April 1992; Dewhirst to DeVorkin, 7 January 1992. Arthur Berry to the Registry Files, 18 October 1902, Cambridge University Archives. Withers, 1929, 342.

17. Russell, 1954, 101–102.

18. Russell, 1954, 103; Spiritual Autobiography, 33. Russell OHI, 1951, records Russell's recitations of English limericks. Russell's penchant for limericks certainly predates Cambridge, however, as it was a favorite recreation of Wilson and of Princetonians generally. Bragdon, 1967, 208; 288; 292; and Russell to Milne, 23 August 1948. Courtesy M. Weston Smith.

19. "Cambridge Minute Books, Minute Books of the $\nabla^2 V$ Club, December 1900–December 1937," entries for October 1902, 15 December 1902, 5 February 1903, 20 November 1903, and 12 May 1904. Archives for the History of Quantum Physics, Reel 38, AIP. Eddington's name first appears in November 1904. Douglas, 1957, 9, notes Eddington's membership and that Jeans was a founder, which is not confirmed in the Minute Books.

20. Spiritual Autobiography, 34.

21. Ibid., 34–35; Milne, 1952, 6; 11.

22. Kushner, 1993, 197; Jeans, 1912.

23. Kushner, 1993, 199.

24. Ibid., 203–204. Burchfield, 1975, 112–115, does not come to this conclusion.

25. Kushner, 1993, 207–208.

26. Darwin, 1898, 343; 346.

27. Russell notes, "1903 Jan–March Darwin, Dynamical Astronomy, Second lecture: Jan 22." Box 70.1, PUL/HNR.

28. Russell notes, "Rotating Figures . . ." 25 November 1902 entry, Darwin course. Box 69.12, PUL/HNR.

29. Russell notes, "R. S. Ball, Planetary theory, 1902." Box 69.11, PUL/HNR.

30. Russell notes, "Figure of a rotating body." 18 November 1902. Box 69.12, PUL/HNR.

31. "Figure of a rotating body." 25 November 1902.

32. Russell to Frank Schlesinger, 15 March 1917. PUL/HNR.

33. Gill, 1900, 477.

34. Russell to Schlesinger, 15 March 1917. PUL/HNR.

35. I thank Andrew Warwick for sharing his insight into the tripos.

36. Smart, 1945.

37. Darwin to Walcott, n.d. Darwin file, CIWA. On the origins of the Carnegie Institution, see Reingold, 1979, 313–341; Bannister, 1979, 85; Miller, 1970; Yochelson, 1994.

38. G. Darwin and R. S. Ball, to the President of the Carnegie Institution, 7 February 1903. Russell file, CIWA.

39. Hinks to Campbell, 21 February 1903. LAUC/WWC; Hinks letterpress book, 250. CUOL.

40. Hinks to Campbell, 21 February 1903. LAUC/WWC.

41. "Notes," 19 February 1903. *Nature* 67:373. Schlesinger was a student of Harold Jacoby at Columbia and then directed the International Latitude Observatory at Ukiah, coming to Yerkes well-prepared to take up astrometric studies. Jacoby, 1908.

42. Hinks to Hale, 21 February 1903. Hinks letterpress book, 244. CUOL.

43. "Research Assistants," *Carnegie Institution of Washington Yearbook* #2 (Washington, 1903), xlvii.

44. Simon Newcomb, "Report on the Case of Dr. Henry N. Russell," 27 February 1903. Russell file, CIWA.

45. The Secretary, CIW, to Russell, 16 March 1903; Russell to The Executive Committee, 28 April 1903. Russell file, CIWA. On the agreement, see Lewis Boss to Daniel C. Gilman, 22 May 1903. Russell file, CIWA.

46. Hinks, 1904, 97. Loewy was the director of the Paris Observatory.

47. Lankford, 1984, 26.

48. Schlesinger, 1899, 242. Loewy, 1901, 396–397.

49. Ball, 1899, 152; Dewhirst, 1982. On Loewy's equatorial Coudé, see Weimer, 1982; King, 1955, 244; 380; [Loewy], 1908, 3; Stratton, 1949, 13. Schlesinger, 1899, discusses how differential optical distortion could in theory be overcome in long-focus refractors.

50. Russell, "Photographic Reductions—Solar Parallax, A. R. Hinks," 20 January. Box 70.3 PUL/HNR. Hinks, 1901, 445.

51. Russell to Gilman, 5 October 1903. Russell file, CIWA.

52. Hinks to Hale and Campbell, 16 April 1903. Hinks letterpress book, 253; 255. CUOL.

53. Hinks to Schlesinger, 17 June 1903. Hinks letterpress book, 262, CUOL. Russell to Gilman, "Report of Progress of work on *Stellar Parallaxes*," 6. Russell file, CIWA.

54. Russell to Gilman, "Report of Progress of work on *Stellar Parallaxes*," 5 October 1903. Russell file, CIWA. In earlier letters to Campbell, Hinks was vague about how the lists would be compiled. Hinks to Campbell and Hale, 21 February 1903. Hinks letterpress book, 244; 250. CUOL.

55. Hinks and Russell, 1905, 784.

56. DeVorkin, 1978, chap. 3; 308.

57. Meadows, 1972; Lockyer, 1902.

58. Russell, 1924. "Course of Addresses by Professor Henry Norris Russell, PhD, delivered in the Physics Building, University of Toronto, upon the applications of modern physics to astronomy." Lecture 8, 146. Box 103.4. See also Russell to Lockyer, 14 January 1911, rough draft. PUL/HNR.

59. Spiritual Autobiography, 35. Hinks to Campbell, 26 August 1903. Hinks letterpress book, 279. CUOL.

60. Eliza Russell's death was not unexpected. "Oyster Bay," *East Norwich Enterprise*, 26 September 1903, report received 22 September. Microfilm/Oyster Bay Library.

61. "Mrs. Eliza H. Russell," *East Norwich Enterprise*, 2–3 October 1903. Microfilm/Oyster Bay Library.

62. Spiritual Autobiography, 35. "Family Information" biographical listing courtesy Margaret Edmondson Olson. Eliza Norris Russell's death was recorded in Abridged Record; frame 9543. Johnston and Reifsnyder, 1990, 72, note Roosevelt's attendance, as does "News Notes," *Oyster Bay Guardian*, 9 October 1903; "Oyster Bay," *East Norwich Enterprise*, 3 October 1903. Microfilm/Oyster Bay Library.

63. Russell to Gilman, "Report . . ." October 5, 1903, 3; "Memoir on Reduction of Photographs by [Turner]'s and Dyson's Methods," n.d., possibly 1903. Box 100.17, PUL/HNR.

64. Lewis Boss to Gilman, n.d., excerpted fragment attached to Russell, "Report . . ." 5 October 1903. Russell file, CIWA.

65. Lewis Boss commentary, quoted in Charles Walcott to Gilman, 15 August 1904; Russell, "Abstract," n.d. ca. 28 June 1904, attached to Russell to "The President and Executive Council of the Carnegie Institution," 29 June 1904. Russell file, CIWA.

66. Russell, "Abstract," 2.

67. Frank Schlesinger later devised even more efficient means of reducing plates. Hoffleit, 1977, 52. [Russell], "Memoir on Reduction."

68. Russell to "The President . . . ," 29 June 1904. Russell file, CIWA. [Russell], "Memoir on Reduction."

69. [Russell], "Memoir on Reduction," 14–15. Russell also decided not to reduce parallaxes to the fundamental frame defined by meridian circle observations.

70. Russell to Wesley, 3 December 1902; 9 December 1902; 26 February 1903. 1902 RAS letters Q–Z, Royal Astronomical Spciety Library.

71. "Meeting of the Royal Astronomical Society," *Observatory* 27:183–185.

72. Russell, "Abstract," 3. Russell to "The President . . . ," 18.

73. Lewis Boss commentary, quoted in Charles Walcott to Gilman, 15 August 1904. Russell file, CIWA.

74. Ibid.

75. Ibid.

76. Campbell to Carnegie Institution, 17 June 1904. LAUC/WWC. Hinks to Campbell, 5 June 1902. Hinks letterpress book, CUOL. Dyson, 1904.

77. Russell, Dugan, and Stewart, 1926, *1*:187. The fate of the Eros campaign will be dealt with in a separate publication.

78. Spiritual Autobiography, 35.

Chapter 5
Return to a New Princeton

1. Russell to Woodrow Wilson, 6 April 1905. Link et al., 1971–1976, *16*:81.

2. HNR jr., 28; MRE #1, 29. Russell to Keener, 9 September 1907. *Duodecennial record* Class of '97. Princeton (1909), 180.

3. Typhoid had a 15 percent mortality rate among those admitted to hospital in London at this time. *Encyclopedia Britannica,* 11th ed., s.v. "Typhoid fever."

4. He finished a column on stellar parallaxes while sailing home. William Maxwell Reed and Garrett P. Serviss wrote the columns for January and February, 1905.

5. Russell to R. S. Woodward, 18 March 1905; 29 September 1906. Russell file, CIWA. Russell, "The Heavens in . . . ," *Sci. Am.* 92:179.

6. Russell to R. S. Woodward, 18 March 1905. Russell file, CIWA.

7. R. W. Pumpelly to Russell, 5 October 1947; 14 May 1948. PUL/HNR. Pumpelly's report is noted in the Russell file, CIWA.

8. Russell to Wilson, 6 April 1905. Link et al., 1971–1976, *16*:81–83; 83.

9. Link et al., 1971–1976, *16*:81.

10. Ibid., 82.

11. Eddington, 1938b, 169. On Kapteyn's methods, see Paul, 1993.

12. Russell to Wilson, 6 April 1905.

13. Kapteyn, 1904, 420–421. On Kapteyn's influence, see Paul, 1993; Turner, 1922. Kapteyn made much the same statements in Kaptyen, 1906, 264.

14. Russell to Wilson, 6 April 1905.

15. Wilson to Lovett, 27 April 1905. Link et al., 1971–1976, *16*:80.

16. Russell to "The President and Executive Committee of the Carnegie Institution," 29 September 1906. Russell file, CIWA. The minutes of the Carnegie Institution trustees note only that "it was not deemed expedient to make an additional grant." "Minutes of the Carnegie Institution Trustees," entry for 10 April 1905. CIWA.

17. Kohler, 1991, 19, examines the debate over priorities at the Carnegie Institution. R. S. Woodward to E. B. Frost, 30 January 1905. YOL/EBF. Forest Ray Moulton of the University of Chicago was similarly cut off, to his great indignation. Woodward to Moulton, 28 September 1906. CIWA/FRM.

18. Russell, 1907b, 89.

19. "Cambridge Observatory Syndics," Minute Book, entry for 26 May 1905. CUOL.

20. Hinks and Russell, 1905.

21. Russell, 1905a, 800.

22. C. W. McAlpin to Woodrow Wilson, 19 November 1903; W. Wilson, "A Memorandum," [ca. 20 November 1903]. Link et al., 1971–197], *15*:55–56; 73.

23. Quotes from Leitch, 1978, 80; 374; Wilson to Lovett, 27 April 1905. See also Bragdon, 1967, 305–306. Russell claimed to have been hired as a preceptor. Russell to E. A. Milne, 23 August 1948, copy courtesy M. Weston Smith.

24. Bragdon, 1967, 306.

25. Minutes of the Trustees of Princeton University, Volume 10, June 1901–January 1908, entry for 21 October 1905, 610–611: report of the Committee on the Curriculum. PUA. Henry Dallas Thompson to Woodrow Wilson, 26 July 1905. Link et al., 1971–1976, *16*:162–163. Bragdon, 1967, 304. By comparison, Young's retirement allowance was some $2,430, paid by the Carnegie Trust. Link et al., 1971–1976, *16*:419.

26. Fine to Wilson, 29 July 1905. Link et al., 1971–1976, *16*:164. Aspray, 1988, 348, notes Fine's offer.

27. Murray, 1935, 345.

28. By 1902, Young was too weak to travel to professional meetings. Young to Campbell, 23 December 1900; 12 October 1903. LAUC/WWC. Young to Hale, 27 May 1903; 29 January 1905. CIT/GEH; Young to Frost, 4 January 1902. UA/GK. Young to Wilson, 6 December 1904, reprinted in Minutes of the Trustees of Princeton University, Volume 10, June 1901–Jan 1908," 518; 540. PUA. Lankford, 1997, describes the multiple entry points to an astronomical career during the period. Veysey, 1965, 443, notes the ubiquity of campus administrators.

29. Young to Russell, 8 August 1905. PUL/HNR.

30. Ibid. Young died three months after his daughter's death, on 3 January 1908.

31. Young to Hale, 31 July 1905; Hale to Young, 19 August 1905. CIT/GEH. Lovett to Magie, 14 November 1902, A. F. West correspondence, Graduate College records. PUA. Minutes of the Trustees of Princeton University, Volume 10, June 1901–Jan 1908, 12 June 1905 entry, 560; 565; Lovett faculty folder, PUA.

32. Lovett contacted John Brashear, Alvan Clark & Sons, and the Warner and Swasey Company. Brashear required $85,000 to build the reflector. He reexamined his estimate several years later for Russell, concluding that it was a gross underestimate. Brashear to Russell, 14 February 1909. See also [Russell], "Notes on the Rebuilding of the Princeton University Observatory," 22 December 1909. Box 100.23, PUL/HNR.

33. Lovett to Woodrow Wilson, 21 November 1906. Link et al., 1971–1976, *16*: 488–489.

34. Young to Hale, 27 May 1903. CIT/GEH.

35. Frost to Young, 11 March 1901. PUL/CAY. Hale to Manley Goodwin, 28 February 1904. Item 28459, Hale Papers, Huntington Library; Hale to Young, 9 November 1899. Letterpress, 7:607, YOL. Reese had his own complaints. Reese to W. S. Adams, 21 April 1901. Adams Papers, Supplement—Box 2: S2.47, HL/WSA.

36. "John Merrill Poor," student records, PUA. Poor published only a few papers on orbits of minor planets.

37. A. Fowler to Young, 16 January 1906. DCA. Mitchell to Hale, 8 April; 6 May; Hale to Mitchell 12 May 1908. AIP/GEH. Wright, 1966, 221.

38. Mitchell, 1905. Russell, 1954, 104.

39. Mitchell to Russell, 14 August [1908] postcard fragment. PUL/HNR. Lovett was never designated director of the observatory, only chairman of the department. Alexander Leitch to Russell, 8 January 1935, Lovett faculty file, PUA.

40. When his father died in 1913, Mitchell left astronomy completely to run the family business and ended up at the Mack Manufacturing Corporation. A. Fowler to C. A. Young, 16 January 1906. AIP/DCA/CAY; A. Fowler to Hale, 17 April 1907. CIT/GEH; R. H. Curtiss to Stebbins, 26 September 1913. UMBL/RHC. Walter Mann Mitchell, Alumni Records, PUA.

41. Dugan faculty folder, PUA. Todd to Dugan, 12 June 1932. PUL/RSD. "President's Report to the Board of Trustees," 1 January 1909, 578. Wilson Papers, PUA.

42. Dugan to Frost, 5 June 1905; Frost to Dugan, 19 June; 28 June 1905; Dugan to Frost, 9 July 1905; 15 July 1905; 30 July 1905; Frost to Dugan, 19 August; 21 August; 22 September; Dugan to Frost, 3 October 1905. The full effect of Hale's departure is described in Frost to Crane, 22 September 1905; and R. S. Woodward to Frost, 30 January; 23 February 1905. *ApJ* correspondence, Box 31:8, YOL/EBF. Osterbrock, 1997a, describes the situation.

43. Dugan to Dean E. P. Puckett, 10 June 1927; Lovett to Dugan, 12 June 1930. PUL/RSD.

44. Logbooks marked "Dugan 1905" and "Dugan 1" with entries in the latter starting on 27 November 1905. Astronomy Department Records. PUA. Dugan was officially hired on 21 October. Minutes of the Trustees of Princeton University, Volume 10, June 1901–Jan 1908, 610–611. PUA.

45. Dugan, 1935, 44.

46. Ibid., 45.

47. Ibid.

48. There was also an equalizing wedge photometer, which Reed and Daniel used between 1902 and 1905. It required a light source. Dugan, 1911; Pickering, 1909a; Drummeter, 1992.

49. Dugan to Campbell, 25 January 1906. LAUC/WWC.

50. Mitchell to Russell, 31 October 1906. PUL/HNR.

51. Dugan to Campbell, 20 September 1907; Campbell to A. O. Leuschner, 30 September 1907, LAUC/WWC.

52. Mitchell to Russell, 31 October 1906. PUL/HNR.

53. Ibid.

54. Dugan to Campbell, 20 September 1907. LAUC/WWC.

55. Dugan to Campbell, 15 February 1909. LAUC/WWC.

56. "Memorandum of Changes in Salaries and Authorizations Passed at Meeting of Finance Committee, June 2, 1908"; "Meeting of Finance Committee, June 8, 1908," [entry dated 8 June 1909]. Woodrow Wilson Presidential Files, file A-WWP-F.16. PUA.

57. Wilson suggested Lovett for the post. Ralph O'Leary, "Rice's Lovett Seeks Truth," typescript draft for the *Houston Post*, 13 April 1953, based upon interviews with Lovett. Lovett file, PUA.

58. Frost to W. F. Magie, 21 January 1908. YOL.

59. Leuschner to Lovett, 31 February 1908. BANC/AOL.

60. Lovett to Leuschner, 4 April 1908. C-B 1016 BANC/AOL.

61. Russell to Edmondson, 24 February 1937. PUL/HNR.

62. Bragdon, 1967, 356.

63. Ibid., 308; chap. 16. Lovett notes the unexpended funds, and his willingness to use them to support Wilson's gubernatorial campaign, in Lovett to Wilson, 23 September 1910. Link et al., 1971–1976, *21*:159.

64. See "Courses Offered but Not Taken by any Graduate Students, first term 1907–1908." Woodrow Wilson Presidential Files, A-WWP-F.10 "Graduate School." PUA.

65. Minutes of the Trustees of Princeton University, Volume 11, entry date 8 June 1909, 55; Henry B. Fine to the Board of Trustees, 31 March 1909. Link et al., 1971–1976, *20*:139.

66. "Memorandum of Changes in Salaries and Authorizations Passed at Meeting of Finance Committee, June 2, 1908." Wilson Presidential Files, PUA.

Chapter 6
Parallaxes, Pedagogy, and the Lives of the Stars

1. "Minutes of the Department of Astronomy, Princeton University, 1910–1917." "Historical box," Princeton University Observatory Library.

2. "Minutes of the Faculty Committee on the Graduate School," 10 November 1913, 22. PUA.

3. Russell, "Notes," on the Rebuilding of the Princeton University Observatory," 22 December 1909 rough draft. Box 100.23, PUL/HNR.

4. Russell, "Notes." Brashear to Russell, 10 December; 14 December 1909; Warner & Swasey Co. to Russell, 20 December 1909. Box 57, Folder 11, PUL/HNR. Russell later told his students that the performance of the Sheepshanks 3-element lens was "beautiful." Russell, "Senior Practical Astronomy," 11 April 1917 entry. PUL/HNR.

5. The decision was not definite until June 1910. Bragdon, 1967, 380–381. Link et al., 1971–1976, *17*:408, entry for 30 September 1907. West and Fine had a complete falling out as a result of this issue. West to Fine, 25 November 1910. A. F. West files, PUA.

6. Russell, "Notes." Thorpe, Myers, and Finch, 1978, 100; 122.

7. "Minutes of the Department of Astronomy," entry for 21 March 1910. Princeton University Observatory.

8. Frost to Dugan, 9 July 1905. *ApJ* correspondence, Box 31:8, YOL/EBF.

9. Dugan, "RT Persei." Box 5, Folder: Manuscripts. PUL/RSD.

10. Russell to "The President and Executive Committee of the Carnegie Institution," 29 September 1906. Russell file, CIWA. The measuring machine is described in Hinks, 1901.

11. Russell, "Determination of Instrumental Constants," entries for 12 March through 23 May 1906. Department of Astronomy records, box 5, "Data Books 20th Century." PUA.

12. Russell to "The President and Executive Committee" 6; Russell, 1906b.

13. Princeton course catalogs, 1905–1909. PUA.

14. Russell took extensive notes on Darwin's 1906 study "On the Figure and Stability of a Liquid Satellite." Russell, 1906a.

15. Upon graduation in June 1908, Daniel won the Thaw Fellowship.

16. Russell and Daniel, 1907c. Russell, "On the Masses of the Stars." Holograph of paper read at the Cambridge meeting of the British Association for the Advancement of Science, 22 August 1904. Box 100.18, PUL/HNR. Paul, 1993, 84.

17. Russell, "On the Probable Distance of Orion," holograph abstract and manuscripts, December 1906. Box 100.19; 100.20. PUL/HNR. Russell to Frost, 11 December 1906. YOL/EBF. An abstract of Russell's work appeared in *Popular Astronomy 15*:444 (1907). Russell underestimated the distance by a factor of 2.5 compared to the modern value. The Astronomical and Astrophysical Society of America was renamed the American Astronomical Society (AAS) in 1914.

18. Frost to Russell, 4 January 1907. YOL/EBF.

19. Russell to George Comstock, 11 January 1909. Comstock had found hints of absorption in Russell's results. Comstock to Russell, 11 November 1907. PUL/HNR. The existence of general absorption in space was then far from accepted. Paul, 1993, chaps. 6; 9. Comstock's differences with Kapteyn are outlined in Comstock, 1907. Comstock, 1908, 33. Russell 1910a, 152.

20. Boxes 78 and 79; esp. 78.7. PUL/HNR.

21. "Accidental Errors," 55–83; quote from 83. Box 79.4 (1). PUL/HNR.

22. Struve and Zebergs, 1962, 190–192; Jones and Boyd, 1971, 380; Pickering, 1886, 535; and Pickering, 1891.

23. Russell to Pickering, 4; 10 April 1908. UA V.630.17.7 Box IV: Russell folder 1908–1910. HUA/ECP. Emphasis in original. Russell's incorrect recollection is in Russell, 1919a, 154. See also Jones and Boyd, 1971, 430–431.

24. Pickering, 1907. Pickering, 1886, 535; 1909a, 75. Pickering created committees of observatory directors he had favored with equipment and support in his attempt to standardize his systems. See, for instance, (1884) "Second Report of the Committee on Standards of Stellar Magnitudes," *Proceedings of the AAAS* 33:29–30. Pickering chaired this committee. DeVorkin, 1981.

25. Russell to Pickering, 4 April; 10 April; 25 April 1908; Pickering to Russell, 22 April; 29 April 1908. PUL/HNR; HUA/ECP.

26. Paul, 1993, 2–3; 121.

27. Nielsen, 1963; Jones and Boyd, 1971, 236–238; DeVorkin 1978, chap. 2.

28. Monck, 1894, quoted in DeVorkin, 1978, 232.

29. Nielsen, 1963. Hertzsprung trained as an engineer with a specialty in photochemistry. Herrmann, 1973; 1976; 1994; and DeVorkin 1978.

30. Hertzsprung, 1905. Hertzsprung to Pickering, 15 March 1906. HUA/ECP.

31. Pannekoek, 1906–1907.

32. Hertzsprung to Pickering, 22 July 1908. HUA/ECP. Noted as well in Hertzsprung, 1908, 380.

33. Clerke, 1903, 204.

34. Paul, 1993. Pickering, 1909c, 3.

35. Pickering to Hertzsprung, 4 August 1908. HUA/ECP; Pickering and Cannon, 1901, 138.

36. Hertzsprung to Pickering, 17 August 1908. HUA/ECP. Hertzsprung to Karl Schwarzschild, 26 August 1908. AIP/KSM. The final manuscript was submitted in December 1908 and was published and received by observatories in the United States in January 1909. Hertzsprung, 1908.

37. He found its apparent diameter to be 0″.05. Hertzsprung, 1906. I am indebted to the late Professor Adriaan Wesselink for insight into Hertzsprung's work. On stellar diameters, see DeVorkin, 1975.

38. For instance, Hertzsprung to Arthur Stanley Eddington, 11 August 1907. Fiche 13/6–7, UA/HER.

39. Quoted from the abstract: ". . . dass so zerstreut zwischen den gewöhnlichen Sternen diese Giganten liegen." *Astronomische Jahresbericht*, 1909, 179 ("these giants are dispersed between the ordinary stars"). Schwarzschild, 1909. Nielsen, 1963, 236; Herrmann, 1994.

40. Russell to Pickering, 24 September 1909. HUA/ECP. Emphasis added.

41. Ibid.

42. The record appears complete. Pickering to Russell, 14 October 1909. PUL/HNR.

43. Pickering to Hale; 13 February; 1 March; 18 May; 17 July; Hale to Pickering 23 February 1909. AIP/GEH.

44. Pickering to Hale, 17 July 1909; Hale to Pickering, 11 March 1910. HL/GEH.

45. Pickering to Frost and Campbell, 2 July 1910. YOL/EBF; LAUC/WWC.

46. Kapteyn to Russell, 7 March; 5 April 1909. PUL/HNR. Russell to Pickering, 18 October 1909. HUA/ECP.

47. Russell to Pickering, 18 October 1909. HUA/ECP. Kapteyn to Russell, 14 March 1911. PUL/HNR.

48. Russell to Pickering, 18 October 1909. HUA/ECP. Hertzsprung, 1908.

49. Russell to Pickering, 18 October 1909. Undated notes from Box 79.1, 270–277. These notes and calculations were written on the verso of copies of a university memo dated April 1910. PUL/HNR.

50. Russell to Pickering, 7 April 1910. HUA/ECP.

51. Russell, 1910c, 878.

52. Russell, 1910e, 883.

53. Ibid.; Nielsen, 1963, 241, suggests that Russell inserted his reference to Hertzsprung after his talk and after he met Schwarzschild, citing Russell to Hertzsprung, 27 September 1910, as his source. Lewis Boss to Russell, 23 August 1910. PUL/HNR.

54. Menu and Conference Notes, 1910 Mt. Wilson. Box 89.11, PUL/HNR.

55. A full account of the deliberations, and information on the history of the union, can be found in DeVorkin, 1981, 40.

56. J. S. Plaskett to Russell, 27 June 1927. PUL/HNR.

57. Paul, 1993.

58. Nielsen, 1963, 235–236. See also Knut Lundmark, *Handbuch der Astrophysik* 5, pt. 1 (Berlin, 1933), 437–441; 437; Hoffleit, *PopAst* 58, nos. 9; 10 (1950); no. 11 (1951); *Harvard reprint no.* 342, 1951, 18. Eddington years later told Hertzsprung, according to Nielsen, that "[o]ne of the sins of your youth was to publish important papers in inaccessible places." This must have been frustrating to Hertzsprung, because he sent Eddington his earliest papers in August 1907. Hertzsprung to Arthur Stanley Eddington, 11 August 1907. Fiche 13/6–7, UA/HER.

59. Schwarzschild's and Hertzsprung's moves to Potsdam were noted, for instance, in "Scientific Notes and News," 1909, *Science* 29:736; 30:914. Hertzsprung was listed seven times in the 1909 *Astronomische Jahresbericht*.

60. Hall, 1900, 18–19.

61. Russell had detected a systematic deviation between his measured parallaxes and those derived from Kapteyn's own mean parallax formula that Kapteyn decided was real, and "the best proof of the reality and accuracy of Russell's parallaxes." Kapteyn to Russell, 14 March 1911. PUL/HNR.

62. Hufbauer, 1991, makes this point regarding the acceptance of Bernard Lyot's work.

63. Russell, 1910a. His final report to the Carnegie, Russell, 1911b, was completed in October 1910, "which has felt like a millstone round my neck." Russell to Edwin Frost, 14 October 1910. YOL/EBF. On astronomers' impressions, see Lundmark, 1922, 148.

64. Mitchell to Russell, 19 January 1921. PUL/HNR. [anon.], 1906, "The Creation of a Star," *Scientific American* 95:323–324.

65. These courses are noted in *Princeton University Course Catalogue*, 1906–1907, 180–181. Russell to Wilson, 6 April 1905. Link et al., 1971–1976.

66. H. N. Russell, "Lecture notes for senior practical astronomy," 14 March 1907 entry. PUL/HNR.

67. Ibid.

68. Ibid.

69. [1910 or earlier]. "Lecture notes: Junior Astronomy." Box 70.11, PUL/HNR.

70. Ibid. This discussion follows DeVorkin, 1977a, 67–68.

71. Russell to Pickering, 5 March 1910. HUA/ECP.

72. Lewis Boss to Russell, 20 December 1910. PUL/HNR.

73. DeVorkin, 1981. 44 n. 37.

74. Lockyer to Russell, 5 October 1910. PUL/HNR.

75. Russell to Lockyer, 14 January 1911. PUL/HNR.

76. Eddington, 1942. Mathematical astronomers like Forest Ray Moulton were finding that much of their work had been anticipated by Ritter's "celebrated series of papers." Moulton to Woodward. 2 February 1909. CIWA/FRM.

77. Russell to Lockyer, 14 January 1911. PUL/HNR. Emphasis in original. The parenthetic remark indicates words crossed out by Russell. James Jeans, review of "Gaskugeln," *ApJ* 30:72–74.

78. Ibid. Russell rewrote this paragraph several times, clearly trying to find a way to say that he believed that spectroscopic criteria were important, but he never could find the right words.

79. DeVorkin and Kenat, 1983a.

80. Russell to Hertzsprung, 27 September 1910, quoted in Nielsen, 1963, 244.

81. [Russell], "A Provisional theory of Stellar Evolution," unpublished manuscript, n.d. ca. late 1910. Box 100.25, PUL/HNR.

82. Ibid., 3–5.

83. Hertzsprung to Russell, 11 October 1910. PUL/HNR. In his first paper in 1905, Hertzsprung agreed with Miss Maury's suggestion that her c-stars formed a collateral evolutionary series: two series proceeding linearly from nebulae. He said the same to Eddington in 1907: Hertzsprung to Arthur Stanley Eddington, 11 August 1907. Fiche 13/6–7, UA/HER.

84. Hertzsprung to Russell, 11 October 1910. PUL/HNR.

85. Russell to Pickering, 3 June 1911. HUA/ECP.

Chapter 7
Building a Life at Princeton

1. HNRjr., 12.

2. Parenthetic remarks added by Keener. Russell to Keener, 9 September 1907. *Duodecennial record of the class of eighteen hundred and ninety-seven*. 1909:180. PUA

3. MRE #1, 16–17; 22–23; #2, 8–9. There are some variations between the two recollections.

4. MRE #1, 43; HNRjr., 28.

5. The *Times* spent most of its ink describing what the women wore and the music. "Prof. Russell Weds Miss Cole," *New York Times*, 25 November 1908, 9.

6. Abridged Record; HNRjr., 28; MRE #1, 4; 26. Stewart, 1958, 312, confirmed the leg injury. In her 1911 will, Ada Norris gave the house to Russell, with some $6,000 each to his two brothers. She added another $3,000 to each brother in 1913, from good stock investments. Ada Louise Norris "Last Will and Testament and Codicil," probated January 30, 1914. Superior Court of New Jersey, Records Center. Courtesy Shirley Maiorino.

7. MRE #1, 16.

8. Russell to Pickering, 29 July 1910; 1 July 1911. Pickering to Russell, 3 August 1910. HUA/ECP. MRE #1, 14.

9. Russell to Pickering, 9 May 1912. HUA/ECP.

10. Russell to Pickering, 12 August 1912. HUA/ECP.

11. Russell to Pickering, 12 August; 1 November 1912. HUA/ECP; PUL/HNR.

12. It had been an attack of eclampsia. HNRjr., 1. *Encyclopedia Britannica*, 11th ed., s.v. "Caesarean Section."

13. Mitchell to Russell, 14 November 1912. PUL/HNR.

14. Russell to Pickering, 1 November 1912. HUA/ECP.

15. Rev. Charles S. Wightman, "Address," *In Memoriam, Reverend Alexander Gatherer Russell*. n.p. Oyster Bay Presbyterian Church archives. Bright's disease is a disorder of the kidneys.

16. His congregation was very aware of his pending death. See George E. Farrar and William L. Swan, in ibid. There is no recorded mention in Russell's, Schlesinger's, or Pickering's papers. Russell incorrectly lists his father's date of death as 1912 in "Abridged Record" without comment beyond its cause. The circumstances of Alexander Russell's death are in Johnston and Reifsnyder, 1990, 76–77. HNRjr., 12.

17. Gordon M. Russell to "Dear People," 5 June 1913. PUL/HNR.

18. Russell to Pickering, 6 February 1914. UAV630.17.5 File 2 1875–1910 Popular—QUI, HUA/ECP.

19. Russell to Pickering, 12 March 1914. HUA/ECP.

20. Russell to "Coal Co.," 16 April 1914. This shopping list was found buried among Russell's research notes, "Effects of Tidal Evolution upon Visual and Spectroscopic Binaries Book I." Box 80.12 Research Notes. PUL/HNR.

21. Data compiled from early editions of the *American Men of Science* by Paul M. Routly, based upon listings and commentary in Visher, 1947, 376–381. Astronomers constituted one of the older groups.

22. H. G. Murray to T. W. Hunt, 12 November 1909. Wilson Presidential Files, PUA, file A-WWP-F.23. Parker D. Handy to "Dear Sir," 12 January 1910. Graduate Council records. PUA.

23. Wilson was then a candidate for New Jersey governor. Bragdon, 1967, 371; 402.

24. Link et al., 1971–1976, *19*:115–116. Jeans, however, had been quite productive: not only had he managed to wed Charlotte Tiffany Mitchell, a member of the prominent New York Tiffany family, but he completed his *Theoretical Mechanics* (1906) and *Mathematical Theory of Electricity and Magnetism* (1908) during his Princeton years.

25. Shenstone OHI, 1979, 6–7. AIP. Roberts, "Memoir," xvi, in Milne, 1952.

26. Bragdon, 1967, 324–325; 328.

27. *Minutes of the Trustees of Princeton University Volume XI*, 484, meeting of 13 April 1911. "Report of the Committee on the Curriculum," Ibid., 10 June 1912 entry, 702. PUA. Russell to Pickering, 12 August 1912. PUL/HNR. Various records indicate that Russell did not become chairman of the department until 1916. As late as 10 November 1913, course offerings were listed under "The Department of Mathematics and Astronomy" in the "Minutes of the Faculty Committee on the Graduate School," 22. PUA.

28. Schlesinger to Russell, 9 August 1912. PUL/HNR.

29. Link et al., 1971–1976, *19*:139: [H. B. Fine], "Departments of Mathematics and Physics," ca. 31 March 1909. PUA.

30. "General Log Book," 23-inch telescope. Box 4, Department of Astronomy records, PUA.

31. Missouri had trouble replacing Seares. R. T. Crawford to A. Ross Hill, 13 April 1910. Missouri file, BANC/DOA.

32. Russell to Pickering, 2 July 1911. HUA/ECP; to Frost, 2 July 1911. Folder 1 #21800, YOL/EBF.

33. Pickering to Russell, 5 July 1911. 1910–1911, Russell folder, HUA/ECP.

34. Russell to Pickering, 2 July 1911. HUA/ECP.

35. Mitchell went to Michigan. Mitchell to R. H. Curtiss, 16 August 1911. UMBL/RHC.

36. R. H. Baker took the job. Curtiss to Mitchell, 13 September 1911. UMBL/RHC. Dugan faculty file. PUA.

37. Mitchell to Curtiss, 5 September 1911. UMBL/RHC.

38. Russell to Pickering, 13 July 1911. HUA/ECP. MRE #1, 29; #2, 1–4.

39. Daniel ended up working at Allegheny as an assistant. Dugan to Campbell, 15 February 1909. LAUC/WWC. Russell to Lewis Boss, 16 December 1909. PUL/HNR.

40. Seares to Crawford, 8 March 1909, emphasis in original; Leuschner to Seares, 29 May 1911; Seares to Leuschner, 3 June 1911. BANC/DOA.

41. "Report of the Committee on the Graduate School," *Minutes of the University Faculty of Princeton University 1902–1914*, entry for 6 May 1912. PUA. Shapley accepted the fellowship on 25 April 1911. O. D. Kellogg to Andrew F. West, 21 February 1911; Shapley, "Application for Graduate Fellowship," graduate file, PUA. "General Log Book," 23-inch telescope. Box 4, Department of Astronomy records, PUA. Campbell to Wheeler, 28 March 1903. LAUC/WWC. Shapley, 1969, 19–22; 26.

42. Shapley, 1969, 31–32. Russell's later students noted similar behavior. See Spitzer and Ostriker, 1997, 505–509.

43. "General Log Book," 23-inch telescope.

44. Shapley graduate file, PUA. Shapley recalls taking very few formal courses. Shapley, 1969, 33.

45. Russell, 1954, 105. Shapley, 1969, 26, recites a slight variation on this, but the flavor is the same.

Chapter 8
Building a Case for Giants

1. On Chamberlin and Moulton's theories, see Brush, 1978; 1996b.

2. Brush, 1980, 49; Russell, 1910d, 187. Brush, 1974.

3. Russell, 1910d, 201–202. I am indebted to Karl Hufbauer for his comments on the style of theoretical astrophysicists.

4. Ibid., 203.

5. Ibid., 207.

6. Russell to Moulton, 17 February; 3 February 1910. PUL/HNR. Frost to Russell, 7 February 1910. Box 48.9, *ApJ* correspondence, YOL/EBF.

7. Russell, 1910b, 883.

8. Bauschinger, 1906, 649–650, quoted in Russell, 1912b, 315.

9. Russell, 1954, 105. Popper, 1967, 88; Wood, 1977, 48; Szafraniec, 1970, 7.

10. For the details, see Szafraniec, 1970.

11. Aitken, 1918, 168. Wood, 1977, 48, notes that it was called the "Russell model." The Russell method, along with its many refinements and extensions, mainly by Russell himself, became the standard mode of analysis for almost fifty years. On its longevity, see Szafraniec, 1970; Wilson and Devinney, 1971.

12. Russell, 1912b, 315; table 1, 333; 1912c, 54 ff.

13. Russell to Pickering, 23 December 1910. HUA/ECP. Russell, 1911a.

14. Russell, 1912d, 576.

15. Russell to Pickering, 4 January 1911. HUA/ECP.

16. Ibid.; Pickering to Russell, 22; 29 June 1911. HUA/ECP.

17. Russell, 1911a, 524–525.

18. Hertzsprung to Russell, 26 September 1912. PUL/HNR.

19. Brown provided the theory, Russell the methods of analysis, and Pickering the data. Russell, 1920a, 105.

20. Kopal, 1986, 158.

21. Russell to Pickering, 8 November; 9 December 1911. HUA/ECP.

22. Russell to Pickering, 9 December 1911. HUA/ECP.

23. Kushner, 1993, 198.

24. Russell to Pickering, 9 December 1911. HUA/ECP. This did not appear in the abstract to his talk. Russell, 1912e, 708.

25. Pickering to C. L. Doolittle, 17 November 1911; 2 January 1912. Box DARR-DRAX, HUA/ECP.

26. Russell, 1912d, 570–571.

27. Russell to Pickering, 6 January 1912, rough draft, PUL/HNR, refers explicitly to Hertzsprung's Pleiades listings, and to the diagrams in Hertzsprung, 1911. Russell may not, however, have seen Rosenberg, 1911, but Hertzsprung's discussion was more complete and fully references Rosenberg's work on the Pleiades and reproduces his diagram, along with others for the Hyades.

28. Russell to Pickering, 6 January 1912. PUL/HNR. Hertzsprung discusses his motives for using clusters in Hertzsprung, 1922, 92. This scenario contradicts the inferences drawn by Spence and Garrison, 1993, who examined only Russell's famous 1914(b) paper.

29. Kapteyn, 1910, 258. Campbell, 1911; Boss, 1912. On Kapteyn's style, see Paul, 1993.

30. Russell, 1912d, 574.

31. Nielsen, 1963, 238; Strand, in Philip and DeVorkin, 1977, 58.

32. Russell, 1912d, 577.

33. Ibid., 579.

34. Aitken to Russell, 15 October 1912. Lewis Boss to Russell, 17 January 1912. Pickering to Russell, 29 March 1912; Russell to Pickering, 30 March 1912; Pickering to Russell, 3 April 1912. PUL/HNR.

35. Shapley, 1913a, 29.

36. Russell to Parkhurst, 12 November 1912, PUL/HNR. Russell, 1913d, 646.

37. Russell, 1913d. On attempts to use the radiation law to predict diameters of stars, see DeVorkin, 1975.

38. Shapley, 1913b.

39. Hale to Pickering, 2 April 1913. HUA/ECP. DeVorkin, 1994, "Origins of the IAU," abstract, August: IAU Netherlands.

40. Russell to Hale, 17 February 1913. AIP/GEH; to Pickering, 24 March 1913. PUL/HNR.

41. Russell to Adams, 15 February; to Schlesinger, 22 February 1913. PUL/HNR.

42. R.T.A. Innes to Russell, 1 September 1913. Campbell to Russell, 16 May 1913; Russell to Campbell, 23 May 1913; Campbell to Russell, 28 May 1913, PUL/HNR.

43. Harlow Shapley graduate file, PUA. Shapley, 1969, 43.

44. Hertzsprung to Russell, 12 June 1913. Shapley to Russell, 24 June 1913. PUL/HNR.

45. Hertzsprung to Russell, 12 June 1913. PUL/HNR.

46. Ibid.

47. Russell, 1913b.

48. Eddington, 1913b, 286.

49. Russell, 1913a, 326.

50. Russell to Campbell, 23 May 1913. PUL/HNR.

51. Russell, 1913a, 327.

52. Ibid. The exact orientation of Russell's diagrams is not known, as the originals have never been recovered.

53. Memorabilia. Box 89.12: Conference Notes, Astronomische Geselleschaft Hamburg 1913: 2 folders. PUL/HNR.

54. Lockyer to Russell, 17 June 1913. PUL/HNR.

55. Eddington to Russell, 28 June 1913, PUL/HNR.

56. Russell to Pickering, 14 February 1913. HUA/ECP.

57. Russell to Pickering, 24 March 1913. See also fragmentary notes and memorabilia, box 89.12;.13, PUL/HNR.

58. Turner, 1913, 382–386. Memorabilia and fragmentary notes, box 89.12. PUL/HNR. Russell, 1947b, 169.

59. Memorabilia and fragmentary notes, box 89.13, PUL/HNR.

60. Turner, 1913, 417.

61. Ibid.

62. Hertzsprung to Adams, 14 November 1914. CIT/GEH.

63. Hertzsprung to Karl Schwarzschild, 6/12 January 1915. Letter courtesy D. Herrmann.

64. Russell to Pickering, 24 March 1913; Russell to Schlesinger, 24 November 1913. PUL/HNR.

65. Russell, 1913e.

66. Russell to Pickering, 21 February 1913, PUL/HNR.

67. Russell to Shapley, 5 August 1913. PUL/HNR.

68. Fowler, 1921, 339.

Chapter 9
At the Theoretical Interface

1. HNRjr., 29.

2. Russell, 1912f, 164–166; 167–168. On the Proctor endowment, see Egbert, 1947, 137.

3. Quote from Kevles, 1978, 70. Burnham, 1971, 253.

4. Russell to Hibben, 13 December 1913. Princeton University, President Hibben folder, PUL/HNR.

5. Pickering to Archibald D. Russell, 16 January 1915. UAV630.17.5 box ROO-SCL, HUA/ECP. Copy in PUL/HNR under A. D. Russell, incoming. On Pickering's funding efforts, see Plotkin, 1978; 1990. A. D. Russell to H. N. Russell, 5 February 1915. PUL/HNR.

6. Since their courses were doubled up and some were not taught, Russell and Dugan were well below the campus average of 11.5 hours. Princeton "Courses of Instruction" for the years 1912–1916. See also 1909–1910 statistics from W. Wilson Papers, Box, File A—WWP—F18b, Magie to Wilson, 24 February 1909, and "Hours of Teaching in Departments, First term 1909–1910." PUA. Russell to Hibben, 13 December 1913. Princeton University, President Hibben folder, PUL/HNR.

7. Shapley to Seares, 27 April; 1 September 1912. Seares to Shapley, 2 October 1912. Box 15.321: "Shapley 1912–1914" folder. HL/FHS.

8. Shapley to Seares, 19 November 1912. Box 15.321., HL/FHS.

9. Seares to Shapley, 19; 27 November 1912. Box 15.321, HL/FHS.

10. Seares to Shapley, 15 March 1913, Box 15.321, HL/FHS.

11. Shapley to Seares, 17 December 1913, Box 15.321, HL/FHS. Shapley to Russell, 14 April 1914. PUL/HNR.

12. Shapley to Russell, 5 May 1914. PUL/HNR. Stebbins, 1910.

13. Russell to Hibben, 3 December 1913. PUL/HNR.

14. Shapley to Russell, 20 May 1914. PUL/HNR.

15. The decision process through which Eddington gained power requires further study, not only as an indicator of changing directions in astronomy, but as a case study of Cambridge politics. Russell to William Hobbs, 21 September 1944. PUL/HNR.

16. Newall, 1911, 349. Eddington, 1914, v. On the mathematicization of statistical astronomy, and Eddington's place in it, see Paul, 1993, 116–117; 136–140.

17. Kapteyn, 1910, 258.

18. Eddington, 1913a, 471. For similar objections, see Puiseux, 1916, 137; Waterman, 1913, 196–199; Charlier, 1917, 389.

19. P. Fox to Russell, 20 November 1913; Russell to Fox, 24 November 1913; Russell to Pickering, 24 November 1913. PUL/HNR.

20. Russell to Schlesinger, 24 November 1913. PUL/HNR.

21. Russell to Pickering, 6 February 1914. PUL/HNR.

22. Russell, 1914b, pagination from *PopAst*.

23. Russell, OHI, 1951.

24. Russell, 1914b, 342–344.

25. Ibid., 342.

26. Ibid., 343; 345.

27. Ibid., 346–347, nn. 1, 2.

28. Ibid., 348.

29. Ibid., 349.

30. Ibid., 350.

31. Ibid., 350–351.

32. Russell, 1914a, 165.

33. Ibid., 171–172; 173.

34. Ibid., 173–174.

35. Eddington to Russell, 1 March 1914. PUL/HNR.

36. Hufbauer, 1981, 284; Kenat, 1987, sec. 5.1.3, 273 ff.

37. Hale, 1908, 20.

38. Rutherford would speak on the elements, T. C. Chamberlin on the origin of the solar system, and other senior scientists would provide essays on evolutionary development from the perspectives of paleontology, mutation theory, genetics, social history, and social thought. Hale to Rutherford, 24 November 1913. CIT/GEH.

39. Campbell to Russell, 28 May 1913. LAUC/WWC.

40. Russell to Campbell, 25 May 1914. LAUC/WWC.

41. Campbell, 1915, 180. Waterman, 1913; Aitken to Russell, 15 October 1912; Adams to Russell, 5 March 1913. PUL/HNR.

42. Fowler, 1915, 392.

43. Hoffleit, 1950, 20–21.

44. Adams, 1913, 89–92. Paul, 1993, 168–169.

45. Adams to Russell, 5 March 1913. PUL/HNR; Hale to Kapteyn, 6 January 1914; Hertzsprung to Adams, 14 May 1914. AIP/CIT/GEH.

46. Hale to Kapteyn, 29 May 1914. AIP/CIT/GEH. Adams and Kohlschütter, 1914, 386.

47. Adams, 1916b; Adams and Joy, 1917.

48. Russell, 1916a, 588.

49. Eddington to Adams, 18 May 1916, quoted in DeVorkin and Kenat, 1983, 108. Russell to Shapley, 15 May 1916. PUL/HNR.

50. Hale to Lockyer, 9 April 1916. AJM/NJL.

51. Adams to Eddington, n.d., CIT/GEH; to Russell, 22 January 1917. PUL/HNR, quoted in DeVorkin and Kenat, 1983, 109–110.

52. Kapteyn to Seares, 7 July 1917. HL/FHS. DeVorkin, 1978, 271. Adams believed that Kapteyn's mean parallax formula "is in serious error for these [giant] stars." Adams to Russell, 25 May 1916. PUL/HNR.

53. Benjamin Boss to Adams, 14 October 1918. Box 1.15, HL/WSA. Kapteyn's disbelief in giants is also expressed in Kapteyn to Hale, 7 October 1918. CIT/GEH.

54. Kapteyn to Russell, 14 March 1911. PUL/HNR. Brashear, 1992; DeVorkin, 1999a.

55. Hertzsprung to Adams, 20 January 1915. CIT/GEH.

56. Schlesinger to Russell, 6 June 1916. PUL/HNR. Dyson's sympathies for the theory can be found in Fowler, 1915, 384; and Dyson to Russell, 31 December 1915. PUL/HNR. Russell was proposed by R. A. Sampson, Alfred Fowler, and Eddington. *MN* 76:447.

57. Eddington's remarks in Fowler, 1915, 387.

58. Eddington to Lockyer, 6 March 1916. AJM/NJL. F. W. Dyson to Russell, 31 December 1915. PUL/HNR.

59. Eddington, 1915b, 392; 396.

60. Hufbauer has noted Schuster's influence on Eddington. Hufbauer, 1981, 284 n. 15.

61. Kenat, 1987, chap. 5, 265–269, provides a detailed discussion of the development of Eddington's theoretical work.

62. Russell, 1913c, Shapley recalls giving a lecture on ways to determine distances to Cepheids while at Princeton, but it appears that he did this assuming they were binaries. Shapley, 1969, 36.

63. Shapley, 1914, 460.

64. Eddington, 1917a, analyzed in Kenat, 1987, 282–285.

65. Eddington, 1917b, 612. Kenat, 1987, 292–293. Eddington, 1926, 397–398. Opacity decreased as the star expanded, allowing the star to radiate its excess energy when it was coolest. Just the opposite took place when the star contracted to minimum volume.

66. Eddington, 1917b, 606.

67. Eddington to Russell, 31 January 1917. PUL/HNR. Kenat, 1987, 299–300.

68. Eddington to Russell, 31 January 1917. PUL/HNR. Eddington, 1917b, 597; 612. Eddington had started with a mean molecular weight of 54, but "[t]he quantum experts Jeans and Lindemann" as well as H. F. Newall advised him that ionization would greatly reduce that number. Eddington, 1938b, 174.

69. Kenat, 1987, 300–303; 315.

70. For a complete discussion, see Hufbauer, 1981, 284–286.

71. Rutherford may have been influenced by his teacher Bickerton, who tried to explain novae by collisions and expansion. Rutherford's suggestion did raise the question of the energy source of the stars. Commentary in Fowler, 1915, 390.

72. Russell to Eddington, 27 July 1917. PUL/HNR.

73. Russell to Eddington, 27 July; Eddington to Russell, 12 August 1917. PUL/HNR.

74. Russell to Hale, 18 September 1917. CIT/GEH. Quoted in Hufbauer, 1981, 287 n. 21.

75. Hufbauer, 1981, 299.

76. Quoted in Kragh, 1995, 94. Among popularizers, see Macpherson, 1922, 261.

77. Fowler, 1921, 350. Russell was first nominated in 1919, along with Annie Cannon and Einstein. Einstein was the choice of the council but the membership balked and the medal was cancelled that year. All three were nominated the next year, but those who resisted relativity also resisted Einstein. Russell won in a compromise vote. *Royal Astronomical Society Council Minutes 11 1915–1928*, 100–101; 123–131. RAS Library. On aspects of this debate, see Clark, 1971, 232; 245–246; Tayler, 1987, 21–22, notes that Annie Cannon was nominated a total of nine times by members of the council but never won the coveted award.

78. Fowler, 1921, 348; 350. On Michelson's stellar interferometer, see DeVorkin, 1975.

79. Tayler, 1987, 22.

80. Fowler, 1921, 336.

81. Ibid., 350.

82. Russell to Shapley, 19 February 1919. PUL/HNR.

Chapter 10
Shifting Allegiance

1. Russell to Schlesinger, 1 July 1914. PUL/HNR.

2. Russell described Elizabeth's condition for the Carnegie Institution of Washington Eugenics Records Office and the Eugenics Society of the United States of America. Russell to Paul Brockett, 22 May 1935. "Abridged Record of Family Traits," NAS, Reel 11: frames 9543 to 9550, PUL/HNR. HNRjr, 1. Russell to Webster, 30 March 1942. SUL/DLW. My sense of the state of knowledge of cerebral palsy during Elizabeth's childhood and typical reactions to having a spastic child comes from George, 1992; Bower, 1993; Phelps, 1966.

3. Russell to Dyson, 27 June 1916. PUL/HNR.

4. Russell, 1914c; DeVorkin and Kenat, 1983b; Shapley, 1915, 459.

5. Bell, 1917.

6. Their ancestors had indeed been neighbors. Russell to Pickering, 3 July, 14 July; Pickering to Russell, 8 July, 17 July 1914. PUL/HNR; HUA/ECP.

7. Phrase taken from the title to Kevles, 1971, 47.

8. Bates, 1965, 121; Cochrane, 1978, 209–215. Kohler, 1991, 73–77; 82–85.

9. Russell hoped to demonstrate that not all scientists were atheists. "Scientific Approach to Christianity I," 1–4, ca. 1916–1917. Box 72.1 (a), PUL/HNR.

10. Ibid., 8; 10–11; 13. Moyer, 1997, 183. Heilbron, 1986, 58, shows that rhetoric like this was a typical scientist's response to accusations that science was "an enemy of faith and family. . . ." At the turn of the century, most physicists would say that physics aimed not at the truth "but only at an exact description of phenomena."

11. Russell to Frank Schlesinger, 15 March 1917. PUL/HNR.

12. Plaskett, 1911, 256.

13. Eric Doolittle to Russell, 17 May 1914. PUL/HNR.

14. Russell to Doolittle, 8 June 1914. See also Box 80.14: "Masses of Stars." PUL/HNR.

15. On how rank-and-file astronomers regarded Burnham, see, for instance, Fox, 1915, "Dedication."

16. Letters ca. 1912–1914, American Philosophical Society file, file 314, HUA/ECP.

17. Pickering, 1909b, 106; General Notes, *PopAst* 17:659. Lankford, 1997.

18. Jones and Boyd, 1971, 250–252; Plotkin, 1978a, 45.

19. Pickering, quoted in Plotkin, 1990, 47. Reingold, 1977; Kohler, 1991, 7.

20. Plotkin, 1978a, 50–51. Pickering, 1909b, 110.

21. Plotkin, 1978a; Reingold, 1968, 239.

22. Plotkin, 1978a, 53; Kohler, 1991, 79–80. Frost felt that prioritizing needs was dangerous, although Hertzsprung wholly agreed that more "common soldiers" were needed to assist the "officers." Frost to Pickering, 1 March 1915. Hertzsprung to Pickering, 24 March; 29 April 1915. HUA/ECP.

23. Pickering to Campbell, 7 November 1916. LAUC/WWC. Schlesinger to Pickering, 22 September; 9 November 1916. HUA/ECP. "Memorandum" attached to "Sub-Committees of the Committee of One Hundred on Scientific Research," reel 12, frame 841, PUL/HNR.

24. Cronon, 1991, 348; Miller, 1996; and Ehrenhalt, 1996, 1; 10, all describe the Chicago of Hale's youth. On Hale as a power broker, see Wright, 1966; Osterbrock, 1993; and Lankford, 1997.

25. Hale to Goodwin, 24 December 1903. HL/GEH.

26. Hale to Goodwin, 28 February 1904. Letter 2859, HL/GEH.

27. Arthur L. Day (Home Sec'y, NAS) to Pickering, 20 May 1913; 22 December 1915. Box 2, No. 322, "National Academy of Sciences"; No. 323, "NAS Committee on Research," HUA/ECP.

28. Cochrane, 1978, 177–178; 194–199; 209–215; Kohler, 1991, 73–77; 82–85. Kohler argues that Hale systematically co-opted the AAAS committees.

29. Hale to Pickering, 19 December 1916. "National Research Council"; "Purpose of the Central Committees Dealing with the Various Branches of Science." NRC chair of the Census Committee, 26 February 1917. 325 file, "National Research Council," HUA/ECP.

30. Hale to Pickering, 19 December 1916, 4; Pickering to Hale, 18 January 1917. 325 file, "National Research Council," HUA/ECP.

31. Russell to Pickering, 11 November 1916. Russell notes, box 80.8, PUL/HNR. Russell planned to collaborate with Herbert Ives, a physicist at Western Electric. Russell to Frost, 11 December 1916. *ApJ* 2:8, YOL/EBF.

32. The earliest draft extant is contained in Pickering to Members of Committee, "Proposed Sketch of Report," 21 February 1917. No. 325A, "NRC Sub-Committee on Astronomy," HUA/ECP. This was very similar to the published version: Pickering, 1917.

33. Russell to Pickering, 14 December 1916. PUL/HNR.

34. Russell to Pickering, 27 February 1917. PUL/HNR.

35. Stebbins to Pickering, 2 March 1917; Campbell to Pickering, 29 March 1917; Pickering to Eichelberger, 3 March 1917; to Comstock, 6 March 1917; to Schlesinger, 6 April 1917. Folder 325A, "NRC Sub-Committee on Astronomy," HUA/ECP.

36. Hale to Pickering, 16 March 1917. CIT/GEH.

37. Ibid. Hale said much the same things to Kapteyn during this time. See Smith, 1982, 67–68.

38. Hale to Pickering, 16 March 1917. CIT/GEH. Wright, 1966, 219–221, recounts Hale's interest in Zeeman's papers prior to his discovery in 1908.

39. Pickering circular letter, 3 April 1917. File 325A: NRC Sub-Committee on Astronomy, HUA/ECP. Pickering to "My dear Professor," 3 April 1917, cover letter to draft "Report of the Astronomy Committee." CIT/GEH; PUL/HNR.

40. Russell appointment letter to NRC committee, 16 February 1917. Reel 11:9668, PUL/HNR.

41. Russell to Pickering, 6 April 1917; Pickering to "My dear Professor," 23 April 1917, attached to "Extracts from Letters." PUL/HNR; Hale to Pickering, 29 May 1917; Pickering to S. W. Stratton, 31 May 1917. HUA/ECP.

42. Pickering to Hale, 20 June 1917. 325A: NRC Sub-Committee on Astronomy, HUA/ECP. Pickering, 1915, 82–85.

43. Pickering to Eliot, 8 August 1917. Box 376, Eliot Papers, UAI.5.150, HUA.

44. Hale to Russell, 16 July 1917. PUL/HNR.

45. Russell to Pickering, 21 July 1917. PUL/HNR.

46. Pickering to Russell, 23 July 1917. HUA/ECP.

47. Russell to Hale, 18 September 1917; Stebbins to Hale, 5 September 1917. Reel 49, NRC files, CIT/GEH.

48. Rough undated notes, fragment, ca. 30 August 1917. R11:9675, PUL/HNR. See also Russell to Hale, 18 September 1917. Reel 49, NRC files, CIT/GEH.

49. Hale to Russell, 25 September 1917. PUL/HNR.

50. Russell to Pickering, 31 October 1917. PUL/HNR; Russell to Hale, 18 September 1917. Reel 49, NRC files, CIT/GEH.

51. Russell to Pickering, 31 October 1917. PUL/HNR.

52. Pickering to Russell, 7 November 1917. PUL/HNR.

53. Russell to Pickering, 22 November 1917. PUL/HNR.

54. Ibid. Emphasis in original.

55. Ibid.

56. Ibid.

57. Pickering to Russell, 28 November 1917. PUL/HNR.

58. Russell to Pickering, 13 August 1918. Folder 313 HUG 1690.6.5, HUA/ECP.

59. Jones and Boyd, 1971, 443.

60. Russell, 1919a, 152.

61. Ibid., 153–154.

62. Ibid., 154–155.

63. Hale to Russell, 24 August; 30 November 1917; 3 January 1918. PUL/HNR.

64. An undated draft in Russell's papers included a tabulation that showed nineteen papers from 1918 and two from 1919 alone. Russell to Hale, 1 February; 13 February 1919; Manuscript, "Problems of Sidereal Astronomy." Box 83.1, PUL/HNR.

65. Russell to Hale, 1 February 1919. IAU File, PUL/HNR.

66. Russell, 1919/1920, 212.

67. Ibid., 273.

68. [1910 or earlier] "Lecture notes: Junior Astronomy." Box 70.11, PUL/HNR.

69. Russell to Eddington, 27 July 1917, PUL/HNR. Quoted in Hufbauer, 1981, 286–287.

70. Russell to Hale, 18 September 1917. PUL/HNR.

71. Russell and Hale pushed to have physicists invited to a formative meeting of the American delegation to the IAU in March 1919. Russell to Hale, 25 February 1919. IAU file, PUL/HNR.

72. Russell, 1919/1920, 214; 275.

73. Aitken to Russell, 16 January 1920. PUL/HNR.

74. Eddington, 1920, 19–20.

75. Seares, 1922a, 252.

76. Struve, 1943, 470; Struve, "The General Needs of Astronomy," 19 January 1955. NSF folder, Berkeley astronomy records, BANC/DOA.

77. Russell, 1920, "Some Problems," 270.

Chapter 11
The Great War

1. On Princeton's prowar stance, see Peterson and Fite, 1957, 3–4. "Dodge Hall Talks," December 1916–January 1917. Box 72.1 (d); "Intellectual Aspects of Religion [Part V] Revealed Religion: Christianity—Possibility of Revelation." Box 72.2 (e), PUL/HNR.

2. Russell to Shapley, n.d. This portion of the rough draft letter was crossed out. Shapley to Russell, 24 June 1913. PUL/HNR. Russell, 1913e, 187.

3. Russell to Frost, 10 February 1917. PUL/HNR.

4. On the clergy's prowar position, see Peterson and Fite, 1957, chap. 11. "common cause," Russell to Frost, 4 May 1917. PUL/HNR.

5. HNRjr., 8.

6. MRE #1, 17.

7. Russell to Hale, 20 January 1917. PUL/HNR.

8. Hale to Russell, 21 January 1917. PUL/HNR. On Hale's wartime organizing, see Kevles, 1978, 109–116.

9. Hale to John M. Clarke, 25 January 1917, cited in Cochrane, 1978, 223 n. 53.

10. Hale to E. C. Pickering, 29 May 1917. HUA/ECP. Kevles, 1978, 109–116.

11. Russell to Frost, 16 March 1917. PUL/HNR. On the huts, see Mayo, 1920, 13–25.

12. Gordon Russell to Henry Norris Russell, 5 April 1917. PUL/HNR. "Vocal Volcano" from Peterson and Fite, 1957, 74.

13. Russell to Frost, 16 March 1917. PUL/HNR. "Revolution in Russia; Czar Abdicates, Michael Made Regent, Empress in Hiding, Pro-German Ministers Reported Slain," front page, *New York Times*, Friday, 16 March 1917.

14. Russell to Frost, 29 March 1917. PUL/HNR. "talking war," Aitken to Campbell, 14 April 1917. LAUC/Aitken file.

15. Russell to C. V. Hibbard, 27 April 1917; and subsequent correspondence between Gordon and Henry. PUL/HNR.

16. Russell to Frost, 4 May 1917. PUL/HNR. On the selective service law, see Ayers, 1919, 17.

17. Cochrane, 1978, 210–213; 222–223; Reingold, 1968.

18. Cochrane, 1978, 231–235. On Trowbridge's war work, see Kevles, 1978, 126–127.

19. On sound-ranging systems, see Kevles, 1978, 127–130.

20. Russell to Hale, 18 September 1917. NRC files, CIT/GEH.

21. Quoted from Russell to Shapley, 18 November 1917. Russell to Pickering, 31 October 1917. PUL/HNR.

22. Russell to Frost, 13 December 1917. YOL/EBF.

23. Russell to Hale, 12 December 1917. CIT/GEH.

24. Russell to Hale, 6 January 1918. John Cole to Russell, 2 September 1917. PUL/HNR.

25. Russell to Hale, 6 January 1918. PUL/HNR.

26. Ibid.

27. Kevles, 1978, 130–131.

28. Russell to Pickering, 29 January 1918. PUL/HNR.

29. Russell to Adams, 15 January [1918], noted as 1917. Roll 96, CIT/GEH.

30. Russell to Hale, 18 January 1918. PUL/HNR. Russell faculty folder, PUA.

31. John Q. Stewart faculty file, PUA.

32. Russell to Hale, 7 January 1918. PUL/HNR.

33. Russell to Millikan, 18; 29 January 1918. NRC file, PUL/HNR.

34. Kevles, 1978, 133–134.

35. Ibid., 134–135. Even General Squier, a member of NACA, favored the NRC. Roland, 1985, 45–47.

36. Kevles, 1978, 134–135.

37. Russell to R. A. Millikan, 18 January 1918. NRC file; to Pickering, 19 March 1918. PUL/HNR. See also "War Records," 25 May 1920, Russell Faculty Folder, PUA.

38. Russell to Pickering, 18 April 1918. HUA/ECP. Mrs. H. N. Russell to Frost, 1 May 1918. YOL/EBF. Russell identified himself variously as "Engineer" and "Consulting and Experimental Engineer." "War Records" file, PUA. Russell to W. S. Adams, 12 May 1918. Roll 96 (R), CIT/GEH. Russell sometimes noted his location in his monthly *Scientific American* columns.

39. Kevles, 1978, 134.

40. Webster to Russell, 25 February 1919. PUL/HNR.

41. Webster to H. Randall, 26 November 1917; Adjutant General of the Army to First Lieut David L. Webster, Aviation Section, Signal Corps, U.S.R., 15 November 1917; G. W. Smith to Webster, 2 November 1917. Box 1, record group SC 131B, SUL/DLW. Webster OHI, 12–13. AIP.

42. Millikan to Air Division, Signal Corps, 26 February 1918. "Orders," 8 March 1918. Box 1, SUL/DLW.

43. Mixed materials, folder 3. Box 1, SUL/DLW.

44. "Individual Flight Reports," 31 August 1918. Folder 3, Box 1, SUL/DLW.

45. Russell heard that Pickering had devised a new azimuthal bearing plate, and wanted to know more about it. Russell to Pickering, 3 November 1918. HUA/ECP

46. Background on aerial navigation in this section is taken from Wright, 1972, 63–73.

47. Russell, Henry Norris, "Report on the Navigation of Aircraft by Sextant Observations," 20 January 1919, 10. Science and Research Department, Bureau of Aircraft Production. Box 5, SUL/DLW. Reprinted, in modified form, as Russell, 1919b. Russell noted the problems using artificial horizon sextants in Russell to Lt. Barrin, 5 December 1918. Reel 2, PUL/HNR.

48. Russell, 1919b, 136. On the problem of the vertical, see Dennis, 1994, 447–449; and MacKenzie, 1990, 66–68. Russell was, unlike most of his astronomical colleagues, fascinated by general relativity. Crelinstin, 1983.

49. Russell, 1919, "Report," 10.

50. They flew HSIL seaplanes, De Havillands (DH-4), and a Curtiss Jenny JN6H. Russell, 1919b, 139–140.

51. Russell refrained from a full description of the device, leaving that to its inventor. See fig. 7, "Bubble Telescope," in Russell, 1919, "Report." SUL/DLW.

52. Russell, 1919b, 140–141. Wright, 1972.

53. Russell, 1919b, 142.

54. Ibid., 144. Wright, 1972, 106.

55. Russell, 1919b, 149; Ault, quoted in Wright, 1972, 107. See also Dutton, 1928, chap. 20.

56. Russell to Shapley, 13 March 1919. PUL/HNR.

57. The drift sight is discussed in Webster to R. H. Young, Finance Division, Patents Department, Bureau of Aircraft Production, 20 January 1919. Box 1, folder 12, SUL/DLW. The patent was granted in July 1922, but they had lost interest in pursuing it. Russell to Webster, 6 July 1922. PUL/HNR.

58. Webster OHI, 16–17. "War Diary, 1918." Box 1, folder 6; Box 2, folder 1, SUL/DLW.

59. Box 1, folder 4, SUL/DLW.

60. Alexander Russell to Henry Norris Russell, 14 September 1918; 28 April 1919; and letters through 1919. PUL/HNR.

61. Alex to Henry, 3 March 1919; Russell to A. Wilmer Duff, 1 May 1936. PUL/HNR.

62. Pickering to Russell, 15 November 1918. HUA/ECP.

63. Millikan, quoted in Burnham, 1971, 309. On how the NRC portrayed itself during wartime, see Kevles, 1978, 117–154; Kohler, 1991, 82–83.

64. Russell to Shapley, 5 May 1919. PUL/HNR. On Russell's "remarkable insight into the problem of interstellar matter and absorption" and his efforts to caution Shapley, see Berendzen, Hart and Seeley, 1976, 77.

65. Webster to Russell, 25 February 1919. PUL/HNR. Russell to Webster, 11 June 1919. SUL/DLW.

66. Webster to Russell, 21 June; 10 July 1919. PUL/HNR.

67. Russell to Eddington, 27 July 1917. PUL/HNR. On the stellar energy problem, see Hufbauer, 1981, 288–293.

68. Russell to Shapley, 19 February 1919. PUL/HNR.

69. Russell to Shapley, 13 March 1919, quoted in Hufbauer, 1981, 295.

70. Conclusions based upon correspondence with Karl Hufbauer.

71. Russell to Webster (unsent draft), 6 May 1919. PUL/HNR. See also Russell, 1919c.

72. Russell to Webster, 9 June 1919, quoted in Hufbauer, 1981, 299.

73. Russell to Hale, 9 June 1919, quoted in Hufbauer, 1981, 298.

74. Hufbauer, 1981, 299.

75. Russell to Adams, 15 March 1917, quoted in DeVorkin and Kenat, 1983a, 109.

Chapter 12
Russell's Turn to Mount Wilson

1. Hoffleit, 1992, 92. Brown, E. W., [1918]. "Memorandum on the Winchester Observatory Addressed to the Board of Managers." Yale University Observatory archival files, courtesy Dorrit Hoffleit.

2. It had an annual income of $24,000, behind Mount Wilson, Harvard, Dudley, Yerkes, and Lick. See Stebbins to Pickering, 2 March 1917. Folder 325A, "NRC Sub-Committee on Astronomy," HUA/ECP. Hoffleit, 1992, 95.

3. Brown, [1918]. "Memorandum." Adams's decision is noted in Osterbrock, 1993, 155.

4. Russell to Hale, 9 June 1919. CIT/GEH.

5. Ibid.

6. Dugan to Russell, 16 June 1919. PUL/HNR.

7. Russell to Webster, 9; 11 June 1919. SUL/DLW.

8. Russell to Hale, 9 June 1919. CIT/GEH.

9. Hale to Russell, 18 June 1919, clipped to Russell letter of 6 June 1919. PUL/HNR. Russell to Hale, 25 June 1919. CIT/GEH.

10. Russell's $3,500 prewar salary, which had been supplemented during the war by a $1,400 stipend, was now officially $5,000 for the academic year. MRE #2, 50; "completely devoted," HNRjr, 1.

11. HNRjr, 1–2.

12. H. H. Turner to Joel Metcalf, 8 February 1919. File 5.160, 1919–1922 Folder No. 14 "Observatory," HUA/ALL. Kapteyn to Hale, 7 February 1919, quoted in Gingerich, 1988, 201. On Shapley's work, see Berendzen, Hart, and Seeley, 1976, 24–38; Smith, 1982.

13. Agassiz reported that a Harvard mathematician "did confess patronizingly that Russell knew as much mathematics as was necessary for an astronomer!" G. R. Agassiz to A. Lawrence Lowell, 7 March 1919. File 5.160, 1919–1922 Folder No. 14 "Observatory," HUA/ALL.

14. Russell to Shapley, 13 March 1919. PUL/HNR.

15. Bailey to Turner, 1 January; 9 January 1920. H. H. Turner folder, UAV630.17.5 box Box TIE-UTI. HUA/ECP.

16. Turner to Bailey, 20 January 1920. HUA/ECP.

17. Ibid.; Bailey to Turner ca. February 1920. HUA/ECP.

18. Frost to H. T. Stetson, 18 December; to Lowell, n.d.; Lewis Boss to Lowell, 10 May; Lowell to Agassiz, 31 October; 18 December; Agassiz to Lowell, 19 November; 17; 26 December, 1919. File 5.160, 1919–1922, Folder No. 14 "Observatory," HUA/ALL.

19. Gingerich, 1988, 203.

20. Russell to Shapley, 31 January 1920. PUL/HNR. "W.H.P." was Edward's younger brother, William H. Pickering. Jones and Boyd, 1971; Strauss, 1994, 45; Sadler, 1990, 62–64.

21. Shapley to Russell, 31 March 1920. PUL/HNR. Smith, 1982, chap. 2; Hoskin, 1976; Berendzen, Hart, and Seeley, 1976, 76–77; 115–116; 134; Hetherington, 1988, 97; 120.

22. Agassiz to Lowell, 28 April 1920, quoted in Gingerich, 1988, 204. In what has come to be known as the "Great Debate" but which historians have shown to be a rather pedestrian discussion, Shapley did not fare so well. Smith, 1982; Hoskin, 1976; Berendzen, Hart, and Seeley, 1976.

23. Russell to Hale, 13 June 1920. PUL/HNR.

24. Hale to Russell, 15 June 1920. PUL/HNR.

25. Russell to Hale, 20 June 1920. PUL/HNR.

26. Ibid.

27. Russell to "My Dear Stanley," 20 June 1920. PUL/HNR, reel 5, in Hale range. On Cobb's career ca. 1920, see White, 1984, 80–83. On administering to family, see Cobb to Charles P. Howland, 23 November 1921. Box 4, folder 88, CL/SC.

28. Russell to "My Dear Stanley," ibid. Russell curiously misquoted the humorist Stephen Leacock, whose character "rode madly off in all directions." Leacock, "Gertrude the Governess," in *Nonsense Novels*, 1912, p. 30, Dover edition.

29. Cobb to Abraham Flexner, 28 March 1925. Box 4, folder 75, CL/SC. Gingerich, 1988, 204–205, gives the impression that Russell tried to say no, but it is unclear what was going on in Russell's mind then.

30. Hale to Russell, 21 June 1920. PUL/HNR. On Hale's state of health during this period, see Wright, 1966, 270–273.

31. Russell to Campbell, 19 July 1920; Campbell to Russell, 27 July 1920. 1910–1923 range, LAUC/WWC. The matter of Harvard's finances is not raised in Jones and Boyd, 1971, nor in Gingerich, 1988, and therefore requires attention.

32. Hale to Russell, 20 October 1920. PUL/HNR.

33. Gingerich, 1988, 205. Hale to Lawrence Lowell, 11 December 1920. CIT/GEH. Agassiz to Lowell, 22 December 1920; Shapley to Lowell, 9 February 1921. File 5.160, 1919–1922 Folder No. 14 "Observatory," HUA/ALL.

34. Hale to Russell, 19 January 1921. PUL/HNR.

35. Russell to Shapley, 25 January 1921. PUL/HNR.

36. Ibid.

37. Russell to Hibben, 26 January 1921. PUL/HNR.

38. Hale to Russell, 20 February 1921.

39. Hale to Rutherford, 1 June 1914. CIT/GEH.

40. Hale, 1897, 311.

41. Hale to Charles E. Mendenhall, 13 January 1905; 3 March 1905; Mendenhall to Hale, 26 March 1905. HL/GEH.

42. Campbell to A. S. King, 13 June 1903; to H. Kayser, 13 June 1903; King to Campbell, 27 July 1903. LAUC/WWC.

43. King to Campbell, 12 June 1903. A. S. King 1903 file, LAUC. Campbell to King, 29 July 1903. Letterbook Volume 80, LAUC/WWC. Hale—King correspondence, CIT/GEH.

44. Colby to "Dear Randalls," 17 March 1915. UMBL/PP.

45. Osterbrock, 1993, 152.

46. Biographical data, HL/PWM.

47. Hale to Merrill, 1 June 1918; Merrill to Hale, 19 August 1918. Box 1GHIJ22, HL/PWM. Emphasis in original.

48. J. H. Moore to Merrill, 27 May 1920. HL/PWM.

49. Merrill to Meggers, 1 March 1920. Box 1, AIP/WFM. Emphasis in original.

50. On Hale's goals in creating Caltech, see Goodstein, 1991, 45; Wright, 1966, 247; Kevles, 1978, 155; Wright, Warnow, and Weiner, 1972, 87.

51. Hale to Kapteyn, 26 September 1917. CIT/GEH.

52. DeVorkin, 1975.

53. Eddington, 1920, 17.

54. DeVorkin, 1975, 12.

55. Ibid., 17, n 46.

56. Russell, 1920b.

57. Russell to Hale, 13 June 1920. PUL/HNR. Moyer, 1997, 261–263, explores the interplay of professional fulfillment and sense of duty in Joseph Henry's life, coming to the same conclusions about how Henry managed to synchronize the two.

58. In 1918, Dugan was still an assistant professor earning $2,500 annually. In January 1920 Russell secured a raise to $3,000, and then in June to $3,500 with the house. Dugan faculty file, PUA.

Chapter 13
Rationalizing Stellar Spectra

1. Russell to William Meggers, 20 January 1927. PUL/HNR.

2. On Adams's report, see Sopka, 1988, 84–86. Russell to C. E. Mendenhall, 4 February 1920; Mendenhall to Russell, 9 February 1920. NRC files, PUL/HNR.

3. Milne, 1924, 95, quoted in DeVorkin and Kenat, 1983a, 110, from which much of this discussion is derived.

4. Russell to Adams, 18 December 1920, quoted in DeVorkin and Kenat, 1983a, 116–117. On Saha's contributions, see DeVorkin, 1993; 1994a.

5. Saha, 1920. DeVorkin, 1993, 167; DeVorkin and Kenat, 1983a; DeVorkin and Kenat, 1983b. J. W. Nicholson, professor of mathematics at the University of London, in an early application of Bohr theory, tried to use the coronal spectrum as a means to deduce theoretically the nature of the structure of the "coronium" atom. Prior to reading Saha's papers, Russell, 1919/1920, 273–274, regarded Nicholson's attempt as a "useful possibility." Nicholson, 1913; 1916. Robotti, 1983.

6. King, 1922, 385; Russell, 1922d.

7. Saha to Hale, 9 July 1921. PUL/HNR.

8. On Saha's fate, see DeVorkin, 1994a.

9. Russell, 1919/1920, 214.

10. Russell, 1921d, 280. Similar rhetoric can be found in Hale, 1922, 68; 75–76; 78.

11. Aitken to Merrill, 12 August 1921. HL/PWM.

12. Fowler and Milne, 1923, 404; 424. See also Milne, 1921, 261; Milne, 1945, 70–71.

13. Walter Baade, quoted in Schwarzschild OHI, 26 August 1991, 10. Bowen OHI, 12.

14. Hale to Russell, 30 September 1921. PUL/HNR.

15. Hale to Hibben, 30 September 1921, clipped to Hale to Mrs. Russell, 4 November 1921. PUL/HNR.

16. Hale to Russell, appended remarks to Hibben, 30 September 1921. PUL/HNR.

17. Hale to Hibben, 30 September 1921, clipped to Hale to Mrs. Russell, 4 November 1921. PUL/HNR.

18. Hibben to Hale, 6 October 1921. PUL/HNR; HL/GEH.

19. Hale to Russell, 24 October 1921. PUL/HNR. Sanford to Slipher, 25 June 1926. LowA/VMS.

20. Merrill to Meggers, 2 March 1922. Box 1, AIP/WFM.

21. Adams to Russell, 5 February 1923. Box 3.49, Russell 1921–1923, HL/WSA.

22. Russell, 1927f, 45–46; 56–57.

23. Russell, 1922a, 356–357.

24. "essentially pragmatic," from Schweber, 1990, 390. Olson, 1967, 406–408; Hickman et al., 1964. I am indebted to David Cassidy and to Marjorie Graham for their insights into Saunders's life. Sopka, 1988, 87, notes how Americans tended to be more experimental and less mathematical than their European counterparts.

25. In 1922, Miguel Catalán discovered that quintet terms in the manganese spectrum exhibited exact numerical relations and could be used for identifying the chemical identities of lines in crowded spectra. Catalán coined the term "multiplets" to denote lines bearing a constant relationship in wavelength to one another. Sánchez Ron, 1994, chap. 3; Meggers, 1951, 7.

26. Born, 1924, 379. On the mounting crisis in physics, see Forman, 1968; 1970; Kronig, 1960; Serwer, 1977; Cassidy, 1979.

27. Schweber, 1990, 386, quoting J. H. Van Vleck. On Heisenberg's model, see Cassidy, 1979; 1992, 111–114; 196, and the references therein. Russell notes his delight with the vector model in Russell to H. D. Babcock, 30 October 1923. PUL/HNR.

28. Kenat and DeVorkin, 1990, 168–169; 174 for L-S coupling. See also Sitterly, 1977. Russell and Saunders, 1925a.

29. Russell to Saunders, 16 November 1923. Saunders to Russell, 21 November 1923. PUL/HNR.

30. Schweber, 1990, 357.

31. Kenat and DeVorkin, 1990, 174–175, refs. 74 and 75.

32. Eckert, 1996, 70–74.

33. Russell, 1924e, 229–230. This abstract was expanded in Russell, 1925a, but was far from complete. McGucken, 1969, and Robotti, 1983, review relevant aspects of the early history of spectral series analysis.

34. Russell, 1925a, 223.

35. Dupree, 1959, 42–43; 391–392.

36. [Meggers], "The Spectroscopic Laboratory of the Optics Division," [statement of the NBS spectroscopy laboratory], 5 June 1919. Meggers to John C. Duncan, 26 October 1917; Campbell to Meggers, 7 April 1919; and letters between A. S. King and Meggers in November 1919. Box 1, AIP/WFM.

37. Meggers to Sommerfeld, 15 June 1923. Box 2, AIP/WFM.

38. Meggers diary entries, 21; 24 June; 24 July 1923; 1 January 1924. AIP/WFM.

39. Meggers diary entry, 21 December 1923. AIP/WFM.

40. Russell to Meggers, 20 July 1923; Meggers to Russell, 25 July 1923. Box 2, AIP/WFM.

41. Russell to Adams, 20; 24 August 1923. HL/WSA.

42. Meggers to Russell, 25 July 1923. Box 2, AIP/WFM.

43. Adams to Russell, 13 January 1921. Box 3.49 Russell 1921–1923, HL/WSA. Sitterly OHI, 27. AIP.

44. Russell to G. K. Burgess, 21 December 1923. AIP/WFM. Mount Wilson staff also endorsed Meggers. Meggers to Charles St. John, 19 January 1920; to Russell, 14 January 1924. AIP/WFM.

45. Meggers to Russell, 14 January 1924. AIP/WFM. Meggers diary entry, 14 December 1923. AIP/WFM.

46. Meggers diary entries, 20 January; 20 February 1928. AIP/WFM.

47. Dunham OHI, 29; 100; Sitterly OHI, 19. Meggers 1928 European Trip Diary, entry for 9 July 1928. AIP/WFM. The IAU is divided into commissions dealing with specialties and specific subjects. Each commission reported on status in the field at the General Assemply once every three years. Blaauw, 1994

48. Dunham OHI, 100. API.

49. Russell to Meggers, 24 January 1924. PUL/HNR.

50. Walters and Burns to Kiess and Meggers, 7 February 1929; Meggers to Burns, 21 February 1929. AIP/WFM.

51. Sitterly, 1977, 29. Russell to Meggers, 12 January 1942. PUL/HNR.

52. First exhibited by Bohr and soon after by W. Grotrian in 1920, these diagrams depicted the energy levels in an atom and the amount of energy each electronic state contains. They became helpful conceptual tools for delineating which transitions were allowed by the various rules then being developed. Russell to F. M. Walters, 19 December 1923. Box 7, NBS correspondence, 1917–1929, AIP/WFM.

53. Russell lecture and research notes, folders 73.13–73.15, ca. 1924–1925. PUL/HNR.

54. On "Normal Science as Puzzle-solving," see Kuhn, 1964, pt. 4.

55. Russell, 1935d, 621.

56. Ibid., 613; 620–621.

57. Russell, 1943c, 211.

58. These are the criteria identified by Kuhn for normal science. Kuhn, 1964, 35–37.

59. Dunham OHI, 100. AIP.

60. Russell to G. C. Winteringer, Princeton Controller, 27 May 1936. PUL/HNR. Russell's "amazing memory" was a subject of discussion among astronomers. See, for instance, Walter Sydney Adams to Otto Struve, 29 May 1943. OS/YOL.

61. Green, 1977, 80. Green later was not certain of this scenario. Green OHI, 21, AIP.

62. Russell to Rhodes Committee, 19 November 1934. PUL/HNR. Green, 1977, 81.

63. Dugan to A. S. King, 5 November 1929. King to Dugan, 4 December 1929. Dugan to King, 22 September 1930. The research assistanship provided $500, and tuition was $100. Dugan to King, 3 April 1930. PUL/RSD. R. B. King OHI, 2–11.

64. Green to Russell, 9 January; 22 May 1939. Russell to Green, 15 May 1939. PUL/HNR.

65. Russell to Shapley, 30 October 1923. PUL/HNR. On the tension between cooperation and competition, see Forman, 1973, 153 n. 7.

66. Meggers to Sommerfeld, 24 January 1924; to Sommerfeld and to H. Kayser, 11 September 1923; to Konen (Bonn), 7 December 1923. AIP/WFM. Meggers to Russell, 21 October 1924. PUL/HNR.

67. Meggers to Russell, 26 November 1924. AIP/WFM; Russell to Meggers, 28 November 1924; 2; 7 February 1925; Ames to Russell, 6 February 1925. PUL/HNR.

68. Russell to Ames, 7 February 1925. PUL/HNR. Members included Saunders, Meggers, Foote, K. Kiess, F. M. Walters, and H. M. Randall of Michigan, as well as Babcock, King, Theodore Lyman, and Robert Millikan.

69. Russell to Ames, 7 February 1925. PUL/HNR.

70. Haramundanis, 1984, 177.

71. Meggers to Russell, 2 March 1949. PUL/HNR.

72. Russell, H. N., 1931, "Report of the Committee on Line Spectra of the Elements," Division of Physical Sciences, National Research Council. Report dated 21 April 1931. Box 104.12, PUL/HNR.

73. White, 1934, 11. On the need for the notation, see Russell to Henry Gale, 20 November 1924. YOL; Russell to Meggers, 20 November 1924; Russell, Shenstone, and Turner to "Dear Sir," June; 19 December 1928. PUL/HNR. Russell, "Autobiographical Statement" (National Academy of Sciences Archives), 23, and Russell, Shenstone, and Turner, 1929, 906. Their report did not achieve total consensus. See Walters and Burns to Kiess and Meggers, 7 February; Meggers to Burns, 21 February; Burns to Meggers and Kiess, 25 February 1929. AIP/WFM.

74. Sitterly OHI, 8–9; 15. AIP. Russell to Adams, 22 May 1925. PUL/HNR. Adams to C. E. Moore, 29 January 1930. Box 48, "Charlotte E. Moore 1928–1932," HL/WSA.

75. Sitterly OHI, 1977, 21–23. AIP.

76. On Moore's value at Mount Wilson, see Wright to Adams, June 1930. LAUC.

77. Moore, 1945, iv.

78. Sitterly OHI, 18–19. AIP. Moore to Wright, 21 April 1931. LAUC. Russell to Adams, 24 June 1932; to Brockett, 23 November 1932; to G. A. Mills, 24 October 1945. PUL/HNR.

79. Russell, 1935d, 28; 51–52 of the manuscript. File 108.17, PUL/HNR. King, 1995.

80. "C.M.S. Alternative report from July 1, 1937 to July 1, 1938." "Charlotte" to "Dear Family," 19 June 1938. PUL/HNR.

81. Adams to Moore [Sitterly], 4 December 1940. HL/WSA.

82. Russell to Meggers, 23 October 1945. PUL/HNR; AIP/WFM. Moore, 1945.

83. Crane, 1972, 27, following Kuhn, 1964, uses the phrase "puzzle-solving" to describe normative practice only.

84. Meggers diary entries, 25 April 1924; 4 January 1926. AIP/WFM.

85. Analysis of relevant indexes by Paul Routly.
86. Russell to Meggers, 7 March 1949. PUL/HNR.
87. On physicists' rationales, see Forman, 1989.
88. Menzel, 1972, 241–242. Sitterly OHI, 27. AIP.

Chapter 14
"A Reconnaissance of New Territory"

1. Haramundanis, 1984, 177.
2. Spitzer, 1977, 6.
3. Gooding notes the importance of tenacity in discovery: "[E]arly success does not obviate the need to establish that the effect is real." Gooding, 1985, 231.
4. Russell, 1914c, quoted in DeVorkin and Kenat, 1983b, 181, from which the bulk of this chapter is derived.
5. Fowler, 1918, 204.
6. Plaskett, 1922; Russell, 1923a.
7. Russell and Compton, 1924. Compton was not happy with this interpretation, however, and asked Russell to remove his name from the paper. But they agreed to dash off a note on it anyway to *Nature* in June 1924. "Compton, K. T. 1925 – 1951" file; entry for 1925, Stewart papers, PUA.
8. Rosseland, 1925a. In contrast to what Bowen asserts in his oral history Rosseland's suggestion may have given Russell and Bowen the idea that led, in 1927, to Bowen's identification of the chief lines in nebulae. The idea was in the air, and Bowen was best equipped to find spectroscopic evidence for the existence of the proper levels. Bowen OHI, 10–11. AIP. Adams to Campbell, 28 November 1924. President's Correspondence File CU-5 1924:496, BANC/UCP. On Bowen's work, see Hirsh, 1979.
9. Shapley to Russell, 2 April 1923, and other letters during this period. PUL/HNR.
10. Dugan to West, 4 March 1921. Menzel file, PUA. Menzel, 1972, 239–240. Russell to Meggers, 21 December 1923. AIP/WFM.
11. Menzel, 1974, 177. AIP. Shapley to Russell, 19 June 1923. PUL/HNR.
12. Haramundanis, 1984, 161–162.
13. Shapley to Russell, 25 October; Russell to Shapley, 30 October 1923. PUL/HNR.
14. Fowler, 1922b, 95–96; 200.
15. Russell to Shapley, 30 October 1923. PUL/HNR.
16. Ibid.
17. See chap. 22, below. Russell to Spitzer, 18 January 1946. PUL/LS.
18. Kidwell, 1984, 15, has explored Payne's viewpoint.
19. Haramundanis, 1984, 164.
20. Russell to Shapley, 19 June 1924. PUL/HNR.
21. Russell to Shapley, 11 October; to Adams, 16 December; to the Committee on Research Fellowships, 26 November, clipped to Russell to Payne, 26 November 1924; Shapley to Russell, 8 October 1924, with a correction in Payne's handwriting. PUL/HNR. Kidwell, 1984, 18, discussing Payne, 1924, 8.
22. Payne to Margaret Harwood, 24 November 1924. SLRC/MH. Payne, 1925b, 183; 1925a, 193–194. DeVorkin and Kenat, 1983b, 185–186.
23. Russell to Payne, 14 January 1925. PUL/HNR, and discussed by Kidwell, 1984, 19–20.
24. Haramundanis, 1984, 177.
25. Payne, 1925a, 197; 1925b, 57; 188.

26. Payne to Margaret Harwood, 9 January 1925. SLRC/MH. Shapley to Russell, 21 April 1925. PUL/HNR. On Plaskett and Payne's differences, see Kidwell, 1984, 18.

27. Struve, 1926, 204–208.

28. Struve and Zebergs, 1962, 220–221.

29. Fowler, 1921, 348. "Toronto Lectures," transcript for 9th lecture, 151. Box 103.4, PUL/HNR. DeVorkin and Kenat, 1983b.

30. Eddington, 1923, 256–257, emphasis in original. On Eddington's need for a low hydrogen abundance, not only to explain why stars have the masses they do, but to keep his models independent of chemical composition, see Eddington, 1926, 245. Russell and Webster, 1922, 181, explore this point.

31. Haramundanis, 1984, 165.

32. Hufbauer, 1981, 301. Eddington, 1920, 18.

33. Russell, 1921d, 289.

34. Eddington, 1924, 332. Hufbauer, "Stellar Structure and Evolution," forthcoming, 4; 7–8.

35. Russell, 1925e, 209.

36. Ibid., 211.

37. DeVorkin and Kenat, 1983b, 196–198, provides source references.

38. Meggers to Russell, 29 April 1925. PUL/HNR. Russell, 1925c.

39. Russell, 1925d, 328; Russell to Meggers, 21 May 1925. PUL/HNR.

40. Russell to Otis L. Crossley, 21 March 1927. PUL/HNR.

41. Stewart, 1922, 311. Stewart faculty file, PUA.

42. Stewart to Russell, 5 January 1920; Russell to Stewart, 27 December 1920; Stewart to Russell, 11 January 1921. PUL/HNR.

43. Stewart, 1923, 187; Stewart, 1924. Russell to Adams, 1 November 1922. PUL/HNR.

44. Rosseland, 1925b, 436.

45. Milne, 1925; "Report of the Council," *MN* 86:232; Milne to Russell, 22 September 1925. PUL/HNR.

46. Menzel, 1972, 237–238.

47. Hearnshaw, 1986, 409; 415, discusses the Schuster-Schwarzschild model of the reversing layer. Stewart, 1925a; Russell and Stewart, 1924. Discussed in [1925], "Report of the Council," *MN* 85:354.

48. Russell to Shapley, 6 February; Shapley to Russell, 21 April 1924. PUL/HNR.

49. Stewart to Russell, 12 April 1924. PUL/HNR.

50. Coolidge quote from Russell to Adams, 4/5 November 1924; 16 December 1924; 18 February 1925; 22 June 1925. PUL/HNR.

51. Russell and Moore, 1925, 1; 12. Russell to Milne, 5 October 1925. PUL/HNR.

52. Russell to Adams, 29 September 1925. PUL/HNR.

53. Adams to St. John, 24 July 1926; St. John to Adams, 4 August 1926. St. John file, HL/WSA.

54. For a fuller discussion, see DeVorkin and Kenat, 1983b, 202.

55. Russell, 1933c, 15. Russell, Adams, and Moore, 1928.

56. Hearnshaw, 1986, 232, sec. 7.18.

57. Russell, Adams, and Moore, 1928, 8.

58. Russell to Eddington, 22 September 1927. PUL/HNR.

59. Russell to Adams, 25 October 1927. PUL/HNR. "Boltzmann" referred to Milne and Fowler's rederivation of Saha's equation, now known as the Saha-Boltzmann equation.

60. Russell and Adams, 1928, 12. Hearnshaw, 1986, 233, notes that Milne and Fowler had suggested using differential analysis, but Russell and Adams were the first to actually carry it out.

61. Russell and Adams, 1928, 35. Russell to Dunham, 11 October 1927. PUL/HNR.

62. Eddington to Russell, 18 September; Russell to Eddington, 15 October; Eddington to Russell, 20 November 1928. PUL/HNR.

63. Eddington to Russell, 20 November 1928. PUL/HNR.

64. Stewart, 1928, 346. An expansion of Stewart's remarks, or the methods he used to support them, have not been found. Russell to Adams, 25 October 1925. PUL/HNR.

65. Russell to Dunham, 27 October 1927. PUL/HNR.

66. Unsöld, 1928, 781; Strömgren, 1951, 224–226; Unsöld, 1927.

67. Unsöld OHI, 13. W. H. McCrea to the author, 11 July 1983, confirmed in Milne, 1930a, 460.

68. Eddington, 1929, 636.

69. Russell to Dugan, 8 March 1928. Box 4, Russell folder, PUL/RSD. [Stewart and Compton], n.d. ca. May 1925. "Research Activities in Astrophysics at Princeton University," Enclosure III, attached to Stewart to Compton 21 May 1925. Stewart to Compton, 5 May 1925. PUL/JQS.

70. Russell's travels were noted in Carl Lampland to V. M. Slipher, 21 June 1928; Lampland diary, entry for 25 July 1928. LowA/COL; letters in Wright files, LAUC.

71. Stratton, 1929, 233–235.

72. Russell thanks Unsöld for lending his lecture notes. Russell to Unsöld, 23 March 1929, cited in Hufbauer, 1991, 104. Russell to Meggers, 22 January 1929. AIP/WFM.

73. Russell and Bowen, 1929, cited in Hufbauer, 1993, ref. 11; Russell, 1929c, 316–317.

74. Russell to Adams, 25 March 1929. PUL/HNR.

75. California Institute of Technology Calendar for January and February, 1929. I am indebted to Karl Hufbauer for pointing out this source.

76. Menzel, 1972, 243. On Lick's eclipse expeditions, see Eddy, 1971. Russell to Aitken, 15 March 1926; Menzel to Russell, 7 September 1926; Russell to Menzel, 27 September 1926. PUL/HNR.

77. Manuscript, "On the Composition of the Sun's Atmosphere," 1. File 106.7. PUL/HNR; Russell, 1929b, 11.

78. Russell, 1929b, 63–64.

79. Ibid., 79.

80. Russell, 1938a, 111. Plaskett to Unsöld, 31 March 1929; Ornstein to Dunham, 22 November 1927. AIP/THD.

81. Unsöld, 1930, 372.

82. Milne, 1930b, 120.

83. Milne, 1929a, 362.

84. Ibid, 363.

85. Wildt, 1939, excerpted in Lang and Gingerich, 1979, 264–267. Wheeler's influence is noted in Schwarzschild OHI, 18 June 1982, 14rd. Redman, 1938, 315.

86. Minnaert, 1965, quoted in DeVorkin and Kenat, 1983b, 214–215. Strömgren, 1940, 218; and Chandrasekhar, 1939; 1951.

87. Russell to Seares, 23 March 1929. PUL/HNR.

88. Russell to Webster, n.d. ca. 1929. PUL/HNR.

89. Russell to Adams, 24 May 1941. PUL/HNR.

90. On the use of power and authority in the American astronomical community, see Lankford, 1997.

91. In 1946, Russell observed that "by far the greater part of the observations have been made in America, while many of the major theoretical advances have come from overseas." Russell, "America's Role in the Development of Astronomy," 2, holo. Box 112.18;.19, PUL/HNR. Russell, 1947.

92. Hiebert, 1986, 434, discusses the importance of making the connection. The motivations leading Russell to prepare addresses like these can be found in Kevles, 1978, 179. I am indebted to Kathy Cooke for her insight into the defenses scientists constructed. Russell to Miss Appleton, 25 November 1924. PUL/HNR. Russell commentary, as reported in James C. Young, "Sees Science Upholding Immortality of Soul," *New York Times*, 17 February 1924, pt. 7, 14.

93. Untitled manuscript accompanying Russell to Rev. Dr. Walter Laidlaw, 1 October 1935, 4. File 108.19, PUL/HNR.

94. Untitled manuscript, 4. File 108.19, PUL/HNR.

95. Russell, 1932b, 300. The original draft was titled "The Relation of Spectroscopy to Other Sciences." File 107.3, PUL/HNR.

96. Russell, 1932b, 300–301.

97. Webster OHI, 12. Kevles, 1978, 159, quoting Webster, attributes the tease to Bragg.

98. Spiritual Autobiography, 42.

99. Candler, 1937, 4; see also the 1964 edition, 3. Saha shared this view, feeling that one could apply the new mechanics to interpret laboratory spectra without mastering its theoretical underpinnings. Saha admitted to a young Subramanyan Chandrasekhar, still studying in Madras in 1930, that he did not understand the "general scope of the Pauli-Principle very much." He disdained overly mathematical constructions; statistical arguments "like Heisenberg's uncertainty principle, always seem to be a bit unconvincing." Saha to Chandrasekhar, 14 July 1930. RUC/SC.

100. Johnson, 1946, v.

101. Russell to Webster, 20 March 1942. SUL/DLW.

Chapter 15
Princeton Astronomy in the 1920s

1. Lankford, 1997, explores the hierarchical nature of the astronomical community.

2. "Keen Rivalry in Astronomy," *Varsity* 43, 14 February 1924, 1. UT Archives. On C. A. Chant's goals, see Jarrell, 1988, chap. 6.

3. Russell to Henry B. Thompson, 10 February 1917. PUL/HNR. Press clippings, Russell faculty folder ca. 1919. PUA. Russell to G. C. Wintringer, 2 January 1920. Princeton Fund, PUL/HNR.

4. Hibben to Russell, 10 March 1922. PUL/HNR. Hibben promised to increase department appropriations from $5,000 to $10,000 per year, so funds could accumulate for the renovation. Hibben to Dugan, 20 February 1922. Research Budget Folder "1920–1939," Department of Astronomy records, PUA. Ron Doel has noted that the delay in gaining the endowment was due to a lingering controversy over the status of the Frick Estate. Private discussion.

5. Russell to Daniel Barringer, 12 December 1925, 2. PUL/HNR. On the low priority of science in the Princeton Fund, see Doel, 1987, 4.

6. Doel, 1987, 1; Kohler, 1991, 209.

7. Kohler, 1991, 211.

8. Ron Doel kindly provided insight into Princeton's institutional cooperative spirit in the 1920s. A sense of this comes as well from Shenstone OHI, 1979.

9. Russell to Barringer, 12 December 1925, 2. PUL/HNR. On the controversy over the nature of the buried mass, and W. F. Magie's and Russell's involvement, see Hoyt, 1987.

10. H. Alexander Smith (Exec Sec) to "My Dear Henry," 2 February 1926; Russell to Smith, 20 April 1926; John Q. Stewart and H. N. Russell, "Astronomy at Princeton," folder 1925–1927. Box II, Astronomy Department Records, PUA. Russell to Henry Gale, 11 March; to Shapley; to Struve, 17 May 1932. PUL/HNR. Miller, 1970; Wright, 1966; and Kohler, 1991, provide insight into funding patterns for astronomy.

11. Shapley, 1927, 772.

12. Thorp, Myers, and Finch, 1978, 182. See Frost, 1933, 60. In a 1925 rating of graduate schools, none of Princeton's departments ranked first in their fields, but astronomy ranked fourth. Hughes, quoted in Thorp, Myers, and Finch, 1978, 180–181. Visher, 1947, 172; 280–281, showed that Princeton lagged behind only California, Chicago, and Virginia in the production of astronomy graduate students, and Russell was named as an "especially stimulating" teacher, garnering the most votes (6) in a 1946 poll.

13. Kohler, 1991, 212.

14. Russell letters, outgoing, in Princeton Department of Astronomy files, ca. 1927. See also a general letter dated 16 December 1925; as well as Russell to Miss Young, 24 April 1927; to Archibald (Archie) A. Gulick 26; 29 November; 14 December 1927; to Hibben, 21 January 1927. PUL/HNR. Class reports, " '97" *Princeton Alumni Weekly*, 15 March 1929. Russell faculty folder, PUA.

15. Russell to Fredrick Delano, 5 April 1927. See also "Harvard College Observatory—Directorship, 1920," file, PUL/HNR.

16. "Research Committee Minutes," Department of Astronomy Records, entry for 28 March 1930. PUA.

17. [Dugan], "Research Budget 1920–1939"; Dugan to G. C. Wintringer, 3 November 1930; Russell to Wintringer, 3 March 1932; Hibben memorandum, 29 April 1932. Budget files. The remounting in 1933 cost some $25,000. "Research-Mounting Telescope A-126 Research-Spectroscopic Tales 1932–1934," Department of Astronomy Records, PUA.

18. Russell to Dugan, 18 March 1925. Controller's files, PUL/HNR

19. Russell to W. Irving Harris (*Princeton Alumni Weekly*), 11 October 1924, and letters through 1930. Princeton Alumni file, PUL/HNR.

20. J. S. Plaskett to Russell, 23 January 1923. PUL/HNR. Starting with the Gold Medal of the RAS in 1921, he subsequently received the Lalande Medal of the French Academy and the Henry Draper Gold Medal of the National Academy of Sciences in 1922. That year Dartmouth became the first to present him with an honorary doctorate of science. He was awarded the Bruce Medal of the Astronomical Society of the Pacific in 1925.

21. Russell to Ginn [C. H. Thurber], 9 November; 16 November 1921. Ginn [Thurber] to Russell, 14; 18 November 1921. PUL/HNR.

22. "Toronto Lectures," transcript for 14th lecture, 29 February 1924, 265. Box 103.5, PUL/HNR. Berendzen, Hart, and Seeley, 1976, 142.

23. Ginn [Thurber] to Russell, 23 May 1924. PUL/HNR.

24. Ginn [Thurber] to Russell, 7 November; 6 December 1924; Russell to Ginn [Thurber], 8 November; 16 December 1924; to Adams, 19 November 1925; to Ginn [Thurber], 14; 30 November 1925, with appended contract; to Ginn [Thurber], 16; 22 December 1925. PUL/HNR. General log book, 23-inch, 1905–1928, Astronomy Department Records. PUA. The chapter on nebulae was in proof when Russell learned of Hubble's work

and was immediately converted to it. See "Note," Russell, Dugan, and Stewart, 1926/1927 (hereafter RDS II), 858; and Berendzen, Hart, and Seeley, 1976, 141.

25. Russell to Ginn [Thurber], 18 January; [E. K. Robinson], 22 January 1925; [Myra deN. Wood], 11 March 1926; [Thurber], 18 March 1926. Stewart to Thurber, 27 April 1926. Dugan to Ginn, 26 May 1926. Ginn [Thurber] to Dugan, 26 May 1926. PUL/HNR.

26. Dugan and Stewart to Thurber, 27 May 1926. PUL/HNR.

27. Stewart to Ginn [Thurber], 14; 16 June 1926. PUL/HNR.

28. Stewart to Ginn [Smith], 14 August 1926. Ginn to Russell, Dugan, and Stewart, 31 August 1926; Secretary to Ginn, 2 September 1926. PUL/HNR.

29. Russell to Ginn, [E. W. Stevens], 26 October 1926. PUL/HNR.

30. Russell to Ginn, [Stevens], 27 October; 1 November 1926. PUL/HNR. Earlier textbooks existed in German: K. Graff, *Astrophysik* (Berlin, 1922), an update of Julius Scheiner's 1912 text *Populäre Astrophysik*.

31. Based upon Russell's share of the royalty, or 3 percent. Russell to Ginn [Thurber], 24 January 1927. PUL/HNR.

32. Promotional material for "Astronomy, Russell-Dugan-Stewart." Ginn, 1927. Author's collection. Ginn file, reel 25, fr. 22687–22690; Russell to Thurber, 9; 20 January 1928. PUL/HNR.

33. C. C. Crump, review of *Astronomy*. *ApJ* 67, 1928, 274–275.

34. C. A. Chant, review of *Astronomy*. *JRASC 21*, 1927, 119–124; 121.

35. Russell to Ginn [George H. Moore], 29 January; to Thurber, 6 March 1931. PUL/HNR.

36. It is quite possible that Russell showed Bowen a draft of his chapter, or the proofs, since the first edition contains a discussion of Bowen's work. As noted in chap. 14, above, both may have gotten the idea from Rosseland. Russell to Thurber, 9; 20 January 1928. PUL/HNR. Bowen OHI, 10. RDS II, 837.

37. Menzel autobiography, chap. 16, p. 2. Gibson Reaves to the author, 24 June 1993. Numerous testimonials are collected in the oral histories contained within the "Sources for History of Modern Astronomy" project, AIP.

38. In 1949, Struve stated that RDS was still the standard text for introductory professional training. Struve, 1949, 385. DeVorkin, 1995.

39. Spitzer OHI, 1977, 31. AIP.

40. Menzel Autobiography, 2–3. AIP.

41. Spitzer OHI, 1977, 30. AIP.

42. Menzel Autobiography, 3–4. AIP.

43. Ibid., 4.

44. Lampland to Slipher, 23 March 1929. LowA/VMS.

45. Dunham OHI, 41–42. AIP.

46. Ibid., 42.

47. Ibid., 41.

48. Ibid., 75.

49. Ibid., 76.

50. Schwarzschild OHI, session 3, 109. AIP.

51. Hale to Frost, 8 February 1922. YOL/EBF. On Hale's breakdowns, see Wright, 1966, 259.

52. King to Birge, 5 July 1921. BANC/RTB.

53. A. S. King to Hale, 22 July 1923. CIT/GEH.

54. A. S. King to Birge, 13 July 1922. BANC/RTB.

55. Adams to Abbot, 9 August 1921. HL/WSA.

56. St. John to Hale, 27 August 1921. St. John file, 59.1031, HL/WSA. On competition and secrecy, see Edge and Mulkay, 1976, 245–248; on antipathy between observers and theorists, 318.

57. St. John et al., 1928.

58. Russell to S. Albrecht, 14 September 1926. DO/SA.

59. Adams to Abbot, 10 May 1923. File 1.5, 1923, HL/WSA.

60. St. John to Adams, 12 June 1928. St. John file 59.1031, HL/WSA.

61. On the origins of the International Education Board Fellowships and their impact, see Kevles, 1978, 191–192.

62. As compiled from listings of Mount Wilson publications, dating from World War I through 1930.

Chapter 16
Stellar Evolution

1. Russell, 1942. 233.

2. Where he stated that he was the first to publish such a diagram in 1913. RDS II, 723–724.

3. Robert B. Sosman to Russell, 14 March 1925; Russell to Sosman, 27 May 1925. PUL/HNR. Harold Jeffreys, *Nature* #2878 (27 December 1924), 934.

4. Russell, "On the Law of Liberation of Energy Within the Stars," typescript, 7 October 1925, 15; 22–23. File 104.10. Russell, 1925b. Russell to Eddington, 22 June; 9–10 October 1925; Jeans to Russell, 2 July 1925. PUL/HNR.

5. Jeans, 1925.

6. Russell, "On the Law of Liberation of Energy Within the Stars," typescript, 7 October 1925, 15; 22–23. File 104.10, PUL/HNR. Hertzsprung suggested that the gaps indicated that the rate of evolution was not constant. Hertzsprung, 1922.

7. Russell to Adams, 19 November 1925. PUL/HNR.

8. Russell, "On the Evolution of the Stars," n.d., ca. 1926. Box 104.26, PUL/HNR.

9. Russell, 1928c, 383.

10. Eddington, 1926, 163.

11. Russell, 1928c, 385–386.

12. Russell to Eddington, 22 September 1927. PUL/HNR.

13. Ibid., and RDS II, 724; 910. Eddington, 1926, 151.

14. Russell, 1931c, 955.

15. Ibid., 958; "Addendum," 147. H. Vogt, 1930, *Veröff. Sternwarte Jena* 8:19. See Russell to RAS, 11 December 1931, and "Note on Vogt's View of Stellar Composition," same date. File 106.27; Russell to Milne, 4 November 1937. PUL/HNR.

16. RDS II, 893. Cowling, 1966, 126; Schwarzschild (private communication) and Tayler, 1996, 362, have pointed out that the Vogt-Russell theorem does not hold up to rigid mathematical inspection for composite stars.

17. RDS II, 916–917.

18. Russell, 1929d, 378–379.

19. Russell to Menzel, 19 December 1927. PUL/HNR.

20. Baker, 1930, 427–428.

21. Hufbauer, forthcoming, "Stellar Structure and Evolution," 6, manuscript.

22. Hufbauer, 1987, 9–11; forthcoming, "Stellar Structure and Evolution," 8–9.

23. Chandrasekhar OHI, 36. AIP. Hufbauer, forthcoming, 8–9. Wali, 1991, 85; 119 et seq. See also the running conversation between Milne and Eddington in 1930 as reported in *Nature 125*:273–274; 453; 489; 708; in *Obs* 53: 113–119; 167–177; 208–211; 238–240; 249–251; 342; as well as throughout volume 90 of the *MN*.

24. *Astronomischer Jahresbericht*, 1930–1931; Strömgren, 1972, 245.

25. The rest of the family, including Lucy, Jean Hetherington, and Elizabeth, left quickly, fearing contagion, and Lucy had to enter college. Henry jr. took weeks to recover from "pleurisy with effusion due to first-infection tuberculosis" in Paris, and then months at home. HNRjr., 18.

26. Milne, 1929b, 17; 24–25; 51–53; Eddington, 1930a; b; d; [D.L.E.], 1931, "Report of the Council," *MN 91*:402.

27. Eddington commentary, 1929, *Obs 52*:349.

28. R. B. King OHI, 12. AIP. Russell notes these lectures as the source of his ideas in Russell to Milne, 21 April 1931. PUL/HNR.

29. Chandrasekhar, 1939, 228; Jeans, 1927, 11, citing Emden's *Gaskugeln*, 96.

30. Milne acknowledged the aid of graduate student computers. Milne, 1930d, 20–29; 55. Cowling, 1985, 5.

31. Cowling, 1930, 100–103, fig. 2; Milne, 1932, 622–624; fig. 1.

32. [1931]. "Minutes of the Meetings," *Proc. Am. Phil. Soc. 70*:vii–xv. Russell to S. B. Nicholson, 16 April 1931; 1 May 1931. PUL/HNR.

33. Russell to Milne, 21 April 1931. PUL/HNR.

34. "Meeting of the Royal Astronomical Society," *Obs 54*:154–155. Cowling OHI, 19. AIP.

35. Russell modified various dimensionless variables Milne had created. Russell, 1931c, 958. Eddington, 1930b, 285. Russell to Eddington, 13 June 1931. PUL/HNR.

36. Dunham to Russell, 17 June 1931. PUL/HNR; Russell, 1931a, excerpted by Milne, 1931, 154; "Meeting of the Royal Astronomical Society," *Obs 54*:100–101; 155.

37. Milne to Russell, 9 May 1931, 8. PUL/HNR.

38. Atkinson to Russell, 6 March 1931. PUL/HNR.

39. Atkinson and Houtermans, 1929; Hufbauer, 1987; forthcoming.

40. Russell, 1931f; Russell, 1933d, 74.

41. Russell and Atkinson, 1931.

42. Morrison established the prize in 1923 in three categories, one being the "best paper in regard to the energy of the sun." "Report of the A. Cressy Morrison Prize Committee for 1930," n.d., ca. 10 December 1930. NYAS/ACM file.

43. Atkinson OHI, 11–18; 27. AIP. In December 1930 Stewart and Condon recommended that Edvard Hugo von Zeipel of the University of Upsala, Sweden, receive the $750 prize for his theory that stars like the Sun were recurrent novae. Stewart and E. U. Condon to R. W. Miner, 10 December 1930. "Report of the A. Cressy Morrison Prize Committee for 1930," n.d., ca. 10 December 1930. NYAS/ACM. For prior winners, see R. W. Miner correspondence with Russell and Dugan, ca. 1925–1928, NYAS/ACM; A. Cressy Morrison to R. S. Dugan, 13; 18 January 1928. PUL/RSD; Russell to A. Cressy Morrison, 18 February 1925; 9 January 1928. PUL/HNR.

44. Russell, 1931c, 953. Strömgren, 1983, 3–4; 1972, 245. See also Goldberg, 1989.

45. Much later Cowling and Martin Schwarzschild would show that Milne's "U-V plane," as it was later called, was extremely useful for fitting integrations from the core and envelope for a full specification of the structure of a star. Cowling, 1966, 130; 1985, 5. For a modern treatment that uses the same nomenclature, see Schwarzschild, 1958, chap. 13.

46. Russell, 1931c, 956; 951.

47. Milne commentary, 10 June 1932 RAS meeting. *Obs* 55:189.

48. Russell, 1936d, 132.

49. Milne to Russell, 27 November 1936. PUL/HNR. Milne notes that it was Russell's work that brought him back to stellar structure, from "a sojourn in the realm of cosmology." Milne to Strömgren, 29 September 1937. Strömgren papers, Aarhus, courtesy Karl Hufbauer.

50. Milne to Chandrasekhar, 26 December 1937, 3–4. RUC/SC. Russell to Milne, 11 December 1937. PUL/HNR. Chandrasekhar, 1987, 83; 90. Wali, 1991, 95–105, et seq.

51. Milne to Chandrasekhar, 26 December 1937. RUC/SC.

52. Russell, 1938a, 112–113. Russell to John C. Cobb, 2 March 1931. PUL/HNR. "computational approach" from Schwarzschild OHI, 18 June 1982, 16–17. AIP. On Bridgman and Slater, see Schweber, 1990. Russell to some extent shared Bridgeman's skepticism regarding the attainability of certainty. Walter, 1990, 173–175.

53. W[oolley], 1936, 318–319; quote from Woolley and Stibbs, 1953, 283.

54. Russell, 1933d, 78.

55. Russell to Eddington, 5 February 1932. PUL/HNR.

56. Eddington to Russell, 23 February; 8 March 1932; Russell to Eddington, 22 March 1932. PUL/HNR.

57. Strömgren, 1983, 3–4; 1972, 245. See also Goldberg, 1989.

58. Strömgren, 1932, 118. Russell to Eddington, 22 March 1932. PUL/HNR. Strömgren, 1983, 3–4; Hufbauer, forthcoming, 10.

59. Russell, 1933d, 68.

60. Ibid., 70–74. Atkinson's theory and its ability to reproduce the relative abundances of the lightest elements is discussed in Hufbauer, forthcoming, 11.

61. Russell, 1933d, 75–76; 79. Russell to Dwight Gray, 26 June 1941. PUL/HNR.

62. Russell, 1935c, 18–19, noted in Wali, 1991, 144.

63. Chandrasekhar, 1934, 377.

64. Cowling, 1966, 121–122.

65. Chandrasekhar OHI, 29. AIP.

66. Russell, 1937a. Wali, 1991, 145.

67. Strömgren, 1933, 247.

68. R-M refers to radius/mass. Chandrasekhar reported that Russell did not acknowledge him at the meeting. Chandrasekhar to Strömgren, 11 June 1933. Strömgren papers, Aarhus, courtesy Karl Hufbauer.

69. Kuiper to Struve, 11 February 1936, Box 2, UA/GK.

70. Struve to Kuiper, 21 March 1936. Box 2, UA/GK.

71. Kuiper, 1937, 176–177.

72. Von Weizsäcker's original fusion mechanism also produced neutrons, which were quickly captured. Strömgren, 1938, 520–521; 533–534. See also Hufbauer, forthcoming, "Stellar Structure."

73. Lucy met George Gardner when he was a first-year student at the Princeton Theological Seminary. They were married at the Seminary Chapel in Princeton. Russell to Abetti, 28 May 1935. PUL/HNR. On the conference, see Hufbauer, forthcoming.

74. George Gardner 1935b. Alumni records, Princeton Theological Seminary. Russell to Olin D. Wannamaker, 6 July 1936. Princeton-Yenching Foundation Folder, PUL/HNR. Russell to G. Abetti, 21 November 1934; 28 May 1935. PUL/HNR. E. H. Roberts to Rev. J. Harry Cotton, 27 January 1936; Cotton to Roberts, 4 February 1936; 6 March 1936. PTS/R.

75. Gamow to Russell (Carnegie file), 10 January 1938; Russell to Gamow, 28 January 1938; to Shapley, 20 January; 3 March 1928; to Struve, 27 June; 8 November 1938; 20 August 1939; to Strömberg, 28 June 1938. PUL/HNR. RDS II, 1938, 917–919. The chapter was completed in March 1938.

76. Gamow to Chandrasekhar, 7 January 193[8]. Chandrasekhar had argued that "it is my opinion that no astrophysicist is more competent to 'help' the physicists than Stromgren is." [Chandra] to Gamow, 29 December 1937. Gamow file, YOL/OS.

77. Russell to Gamow, 28 January; 14 February 1938. PUL/HNR. I am indebted to Karl Hufbauer for his clarification of Gamow's early work.

78. T. E. Sterne to Shapley, 4 December 1938. HUA/HSDC. Emphasis added. Marshak OHI, 33. AIP.

79. Oort, 1939, *Transactions International Astronomical Union* 6:291. Hufbauer, forthcoming, 21; Russell, 1938a.

80. McCrea to Chandrasekhar, 10 August 1938. Folder 21.15, RUC/SC. Russell, 1938b. On Lyot, see Hufbauer, 1994.

81. Gamow to Struve, 17 September 1938. YOL/OS.

82. Gamow to Russell, 8 October; Russell to Gamow, 24 October 1938. PUL/HNR.

83. Russell to A. van Maanen, 28 January 1939; to Shapley, 27 January 1939, 4. PUL/HNR. Russell, 1937a, 49; 74; 77.

84. Russell, 1937a, 77. Hufbauer, 1981.

85. Russell to Conklin, 8 October 1938. PUL/HNR.

86. Russell to A. S. King, 8 February 1939. PUL/HNR. Also implied from Russell's remarks. Russell, 1939d, 295.

87. Russell to A. S. King, 20 February 1939. PUL/HNR. Shapin, 1994, 414, suggests the importance of face-to-face contact in knowledge production.

88. This must have been apparent to Eddington as well. See Eddington, 1938a, 140. Bethe, 1939, reprinted in Lang and Gingerich, 1979, 320–338. See 334–336. On the importance of expertise in establishing trust, see Shapin, 1994, 412.

89. Russell, 1939d, 295.

90. Russell to The Editor of the New York Times, 18 February 1939; to V. M. Slipher, 21 February 1939. PUL/HNR.

91. Russell, 1939a; Russell to Roy Waldo Miner (New York Academy of Sciences), 7 November; 11 November 1940; Miner to Russell, 6 November 1940. PUL/HNR. The awarding of the $300 Lewis Prize to Russell was noted in E. G. Conklin to Shapley, 21 March 1939; Shapley to Conklin, 23 March 1939. Box 28, UA V 630.22 HUA/HSDC.

92. Russell, 1939d, 306; 1941c. Although Shapley ran the conference, the academy sent the speakers list to Russell for approval before anyone was invited. Miner to Russell, 5 October 1939. PUL/HNR.

93. Russell, 1942, 233.

94. DeVorkin and Kenat, forthcoming.

95. Quoted in Osterbrock, 1997b, 298. Russell, 1948a.

Chapter 17

Binary Stars and the Formation of the Solar System

1. Steins, 1936, 107; Schwarzschild OHI, 18 June 1982, 6. NASM

2. Sitterly, 1977, 28. Russell to K. Steins, 21 July 1933. Box 27:32, PUL/HNR. I am indebted to Peggy Kidwell for alerting me to Steins's survey, and to her interpretative essay: Kidwell, 1990.

3. Russell to Bok, 17 October 1946. PUL/HNR. Emphasis added.

4. Brown et al., 1922, 9.

5. Russell, 1922b, 241; Russell, 1923b; Russell, Adams, and Joy, 1923; Russell, 1928b, 89, unchanged from 25 November 1922 holograph, "On the Determination of Dynamical Parallaxes," 1. 102.30, PUL/HNR.

6. Seares, 1922b, 165–167.

7. Russell, 1922b, 238.

8. Russell to Aitken, 22 November 1922. PUL/HNR. Russell, 1922b, 238.

9. Russell to Aitken, 11 April 1923. PUL/HNR. Russell, 1928b, 95.

10. Redman, 1927; 1928.

11. Aitken, 1933, 131–132.

12. Russell to Struve, 27 June 1938. YOL/OS. Strömberg meanwhile published interim studies and also suggested that he and Russell delay publication until they could meet face-to-face in Pasadena. Russell to Adams, 24 November 1939; to Seares, 18 June; 31 October; 14 December 1939; 13; 22; 24 May 1940. PUL/HNR.

13. Russell and Moore, 1933; 1938; 1940a; 1940b. Russell to Struve, 8 November 1938. PUL/HNR.

14. Russell and Moore, 1940a, vii.

15. Russell to Eddington, 30 March 1936; to Gamow, 14 February 1938; to Shapley, n.d. PUL/HNR.

16. N. C. Dunér, noted in Struve and Zebergs, 1962, 300. Tscherny, 1927; 1929; Walter, 1933, 1 n. 2.

17. Russell to Menzel, 19 December 1927. PUL/HNR. Blagg, 1928; Tscherny, 1928.

18. Russell manuscript, "On the Advance of Periastron in Eclipsing Binaries," 8, crossed out line. File 106.1, PUL/HNR. Russell, 1928a.

19. Dugan and Waterfield, 1928, 18. Quote from Dugan, 1929, "The Role of the Eclipsing Binary," AAAS talk, Albuquerque, 1929. PUL/RSD.

20. Dugan to Nijland, 21 May 1930. PUL/RSD. Redman, 1930, 759.

21. Russell and Dugan, 1930, 215.

22. Walter, 1933, 3; 14, cited in Kopal, 1938, 448. Kopal to Chandrasekhar, 3 November 1935. RUC/SC.

23. Kopal, 1938, 448; Cowling, 1938, 734; 741; 744. See also Wood, 1977, 48; Cowling, 1985, 10; and Cowling OHI, 28–30. AIP.

24. Sterne, 1939.

25. Ibid., 672. Sterne's influence on Russell's thinking can be found in [Russell], "Notes on Ellipticity in Eclipsing Binaries," folder 1 of 2, June 1939; folder 2, December 1938. 109.23; 109.24, PUL/HNR.

26. Russell, 1939b, 641; 674.

27. Doel, 1996, 41.

28. Russell to Donald Lee Cyr, 18 November 1941. PUL/HNR.

29. Brush, 1996b, 9–11; and Dick, 1996, 53–54; 178. Doel, 1996, 33–42, provides a detailed overview of Russell's contributions to cosmogony.

30. Brush, 1996b, 5–9; 60–67; Dick, 1996, 166–172; Brush, 1978; Burchfield, 1975.

31. Russell, 1924d, 410. Russell to W. D. MacMillan, 11 December 1924, 3. PUL/HNR. Brush, 1996b, 68–74.

32. Russell to L. H. Adams, 25 February 1927. PUL/HNR. Russell, 1929a, 7.

33. Russell, 1931g, 92. Brush, 1996b, 79–80.

34. S. A. Mitchell to Russell, 21 February 1933; Russell to Mitchell, 24 February 1933. University of Virginia file, PUL/HNR.

35. Doel, 1996, 33–39.

36. Russell to V. M. Slipher, to Carl Lampland, 10 April; to Dugan, 11 April 1934. PUL/HNR.

37. Russell to Eddington, 27 February; to Jeffreys, 19 March; 18 April 1934. PUL/HNR.

38. Russell to Mitchell; to J. C. Newcomb, 2 May: to Meggers, 10 May; to V. M. Slipher, 14 May; to Dr. Alexander Lambert (family physician in New York), 11 June 1934. PUL/HNR.

39. Russell to Meggers, 10 May; to V. M. Slipher, 14 May 1934. PUL/HNR.

40. Frank Edmondson OHI, 26. AIP.

41. Russell, 1935e, 93.

42. Ibid., 134.

43. Ibid., 137–139. This was the least accepted of Russell's many speculations. Luyten abhorred any "external theory aided by the dim past of an expanding universe." Luyten, 1938, 85. Doel, 1996, 40–41, discusses the implications of convergent timescales.

44. Doel, 1996, 40; 43.

45. Russell to W. M. Smart, 17 May 1934. PUL/HNR. Eisenhart to Mr. Lewis Walters, 17 May 1934. Walters Alumni file, courtesy Nanci Young, PUA.

46. Russell to E. W. Brown, 12 October 1935. PUL/HNR.

47. Russell to Adams, 4 January; to Smart, 10 February 1936. PUL/HNR.

48. Lyttleton, 1936b, 559.

49. Russell to Adams, 14 March 1936. PUL/HNR.

50. Russell to R. H. Simpson [Commonwealth Fund Fellowships], 2 March 1936. Russell to Smart; to Eddington, 17 March; to The Vice Chancellor of Cambridge University, 20 March 1936. PUL/HNR.

51. Lyttleton, 1936b, 559–560; 563–567.

52. Russell to Smart, 9 April 1936. PUL/HNR.

53. Russell to Eddington, 23 February 1937. PUL/HNR; W. H. Wright to Russell, 2 June 1936. Russell file, LAUC/WHW.

54. Smart to Russell, 20 May 1936. PUL/HNR.

55. Russell, 1936b, 205. "Astronomy" (for the American Year Book 1936), December 1936. File 109.6, PUL/HNR; Russell, 1936a, 267. Robertson, 1937, 423; 425. Brush, 1996b, 10.

56. Luyten to Russell, 28 January, 24 February 1937. PUL/HNR.

57. Luyten was then applying for Guggenheim and American Philosophical Society support. Luyten to Russell, 28 January, 24 February 1937. Russell to Luyten 12 February 1937. PUL/HNR.

58. Luyten and Hill, 1937, 109–111. Luyten to Russell, 4 April 1937. PUL/HNR.

59. Russell to Luyten, 27 April; 14 May; Luyten to Russell, 5 May; 12 May. PUL/HNR.

60. Russell to Luyten, 27 May; Luyten to Russell, 17; 27 June 1937, with appended manuscript "On the Origin of the Solar System," 11. PUL/HNR.

61. Lyttleton to Russell, 31 May 1937. PUL/HNR.

62. Russell to Lyttleton, 17 June 1937. PUL/HNR.

63. Russell to Luyten, 2 July; 13 July 1937. Luyten to Russell, 22 September; quoted from Russell to Luyten, 2 October 1937; to Russell, 8 November 1937. PUL/HNR. Luyten and Hill, 1937, 482.

64. Luyten, 1938, 84–85. Brush, 1978, 86–97 has noted how Americans favored monistic theories. Both Doel, 1996, 39–40; 116–117 and Dick, 1996, chap. 4, explore this factor's broader implications.

65. Brush, 1996b, 79–81. Kuiper to Struve, 11 February 1941. YOL/OS.

66. Lyttleton, 1939a; 1939b.

67. Spitzer OHI, April 1977, 31; 36. AIP.

68. Ibid., 17–19; 26; quote from 28.

69. Spitzer files, Box 9, Astronomy Department Records, PUA. Spitzer, 1939a. Spitzer OHI, April 1977, 28–29; 32. AIP.

70. Russell to Dean of Graduate School, Yale University, 27 February 1939. He said much the same to Harvard. PUL/HNR. On Spitzer's succession at Princeton, see chap. 22, below.

71. Spitzer OHI, April 1977, 37; 10 May 1978, 56. AIP.

72. Russell to Spitzer, 18 May 1939. PUL/HNR. Spitzer OHI, May 1978, 56–57. AIP.

73. Russell to Spitzer, 18 May 1939, PUL/HNR; Spitzer, 1939b, 687–688.

74. Russell, 1940, 141. Russell to Spitzer, 18 May 1939. PUL/HNR.

75. Russell to Struve, 10 June 1940. PUL/HNR.

76. Lyttleton to Struve, remarks for the reviewer, 13 July 1940; Luyten to Russell, 14 February 1941; Russell to Lyttleton, 9 April 1940. PUL/HNR.

77. Luyten to Russell, 14 February; Russell to Luyten, 17 February 1941. PUL/HNR.

78. Russell to Struve, 11 March; 28 March; Spitzer to Russell, 25 March 1941; Struve to Lyttleton, 1 April 1941. PUL/HNR.

79. Lyttleton, 1941a; 1941b, 225; 1942.

80. Russell to Spitzer, 20 October 1941, appended to Spitzer to Lyttleton, 18 October 1941. PUL/HNR.

81. Luyten, 1941; 1942, 483 n. 4.

82. Lankford and Slavings, 1996, and Lankford, 1997, explore this characteristic of the American astronomical community.

83. Eddington, 1938b, 172.

84. Russell to Chandrasekhar, 26 January 1945. PUL/HNR.

85. "Dodge Hall Discussion November 1917," Box 72.2 (b), PUL/HNR.

86. RDS 1, 468, quoted in Dick, 1996, 54.

87. Russell, "Life in Space and Time," manuscript draft, identical to that published in E. V. Cowdry, ed., *Human Biology and Racial Welfare* (New York, 1930). Box 105.16, PUL/HNR.

88. Russell, review of Eddington, *The Nature of the Physical World*. Manuscript draft sent to editor of the *Yale Review*, 29 March 1929. Box 106.11, PUL/HNR.

89. "Dodge Hall Discussion November 1917" session on "Some Intellectual Problems of Religion, viewed from The Standpount of Physical Science." Box 72.2 (d), PUL/HNR.

90. Dick, 1996, chap. 2, identifies the correlation.

91. Russell, "The Birth of our Planet," 14 April 1941. File 110.19; Russell to the United States Rubber Company (NYC), 8 May 1945; "Broadcast: Are the Planets Habitable?" Manuscript file, box 112, folder 2, PUL/HNR.

92. Russell, 1943a, 19.

93. Ibid.

Chapter 18
The Royal Road

1. Stewart, 1958, 312. HNRjr., 29–30. MRE #1, 69–70.

2. Russell, 1938a, 108–109.

3. Ibid., 112–113.

4. Ibid., 111.

5. Schweber, 1986; 1990; Eckert, 1996, 69–71.

6. Russell, 1938a, 112–113.

7. "General Notes," File 107.5; Russell to Aitken, 29 April 1933. PUL/HNR.

8. "General Notes." On computing's requiring certain "tacit, unwritten, and non-verbal" craft skills, see Ceruzzi, 1991, 239.

9. Szafraniec, 1970, 13. Russell to Aitken, 13 October 1916. PUL/HNR. Aitken, 1918, 168. Pinch, 1986, 151, examines the work of a theorist, who like Russell tested theory not only against nature but against social factors.

10. President Avila Camacho donated $1,000 and presented Russell with a heraldic decoration. Shapley to Dodds, 5 April; to Arthur W. Butler, 26 December 1945; Boyce and Meggers to J. E. Mack, n.d.; Shapley to the Ambassador of Mexico, 17 December 1945; HCO Project Files, "HNR Fund" UAV 630.22.5 Box 2, HUA/HSDC. On the 1942 Mexican Conference, see UAV 630.22.5 Box 1. HUA/HSDC.

11. Crawford also opposed naming the lectureship for a living astronomer, on the same grounds as he objected to the renaming of "Hoover Dam," but added: "I take second place to no one in admiration of your work and in recognizing you as the foremost living American astronomer . . . I can well believe that this action of the Council is quite an embarrassment to you." R. T. Crawford to Russell, 6 November 1945; Russell to Crawford, 13 November 1945. LAUC/RTC. Wright to Shapley, 12 November 1946; Shapley to Wright, 18 November 1946. The original goal was $25,000. Shapley to Committee, 30 June 1945; Shapley to McLaughlin, 24 June 1945. "HNR Fund" UAV 630.22.5 Box 2. HUA/HSDC.

12. Russell, 1932a, 199; 1935e, 19.

13. C. H. Payne to F. H. Seares, 18 February 1931. HL/FHS. Plans had to be ratified by both Shapley and Russell. C. H. Payne to Dugan, 14 January 1933. PUL/RSD. Wright, 1987, chap. 13, 61. Payne's was a staff position at the observatory without faculty rank until 1938. On Payne's advancement at Harvard, see Kidwell, 1984, 25–27.

14. Kopal, 1986, 172–173.

15. Ibid., 151–152. See E. J. Fisher (Institute of International Education, NYC) to "Director Lick," 15 February 1938. LAUC. Kopal to W. S. Adams, 23 March 1938. 38.661: Zdeněk Kopal folder, HL/WSA. See also Kopal folder, Box 21A, HUG/HSPC.

16. Kopal to Shapley, 23 July 1941; 17 August 1946 [year uncertain]. Kopal file, box 21A, HUG 4773.10, HUA/HSPC. Kopal's career planning was not unique among Europeans. See Hufbauer, 1993, 205.

17. Shapley to Drury, 21 May; 11 June; 23 June; Duggan to Shapley, 7 August; Shapley to Drury, 28 August 1941. Kopal folder, UAV 630.22 "Hoc—K 1940–1950," HUA/HSDC.

18. Rough notes marked "Eclipsing Variables, Unpublished Paper by Z. Kopal (lent by Shapley Dec 1938)." Folder 72.17 (e), PUL/HNR.

19. Russell to Shapley, 19; 23 December 1938; Shapley to Russell, 23 February; 8 March 1939. PUL/HNR. On Prager's fate in Europe, see Blaauw, 1994.

20. Schwarzschild to Chandrasekhar, 14 January 193[9]. File 28.13, RUC/SC.

21. Shapley later assured Kopal that Sterne had no right to his anger. Shapley to Kopal, 8 March 1941. Kopal folder, Box 21A, HUG 4773.10, HUA/HSDC.

22, Russell to Struve, 8 November 1939, PUL/HNR, repeated his misquote of Leacock's character, who "rode madly off in all directions." Leacock, 1912, 30. Russell to Shapley, 4 January 1939; to V. M. Slipher, 17 November; to Shapley, 23 December 1938; PUL/HNR.

23. Russell to Shapley, 27 January 1939. PUL/HNR.

24. Russell to Shapley, 2 February 1939. PUL/HNR.

25. Kuiper described Shapley's "*criminal spirit with respect to science.*" Emphasis in original. Kuiper to Carl Seyfert, 26 January 1938. Kuiper file, YOL/OS.

26. Russell to Shapley, 3; 27 February; 3; 7; 10; 30 March 1939. PUL/HNR.

27. Kuiper to Hertzsprung, 26 March 1940. UA/GK.

28. Kopal, 1986, 186; "Russell-worship" at Harvard is noted on 182. These recollections are born out by contemporary correspondence.

29. Kopal to Russell, 29 October 1943. PUL/HNR. Kopal to Russell, n.d., ca. fall 1943. Kopal folder, UAV 630.22 "Hoc—K 1940–1950," HUA/HSDC.

30. Russell to Adams, 7 February 1941. PUL/HNR. Others Russell identified included Merrill, Pierce, and Sitterly on the Princeton side, with Kopal, Gaposchkin, and the Milton Bureau on the Harvard side.

31. Kopal to Russell, 10 March 1942. PUL/HNR.

32. Kopal to Russell, 24 March 1941; 22 January 1942. PUL/HNR.

33. Adams to Russell, 31 October 1940, HL/WSA; Russell to Shapley, 13 May; 20 June 1941. PUL/HNR. Vakalopoulos, 1985. Goralski, 1981, entry for 9 April 1941.

34. Russell to Kopal, 26 November 1943. PUL/HNR. Kopal to Shapley, n.d., ca. fall 1943. Kopal folder, UAV 630.22 "Hoc—k 1940–1950," HUA/HSDC.

35. Shapley to Russell, 4 February 1944. PUL/HNR. Kopal's autobiography devotes considerable space to Russell, "based on a personal acquaintance of almost twenty years in the last quarter of Russell's life (and backed by extensive correspondence now on deposit with the Manchester University Library)." Unfortunately, when the author visited the Ryland Library specifically to examine this file, after months of preparatory letters, the range containing correspondence with Russell ("r") could not be located. The rest of Kopal's papers are very rich. Kopal, 1986, 168; 180–181; 210.

36. Kopal to Shapley, 8 April 1945. "HNR Fund" folder, "HCO Project Files," UAV 630.22.5 Box 2. HUA/HSDC.

37. Russell to Shapley, 9 October 1943. PUL/HNR. The addition was Kopal's development of alpha functions.

38. Russell to Shapley, 2 February 1944. PUL/HNR. Contrary to Kopal's recollection, it took quite some time to convince Russell to write the introduction. Kopal to Shapley, 8 April 1945. "HNR Fund" folder, "HCO Project Files," UAV 630.22.5 Box 2. HUA/HSDC. Kopal, 1986, 208.

39. Quotes from Edge and Mulkay, 1976, 234 ff., who examine competition and cooperation in radio astronomy, including the role of career advancement. They reflect continually on Hagstrom's (1965, 70) study of competition.

40. Edge and Mulkay, 1976, 137.

41. Shane to Sproul, 4 February 1946; Russell to Shane, 18 March 1946; LAUC/Morrison Folder, 1937–1951. G. Kron OHI, 37–38, AIP; DeVorkin, 1985, 1214.

42. Bowen OHI, 41. AIP.

43. Kopal to Russell, 15 December 1943. PUL/HNR.

44. Owens explores the differential analyzer as a symbol. Owens, 1986, 85 ff.

45. Kopal to Russell, 15 December 1943. PUL/HNR. On the machine builders as protagonists for a new world, see Owens, 1986; 1995, 13. Kopal, 1986, 217.

46. Russell to W. J. Eckert, 12 March; 8 June; 14; 20 December 1937. PUL/HNR. On the bureau, see Eckert, 1937. In 1939 the bureau was named the Watson Computing Bureau.

47. Owens, 1986, 93–94; 1995, 13. Ceruzzi, 1991, 238.

48. Russell to Kopal, 9 April 1941; "Eclipsing Variables," Folder 4, Booklet VI, October through November 1941. File 72.18, PUL/HNR. Edge and Mulkay (1976, 174–175) found, in like manner, that competing groups committed themselves to a partcular technology or reduction technique to resolve an "internal" dispute, rather than to solve a specific astrophysical problem.

49. Russell to Kopal, 2 February 1944. PUL/HNR. On de-skilling and automation in industry, see Noble, 1984, 40–41.

50. Struve, 1946, 329–330; 339.

51. Russell to Shapley, 26; 27 March 1946. PUL/HNR.

52. Russell to Charlotte Moore Sitterly, 29 April 1946. PUL/HNR.

53. Russell to Shapley, 14 February 1946. PUL/HNR.

54. Russell to Kopal, 27 March, 4 May 1946. PUL/HNR.

55. Russell to Mrs. Sergei Gaposchkin, 29 September 1945. Russell to Stanley Cobb, 31 March; to Shapley, 26 March 1946. PUL/HNR.

56. *Bulletin of the Panel on the Orbits of Eclipsing Binaries No. 1*, May 1946, 2. Kopal file, LAUC/GEK.

57. Whitford to Kron, 17 September 1946. Whitford file, LAUC/GEK. Kron, Gerald E., 1946. "Reviews," *PASP* 58:272; Henk van de Hulst, 1946, "Reviews," *ApJ* *104*:463–464. Years later Kopal recalled only one positive statement, quoting the first line of André Danjon's review in the *Annales d'Astrophysique*. André Danjon, 1946, "Bibliographie," *Annales d'Astrophysique* 9:234. Kopal, 1986, 209.

58. A. Beer, 1947, "Eclipsing Binaries," *Obs* 67:150–152. L[eon] C[ampbell], 1946, "Book Review," *Popular Astronomy* 54:384. Merrill, 1947, "Books and the Sky," *Sky and Telescope* 6:14–15. See also Struve, 1945, 381.

59. Whitford to Kopal, 24 November 1947. Whitford file, LAUC/GEK. DeVorkin, 1985, 1215–1217.

60. "Eclipsing Variables Lecture Notes," starting 12 February 1946. File 73.2, PUL/HNR.

61. Gibson Reaves to the author, 24 June 1993.

62. Trans Novas to Russell, 11 May; Russell to Trans Novas, 17 May 1946. Caroline University File, PUL/HNR. Dean of the Science Faculty of the Caroline University, Prague, to Struve, 11 May 1946; Struve to Dean of the Faculty, 21 May 1946. YOL/OS. Kopal, 1986, 215–216, appears to have been unaware of Russell's endorsements.

63. Kopal to Russell, 28 October 1945. PUL/HNR. Kopal, 1986, 208–209.

64. Kopal, 1946, 13. Kopal's remark was picked up by reviewers. Beer, 1947, "Eclipsing Binaries," *Obs* 67:150–152.

65. Russell, 1948b, 197–198.

66. Ibid., 205; 207–208.

67. These included his IAU Commission presidency and his NRC committee chair. Russell to Shapley, 11 September 1948. "Roc—S" file, UAV 630.22. HUA/HSDC. Russell to Jan Oort, 10 November 1947; to Members of Commission 35, 11 July 1947. IAU Commission Files, PUL/HNR. These counter Kopal's allegations. Kopal, 1986, 212–213.

68. Whipple to Kopal, 31 March 1950; Kopal to Shapley, 4 April 1950. "Hoc—K 1940–1950" file; Shapley to Russell, 16 April; Russell to Shapley, 22 April 1950. "Roc—S" file, UAV 630.22. HUA/HSDC.

69. Russell to Shapley, 20 November 1950. PUL/HNR.

70. Russell to Shapley, 19 May 1950. "Roc—S 1940–1950" file, UAV 630.22. HUA/HSDC. Kopal to Russell, 7 May 1950; to Struve, 13 May 1950. "Hoc—K 1940–1950"

UAV 630.22, HUA/HSDC. On the fate of the center, see Owens, 1995, 3; 1986, 85. Kopal to Chandrasekhar, 14 June 1950. File 19.11 "Kopal, Zdeněk, 1935–1968," RUC/SC.

71. Doel, 1996, 163.

72. Kopal, 1986, 220–224. Kopal to Chandrasekhar, 14 June 1950. File 19.11 "Kopal, Zdeněk, 1935–1968," RUC/SC. Shapley to Kopal, 7 July; Kopal to Shapley, 9 July; 18 August 1950. "Hoc—K 1940–1950" file, UAV 630.22. HUA/HSDC.

73. Shapley to Russell, 19 September; Russell to Shapley, 22 September 1950. "Roc—S 1940–1950" file, UAV 630.22. HUA/HSDC.

74. Kopal, 1986, 186; 182–183.

75. Ibid., 424–425.

76. Kopal, 1950, 180–181.

77. Irwin, 1951, 156. See also R. L. B[aglow], 1951, [Review of Publications] *JRASC* 45:136–137.

78. From an analysis by Alex Magoun of the *Science Citation Index* summary volumes through 1962.

79. Huffer and Eggen, 1947a; 1947b.

80. Huffer to Kopal, 1; 31 August 1951. Binder L1, RUM/ZK. Huffer and Kopal, 1951.

81. Dirk Brouwer to Kopal, 17 December 1951. Box B Folder "Brit-Bux," RUM/ZK.

82. Katherine Kron to author, 19 February 1995, with appended letters, Lenouvel to Katherine Kron, 18 May 1953; K. Kron to Lenouvel, 3 April 1953. Copies in LAUC/Kron Collection.

83. Binnendijk, 1960, 69; 163; 165; quote from 258; 272.

84. This is in contrast to the interdisciplinary rivalries Doel has explored in solar system astronomy. Doel 1996, chaps. 4–5; esp. 150.

Chapter 19
A Summer Place

1. Giclas, "Reminiscences" draft, 12–13. LowA. On the Lowell Observatory, see Hoyt, 1976; 1980; Putnam, 1994; Strauss, forthcoming.

2. Lampland Diary, 11 May 1921 entry, LowA/COL; V. M. Slipher to R. L. Putnam, 21 March 1928, LowA/VMS.

3. [Russell], 1907a, 25–26. On Lowell as a threat, and the reaction, see Strauss, forthcoming. On the canal and water vapor controversy, Hoyt, 1976; DeVorkin, 1977b.

4. H. N. Russell '97 Student notes, Fall 1895, little brown notebook "Astronomy." Box 69.3, PUL/HNR.

5. C. A. Young, *Manual of Astronomy* (1902, rev., 1904), 371.

6. H. N. Russell, "The Heavens in March, 1901," quoted in Hoyt, 1976, 95.

7. Russell to Abbot, 23 June 1915. PUL/HNR.

8. Russell to Frost, 28 January 1916. PUL/HNR. Russell, 1916c.

9. Russell to Barnard, 5 February 1916. PUL/HNR.

10. Shapley to Russell, 26 April 1927; Putnam to V. M. Slipher; to Russell, 31 May 1927. LowA/VMS; PUL/HNR. Putnam to Shapley, 17 June 1927. Box 15, HUA/HSDC.

11. Putnam to Russell, 5 May; Russell to Putnam, 28 May 1927. PUL/HNR. Shapley to Putnam, 15 June; Putnam to Shapley, 17 June 1927. Box 15, HUA/HSDC. V. M. Slipher to Putnam, 15 June; Putnam to V. M. Slipher, 21 June 1927. LowA/VMS. Russell to V. M. Slipher, 28 May 1931. PUL/HNR.

12. Lampland diary, entries for 7 June 1927 and 6 July 1927. LowA/COL. Slipher to Duncan, July 1927, n.d. LowA/VMS. Russell to Slipher, 28 May 1931. PUL/HNR; LowA/VMS. MRE #1, 11–12.

13. Lampland diary, 8 July 1927. LowA/COL. Slipher to Adams, 12 July 1927. LowA/VMS.

14. Lampland diary entries, 8 July 1927 to 4 August 1927. LowA/COL.

15. Putnam to V. M. Slipher, 18 March 1929. LowA/RLP.

16. P. I. Wold to Slipher, 13 May 1926; Slipher to Bennett, 7 August 1926; to Putnam, 22 August 1927. LowA/VMS. Smith, 1994.

17. Putnam to Slipher, 31 August; 12 December; 31 December 1927. LowA/VMS. Dugan to Lampland, 1 October 1928. LowA/COL.

18. Hoyt, 1980, 151–152. Russell to Pettit, 25 January 1923. PUL/HNR.

19. Russell to Lampland, 3 March 1926. PUL/HNR. Lampland to Miller, 19 February 1928. LowA/COL. Slipher to Putnam, 18 February; Putnam to Slipher, 28 February; Slipher to Putnam, 5 March 1928. LowA/VMS.

20. Lampland to V. M. Slipher, 3 March 1929. LowA/VMS.

21. Lampland to Slipher, 10, 7 March 1929; Diary entry, 14 March 1929. LowA/VMS/COL.

22. Lampland to Slipher, 23 March 1929. LowA/VMS.

23. Shapley to Russell, 22 May 1929. PUL/HNR. Quote from Lampland to Slipher, 9 April 1929. LowA/VMS.

24. Slipher to Putnam, 5 August 1929. LowA/VMS.

25. Stewart to the Graduate Dean, 16 June 1930. PUL/JQS. Stewart to Lampland, 16 June 1930. LowA/COL. Bennett to Dugan, 25 October 1930; 9 November; Dugan to Bennett, 27 October 1930. PUL/RSD. Russell to Bennett, 19 May 1931. PUL/HNR.

26. Slipher to Cogshall, 29 May 1930. LowA/VMS.

27. Slipher to Putnam, 31 December 1928. LowA/VMS. On Slipher's attitude, see Smith, 1994; and Hoyt, 1980, 178.

28. Russell to Lampland, 30 December 1930. PUL/HNR. Tombaugh to Slipher, 1 November 1932. LowA/VMS. On Tombaugh's later work, see Hoyt, 1980, chap. 12.

29. Zorn, 1988. Adel, 1933. Adel OHI, 15; 17–19; 47.

30. "Report of the Committee on the Henry Draper Fund," 3 November 1932. Reel 11, NAS folder. Russell to Aitken, 3 November 1932; to Slipher, and Campbell, 10 April 1934. PUL/HNR.

31. On the discovery of Pluto, see Hoyt, 1980. Slipher to Miller, 8 March 1930. LowA/VMS.

32. Hoyt, 1980. 226–228; Shapley to Slipher, 20 March 1930. LowA/VMS.

33. Slipher to Putnam, 1 May 1930, quoted in Hoyt, 1980, 229.

34. On the itinerary, see HNRjr, 16–17; MRE #1, 13; Russell to Breasted, 22 June 1929. PUL/HNR.

35. Russell, 1930c, 21–22.

36. Ibid.; Russell, 1930b, 364.

37. HNRjr, 18. MRE #1, 32.

38. Russell, 1930m, 446.

39. Russell, 1931e, 91.

40. Brown to Russell, 7 February 1931. PUL/HNR.

41. Ibid.

42. Appended to Russell, 1931b, 303. Russell to Bennett, 19 May 1931. PUL/HNR.

43. Brown to Putnam, 17 March 1930. Pluto folder, LowA/RLP. The question of the accidental nature of the discovery has lingered as an issue among dynamical astronomers. Kuznik, 1993.

44. Russell to Slipher, 28 May 1931. PUL/HNR.

45. Russell to Slipher, 31 October 1930; to Hubble, 21 October 1932. PUL/HNR.

46. Russell to Jean Hetherington, 7 May 1934; 29 January 1938; to Wynn Armstrong, 18 February 1937. PUL/HNR. MRE #1, 53–55.

47. MRE #2, 50.

48. Russell to Adams, 3 May 1933. The British Association had asked him to attend as representative of the AAAS, since he was the newly elected president. Russell to British Association, 17 May 1933. Russell had been thinking about renting a car to drive around England that summer but decided to play it safe, "in the face of utter uncertainty." Russell to H. H. Plaskett, 27 April 1933. PUL/HNR. On paying the bills, and family income, see MRE #1, 40–41.

49. HNRjr., 7.

50. MRE #2, 40–41.

51. HNR jr, 30–31. S. Chandrasekhar, conversation with author, June 1992.

52. MRE #2, 42.

53. Russell to Dr. Charles Manly Griffiths, 19 September 1933; to Dr. Alexander Lambert, 3 October 1933. PUL/HNR. Bower, 1993, 35; 50. Phelps, 1966, 84.

54. HNRjr., 2. Elizabeth still spent winters in Tucson. Russell to Seares, 15 June 1939; Russell to G. Abetti, 14 October 1940. PUL/HNR. On Roosevelt's asthma, see McCullough, 1981.

55. Cogshall to Slipher, 15 February 1931. LowA/VMS.

56. F. K. Edmondson OHI, 9–11.

57. Ibid., 26–27.

58. Henry jr. was also there. Slipher to Cogshall, 1 June 1934. LowA/VMS; F. K. Edmondson OHI, 23–25.

59. F. K. Edmondson OHI, 25. Conversations with Henry Giclas, 5 January 1993. Russell to G. Abetti, 21 November 1934. PUL/HNR.

60. F. K. Edmondson OHI, 28. Russell was thinking of Harvard even before their marriage. Russell to G. Abetti, 21 November 1934. PUL/HNR.

61. Russell to G. Abetti, 21 November 1934. PUL/HNR.

62. Russell to Putnam, 11 December 1934. PUL/HNR. F. K. Edmondson OHI, 29–30.

63. Russell to K. P. Williams, 20 November 1934. PUL/HNR.

64. Russell to Williams, 20 November 1934; 24 December 1934; to Edmondson, 1 February; 4 March 1935. Shapley to Russell, 12 February 1935. PUL/HNR.

65. *Princeton Alumni Weekly*, 6 December 1935. The Coles had acquired the Jamestown property after they sold the Southport home in 1920, and by the 1930s it was used regularly by the Russells.

66. Lampland to Edmondson, 2 November 1935. LowA/COL.

67. Mitchell to Russell, 13 February; Russell to Mitchell, 15 February 1937. PUL/HNR. Van de Kamp OHI #2, 114; DeVorkin, 1984b, 55. Mitchell to Edmondson, 13 February 1937. Edmondson file, PUL/HNR.

68. F. K. Edmondson OHI, 36–38. Russell to Edmondson, 15; 17 February 1937. PUL/HNR.

69. Russell to Edmondson, 17 February 1937, 4. PUL/HNR.

70. Russell to Mitchell, 10 March; to Williams, 12 March 1937. PUL/HNR. Slipher to Cogshall, 10 June 1937. LowA/VMS.

71. Russell said much the same to Shapley about his own staff. Russell to Shapley, 3 June 1937. PUL/HNR.

72. Edmondson to C. O. Lampland, 14 July 1937. LowA/COL.

73. Ibid.

74. Ibid.

75. Ibid.

76. Russell to Slipher, 21 June 1937. PUL/HNR.

77. Russell to R. H. Curtiss, 7 March 1927. PUL/HNR. Curtiss had been especially anxious to have Russell attend, to attract attention and support. Milne did not attend until 1929.

78. Russell to Slipher, 20 January 1931. PUL/HNR.

79. Russell to Campbell, 3 February 1933. PUL/HNR. Campbell to Slipher, 6 February 1933. Campbell folder, LowA/VMS.

80. Jarrell, 1988.

Chapter 20
Influencing Institutions and the Profession

1. Van de Kamp OHI, 1977, 47. AIP. Lankford, 1997, 204–206, explores the influence that "the Generals" had on American astronomy. On the origins of the Neighborhood Club, sometimes called "The Neighbors," see Schlesinger to Russell, 30 November 1919; 27 May 1920; 18 September 1920. PUL/HNR.

2. "News and Views," *Nature 139*:1099 (26 June 1937); "Signal Honor for Professor Russell," *Scientific American 157*:182–183 (September 1937); "American Astronomical Society Reports Progress," *Sky* 3:3; 16–19; 28 (March 1939); Shapley, 1957, 260; van de Kamp OHI, 1977, 48. AIP.

3. Lankford, 1997, 182–183.

4. Struve to Markowitz, 8; 22 October 1932; 27 February 1933. YOL/OS.

5. Osterbrock, 1997a, 115, notes that the choice of a director was usually a management prerogative. Fewer than two hundred people showed up to AAS meetings in the 1930s. DeVorkin and Routly, 1999.

6. Jesse Greenstein to Russell, 26 December [1940], PUL/HNR, responding to Russell, 1941b. Struve to Stebbins, 23 August 1941. YOL/OS.

7. On Hale's goals, see Kevles, 1968, 431–432. Ron Doel (1985) has found that Russell frequently discussed academy policy issues with his campus colleague Augustus Trowbridge.

8. Kevles, 1968, 435. On the fund, see Davis and Kevles, 1974; Dupree, 1957, 338–343; Kevles, 1978, 185–187.

9. "Scientific Members of the Trustees of the National Research Fund as Representative of Fields of Research," NRF incoming file, PUL/HNR. Doel, 1985.

10. Hale to Russell, 13 November 1928. NRF incoming file, PUL/HNR.

11. Davis and Kevles, 1974, 215. Russell to T. H. Morgan; to Ostwald Veblen, 16 May 1928. NRF file, PUL/HNR.

12. Russell to "My Dear . . . ," 30 April 1928. NRF file, PUL/HNR. Letters went to Shapley (Harvard), Schlesinger (Yale), Brown (Yale), Curtis (Allegheny), Curtiss (Michigan), Frost (Yerkes), Stebbins (Wisconsin), Slipher (Lowell), Adams (Mount Wilson), Ait-

ken (Lick), Abbot (Smithsonian), Mitchell (Virginia), Leuschner (Berkeley), Miller (Swarthmore), and to Hale for his information.

13. Untitled report, 10, attached to Russell to T. H. Morgan, 16 May 1928. NRF file, PUL/HNR. Adams reported that the Carnegie Institution already provided sufficient funds for their work, and so for the moment they would not propose.

14. Ibid., Untitled report. Lankford, 1997, 215–219, analyzes the first round of responses in detail.

15. Incoming correspondence, ca. early May through June, 1928. NRF file, PUL/HNR. Russell to Morgan, 16 May 1928. PUL/HNR. Both Leuschner and Curtiss (at Michigan) ran large graduate programs in astronomy. Leuschner to Curtiss, 5 June 1928; to Russell, 22 May 1928. NRF incoming file, PUL/HNR.

16. Hale to Morgan, 13 June 1928; Russell to Vernon Kellogg, 11 October 1928; to Abbot, Schlesinger, Shapley, and Hale, 31 October 1928. Campbell to Russell, 7 November 1928. NRF file, PUL/HNR.

17. Russell to Hale, 28 November 1928. PUL/HNR.

18. Davis and Kevles, 1974, 214, considered a host of other factors as well in their analysis. There is no evidence that Russell was ever concerned with the failure of the fund. Leuschner to Millikan, 12 April 1929. NRF incoming file, PUL/HNR.

19. V. M. Slipher to Russell, 6; 12 May 1928; Stebbins to "Members of the Section," 19 November 1932. NAS files, incoming, PUL/HNR.

20. Shapley to Russell, 24 July 1928. PUL/HNR. Legible copy in Box 17.124, UAV 630.22. HUA/HSDC. Shapley to Schlesinger, 16 August 1928, Reel 29, AIP/Yale/FS.

21. Putnam to V. M. Slipher, 12 September 1928. Putnam file, LAUC/VMS. The travails surrounding Hale's big telescope have been examined in Wright, 1966, 387–399. On the general policies of the Rockefeller Foundation and Hale's relationship with it, see Kohler, 1991, 233–235.

22. "A Great Opportunity," *New York Times*, 7 November 1926, pt. 2, p. 8.

23. Shapley to Russell, 24 July 1928. PUL/HNR.

24. On Shapley's general influence at the foundation, see Kohler, 1991, 149; 215; 256. On Shapley's specific advice regarding priorities in astronomy, see Shapley to Wickliffe Rose, 25 March 1925; to Rose, 20 November 1926; Rose to Shapley, 20 November 1926; T. B. Applegate to Shapley, 6 May 1927; letters, Shapley to Rose, March–October, 1927. This included an omnibus Harvard proposal equal in size to the gift to Caltech. Shapley to Rose, 15 May 1928. Box 16, UAV 630.22. HUA/HSDC.

25. Russell, "Memorandum on the Administration of a Great Telescope," 20 April 1928, 1; 4. Marked "rec'd" at the carnegie Institution on 30 April. Reel 31, CIT/GEH. Draft dated 28 April in PUL/HNR.

26. Quoted in Anderson to Hale, 26 September 1928. CIT/GEH.

27. Hale to Arnett, 22 September 1928. CIT/GEH.

28. Russell, "Note on the Site of the 200-inch telescope," appended to Russell to Hale, 27 September 1928. PUL/HNR.

29. Emphasis in original. Russell to Hale, 27 September 1928. PUL/HNR. Russell to Shapley, 23 August 1928. Box 17.124 UAV 630.22, HUA/HSDC. I am indebted to Joanne Palmieri for alerting me to this letter.

30. Telegram, Anderson to Hale, 28 September 1928. CIT/GEH.

31. "Scientific Approach to Christianity III," Box 72.1 (b), PUL/HNR. Shapley to Russell, 26 September 1928; Russell to Shapley, 3 October 1928. PUL/HNR. A slightly altered version was sent to Shapley, Box 17.124, UAV 630.22, HUA/HSDC. When Shapley backed down, Russell endorsed the Harvard-Leiden proposal for South Africa, stating that

it was "entirely to be commended as much important work is urgently needed in the southern hemisphere." Russell to Trevor Arnett, 3 October 1928. PUL/HNR. Kipling, 1914, "A Song in Storm." MRE #1, 36.

32. J. A. Anderson to Hale, 8 November 1928; Shapley to Hale, 17 November 1928. CIT/GEH.

33. Russell, 1929f; Russell to Hale, 26 November 1928. CIT/GEH.

34. Hale to Adams, 17 April 1930. AIP/GEH.

35. Aitken, 1933, 127–128.

36. Menzel, Autobiography, 202–209.

37. Aitken to Dugan, 15 March 1927. LAUC/RGA. Aitken and Leuschner blocked Menzel from encouraging graduate students to do astrophysical theses. Aitken to Leuschner, 26 November 1929. LAUC/RGA.

38. Quote from "Shane" interview, 19 September 1939; "Aitken" interview, 18 October 1939. Listing "A." Folder "UC Berkeley Lick Observatory Selection of Director for 1942," BANC/RTB. Birge, "History," vol. 2, pt. 7, 29. BANC/RTB; Menzel, Autobiography, 202–203.

39. Menzel, Autobiography, 209.

40. Menzel recalls that he had to publish his theory in the *Monthly Notices*, although it appeared elsewhere: Autobiography, 210; 222–227. Menzel, 1931; Menzel, 1972, 241.

41. Osterbrock, Gustafson, and Unruh, 1988, 171; 206–210. Menzel, Autobiography, 212–214.

42. Russell to Menzel, 23; 28 January 1931; 3 May; 27 June 1932. PUL/HNR. Adams to Gale, 24 March 1932. HL/WSA. Menzel, Autobiography, 232.

43. Goldberg, 1989; Russell-Menzel correspondence, 1932–1934. PUL/HNR.

44. Aitken to Kuiper, 27 January 1932; 17 January; 7 March 1933. LUAC/RGA. Wright to Sproul, 3 July 1935. U.C. Office of the President 1935 folder, LAUC/WHW.

45. Cruikshank, 1993, 260. Osterbrock, Gustafson, and Unruh, 1988, 200–202.

46. Kuiper to Bok, 21 January 1935; Hertzsprung to Kuiper, 2 April 1936. UA/GK. Translated from the Dutch by Henk Boute. Osterbrock, Gustafson, and Unruh, 1988, 202. Shapley to Birge, 12 November 1940. Shapley folder; R. T. Birge file, "U.C. Berkeley, Lick Observatory, Selection of Director for 1942," BANC/RTB.

47. Russell to Kuiper, 12 December 1934. PUL/HNR. UA/GK.

48. Bok to Kuiper, 17; 26 January 1935. UA/GK. Boute translation.

49. Wright to Leuschner, 10 October 1934. LAUC/WHW. Sproul to Aitken, 14 February 1935; Aitken to Sproul, 18 February; 25 February 1935. Aitken microfilm, LAUC.

50. Aitken to Sproul, 8 March 1935. LAUC.

51. Kuiper to J. van der Bilt, 18 January 1935. UA/GK. Boute translation.

52. Kuiper to Russell, 19; 20 February 1935. PUL/HNR. On Wyse's appointment, see Kuiper to Hertzsprung, 8 March 1935. UA/GK.

53. Russell to Aitken, 23 February 1935. PUL/HNR; Aitken to Russell, 1 March 1935. Russell file, LAUC/RGA.

54. Aitken to H. A. van Coenen Torchiana, 29 March 1935. Kuiper folder, LAUC/RGA. On nativism, see Higham, 1955, 4.

55. Russell to Shapley, 23; 27 February 1935; Shapley to Russell, 25 February 1935. PUL/HNR.

56. Russell to Shapley, 23 February 1935; Shapley to Russell, 25 February 1935; Russell to Leuschner, 25 February 1935. PUL/HNR.

57. Russell, 1936e, 314.

58. Adel OHI, 52; Shapley to Russell, 26 April 1935. PUL/HNR; Lyman Spitzer, "Newton Lacy Pierce 1905–1950," 3. Typescript for *Popular Astronomy*.

59. Dunham to Russell, 13 January 1935. AIP/THD.

60. Adams to Gale, 13 June 1931. HL/WSA.

61. Ibid.

62. Gale to Adams, 22 May 1931; Adams to Gale, 1 June 1931. HL/WSA. Slipher to Fox, 12 January 1932. LowA/VMS.

63. Russell to Gale, 25 May 1931; to Shapley, 11 March 1932. PUL/HNR. Osterbrock, 1997a, 111–112; 121.

64. Russell to Shapley, 11 March; to Gale, 9 March 1932. PUL/HNR.

65. Russell to Gale, 11; 18 March; to Shapley; to Struve, 17 May 1932. PUL/HNR.

66. On Hutchins's support, see Fermi, 1971, 76–77. On Gale's xenophobia, see Gale to Saunders, 18 November 1922, quoted in DeVorkin, 1994a.

67. Struve to Shapley, 18; 29 February; 8 March 1932; to Russell, 13 March; 10 May 1932. PUL/HNR. Quote from 29 February.

68. Struve, 1935, 485–486. See also Struve, 1931. His Guggenheim Fellowship is noted in "Reports of Observatories, 1928–1929," *PopAst* 38:344 ff.

69. Struve, draft "Memorandum," 4 February 1941, attached to Struve to Hutchins, 19 February 1941. YOL/OS.

70. Struve to Unsöld, 21 February 1935. AIP/OS/YOL.

71. Struve to Kuiper, 19 December; Kuiper to Struve, 23 December 1935. UA/GK. Morgan to Struve, 29 November 1935. AIP/OS/YOL. Russell to Struve, 3 January 1936. PUL/HNR.

72. Shapley to Russell, 20 March 1935. PUL/HNR. Shapley to Chandrasekhar, 27 November 1934. RUC/SC.

73. Wali, 1991, 162–165.

74. Russell to Struve, 29 January 1936. PUL/HNR.

75. Struve to Russell, 3 February 1936; Russell to Struve, 5 February 1936. PUL/HNR.

76. Struve to Kuiper, 5 February 1936. "Confidential G. P. Kuiper—Otto Struve" 1935–1950, UA/GK. Shapley to Struve, 9 January 1936. HUA/HSDC. Gale to Struve, 28 January 1936. AIP/OS/YOL.

77. Hutchins to Struve, 15 January 1936. AIP/YOL/OS. Struve to Russell, 3 February 1936. PUL/HNR; Struve to Kuiper, 5 February 1936. UA/GK. I am indebted to Karl Hufbauer for commentary on Atkinson.

78. Struve to Kuiper, 21 March 1936. AIP/OS/YOL.

79. Kuiper to Struve, 8 February; 1 March 1936. UA/GK.

80. Russell to William Smart, 9 April 1936. PUL/HNR.

81. Kuiper to Struve, 1 March 1936. UA/GK. Wali, 1991, 164–165. Chandrasehkar OHI, 1977, 58. AIP.

82. Russell to Adams, 18 January 1937. PUL/HNR. Struve to Gale, 3 February 1936. AIP/YOL/OS. Struve's reluctance to hire Chandrasekhar is noted in Wali, 1991, 235–236.

83. Birge to Sproul, n.d., draft, ca. November 1934. Lick Observatory folder, carton 2, BANC/RTB.

84. Adams to Birge, 30 September 1940. BANC/RTB.

85. Ibid.

86. [Birge], Leuschner interview, entry for 11 November 1940, 59–62. "U.C. Berkeley Lick Observatory Selection of Director for 1942," BANC/RTB.

87. Russell to Birge, 4 October 1940; 16 October 1940. BANC/RTB. Kohler, 1991, chap. 13, has explored the role of new instrumentation in how research agendas were defined in areas experiencing rapid change.

88. Russell to Birge, 4; 16; 28 October 1940. BANC/RTB.

89. Struve to Kuiper, 27 February 1940. YOL/OS.

90. Adams–V. Bush correspondence, August 1939 through September 1940. Adams Papers, HL/WSA.

Chapter 21
Astronomical Isolationism

1. Nativism is used here to mean "intense opposition to an internal minority on the ground of its foreign connections." Higham, 1955, 4.

2. Strömgren, 1932, 118; 1933.

3. Kuiper to Chandrasekhar, 15 March 1936. 19.21: Kuiper folder 2: 1936–1950, RUC/SC. Kuiper to Hertzsprung, 14 March 1936. Box 29a, "Hertzsprung, Ejnar" 1929–1957, UA/GK. Milne, 1930c, 204. Eddington, 1926, 163.

4. Kuiper to Chandrasekhar, 15 March 1936. 19.21: Kuiper folder 2: 1936–1950, RUC/SC. Similar arguments are made in Kuiper to Hertzsprung, 14 March 1936. Box 29a, "Hertzsprung, Ejnar" 1929–1957, UA/GK. Luyten, 1987, 29.

5. Ibid., Kuiper to Chandrasekhar, 15 March 1936.

6. Russell to Adams, 31 October 1932. PUL/HNR. Russell OHI, 1951

7. Kuiper to Hertzsprung, 14 March 1936; Hertzsprung to Kuiper, 2 April 1936. Box 29a, "Hertzsprung, Ejnar" 1929–1957, UA/GK.

8. Hertzsprung, 1937, 309. Hertzsprung's tenure at Lick is discussed in Herrmann, 1994, 159–160. Translated by Carola Jeschke.

9. Hertzsprung, 1937, 309; 312.

10. Morgan to Chandrasekhar, n.d. "3 PM Thursday." File 22.14: "W. W. Morgan," RUC/SC.

11. Hertzsprung to Kuiper, 7 March 1935. UA/GK. Translated by Henk Boute.

12. Kuiper to Hertzsprung, 14 March 1936. UA/GK. Boute translation. On Dutch cultural imperialism, see Pyenson, 1989, 266–273.

13. Wright to Sproul, 16 April 1936. U.C. Office of the President file, LAUC/WHW. I am indebted to Dorothy Schaumberg and Don Osterbrock for this reference.

14. Higham, 1955, 324; Daniels, 1990, quoted on 265; 282–284; 295–296; Fermi, 1971, 24–27.

15. Ph.D. production data from Berendzen and Moslen, 1972, 48, fig. 2. Fermi, 1971, 29; Daniels, 1990, 289. Higham, 1955, 329, describes how nativism is linked to domestic economics.

16. Kevles, 1978, 281–282, notes the discomfiture of the displaced elite.

17. Fermi, 1971, 295–297. In the late 1930s the American Astronomical Society contained fewer than six hundred members. Only a few hundred were active observatory-based researchers. Meetings rarely drew more than one hundred people. DeVorkin and Routly, 1999, 128, fig. 1. Clemence and Jenkins, 1948.

18. ". . . risky," Russell, "Memorandum on the Observatory and Department of Astronomy," 10 January 1934, 6. Princeton University, Dodds file; "New England," Russell to Shapley, 10 November 1936. PUL/HNR. In 1916, Russell accepted a Jew as a graduate student, on Frost's urging, but had to be assured by Frost that the young man would not be a social pariah and would not attempt any "outside activities of a money-making na-

ture," since Russell was proposing him for the Thaw Fellowship. Frost to Russell, 9 June; 11 July; 24 July 1916; Russell to Frost, 14 June; 17 July. YOL/EBF; Frost to Russell, 16 December 1916; Russell to Hibben, ca. late 1916. PUL/HNR.

19. Slipher to Russell, 8 February 1939. PUL/HNR.

20. Weiner, 1969, 213–217, discusses the Emergency Committee in Aid of Displaced German Scholars and identifies Veblen and Ladenburg as leading figures. Fermi, 1971, 76–77.

21. Schlesinger to Adams, 8 March 1934; Adams to F. Stratton, 24 August 1935. HL/WSA; Struve to Cattell, 17 March 1938. AIP/OS/YOL. Weiner, 1969, 215. DeVorkin, 1984b.

22. Russell to E. R. Clinchy, 21 November 1938, and appended remarks. File 11:9648; Russell to Robert A. Ashworth, 7 November 1938. PUL/HNR.

23. Russell to Guthnick, 7 February 1938. PUL/HNR.

24. Russell took Gaposchkin's name off the list for secretary, arguing that her considerable "professional and domestic obligations" kept her "working nearly to the limit" as one of the country's best and most conscientious astrophysicists. Therefore she should not be asked to take on the added obligation of what was essentially a clerical post. Russell left unsaid the fact that Payne was essential to the management of Russell's double star interests at Harvard. Russell to Struve and Nassau, 26 October; 7; 17 November; to Struve, 1 November; to John C. Duncan, 23 November 1938. Reel 1, frame 816 ff.; "American Astronomical Society Nominating Committee" file, PUL/HNR. Hereafter "AAS Nom. Com."

25. Aitken to Russell, 5 December; Duncan to Russell, 6 December 1938. "AAS Nom. Com.," PUL/HNR.

26. Russell to Aitken, 7 December 1938. "AAS Nom. Com.," PUL/HNR.

27. Ibid.

28. Ibid.

29. [R. T. Birge], Listing "A," 19 September entry. "U.C. Berkeley Lick Observatory Selection of Director for 1942," BANC/RTB.

30. Russell to Aitken; to Duncan, 7 December; Aitken to Russell, 10 December 1938. "AAS Nom. Com.," PUL/HNR.

31. Aitken to Russell, 10 December 1938. PUL/HNR.

32. Russell to Struve and Nassau, 8 December 1938. "AAS Nom. Com.," PUL/HNR.

33. Russell to Aitken, 13 December 1938; Duncan to Russell, 22 December 1938. "AAS Nom. Com.," PUL/HNR.

34. Duncan to Russell, 22 December 1938. "AAS Nom. Com.," PUL/HNR.

35. Russell to Aitken, 4 January 1939. "AAS Nom. Com.," PUL/HNR.

36. Struve to Russell, 7 January 1939. "AAS Nom. Com.," PUL/HNR. Dean Henry Gale told Struve: "As you probably know, the general belief down town is that we are a communist center at the University." Struve polled his staff to find that 3 were voting for Landon, 2 for Roosevelt, and none for Browder or Thomas. Gale to Struve, 12 October 1936. U.C. Office of the Dean of Physical Sciences file, YOL/OS.

37. Emphasis in original. Aitken to Russell, 10 January 1939. "AAS Nom. Com.," PUL/HNR. Luyten was born in Java. Luyten, 1987, 2.

38. Emphasis in original. Russell, "A Natural Scientist Evaluates Intellectual Freedom." Lincoln's Birthday address, Alexander Hall, 12 February 1939. File 109.18, PUL/HNR. Russell to Paula Jacobs, 2 March 1939; to Franz Boas, 8 March 1939; 28 May 1940. PUL/HNR. On Boas and the American Committee, see Kuznick, 1987, chaps. 6–7; and on the Lincoln's Birthday event, 188.

39. Russell, 1939c, 479–480. Russell's "the devil incarnate" was a quote from presumably a Robert Atkinson letter.

40. Ibid.

41. Ibid., 481.

42. Russell to Dr. D. W. Anthony (NAACP branch chairman), 16 March 1939. PUL/HNR.

43. Russell, 1940, "The Price of Intellectual Freedom," holograph. Box 110.6, PUL/HNR. Adams to Schlesinger, 23 August 1939. File 59.1045, HL/WSA. Robert Marshak OHI, 47. AIP. Weiner, 1969, 223. McCutcheon, 1991.

44. Russell to Gerald Pirsig, 18 January 1941. PUL/HNR. Struve to Birge, 4 November 1940. BANC/RTB; to Gale, 12 December 1938. U.C. Office of the Dean file, YOL/UC.

45. Russell to Cattell, 1 October 1941. PUL/HNR.

46. Kevles, 1978, 219; Weiner, 1969, 226 n. 71; Coben, 1971. I am indebted to Sam Schweber and Paul Forman for several very helpful discussions about the character of the American physics community during this period.

47. Russell to Luyten, 22 May 1940; Luyten to Russell, 20; 25 May 1940. PUL/HNR. On the peace resolution, see Butler, 1994, 30–34; Kuznik, 1987, 245–246.

48. Russell to Adams, 26 May 1941. PUL/HNR.

49. Russell to Bloom, 20 January 1941; Russell to Webster, 22 September 1941. PUL/HNR.

50. Russell to Boas, 8; 29 March 1939; 23 May 1940; to Louis Bromfield, 14 March 1939; to Bertha Josselyn Foss, 14 April 1939. PUL/HNR. Kuznick, 1987, chap. 6. Russell enthusiastically voted for Wendell Willkie in 1940 and contributed to his campaign. Russell to A. H. Wilson, 28 June 1940; to Joseph A. Bower, 5 July 1940. PUL/HNR.

51. Russell to Webster, 13 October 1941. PUL/HNR.

52. Russell remained on the committee until 1952, happy to be useful. Chandrasekhar to Kuiper, 12 August 1943. UA/GK. Russell's committee records are in Department of Astronomy Records, box II, section 1, "Miscellaneous Administrative Correspondence and Budgets, 1920–1946," PUA/A; Harold Urey papers, MSS44 Box 91, folder 22, UCSD/HL. The committee consisted, at various times during and after the war, of H. H. Dryden, J. W. Beams, G. B. Kistiakowsky, I. I. Rabi, H. C. Urey, John von Neumann, Theodore von Karman, P. L. Alger, Bernard Lewis, and A. W. Hill. Leslie E. Simon to Arlie V. Bock, 15 November 1948. Reel 22, PUL/HNR.

53. Russell to Wigner, 5 July 1946. PUL/HNR.

54. Russell, 1947b, 166; 170; 172.

55. Ibid., 172–173.

56. Russell to E. Wigner, 8 July 1946. PUL/HNR. On Russell's affinity for Kipling's philosophy, see MRE #1, 36.

Chapter 22
Searching for a Replacement

1. MRE #1, 63; #2, 52.

2. Obituary notices filed at the Presbyterian Historical Society, MI 51-R913g. Russell to Adams, 1 February; to V. M. Slipher, 3 February; to Aitken, 16 February 1933; Russell to New Jersey Tax Bureau, 24 February 1933. PUL/HNR.

3. Russell to Wynn Armstrong, 27 February 1934. PUL/HNR.

4. Russell to Dugan, 24 February 1934. PUL/HNR.

5. Miscellaneous files and clippings, box 40, PUL/JQS. Doel, 1985, 3.

6. Russell regarded continued Einstein shift work as a useful way to keep astronomy before the public. Stewart to Frank Schlesinger, 11 May 1937. Box 30, Folder S, PUL/JQS.

7. Stewart to K. T. Compton, 5 May 1925; "Memorandum for Professor K. T. Compton," 11 June 1929. Box 11, PUL/JQS. Stewart to Robert R. Williams, 13 May 1948. Research Corporation folder, PUL/JQS.

8. Russell to Charles Dollard (Carnegie Corporation of New York), 3 February 1942. PUL/HNR. Russell did respect some of Stewart's work, notably Stewart, 1941. Russell to Luyten, 17 May 1940. PUL/HNR.

9. These included Serge Korff and Menzel. Stewart, "Memorandum," 11 June 1929. Box 11, PUL/JQS.

10. Russell to Adams, 26 April 1934. PUL/HNR.

11. Russell, "Memorandum on the Observatory and Department of Astronomy," 10 January 1934, 1–4; 7; 6. Princeton University, Dodds file, PUL/HNR.

12. Ibid., 6.

13. Russell to Dugan, 24 February 1934; to Adams, 26 April 1934. PUL/HNR.

14. Russell to Adams, 9 April 1934. PUL/HNR.

15. Russell to Adams, 26 April 1934. PUL/HNR.

16. Adams to Russell, 10; 31 May 1934. PUL/HNR. Dunham to Russell, 10 August 1934. AIP/THD.

17. Adams to Russell, 20 March; 30 April 1935. PUL/HNR. Adams to Dugan, 1 April 1935; Note, 8 August 1934. Wildt folder, file 72.1284, HL/WSA.

18. [Stewart] to Dunham, 5 March 1935. AIP/THD; Dunham to Adams, 21 March; 18; 27 April 1935. File 18.322, HL/WSA.

19. Russell to Dunham, 26 December 1935. PUL/HNR; Adams to Russell, 16 January; quote from Adams to Merriam, 29 September 1936. File 8.114, HL/WSA.

20. Dunham to Adams, 20 May 1936. Dunham file 18.322, HL/WSA.

21. Plaskett to Dunham, 17 March 1936. AIP/THD.

22. Dunham to Plaskett, 23 April 1936. Plaskett to Dunham, 12 October; 26 November 1936. AIP/THD.

23. Dunham also never accepted Plaskett's offer. Dunham to Plaskett, 16 November; 13; 21 December 1936; telegram, Plaskett to Dunham, 29 December 1936. Russell to Dunham, 24 November 1936. AIP/THD.

24. Struve also noted that Dunham feared teaching and wanted to be in contact with theorists like Strömgren and Chandrasekhar, and that he was frustrated that at Mount Wilson, the only theory was of the "expanding universe." Struve memo, 23 June 1937; Struve to Gale, 14 October 1937. YOL/OS; Adams to Gale, 9 July 1937. HL/WSA.

25. Telegram, Russell to Dunham, 26 October 1936. AIP/THD.

26. Russell to Adams, 29 October; to Dunham, 24 November 1936. PUL/HNR.

27. Russell to Dodds, 20 January 1937. PUL/HNR.

28. Atkinson to Russell, 20 March; Russell to Atkinson, 20 March 1935. PUL/HNR. Comments on Atkinson based upon discussions with Karl Hufbauer.

29. Russell to Dodds, 20 January; 3 February 1937; to Root, 28 January; to Atkinson, 29 January; 4; 9 February. "Russell 1931–1939," folder, Astronomy Department Records, PUA. Russell to Adams, 12 February 1937. PUL/HNR.

30. Atkinson to Russell, 8 April 1937. PUL/HNR; to Dugan, 8 April 1937. Box 5, Astronomy Department Records, PUA.

31. Russell to Atkinson, 5 July 1940. AIP/RA. Russell was angered by the petitions of the American Association of Scientific Workers. Butler, 1994, 30–34; Kuznik, 1987, 245–246, identify Princeton's reaction.

32. HNRjr, 23.

33. Spitzer, "Newton Lacy Pierce 1905–1950," typescript for *Popular Astronomy*, 3.

34. Russell to Dodds, 28 August 1937. General Budget 1928–1939 folder, Astronomy Department Records, PUA.

35. Russell to l'Académie des Sciences, 9 February; to M. A. Lacroix, 14 December 1939; "Presidential Address," Paris Colloquium on Astrophysics, July 17–23, 1939. Box 110.1, PUL/HNR. Russell to Dodds, 31 July 1940. Princeton, Dodds file; Robert Root to Russell, 3 August 1940. Dugan file, PUA. The Russells were in Bordeaux as late as the 18th, when Russell took the time to witness the signing of a will. Margaret Hamilton to Russell, 22 November 1945. PUL/HNR

36. Russell to Wildt, 22 June 1939. PUL/HNR. Russell, "Form Y," 9 December 1940. Wildt faculty file. Salary schedule, Pierce faculty file. PUA. Lyman Spitzer and Martin Schwarzschild, discussions with the author, August 1996.

37. Russell to Dean of the Graduate School (Eisenhart), 4 October 1935. Princeton Graduate School file, Eisenhart folder, PUL/HNR. Rosseland had sent an endorsement for a graduate student.

38. Hufbauer, 1996.

39. "Svein Rosseland," chronology, typescript, courtesy Karl Hufbauer.

40. Russell to Struve, 27 September 1940; to Adams 28; 31 October; 14 December 1940. Adams to Russell, 31 October 1940. PUL/HNR.

41. Russell to Adams, 14 December 1940; 4 January 1941. PUL/HNR. Rosseland to Shapley, 11 January 1941. "Roc-S" file, UAV 630.22, HUA/HSDC.

42. Alexander Leitch to Honorable Breckinridge Long, 5 January 1941. Rosseland folder, PUA.

43. Russell to Adams, 7 February 1941. PUL/HNR.

44. Annette O. Dugan to Adams, 20 September 1940. File 18.316, HL/WSA: Russell to Mrs. Dugan, 7 January 1941. PUL/HNR.

45. Russell to Prof. James B. Pratt, 9 October 1941; to Webster, 22 September 1941. PUL/HNR. Accounts of the Rosseland's escape vary. Shapley to Russell, 28 March 1941. Rosseland folder, PUA.

46. Russell to Atkinson, 30 October 1941. PUL/HNR. On their itinerary and experiences, see Russell to Adams, 3; 7 July 1941; Leitch, entry for August 8, 1941 in Rosseland folder, PUA; Russell, 1954, in Philip and DeVorkin, 1977, 112–113.

47. Rosseland to Shapley, 24 September 1941; Shapley to Rosseland, 15 September. "Roc—S" file, UAV 630.22, HUA/HSDC. Russell to Atkinson, 30 October 1941. PUL/HNR.

48. Russell to Adams, 20 October 1941; Russell to Ladenburg, April 1942, n.d., PUL/HNR.

49. Frank Aydelotte to Chandrasekhar, 18 July 1941; Chandrasekhar to Aydelotte, 23 July 1941. Institute for Advanced Study Archives. Russell to Adams, 25 July 1941; 8 August 1941. PUL/HNR.

50. Russell to Adams, 11 November 1941; to Atkinson, 30 October 1941. PUL/HNR. Chandrasekhar OHI, 92–93. AIP.

51. Chandrasekhar to Kuiper, 20 October 1941. Box 28a, UA/GK.

52. Chandrasekhar to Kuiper, 20 October 1941. UA/GK. "Notice," 7 October 1941. "Observatory Journal Club: a set of informal colloquia at the Observatory of Instruction," Folder 22.54, PUL/HNR. This notice had attached titles and speakers through the fall.

53. Chandrasekhar OHI, 92–93. AIP.

54. Ibid.; Chandrasekhar to Kuiper, 20 October 1941. UA/GK.

55. Russell to Adams, 20 October 1941. PUL/HNR; "Notice," *Princeton Herald*, 7 November 1941. Rosseland file, PUA. Kuiper to Chandrasekhar, 14 November 1941. Folder 19.21, Kuiper folder 2: 1936–1950, RUC/SC.

56. Chandrasekhar OHI, 92–93. AIP.

57. Ibid., 94. On Russell's speaking constantly at meetings, see Shenstone OHI, 54, AIP; conversation with Atkinson, April 1977.

58. Russell to Franklin Roach, 20 October 1941. PUL/HNR.

59. Marshak to Chandrasekhar, 11 November 1941. Folder 21.10 "Marshak 1938–1969," RUC/SC.

60. Russell to Dean Root, 18 October; 18 November; to Shapley, 11 December; to Willis, 19 December 1941. Princeton University-Secretary. PUL/HNR. Rosseland, however, found his early studies under Vilhelm Bjerknes "uninteresting." Friedman, 1989, 180.

61. Russell to S. A. Mitchell, 8 May through 30 November 1942; to Shapley, 8 May 1942; to Dodds, n.d., ca. 23 September 1942. Princeton University President—Dodds, file, PUL/HNR. Rosseland to Shapley, 28 December 1942. "Roc—S" file, UAV 630.22, HUA/HSDC. Wildt file, PUA. Dorothy Davis Locanthi OHI. AIP.

62. Russell to President Dodds, 15 May 1942. Astronomy Department records, box II, section 1: "Miscellaneous Administrative Correspondence and Budgets, 1920–1946," PUA/A.

63. Russell to Shapley, 7 June 1943. PUL/HNR. Rosseland to Shapley, 29 June 1943. "Roc—S" file, UAV 630.22. HUA/HSDC.

64. Russell to Shapley, 22 June 1943. PUL/HNR.

65. "The Present Situation of the Department of Astronomy," 16 May 1944; Budget sheets, 1944–1945. Astronomy Department records, PUA/A.

66. Rosseland to Shapley, 29 May 1945. "Roc—S" file, UAV 630.22, HUA/HSDC. Russell to Apgar, Princeton Purchasing Dept, 7 July 1944. Reel 13:11905; Russell to F. S. Osborne, Princeton Public Relations, 7 February 1945. Reel 13:11533, PUL/HNR.

67. Russell to Hugh S. Taylor, 16 January 1946; confirmed in 22 May 1946 letter to Committee on Committees, Elmer G. Butler; Russell to Mills (Controller), 5 February 1946; Russell to Oort, 22 January 1946. PUL/HNR.

68. Russell to Mills (Controller), 5 February 1946; Russell to Abetti, 11 December 1945. PUL/HNR.

69. Russell to L. P. Eisenhart, 22 November 1943; Russell to Shapley, 28 November 1945; 10 January 1946; to Kenneth Condit and George Brakeley, 15 December 1945. PUL/HNR.

70. Struve to Russell, 15 January 1945. Dodds had asked all academic departments to form external review committees. Russell to Dodds, 21 June 1941. Princeton University President-Dodds file; Russell to Struve, 9 February 1945. PUL/HNR.

71. Russell to President Dodds, 14 December 1945. PUL/HNR. Minutes of the Scientific Research Committee, 2 October 1945. "Chandrasekhar 1946 Astronomy" folder, Astronomy Department Records, PUA/A. Spitzer to Russell, 13 January 1946. PUL/LS.

72. Russell to Spitzer, 18 January 1946. PUL/LS. Shapley's enormous breadth of activity, both as social activist and observatory director, has yet to be appreciated. Glimpses

can be gleaned from Haramundanis, 1984; Kuznick, 1987; Kevles, 1978; Diamond, 1992; England, 1982; Doel, 1996, 27–29; and DeVorkin, 1984b.

73. Russell to Shapley, 10 January 1946; to Taylor, 8 February 1946; to Shapley, 18 March 1946. PUL/HNR.

74. Russell to Stanley Cobb, 21 March 1946. PUL/HNR.

75. Russell to Shapley, 14 February 1946. PUL/HNR.

76. Russell to Shapley, 26 March 1946. PUL/HNR.

77. Shapley to Russell, 29 April 1946. PUL/HNR.

78. Ibid.; Russell to Shapley, 8 May 1946. PUL/HNR.

79. Russell to Shapley, 8 May 1946. PUL/HNR.

80. Russell to Shapley, 31 May 1946; to Chandrasekhar, 12 June 1946; to Bowen, 28 June 1946. PUL/HNR.

81. Russell to Taylor, 22 July 1946. PUL/HNR.

82. Russell to Taylor, 12 August 1946; to Chandrasekhar, 29 August 1946. PUL/HNR.

83. "Harlow Shapley," FBI File number 100–16321, "Internal Security" files, Boston office, 23 July 1946, Abstract, 4. FBI File 100–341825. On Hoover's thinking regarding Shapley, see Diamond, 1992, 293 n. 13.

84. Shapley to Russell, 4 August 1946; Russell to Shapley, 5; 9 August 1946. PUL/HNR.

85. Russell to Bowen, 18 October 1945; to Shane, 11 December 1945; Shapley to Russell, 19 August 1946. PUL/HNR. Osterbrock, 1992.

86. Wali, 1991, 200.

87. Russell to Shapley, 22 August 1946. PUL/HNR.

88. Russell commentary in Wigner, 1947, 72.

89. Russell to Shapley, 22 August 1946. PUL/HNR.

90. Hutchins to Russell, 31 October 1946. University of Chicago file, PUL/HNR.

91. Spitzer OHI, 8 April 1977, 39. AIP.

92. Telegram, Shapley to Russell, 29 October 1946; Russell to Spitzer, 30 October 1946. PUL/HNR.

93. Russell to Shapley, 30 October 1946. PUL/HNR.

94. Spitzer to Goldberg, 24 July 1946. PUL/LS. On Spitzer's plan at Yale, see DeVorkin, 1992, 208–215; 219 n. 77.

95. Spitzer OHI, April 1977, 39, AIP; Brouwer to Spitzer 12 August 1946. PUL/LS. Schwarzschild OHI, 16 December 1977, 101–102, AIP. Spitzer, "Commentary," 511, in Spitzer and Ostriker, 1997.

96. Schwarzschild OHI, 16 December 1977, 101–102. AIP.

97. Spitzer OHI, May 1978, 71. AIP.

98. Struve to Russell, 12 November 1946. PUL/HNR.

99. Spitzer to Shapley, 1 November 1946. PUL/LS. On Spitzer's consulting, see DeVorkin, 1992.

100. "Proposed Long-Range Plan for the Princeton University Observatory," 1, attached to Spitzer to Shapley, 1 November 1946. PUL/LS. Spitzer's proposal and commentary based upon it have been reprinted in Spitzer and Ostriker, 1997.

101. Spitzer, "Proposed Long-Range Plan," 2. PUL/LS.

102. Russell to Spitzer, 14 November 1946. PUL/HNR.

103. Russell to Shapley, 14 November 1946; Shapley to Russell, 18 November 1946. PUL/HNR.

104. Lyman to "Dear Honorary Uncle," 2 November 1946; Lyman to "Dear Uncle Henry," 15 November 1946; Russell to "Dear Lyman," 6 December 1946. PUL/HNR.

105. Hugh S. Taylor, "Memorandum to President Dodds concerning the Dept. of Astronomy," 10 December 1946. Spitzer Personnel file, PUA; Dodds to Spitzer, 19 December 1946. Personal File; Taylor to Spitzer, 16 January; Spitzer to Dodds, 17 January; Dodds to Spitzer, 10 February 1947. PUL/LS. Russell to Spitzer, 15 February 1947. PUL/HNR.

106. Russell to Spitzer, ca. April 1947. Russell to Alexander Leitch, 18 April, 1947. Princeton University Secretary, Leitch file, PUL/HNR.

107. Russell to Spitzer, 29 April; 16 May; 25 May; 2 June 1947. PUL/HNR.

108. Russell to Spitzer, 3 July 1947. PUL/HNR.

109. Russell to Spitzer, 23 June; 3 July 1947. PUL/HNR.

110. Russell to Shapley, 12 December 1948. "Roc—S" file, UAV 630.22, HUA/HSDC. Edmondson to V. M. Slipher, 4 December 1948. LowA/VMS. Gibbs to Russell, 4 February 1949; NRC files; Russell to Danjon, 9 November 1950. PUL/HNR.

111. Spitzer OHI, 27 November 1991, 7–8. Schwarzschild to the author, 16 August 1996.

112. Review, 18 September 1950. Stewart file, PUA. Conversation with L. Spitzer; Spitzer endorsed Pierce's efforts, but Kron and Whitford felt that the design was premature. Spitzer OHI, 27 November 1991, 7–8. "Memorial," Pierce faculty file, PUA. Wood et al., 1951, vi.

113. Russell to Bowen, 14 March 1949. PUL/HNR. Schwarzschild to the author, 16 August 1996.

114. DeVorkin and Kenat, forthcoming.

Chapter 23
Russell's Universe

1. Russell to Webster, 30 March 1942. SUL/DLW. On Webster's fate at Stanford, see Leslie and Hevly, 1985; Leslie, 1987; Lowen, 1997; Galison, Hevly, and Lowen, 1992.

2. Russell to Webster, 30 March 1942. SUL/DLW.

3. Ibid.

4. Russell to Finkelstein, 11 July 1949. PUL/HNR.

5. Russell to Webster, 30 March 1942. SUL/DLW.

6. Leacock, 1912, 30.

7. Shapley, 1958, 361.

8. Russell, 1916, 304; 1918, 192. On Slipher's work, see Smith, 1982, 17–22.

9. Doel, 1996, chap. 1.

10. Gooding, 1985; DeVorkin, 1994b.

11. Haramundanis, 1984, 162. On Russell's view of the borderline between physics and astronomy, see Russell, 1927f, 45.

12. Gavroglu and Simoes, 1994, 110; Servos, 1990, 322–323.

13. Margaret Deland to Russell, 14 November 1928; Russell to Deland, 30 November 1928. PUL/HNR.

14. Russell to Deland, 30 November 1928. PUL/HNR.

15. Swings, 1968, 257–258.

16. For a contrasting view, see Lankford, 1997, 286.

17. HNRjr, 3. Russell to John D. Fitzgerald, 19 November 1934; to Leo Goldberg, 10 May 1947. PUL/HNR.

18. Hollinger, 1996, 66. HNRjr, 3.

19. Russell, 1947b, 166; 172–173; 174; 176. Robert Merton, 1942, "The Normative Structure of Science," 275–276, quoted in Shapin, 1994, 413.

20. Stewart, 1958, 312. HNRjr., 29–30; MRE #1, 69–70.

21. Shapley, 1958, 361–362. MRE #2, 12. Russell died of an acute myocardial infarction and is buried in the Princeton Cemetery of the Nassau Presbyterian Church, in sight of the stately Norris family marker where his mother and aunt are buried and where Jean Hetherington lies next to Gordon Russell. Lucy May and Elizabeth Russell are with Russell.

• GLOSSARY OF KEY TERMS •

Bohr atomic theory: developed by Niels Bohr in 1913, envisions the atom as a nucleus of positive charge surrounded by orbiting electrons, each constrained to occupy specific energy states, and resembling, to spectroscopists of Russell's generation, a tiny well-ordered solar system. Movement (transitions) of electrons between energy states requires energy (absorption of photons) or releases energy (emission of photons). By the 1930s, the nuclear component was envisioned as a collection of protons and neutrons, and the electrons moved in a probabilistic cloud of energy states.

Effective temperature: the temperature of a perfect radiator (or black body, according to Planck's law of radiation) whose diameter and distance are equal to those of the object in question.

Ionization potential: in the Bohr atom, the minimum amount of energy (supplied by photons) required to allow an electron, sitting in the lowest (ground) state, to escape entirely from that atom. Ionized atoms have different spectroscopic properties from neutral atoms.

Magnitude, absolute: (see magnitude, apparent) numerical system used to indicate the "actual" brightnesses of stars, if seen from a standard distance in space (10 parsecs, or 32.6 light years). Absolute magnitude can be related to the luminosity of the object, the amount of energy it emits over all wavelengths. There are also mathematical descriptions based upon Planck's radiation law, equating the luminosity, linear diameter, distance, and temperature of a star.

Magnitude, apparent: a numerical system used by astronomers to indicate the observed relative brightnesses of stars. Larger numbers indicate fainter objects.

Mass-luminosity law: first detected empirically by Hertzsprung, Russell, and others from double-star statistics, and then put on a rational basis by Eddington in 1924, it is the realization that there is a definite relationship between the mass of a star and its total energy output, or its luminosity. The relationship is not linear; roughly, luminosity varies as the cube of the mass of a star (more precisely its mass raised to the 3.5 power). The historically important issue is that both giants and dwarfs were found to adhere to this law, which meant that both behave as perfect gases.

Parallax, solar: the parallactic shift of the Sun as seen simultaneously from different parts of Earth. The solar parallax became synonymous with the size of Earth's orbit and was used rhetorically in schemes to determine the scale of the solar system and the value of the astronomical unit, the average distance of Earth from the Sun.

Parallax, stellar: the apparent shift in position of a nearby star with respect to background stars, caused by the annual motion of Earth around the Sun. First observed through the use of graduated visual micrometers on equatorial refractors and by meridian circle observations, which referred all positions to a fundamental frame. Stellar parallaxes are measures of distance, since the magnitude of the shift is inversely proportional to the distance of the object.

Perfect radiator (black body): A body that radiates energy according to Planck's radiation law. It absorbs all incident radiation it receives, and so in the absence of external radiation would appear absolutely black.

Planck's radiation law: At the turn of the century Max Planck developed a scheme to describe energy as quantized, implying that there was a fundamental unit of energy in Nature. These energy bundles (or packets) were called photons, and the amount of energy they possessed was a function of the frequency (vibrational energy state) of the photon. Planck also developed an expression to describe the total radiation emitted by a perfect radiator, embodying the characteristics of radiating bodies over all frequencies. This expression showed how the energy radiated by an ideal radiator could be equated with its temperature and the geometry of the radiating body.

Proper motion: cumulative motion across the line of sight, first detected through comparison of the positions of stars in old catalogs with modern positions.

Radial velocity: motion of an object in the line of sight, determined astronomically through the measuring of how much absorption or emission lines in the spectrum of the object are shifted to the red or to the blue end of the spectrum.

Spectral classification: starting in the 1860s, as soon as physicists realized that the spectrum of a luminous object could reveal both chemical composition and physical constitution, astronomers began to catalog the spectroscopic appearances of stars. By the end of the century there were over a dozen competing systems of spectral classification, each based upon how the patterns of absorption or emission lines seen in the spectra changed with the color of the star, or how they merely fit into a continuous sequence. Starting in the late 1880s, working under E. C. Pickering's direction, first Williamina P. Fleming and then Annie J. Cannon led teams of women to classify a vast number of stars by their spectra, using at first a simple alphabetical scheme of the first fifteen letters of the alphabet (A, B, C, . . . O, with the exception of J) based upon the intensity of the hydrogen lines of the Balmer Series. This empirical system was soon replaced by one embodying evolutionary considerations by rearranging the sequence to indicate linear descent from nebulae (O, B, A, F, G, K, M) and combining various redundant classes. This, with numerical subdivisions, became the familiar Harvard system. In spite of Harvard's enormous commitment, in 1900 most astronomers referred to the simple numerical systems of A. Secchi and H. C. Vogel, which classified into three or four major types (by Roman numeral). The modified Cannon system produced discrete classifications for stars (i.e., F0, F3, F5, F8). This is why Russell's early diagrams take on a "quantized" appearance (see fig 9.2). Color Index, which is what Hertzsprung typically used, was a continuous numerical system, which is why his "gaps" were real and Russell's were illusory.

Variable stars: stars that vary in light, periodically or erratically. Searching out the cause of the variation has since led to the recognition of two major types: eclipsing binaries and intrinsic variables. The former are systems of stars whose orbits happen to be in our line of sight, so that one star periodically eclipses the other, resulting in an overall diminution of light. In the historical literature, these stars tended to be called "eclipsing variables" at first. Some indeed are eclipsing systems containing intrinsic variables. Most periodic intrinsic variables are now thought to be caused by some form of pulsation mechanism, where the diameter and the surface temperature of the object both vary.

• BIBLIOGRAPHY •

General Sources

Adams, Walter S. 1913. "Note on the Relative Intensity at Different Wave-Lengths of the Spectra of Some Stars Having Large and Small Proper Motions. *ApJ* 39: 89–92.

———. 1916a. "Investigations in Stellar Spectroscopy. I. A Quantitative Method of Classifying Stellar Spectra." *PNAS* 2:143–147.

———. 1916b. "Investigations in Stellar Spectroscopy. IV. Spectroscopic Evidence for the Existence of Two Classes of M Type Stars." *PNAS* 2:157–163.

———. 1928. "The Past Twenty Years of Physical Astronomy." *PASP* 40:213–228.

———. 1933. "Astrophysics and the Ionization Theory." *PASP* 45:215–226.

Adams, Walter S., and A. H. Joy. 1917. "The Luminosities and Parallaxes of Five Hundred Stars, First List." *ApJ* 46:313–339.

Adāms, Walter S., and Arnold Kohlschütter. 1914. "Some Spectral Criteria for the Determination of Absolute Stellar Magnitudes." *ApJ* 40:385–398.

Adel, Arthur. 1933. "The Carbon Dioxide Molecule: Its Infrared Spectrum and Mechanical Structure." Ph.D. diss., University of Michigan.

Aitken, Robert Grant. 1918. *The Binary Stars*. New York: D. C. McMurtrie.

———. 1933. "Notes on the Progress of Astronomy in the Year 1932–1933." *PASP* 45:127–137.

Aspray, William. 1988. "The Emergence of Princeton as a World Center for Mathematical Research. 1896–1939." In *History and Philosophy of Modern Mathematics*, edited by William Aspray and Philip Kitcher, 346–366. Minneapolis: University of Minnesota Press.

Atkinson, R. d'E., and Fritz Houtermans. 1929. "Transmutation of the Lighter Elements in Stars." *Nature* 123:567.

Ayers, Leonard P. 1919. *The War with Germany: A Statistical Summary*. Washington: U.S. Government Printing Office.

Baker, Robert H. 1930. *Astronomy*. New York: D. Van Nostrand.

Ball, Robert S. 1899. "Preliminary Description of the New Photographic Equatorial of the Cambridge Observatory." *MN* 59:152–155.

Bannister, Robert C. 1979. *Social Darwinism: Science and Myth in Anglo-American Social Thought*. Philadelphia: Temple University Press.

Bates, Ralph S. 1965. *Scientific Societies in the United States*. 3d ed. New York: Pergamon.

Bauschinger, Julius. 1906. *Die Bahnbestimmung der Himmelskörper*. Leipzig: W. Engelmann.

Becker, Barbara J. 1993. "Eclecticism, Opportunism, and the Evolution of a New Research Agenda: William and Margaret Huggins and the Origins of Astrophysics." Ph.D. diss., The Johns Hopkins University.

Bell, Louis. 1917. "The Physical Interpretation of Albedo I." *ApJ* 45:1–29.

Berendzen, Richard, ed. 1972. *History of and Education in Modern Astronomy*. Annals of the New York Academy of Sciences 198.

Berendzen, Richard, Richard Hart, and Daniel Seeley. 1976. *Man Discovers the Galaxies*. New York: Science History Publications.

Berendzen, Richard, and Mary Treinen Moslen. 1972. "Manpower and Employment in American Astronomy." In Berendzen, 1972, 46–65.

Bethe, Hans A. 1939. "Energy Production in Stars." *PhysRev* 55:434–456.

Binnendijk, Leendert. 1960. *Properties of Double Stars*. Philadelphia: University of Pennsylvania Press.

Birge, R. T. N.d. "History of the Physics Department, University of California, Berkeley." Unpublished manuscript. BANC/RTB.

Blaauw, Adriaan. 1994, *History of the IAU*. Dordrecht: Kluwer.

Blagg, Mary A. 1928. "Discussion of Some Further Observations of [Beta] Lyrae." *MN* 88:162–174.

Boltzmann, Ludwig. 1992. "A German Professor's Trip to El Dorado." *PhysTod* 45: 44–51.

Born, Max. 1924. "Über Quantenmeckanik." *ZPhys* 26:379–95.

Boss, Benjamin. 1912. "Systematic Motions of the Stars Arranged according to Type." *AJ* 27:92–94.

Bower, E. 1993. "Physiotherapy for Cerebral Palsy: A Historical Review." In *Clinical Neurology* 2, edited by D. Ward, 29–54.

Bowler, Peter J. 1985. "Scientific Attitudes to Darwinism in Britain and America." In Kohn, 1985, 641–681.

Bowman, Eleanor. 1981. *The Present Day Club*. Princeton: The Present Day Club.

Bozeman, Theodore Dwight. 1977. *Protestants in an Age of Science: The Baconian Ideal and Ante-bellum American Religious Thought*. Chapel Hill: University of North Carolina Press.

Bragdon, Henry Wilkinson. 1967. *Woodrow Wilson: The Academic Years*. Cambridge: Harvard University Press.

Brashear, R. S. 1992. "Timing Is Everything: World War I and the Spectroscopic Parallax." AAS talk, manuscript, Atlanta, January 1992.

———. 1993. "A Brewing Storm: Hale, Gale, and Researcher Selection at the Mount Wilson Observatory." AAS talk abstract, Phoenix, January 1993.

Brock, W.H. 1969. "Lockyer and the Chemists: The First Dissociation Hypothesis." *Ambix* 16:81–99.

Browne, Janet. 1995. *Charles Darwin: Voyaging, A Biography*. Princeton: Princeton University Press.

Bruce, Robert. 1987. *The Launching of Modern American Science 1846–1876*. New York: Knopf. Reprinted, Ithaca: Cornell, 1988.

Brush, Stephen G. 1978. "A Geologist among Astronomers: The Rise and Fall of the Chamberlin-Moulton Cosmogony." *JHA* 9:1–41; 77–104.

———. 1980. "Poincaré and Cosmic Evolution." *PhysTod* 33:42–49.

———. 1996a. *Transmuted Past*. Cambridge: Cambridge University Press.

———. 1996b. *Fruitful Encounters*. Cambridge: Cambridge University Press.

Burchfield, Joe D. 1975. *Lord Kelvin and the Age of the Earth*. New York: Science History Publications. Revised, Chicago: University of Chicago Press, 1990.

Burnham, John C. 1971. *Science in America: Historical Selections*. New York: Holt, Reinhart and Winston.

Butler, Loren. 1994. "Robert S. Mullikan and the Politics of Science and Scientists. 1939–1945." *HSPS* 25:25–45.

Butler, Orville Roderick. 1993. "The Birth of American Astrophysics: The Development of a Science in Its Cultural Context." Ph.D. diss., Iowa State University.

Camerini, Jane. 1997. "The Power of Biography." *Isis* 88:311.

Campbell, William Wallace. 1906. "The Problems of Astrophysics." *International Congress of Arts and Science* 6:446–469. New York: University Alliance.

———. 1911. "Some Peculiarities in the Motions of the Stars." *Lick Observatory Bulletin* 6, no. 196:125–135.

———. 1913. *Stellar Motions*. New Haven: Yale University Press.

———. 1915. "The Evolution of the Stars and the Formation of the Earth." *Popular Science Monthly* 87 (September 1915): 209–235; *Scientific Monthly*, 1 October 1915, 1–17; November. 177–194; December, 238–255.

Candler, Albert C. 1937. *Atomic Spectra and the Vector Model*. Cambridge: Cambridge University Press.

[Carnegie Institution of Washington]. 1922. "Staff of Investigators for Year 1921." *Carnegie Yearbook No. 20*. Washington, D.C.

Carson, C., and S. Schweber. 1994. "Recent Biographical Studies in the Physical Sciences." *Isis* 85:284–292.

Cassidy, David. 1979. "Heisenberg's First Core Model of the Atom: The Formation of a Professional Style." *HSPS* 10:187–224.

———. 1992. *Uncertainty: The Life and Science of Werner Heisenberg*. New York: W. H. Freeman.

Ceruzzi, Paul E. 1991. "When Computers Were Human." *Annals of the History of Computing* 13:237–244.

Chandrasekhar, Subrahmanyan. 1934. "Stellar Configurations with Degenerate Cores." *Obs* 57:373–377.

———. 1939. *An Introduction to the Study of Stellar Structure*. Reprinted, New York: Dover, 1957.

———. 1951. "The Structure, the Composition, and the Source of Energy of the Stars." In Hynek, 1951, chap. 14.

———. 1987. *Truth and Beauty: Aesthetics and Motivations in Science*. Chicago: University of Chicago Press.

Charlier, C.V.L. 1917. "Monistic and Dualistic Conceptions of the Stellar Universe." *Obs* 40:387–389.

Christianson, Gale E. 1995. *Edwin Hubble: Mariner of the Nebulae*. New York: Farrar, Straus and Giroux.

Clark, Ronald W. 1971. *Einstein: The Life and Times*. New York: World.

Clemence, G. M., and L. Jenkins. 1948. *The Astronomical Journal General Index*. New Haven.

Clerke, Agnes Mary. 1887. *A Popular History of Astronomy during the Nineteenth Century*. Edinburgh: Adam and Charles Black.

———. 1903. *Problems in Astrophysics*. Edinburgh: Adam and Charles Black.

Coben, Stanley. 1971. "The Scientific Establishment and the Transmission of Quantum Mechanics to the United States." *American Historical Review* 76:442–466.

Cochrane, Rexmond C. 1978. *The National Academy of Sciences: The First Hundred Years. 1863–1963*. Washington D.C.: National Academy.

Comstock, George C. 1907. "The Luminosity of the Fixed Stars." *AJ* 25:169–175.

Comstock, George C. 1908. "Proper Motions of Faint Stars." *Publications of the Washburn Observatory 12*, pt. 1:1–34.

Cooley, L. C., R. S. Campbell, R. O. Kirkwood, and R. F. Sterling, eds. 1897. *The Nassau Herald of the Class of 1897*. Princeton: Princeton University Press.

Cowling, T. G. 1930. "On a Point-Source Model of a Star." *MN 91*:91–108.

———. 1938. "On the Motion of the Apsidal Line in Close Binary Systems." *MN* 98:734–744.

———. 1966. "The Development of the Theory of Stellar Structure." *QJRAS* 7:121–137.

———. 1985. "Astronomer by Accident." *ARAA* 23:1–18.

Crane, Diana. 1972. *Invisible Colleges*. Chicago: University of Chicago Press.

Crawford, Elizabeth, John. L. Heilbron, and Rebecca Ullrich. 1987. *The Nobel Population*. Berkeley: Office for the History of Science and Technology.

Crelinstin, Jeffrey. 1983. "William Wallace Campbell and the 'Einstein Problem': An Observational Astronomer Confronts the Theory of Relativity." *HSPS 14*:1–91.

Cronon, William. 1991. *Nature's Metropolis: Chicago and the Great West*. New York: Norton.

Cruikshank, Dale P. 1993. "Gerard Peter Kuiper." *BMNAS 62*:259–295.

Cuyler, Theodore L. N.d. *Christianity in the Home*. New York: Baker and Taylor.

Daniels, Roger. 1990. *Coming to America: A History of Immigration and Ethnicity in American Life*. New York: HarperCollins.

Darwin, George Howard. 1898. *The Tides and Kindred Phenomena in the Solar System*. 1898. Reprinted, London: W. H. Freeman, 1962.

———. 1900. "Address, Delivered by the President, Professor G. H. Darwin, on Presenting the Gold Medal of the Society to M. H. Poincaré." *MN 60*:406–415.

———. 1906. "On the Figure and Stability of a Liquid Satellite." *Philosophical Transactions of the Royal Society Series* A *206*:160–248.

Davis, Lance E., and Daniel J. Kevles. 1974. "The National Research Fund: A Case Study in the Industrial Support of Academic Science." *Minerva 12*:207–220.

Davis, W. P., S. M. Palmer, A. S. Wrenn, and P. R. Colwell, eds. 1912. *Quindecennial Record of the Class of Eighteen Hundred and Ninety-Seven*. Princeton: Princeton University Press.

Dennis, Michael Aaron. 1994. " 'Our First Line of Defense': Two University Laboratories in the Postwar American State." *Isis 85*:427–455.

DeVorkin, David H. 1975. "A. A. Michelson and the Problem of Stellar Diameters." *JHA* 6:1–18.

———. 1977a. "The Origins of the Hertzsprung-Russell Diagram." In Philip and DeVorkin, 1977, 61–77.

———. 1977b. "W. W. Campbell's Spectroscopic Study of the Martian Atmosphere." *QJRAS 18*:37–53.

———. 1978. "An Astronomical Symbiosis: Stellar Evolution and Spectral Classification (1860–1910)." Ph.D. diss., University of Leicester.

———. 1981. "Community and Spectral Classification in Astrophysics: The Acceptance of E. C. Pickering's System in 1910." *Isis* 72:29–49.

———. 1982. "Venus 1882: Public, Parallax, and HNR." *Sky and Telescope* 64:524–526.

———. 1984a. "Stellar Evolution and the Origins of the Hertzsprung-Russell Diagram in Early Astrophysics." In *Astrophysics and Twentieth-Century Astronomy to 1950*, edited by M. A. Hoskin and O. Gingerich, 90–108. Cambridge: Cambridge University Press.

———. 1984b. "The Harvard Summer School in Astronomy." *PhysTod* 37:48–55.

———. 1985. "Electronics in Astronomy: Early Applications of the Photoelectric Cell and Photomultiplier for Studies of Point Source Celestial Phenomena." *PIEEE* 73:1205–1220.

———. 1992. *Science with a Vengeance*. New York: Springer-Verlag.

———. 1993. "Saha's Influence in the West: A Preliminary Account." In *Meghnad Saha Birth Centenary Commemoration Volume*, edited by S. B. Karmohapatro, 154–202. Calcutta: Saha Institute of Nuclear Physics.

———. 1994a. "Quantum Physics and the Stars IV: Meghnad Saha's Fate." *JHA* 25:155–188.

———. 1994b. "A Fox Raiding the Hedgehogs: How Henry Norris Russell Got to Mount Wilson." In *The Earth, the Heavens and the Carnegie Institution of Washington*, edited by G. Good, 103–111. Washington, D.C.: American Geophysical Union.

———. 1995. "Textbooks and Training: A Literature for Theoretical Astrophysics." Manuscript talk, American Astronomical Society, Tucson, January 1995.

———. 1999a. " 'The Discovery and Exploitation of Spectroscopic Parallaxes' Revisited." In *Anni Mirabiles: A Symposium Honoring the Ninetieth Birthday of Dorrit Hoffleit*, edited by A. G. Davis Philip et al., 17–23. Schenectady: L. Davis Press.

———, ed. 1999b. *The American Astronomical Society's First Century*. New York: American Institute of Physics.

DeVorkin, David H., and R. Kenat. 1983. "Quantum Physics and the Stars I: The Establishment of a Stellar Temperature Scale." *JHA* 14:102–132.

———. 1983. "Quantum Physics and the Stars II: The Abundances of the Elements in the Atmospheres of the Sun and Stars." *JHA* 14:180–222.

———. Forthcoming. "Stellar Evolution. 1938–1955." In *General History of Astronomy 4b*, edited by M. A. Hoskin and O. Gingerich. Cambridge: Cambridge University Press.

DeVorkin, David H., and Paul Routly. 1999. "The Modern Society: Changes in Demographics." In DeVorkin, 1999b, 122–136.

Dewhirst, David W. 1982. "A Note on Polar Refractors." *JHA* 13:119–120.

Diamond, Sigmund. 1992. *Compromised Campus*. Oxford: Oxford University Press.

Dick. Steven J. 1996. *The Biological Universe*. Cambridge: Cambridge University Press.

Dillenberger, John. 1988. *Protestant Christianity*. New York: Macmillan.

Dillenberger, John, and Claude Welch. 1954. *Protestant Christianity Interpreted through Its Development*. New York: Scribner's.

Doel, Ronald E. 1985. "Department of Astronomy, Princeton University 1918–1932." Manuscript draft.

———. 1987. "Philanthropy and Changing Patterns of Research: Astronomy, Biology, Chemistry and Physics. 1918–1932." Transcript of a talk, Philadelphia, April 1987.

———. 1996. *Solar System Astronomy in America*. Cambridge: Cambridge University Press.

Douglas, A. Vibert. 1957. *The Life of Arthur Stanley Eddington*. London: Thomas Nelson.

Drummeter, L. F., Jr. 1992. "Dugan's Polarization Photometer and Other Mysteries." Manuscript, National Museum of American History, physical sciences division files.

Dugan, Raymond Smith. 1911. "Photometric Researches: The Algol-System RT Persei." *Contributions from the Princeton University Observatory No. 1*. Princeton: Princeton University Observatory.

———. 1930. "Why We Observe Double Stars." *PopAst* 38:392–395.

———. 1935. "Princeton's New Observatory." *PASP* 47:43–45.

Dugan, Raymond Smith, and W.F.H. Waterfield. 1928. "The Periods of Y Cygni." *Harvard Observatory Bulletin* 856:18–23.

Dupree, A. Hunter. 1957. *Science in the Federal Government: A History of Policies and Activities to 1940*. Reprinted, Baltimore: Johns Hopkins University Press, 1986.
———. 1959. *Asa Gray*. Cambridge: Harvard University Press. Reprinted, Baltimore: Johns Hopkins University Press, 1988.
———. 1986. "Christianity and the Scientific Community in the Age of Darwin." In *God and Nature: Historical Essays on the Encounter between Christianity and Science*, edited by David C. Lindberg and Ronald L. Numbers, 351–368. Berkeley and Los Angeles: University of California Press.
Dutton, Cmdr. Benjamin. 1928. *Navigation and Nautical Astronomy*. Annapolis: United States Naval Institute.
Dyson, F. W. 1904. "Eros and the Solar Parallax." *Obs* 27:131.
Eastman, J. R. 1892. "The Neglected Field of Fundamental Astronomy." *PAAAS* 41:17–32.
Eckert, Michael. 1996. "Theoretical Physicists at War: Sommerfeld Students in Germany and as Emigrants." In Forman and Sánchez-Ron, 1996, 69–86.
Eckert, W. J. 1937. "The Astronomical Hollerith Computing Bureau." *PASP* 49:249–253.
Eddington, Arthur Stanley. 1913a. "Some Problems of Astronomy. XII. The Distribition of the Spectral Classes of Stars." *Obs* 36:467–471.
———. 1913b. "Meeting of the Royal Astronomical Society." *Obs* 36:277–290.
———. 1914. *Stellar Movements and the Structure of the Universe*. London: Macmillan.
———. 1915a. "The Dynamics of a Globular Stellar System, Second Paper." *MN* 75:366–376.
———. 1915b. "The Relation between the Velocities of Stars and Their Brightness." *Obs* 38:392–396.
———. 1917a. "The Pulsation Theory of Cepheid Variables." *Obs* 40:290–293.
———. 1917b. "Further Notes on the Radiative Equilibrium of the Stars." *MN* 77:596–612.
———. 1920. "The Internal Constitution of the Stars." *Nature* 106:14–20.
———. 1923. "The Interior of a Star." *Proceedings of the Royal Institution* (February 23). Reprinted in *Royal Institution Library of Science: Astronomy* 2, edited by B. Lovell, 256–257. Barking: Elsevier, 1970.
———. 1924. "On the Relation between the Masses and Luminosities of the Stars." *MN* 84:308–332.
———. 1926. *The Internal Constitution of the Stars*. Cambridge: Cambridge University Press.
———. 1927. *Stars and Atoms*. Oxford: Oxford University Press.
———. 1929. "The Formation of Absorption Lines." *MN* 89:620–636.
———. 1930a. "The Effect of Boundary Conditions on the Equilibrium of a Star." *MN* 90:279–284.
———. 1930b. "On Professor Milne's Treatment of the Mass-Luminosity Problem." *MN* 90:284–286.
———. 1930c. "The Connection of Mass with Luminosity for Stars." *Obs* 53:208–211.
———. 1930d. "The Effect of Stellar Boundary Conditions: A Reply." *MN* 90:808–809.
———. 1932. "The Decline of Determinism." *Mathematical Gazette* 16:66; Reprinted in *Annual Report of the Smithsonian Institution for 1932*, 141–157. Washington, D.C., 1933..
———. 1938a. "Constitution of the Stars." *Annual Report of the Smithsonian Institution for 1937*, 131–144. Reprinted from *Scientific Monthly* 43:385–395.
———. 1938b. "Forty Years of Astronomy." *Discovery* 1, n.s. (July): 167–178.

———. 1942. "Robert Emden." *Obs 102*:77.
Eddy, John A. 1971. "The Schaeberle Forty-foot Eclipse Camera of Lick Observatory." *JHA* 2:1–22.
Edge, David O., and Michael Mulkay. 1976. *Astronomy Transformed: The Emergence of Radio Astronomy in Britain*. New York: John Wiley.
Egbert, Donald Drew. 1947. *Princeton Portraits*. Princeton: Princeton University Press.
Ehrenhalt, Alan. 1996. "A Town Second to None." *Washington Post Book World* 26:1; 10.
Emden, R. 1907. *Gaskugeln: Anwendungen der Mechanischen Wärmetheorie auf Kosmologische und Meteorologische Probleme*. Leipzig: Teubner.
England, J. Merton. 1982. *A Patron for Pure Science*. Washington, D.C.: National Science Foundation.
Fermi, Laura. 1971. *Illustrious Immigrants*. Chicago: University of Chicago Press.
Fleming, Donald. 1969. "Émigré Physicists and the Biological Revolution." In *The Intellectual Migration: Europe and America. 1930–1960*, edited by Donald Fleming and Bernard Bailyn, 152–189, Cambridge: Harvard University Press.
Forman, Paul. 1968. "The Doublet Riddle and Atomic Physics circa 1924." *Isis* 59:156–74.
———. 1970. "Alfred Landé and the Anomalous Zeeman Effect. 1919–1921." *HSPS* 2:153–261.
———. 1973. "Scientific Internationalism and the Weimar Physicists: The Ideology and Its Manipulation in Germany after World War I." *Isis 64*:151–180.
———. 1987. "Behind Quantum Electronics: National Security as Basis for Physics Research in the United States. 1940–1960." *HSPS 18*:149–229.
———. 1989. "Social Niche and Self-Image of the American Physicist." In *The Restructuring of Physical Sciences in Europe and the United States 1945–1960*, edited by Michelangelo De Maria, Mario Grillo, and Fabio Sebastini, 96–104. Singapore: World Scientific.
Forman, Paul, and José M. Sánchez-Ron, 1996. *National Military Establishments and the Advancement of Science and Technology*. Dordrecht: Kluwer.
Fowler, Alfred. 1915. "Spectral Classification of Stars and the Order of Stellar Evolution." *Obs* 38:379–392.
———. 1918. [Untitled review.] *JBAA* 28:197–204.
———. 1921. "Address Delivered by the President, Professor A. Fowler, on the award of the Gold Medal to Professor Henry Norris Russell." *MN 81*:334–350.
———. 1922a. *Report on Series in Line Spectra*. London: The Physical Society.
———, ed. 1922b. 1922. *Transactions of the International Astronomical Union 1*. London: Imperial College.
Fowler, R. H., and E. A. Milne. 1923. "The Intensities of Absorption Lines in Stellar Spectra, and the Temperatures and Pressures in the Reversing Layers of Stars." *MN* 83:403–424.
Fox, Philip. 1915. "General Account of the Dearborn Observatory." *Annals of the Dearborn Observatory of Northwestern University*. Vol. 1. Evanston: Northwestern University.
Friedman, Robert Marc. 1989. *Appropriating the Weather: Vilhelm Bjerknes and the Construction of a Modern Meteorology*. Ithaca: Cornell University Press.
Frost, Edwin B. 1910. "Charles Augustus Young." *BMNAS* 7:91–114.
———. 1933. *An Astronomer's Life*. Boston: Houghton-Mifflin.
Galison, Peter, Bruve Hevly, and Rebecca Lowen. 1992. "Controlling the Monster: Stanford and the Growth of Physics Research. 1935–1962." In *Big Science*, edited by Peter Galison and Bruce Hevly, 47–77. Stanford: Stanford University Press.

Gavroglu, Kostas, and Ana Simoes. 1994. "The Americans, the Germans and the Beginnings of Quantum Chemistry." *HSPS* 25:47–110.

George, Mark S. 1992. "Changing Nineteenth Century Views on the Origins of Cerebral Palsy: W. J. Little and Sigmund Freud." *Journal of the History of Neurosciences* 1:29–37.

Gill, David. 1900. "An Astronomer's Work in a Modern Observatory." In *Essays in Astronomy*, edited by E. S. Holden, 475–495. New York: D. Appleton. Reprinted from *Proceedings of the Royal Institution of Great Britain*, 1890–1892, and reprinted in *Royal Institution Library of Science 1*, edited by B. Lovell, 362–377 (1970).

Gingerich, Owen. 1988. "How Shapley Came to Harvard or, Snatching the Prize from the Jaws of Debate." *JHA* 19:201–207.

———. 1992. *The Great Copernicus Chase and Other Adventures in Astronomical History*. Cambridge, Mass.: Sky Publishing Corporation.

Goldberg, Leo. 1989. "Quantum Mechanics at the Harvard Observatory in the 1930s." Manuscript draft in *Problems in Theoretical Physics and Astrophysics: A Collection of Essays dedicated to V. L. Ginzburg on his Seventieth Birthday*. Moscow.

Golinski, Jan. 1990. "The Theory of Practice and the Practice of Theory: Sociological Approaches in the History of Science." *Isis* 81:492–505.

Gooding, David. 1985. " 'He who proves, discovers': John Herschel, William Pepys and the Faraday Effect." *Notes and Records of the Royal Society of London* 39:229–244.

Goodstein, Judith R. 1991. *Millikan's School*. New York: Norton.

Goodrich, H. B., and R. H. Knapp, 1952. *Origins of American Scientists*. University of Chicago Press.

Goralski, Robert. 1981. *World War II Alamanac 1931–1945*. New York: Putnam.

Green, Louis Craig. 1937. "The Iron Spectrum in the Far Ultra-violet from 2300 to 600 Angstroms." Ph.D. diss., Princeton University, May 1937.

———. 1977. "Some Recollections of Henry Norris Russell." In Philip and DeVorkin, 1977, 79–81.

Gregory, Frederick. 1986. "The Impact of Darwinian Evolution on Protestant Theology in the Nineteenth Century." In *God and Nature: Historical Essays on the Encounter between Christianity and Science*, edited by David C. Lindberg and Ronald L. Numbers, 369–390. Berkeley and Los Angeles: University of California Press.

Hagstrom, W. O. 1965. *The Scientific Community*. New York: Basic Books. Reprinted, 1975.

Hale, George Ellery. 1897. "The Aim of the Yerkes Observatory." *ApJ* 6:310–321.

———1902. "Stellar Evolution in the Light of Recent Research." *Popular Science Monthly* 60:291–313.

———. 1908. *The Study of Stellar Evolution: An Account of Some Recent Methods of Astrophysical Research*. Chicago: University of Chicago Press.

———. 1922. *The New Heavens*. New York: Scribner's.

Hall, Asaph, Jr. 1900. "On the Teaching of Astronomy in the United States." *Science*, n.s., 12:15–20.

Hankins, Thomas L. 1979. "In Defence of Biography: The Use of Biography in the History of Science." *History of Science* 17:1–16.

Haramundanis, Katherine, ed. 1984. *Cecilia Payne-Gaposchkin: An Autobiography and Other Recollections*. Cambridge: Cambridge University Press.

Hearnshaw, J. B. 1986. *The Analysis of Starlight: One Hundred and Fifty Years of Astronomical Spectroscopy*. Cambridge: Cambridge University Press.

Heilbron, John L. 1986. *The Dilemmas of an Upright Man, Max Planck as a Spokesman for German Science*. Berkeley and Los Angeles: University of California Press.
Hentschel, Klaus. 1993. "The Discovery of the Redshift of Solar Fraunhofer Lines by Rowland and Jewell in Baltimore around 1890." *HSPS* 23:219–277.
Herrmann, D. B. 1973. *The History of Astronomy from Herschel to Hertzsprung*. Translated by K. Krisciunas. Cambridge: Cambridge University Press, 1984.
———. 1976. "Ejnar Hertzsprung, Zur Strahlung der Sterne." *Ostwald's Klassiker*, no. 255. Leipzig.
———. 1994. *Ejnar Hertzsprung, Pioneer der Sternforschung*. Berlin: Springer-Verlag.
Herschel, Mrs. John. 1879. *Memoir and Correspondence of Caroline Herschel*. 2d ed. London: John Murray.
Hertzsprung, Ejnar. 1905. "Zur Strahlung der Sterne." *Zeitschrift für wissenschafliche photographie* 3:449.
———. 1906. "Über die optische Stärke der strahlung des schwarzen Körpers und das minimale Lichäquivalent." *Zeitschrift für wissenschaftliche photographie* 4:43–54.
———. 1908. "Über die sterne der Unterabteilungen c und ac nach der Spektralklassifikation von Antonia C. Maury." *AN* 179:373–380.
———. 1911. "Über die Verwendung Photographischer Effektiver Wellenlängen Zur Bestimmung von Farbenäquivalenten." *Potsdam Publications* 22, pt. 1:1–21.
———. 1922. "Remark on the Relation between Colour, Proper Motion and Apparent Magnitudes of the Stars." *BAN* 17:91–92.
———. 1937. "On Collaboration in Astronomy." *PASP* 49:309–312.
Hetherington, Norris S. 1988. *Science and Objectivity*. Ames: Iowa State University Press.
Hevly, Bruce. 1996. "The Heroic Science of Glacier Motion." *Osiris* 11:66–86.
Hickman, R. W., F. V. Hunt, O. Oldenberg, and E. C. Kemble. 1964. "Frederick Albert Saunders." *Harvard University Gazette*, 2 May, 188–189.
Hiebert, Erwin N. 1986. "Modern Physics and Christian Faith." In *God and Nature: Historical Essays on the Encounter between Christianity and Science*, edited by David C. Lindberg and Ronald L. Numbers, 424–447. Berkeley and Los Angeles: University of California Press.
Higham, John. 1955. *Strangers in the Land: Patterns of American Nativism. 1860–1925*. New Brunswick, N.J.: Rutgers University Press. 2d ed., 1994.
Hinks, Arthur R. 1899. "Note on the Construction and Use of Réseaux." *MN* 59:530–532.
———. 1901. "The Cambridge Machine for Measuring Celestial Photographs." *MN* 61:444–458.
———. 1904. "Eros and the Solar Parallax." *Obs* 27:97–101.
Hinks, Arthur R., and Henry Norris Russell. 1905. "Determinations of Stellar Parallax from Photographs Made at the Cambridge Observatory. Introductory paper." *MN* 65:775–787.
Hirsh, Richard F. 1979. "The Riddle of the Gaseous Nebulae." *Isis* 70:197–212.
Hoch, Paul K. 1983. "The Reception of Central European Refugee Physicists of the 1930s: USSR, UK, USA." *Annals of Science* 40:217–246.
Hoeveler, J. David, Jr. 1981. *James McCosh and the Scottish Intellectual Tradition*. Princeton: Princeton University Press.
Hoffleit, Dorrit. 1950. "The Discovery and Exploitation of Spectroscopic Parallaxes." *PopAst* 58, nos. 9; 10; 59, no. 1. *Harvard Reprint* 342.
———. 1977. "H.N.R. as a Pioneer in Trigonometric Parallaxes." In Philip and DeVorkin, 1977, 51–54.

Hoffleit, Dorrit. 1992. *Astronomy at Yale 1701–1968*. New Haven: Connecticut Academy of Arts and Sciences.

Holden, E. S. 1898. "The Teaching of Astronomy in the Primary and Secondary Schools and in the University." In *Report of the Commission of Education for the Year 1897–1898 Volume 1*, 869–892. Washington, D.C.

Hollinger, David A. 1996. *Science, Jews, and Secular Culture*. Princeton: Princeton University Press.

Holmes, Larry. 1981. "The Fine Structure of Scientific Creativity." *History of Science 19*:60–70.

Holton, Gerald. 1978. *The Scientific Imagination: Case Studies*. Cambridge: Cambridge University Press.

Hoskin, M. A. 1976. "The 'Great Debate': What really happened." *JHA* 7:169–182.

———. 1979. "Goodricke, Pigott and the Quest for Variable Stars." *JHA 10*:23–41.

Hovencamp, Herbert. 1978. *Science and Religion in America 1800–1900*. Philadelphia: University of Pennsylvania Press.

Howe, Herbert A. 1901. "Astronomical Books for the Use of Students." *PopAst* 9:169–176.

Hoyt, William Graves. 1976. *Lowell and Mars*. Tucson: University of Arizona Press.

———. 1980. *Planets X and Pluto*. Tucson: University of Arizona Press.

———. 1987. *Coon Mountain Controversies*. Tucson: University of Arizona Press.

Hufbauer, Karl. 1981. "Astronomers Take Up the Stellar-Energy Problem. 1917–1920." *HSPS 11*:277–303.

———. 1987. "Solutions Proposed to the Energy Problem before 1938." Typescript, American Physical Society, 21 April 1987.

———. 1991. *Exploring the Sun: Solar Science Since Galileo*. Baltimore: Johns Hopkins University Press.

———. 1993. "Breakthrough on the Periphery: Bengt Edlén and the Identification of the Coronal Lines. 1939–1945." In *Center on the Periphery: Historical Aspects of Twentieth-Century Swedish Physics*, edited by Svante Lindqvist, 199–237. Cambridge, Mass.: Science History Publications.

———. 1994. "Artifical Eclipses: Bernard Lyot and the Coronagraph. 1929–1939." *HSPS 24*:337–394.

———. 1996. "Tools for a New Specialty: Svein Rosseland and Theoretical Astrophysics. 1926–1940." Typescript.

———. Forthcoming. "Stellar Structure and Evolution [1924–1938]." In *General History of Astronomy 4b*, edited by, M. A. Hoskin and O. Gingerich. Cambridge: Cambridge University Press.

Huffer, C. M., and Olin J. Eggen. 1947a. "A Photoelectric Study of the Eclipsing Variable TX Ursae Majoris." *ApJ 105*:117–221.

———. 1947b. "A Photoelectric Study of the Eclipsing Variable AR Aurigae." *ApJ 105*:217–221.

Huffer, C. M., and Zdeněk Kopal. 1951. "Photoelectric Studies of Five Eclipsing Binaries." *ApJ 114*:297–330.

Hussey, William J. 1898. "Elements of the Minor Planet DQ." *AJ 19*:120.

———. 1900. "Notes on the Progress of Double-Star Astronomy." *PopAst 12*:91–103.

———. 1912. "A General Account of the Observatory." Publications of the Astronomical Observatory of the University of Michigan, 1. Ann Arbor.

Hynek, J. A., ed. 1951. *Astrophysics: A Topical Symposium*. New York: McGraw-Hill.

[International Astronomical Union.] *Transactions of the I.A.U.* 5; 6; 7; 8; 9; *10* (1935 through 1958).

Irwin, J. B. 1951. Review of *The Computation of Elements of Eclipsing Binary Systems*. *PASP* 63:155–157.
Jacoby, Harold. 1908. "John Krom Rees." *PopAst* 16:639–642.
Jarrell, Richard A. 1988. *The Cold Light of Dawn: A History of Canadian Astronomy*. Toronto: University of Toronto Press.
Jeans, James H. 1909. Review of *Gaskugeln*. *ApJ* 30:72–74.
———. 1910. "On the 'Kinetic Theory' of Star Clusters." *MN* 74:109–112.
———. 1912. "George Howard Darwin." *Dictionary of National Biography, 1912–1921*, vol. 24, edited by H.W.C. Davis and J. R. H. Weaver, 144–147. London: Oxford University Press.
———. 1925. "A Theory of Stellar Evolution." *MN* 85:914–915.
———. 1927. "Recent Developments of Cosmical Physics." *JRASC* 21:7–34.
Johnson, Deryl Freeman. 1968. "The Attitudes of the Princeton Theologians toward Darwinism and Evolution from 1859–1929." Ph.D. diss., University of Iowa.
Johnson, R. C. 1946. *Atomic Spectra*. London: Methuen.
Johnston, William Davison, and Richard W. Reifsnyder. 1990. *A Pilgrimage of Faith: The History of the First Presbyterian Church of Oyster Bay, New York. 1844–1989*. Oyster Bay: The First Presbyterian Church of Oyster Bay.
Jones, B. Z., and L. G. Boyd. 1971. *The Harvard College Observatory 1839–1919*. Cambridge: Harvard University Press.
Kapteyn, J. C. 1904. "Statistical Methods in Stellar Astronomy." *International Congress of Arts and Sciences* 6 (1906): 396–425. New York: University Alliance.
———. 1906. "Star Streaming." *Report of the British Association for the Advancement of Science, Section A 1905*, 257–265.
———. 1910. "On Certain Statistical Data Which May Be Valuable in the Classification of the Stars in the Order of Their Evolution." *ApJ* 31:258–269.
Kargon, Robert H. 1982. *The Rise of Robert Millikan: Portrait of a Life in American Science*. Ithaca: Cornell University Press.
Keeler, James E. 1897. "The Importance of Astrophysical Research and the Relation of Astrophysics to Other Physical Sciences." *ApJ* 6:271–288.
Keener, John Henry, ed. 1900. *Triennial Records of the Class of 1897*. Princeton: Princeton University Press.
———. 1903. *Princeton University Sexennial Record of the Class of Eighteen Ninety-Seven*. Princeton: Princeton University Press.
———. 1909. *Duodecennial Record of the Class of Eighteen Hundred and Ninety-Seven*. Princeton: Princeton University Press.
Kenat, Ralph Carl, Jr. 1987. "Physical Interpretation: Eddington, Idealization and the Origin of Stellar Structure Theory." Ph.D. diss., University of Maryland.
Kenat, Ralph Carl, Jr., and David H. DeVorkin. 1990. "Quantum Physics and the Stars III: Towards a Rational Theory of Stellar Spectra." *JHA* 21:157–186.
Kennedy, A. M. 1947. *Princeton University 1897 Fiftieth Anniversary Book*. Princeton: Princeton University Press.
Kevles, Daniel J. 1968. "George Ellery Hale, the First World War, and the Advancement of Science in America." *Isis* 59:427–437.
———. 1971. " 'Into Hostile Political Camps': The Reorganization of International Science in World War I." *Isis* 62:47–60.
———. 1978. *The Physicists*. New York: Vintage.
Kidwell, Peggy Aldrich. 1981. "Prelude to Solar Energy: Pouillet, Herschel, Forbes and the Solar Constant." *Annals of Science* 38:457–476.

Kidwell, Peggy Aldrich. 1984. "An Historical Introduction to 'The dyer's hand.' " In Haramundanis, 1984, 11–37.
———. 1990. "American Scientists and Calculating Machines—From Novelty to Commonplace." *Annals of the History of Computing* 12:31–40.
King, Arthur S. 1922. "Electric Furnace Experiments Involving Ionization Phenomena." *ApJ* 55:380–385.
King, Henry C. 1955. *The History of the Telescope*. High Wycombe, Bucks.: Charles Griffin. Reprinted, New York: Dover, 1979.
King, Robert B. 1995. "Arthur S. King. 1876–1957." *BMNAS* 68:1–15.
Klinkerfues, Wilhelm. 1899. *Theoretische Astronomie*. Braunscheig: F. Vieweg, 1871. Expanded by H. Buchholz, 1899.
Kohler, Robert E. 1991. *Partners in Science, Foundations and Natural Scientists 1900–1945*. Chicago: University of Chicago Press.
Kohn, David, ed. 1985. *The Darwinian Heritage*. Princeton: Princeton University Press.
Kopal, Z. 1938, "On the Motion of the Apsidal Line in Close Binary Systems." *MN* 98:448–458.
———. 1946. *An Introduction to the Study of Eclipsing Variables*. Cambridge: Harvard University Press.
———. 1950. *The Computation of Elements of Eclipsing Binary Systems*. Harvard Observatory Monograph No. 8.
———. 1986. *Of Stars and Men, Reminiscences of an astronomer*. Bristol: Adam Hilger.
Kragh, Helge. 1987. *An Introduction to the Historiography of Science*. Cambridge: Cambridge University Press.
———. 1995. "Cosmology between the Wars: The Nernst-MacMillan Alternative." *JHA* 26:93–115.
Kronig, R. 1960. "The Turning Point." In *Theoretical Physics in the Twentieth Century: A Memorial Volume to Wolgang Pauli*, edited by M. Fierz and V. F. Weisskopf, 5–39. New York: Interscience Publishers.
Kuhn, Thomas S. 1964. *The Structure of Scientific Revolutions*. Chicago: University of Chicago Press. 1962.
Kuiper, Gerard. 1937. "On the Hydrogen Content of Clusters." *ApJ* 86:176–197.
Kushner, David. 1993. "Sir George Darwin and a British School of Geophysics." *Osiris* 8:196–224.
Kuznik, Frank. 1993. "Is Something Out There?" *Air and Space/Smithsonian* 8:47–53.
Kuznik, Peter J. 1987. *Beyond the Laboratory: Scientists as Political Activists in 1930s America*. Chicago: University of Chicago Press.
Lang, Ken R., and Owen Gingerich. 1979. *A Sourcebook in Astronomy and Astrophysics 1900–1975*. Cambridge: Harvard University Press.
Lankford, John. 1981. "Amateurs and Astrophysics: A Neglected Aspect in the Development of a Scientific Specialty." *Social Studies of Science* 11:275–303.
———. 1983. "Photography and the Long-Focus Refractor: Three American Case Studies. 1885–1914." *JHA* 14:77–91.
———. 1984. "The Impact of Photography on Astronomy." In *Astrophysics and Twentieth Century Astronomy to 1950, Pt. A.*, edited by, Owen Gingerich, 16–39. Cambridge: Cambridge University Press.
———. 1997. *American Astronomy: Community, Careers, and Power. 1859–1940*. Chicago: University of Chicago Press.
Lankford, John, and Ricky L. Slavings. 1996. "The Industrialization of American Astronomy. 1880–1940." *PhysTod* 49:34–40.

Leacock, Stephen. 1912. *Nonsense Novels*. Reprinted, New York: Dover. 1971.
Lears, T. J. Jackson. 1981. *No Place of Grace: Antimodernism and the Transformation of American Culture 1880–1920*. New York: Pantheon.
Lefschetz, Solomon. 1969. "Luther Pfahler Eisenhart." *BMNAS 40*:73.
Leitch, Alexander. 1978. *A Princeton Companion*. Princeton: Princeton University Press.
Leslie, Stuart W. 1987. "Playing the Education Game to Win: The Military and Interdisciplinary Research at Stanford." *HSPS 18*:55–88.
Leslie, Stuart W., and Bruce Hevly. 1985. "Steeple Building at Stanford: Electrical Engineering, Physics, and Microwave Research." *PIEEE* 73:1169–1180.
———Link, Arthur S. 1971. *The Higher Realism of Woodrow Wilson*. Nashville, Tenn.: Vanderbilt University Press.
Link, Arthur S., et al., eds. 1971–1976. *The Papers of Woodrow Wilson*. Princeton: Princeton University Press.
Livingstone, David R. 1992. "Darwinism and Calvinism, The Belfast-Princeton Connection." *Isis* 83:408–428.
Lockyer, Norman. 1902. *Catalogue of Four Hundred and Seventy of the Brighter Stars*. London: Solar Physics Committee.
Loetscher, Lefferts A. 1967. "New Vitality in Church and Nation." In *The First Presbyterian Church of Princeton*, edited by Arthur S. Link, 29–62. Princeton: Princeton University Press.
Loewy, M. 1901. "Report of the International Astronomical Conference Held at the Paris Observatory in July, 1900." *PopAst* 9:393–400.
———. 1908. "Maurice Loewy." *PopAst 16*:1–9.
Longfield, Bradley J. 1991. *The Presbyterian Controversy, Fundamentalists, Modernists and Moderates*. New York: Oxford University Press.
Lowen, Rebecca. 1997. *Creating the Cold War University*. Berkeley and Los Angeles: University of California Press.
Lundmark, Knut. 1922. "The Parallaxes of Stars Derived from Spectral Class and Apparent Magnitude." *PASP* 34:147–155.
———. 1932/1933. "Luminosities, Colours, Diameters, Densities, Masses of the Stars." *Handbuch der Astrophysik* 5, pt. 1. chap. 4; 1933. pt. 2. Chap. 4. Berlin: Springer-Verlag.
Luyten, W. J. 1938. "On the Origin of the Solar System." *Obs 61*:83–85.
———. 1939. "On the Origin of the Solar System." *MN* 99:692–696.
———. 1940. "Again the Origin of the Solar System." *Obs* 63:72–75.
———. 1941. "The Stream Motions of 92,656 Stars." *ApJ* 93:250–266.
———. 1942. "On the Origin of the Solar System." *ApJ* 96:482–483.
———. 1987. *My First Seventy-Two Years of Astronomical Research: Reminiscences of an Astronomical Curmudgeon*. Private publication.
Luyten, W. J., and E. L. Hill. 1937. "On the Origin of the Solar System." *Obs* 60:109–111; *ApJ* 86:470–482.
Lyttleton, Raymond A. 1936a. "The Solar System and Its Origin." *Nature* 137:664.
———. 1936b. "The Origin of the Solar System." *MN* 96:559–568.
———. 1939a. "On the Rotations of the Planets." *MN* 99:181–187.
———. 1939b. "On the Fission of a Rotating Fluid Mass." *MN* 99:567–568.
———. 1940a. "On the Origin of the Solar System." *MN 100*:546–553.
———. 1940b. "On the Origin of the Planets." *Obs* 63:206–213.
———. 1941a. "On the Theory of the Origin of the Planets." *ApJ* 93:267–274.
———. 1941b. "On the Origin of the Solar System." *MN 101*:216–226.

Lyttleton, Raymond A. 1942. "On the Formation of Planets." *ApJ* 96:155–156.
Lyttleton, Raymond A., and F. Hoyle. 1940. "The Evolution of the Stars." *Obs* 63:39–43.
MacDonald, Ruth K. 1989. *Christian's Children: The Influence of John Bunyan's The Pilgrim's Progress on American Children's Literature*. New York: Peter Lang.
MacKay, Donald M., ed. 1965. *Christianity in a Mechanistic Universe and Other Essays*. Chicago: Inter-Varsity Press.
MacKenzie, Donald. 1990. *Inventing Accuracy: An Historical Sociology of Nuclear Missile Guidance*. Cambridge: MIT Press.
Macpherson, Hector. 1922. "Herschel's World View in the Light of Modern Astronomy." *Obs* 45:254–261.
Mayo, Katherine. 1920. *"That Damn Y"*. Boston: Houghton-Mifflin.
McCosh, James. 1890. *The Religious Aspect of Evolution*. New York: Scribner's Sons.
McCrea, W. H. 1979. "[Einstein's Relationships with the Royal Astronomical Society]." *Obs* 99:105–107.
McCullough, David. 1981. *Mornings on Horseback*. New York: Simon and Schuster.
McCutcheon, R. A. 1991. "The 1936–37 Purge of Soviet Astronomers." *Slavic Review* 1:100–117.
McGucken, William. 1969. *Nineteenth Century Spectroscopy*. Baltimore: Johns Hopkins Press.
Meadows, A. J. 1966. "The Discovery of an Atmosphere on Venus." *Annals of Science* 22:117–127.
———. 1972. *Science and Controversy: A Biography of Sir Norman Lockyer*. Cambridge: MIT Press.
Meggers, William F. 1951. "Multiplets and Terms in Technetium Spectra." *Journal of Research of the National Bureau of Standards* 47:7–14.
Melnick, Edward K. 1988. "Organizational Culture within an Independent Day School: Formation and Functions." Ed.D. diss., Hofstra University.
Menzel, Donald H. 1931. "Pressures at the Base of the Chromosphere: A Critical Study of Milne's Theories." *Obs* 54:133–134.
———. 1972. "The History of Astronomical Spectroscopy." In Berendzen, 1972, 225–244.
Merrill, John E. 1950. "Tables for Solution of Light Curves of Eclipsing Binaries, Coefficient of Limb Darkening x = 0.0." *Contributions from the Princeton University Observatory No. 23*. Princeton: Princeton University Observatory.
———. 1953. "Nomographs for Solution of Light Curves of Eclipsing Binaries." Contributions from the Princeton University Observatory No. 24. Princeton: Princeton University Observatory.
———. 1957. "Avenues of Progress in Interpretation of Light Curves." *JRASC* 51:23–28.
Merrill, Paul. 1923. "A Research Career in Astronomy." *Science* 57:546–548.
Miller, Donald L. 1996. *City of the Century: The Epic of Chicago and the Making of America*. New York: Simon and Schuster.
Miller, Howard S. 1970. *Dollars for Research: Science and Its Patrons in Nineteenth-Century America*. Seattle: University of Washington Press.
Milne, E. A. 1921. "Ionisation in Stellar Atmospheres." *Obs* 44:261–269.
———. 1924. "Recent Work in Stellar Physics." *Proceedings of the Physical Society, London* 36:94–113.
———. 1925. "The Width of Fraunhofer Lines: A Reply to Professor Stewart." *MN* 85:739–750.

———. 1929a. "Anomalous Effects in Astrophysics." In "Correspondence," *Obs* 52:358–363.
———. 1929b. "The Masses, Luminosities, and Effective Temperatures of the Stars." *MN* 90:17–54.
———. 1930a. "The Structure and Opacity of a Stellar Atmosphere." [Bakerian Lecture.] *Philosophical Transactions* 228:421–461.
———. 1930b. "Temperature Gradients and Molecular Weight in the Sun's Atmosphere." *Obs* 53:119–120.
———. 1930c. "Thermodynamics of the Stars." In *Handbuch der Astrophysik III*, pt. 1, chap. 2. Springer-Verlag.
———. 1930d. "The Analysis of Stellar Structure." *MN* 91:4–55.
———. 1931. "Meeting of the Royal Astronomical Society." *Obs* 54:100–101; 154–155.
———. 1932. "The Analysis of Stellar Structure, II." *MN* 92:611–639.
———. 1945. "Ralph Howard Fowler 1889–1944." *Obituary Notices of the Fellows of the Royal Society* 5:61–78.
———. 1952. *Sir James Jeans: A Biography*. Cambridge: Cambridge University Press.
Mitchell, Walter Mann. 1905. "Researches in the Sun-Spot Spectrum Region F to a." *ApJ* 22:4–41.
Monck, W.H.S. 1894. "The Spectra and Colours of Stars." *Journal of the British Astronomical Association* 5:418–419.
Moore, Charlotte E. 1932. *A Multiplet Table of Astrophysical Interest*. Princeton: Princeton University Observatory. Abstract: *Publications of the American Astronomical Society* 7:97.
———. 1933. *Atomic Lines in the Sun-Spot Spectrum*. Princeton: Princeton University Observatory.
———. 1945. *A Multiplet Table of Astrophysical Interest, Revised Edition. Contributions from the Princeton University Observatory* No. 20; NBS Technical Note No. 36. Washington, D.C.: Government Printing Office.
Moore, Charlotte E., and H. N. Russell. 1926. "On the Winged Lines in the Solar Spectrum." *ApJ* 63:1–12.
Moyer, Albert. 1992. *A Scientist's Voice in American Culture*. Berkeley and Los Angeles: University of California Press.
———. 1997. *Joseph Henry: The Rise of an American Scientist*. Washington, D.C.: Smithsonian.
Murray, H. G. 1935. "Gentlemen-Scholars of 1890, No. 11—Charles Augustus Young." *Princeton Alumni Weekly*, 1 February, 345.
Newall, H. F. 1911. "Stellar Spectroscopy in 1910." *MN* 71:346–353.
Nicholson, J. W. 1913. "The Physical Interpretation of the Spectrum of the Corona." *Obs* 36:103.
———. 1916. "The Nature of the Coronium Atom." *MN* 76:415.
Nielsen, Axel V. 1963. "History of the Hertzsprung-Russell Diagram." *Centaurus* 9:219–252.
Noble, David F. 1984. *Forces of Production*. New York: Knopf. Reprinted, Oxford: Oxford University Press, 1986.
Nordholt, J. W. Schulte. 1991. *Woodrow Wilson*. Translated by Herbert H. Rowen. Berkeley and Los Angeles: University of California Press.
Norris, Edwin Mark. 1917. *The Story of Princeton*. Boston: Little Brown.
North, John. 1995. *Astronomy and Cosmology*. New York: W. W. Norton.

Numbers, Ronald L. 1977. *Creation by Natural Law: Laplace's Nebular Hypothesis in American Thought*. University of Washington Press.
Olson, Harry F. 1967. "Frederick Albert Saunders." *BMNAS* 39:403–416.
Oppolzer, Theodore R. von. 1880. *Lehrbuch zur Bahnbestimmung der Kometen und Planeten*. Leipzig: Engelmann.
Osterbrock, Donald E. 1984. "The Rise and Fall of Edward S. Holden: Part I." *JHA 15*:81–127.
———. 1984. *James E. Keeler: Pioneer American Astrophysicist and the Early Development of American Astrophysics*. Cambridge: Cambridge University Press.
———. 1986. "Failure and Success: Two Early Experiments with Concave Gratings in Stellar Spectroscopy." *JHA 17*:118–129.
———. 1990. "Armin O. Leuschner and the Berkeley Astronomy Department." *Astronomy Quarterly* 7:95–115.
———. 1992. "The Appointment of a Physicist as Director of the Astronomical Center of the World." *JHA* 23:155–165.
———. 1993. *Pauper and Prince: Ritchey, Hale, and Big American Telescopes*. Tucson: University of Arizona Press.
———. 1997a. *Yerkes Observatory 1892–1950*. Chicago: University of Chicago Press.
———. 1997b. "Walter Baade, Observational Astrophysicist (3): Palomar and Göttingen 1948–1960 (Part A)." *JHA* 28:283–316.
Osterbrock, Donald E.,J. R. Gustafson, and W. J. S. Unruh. 1988. *Eye on the Sky, Lick Observatory's First Century*. Berkeley and Los Angeles: University of California Press.
Owens, Larry. 1986. "Vannevar Bush and the Differential Analyzer: The Text and Context of an Early Computer." *Technology and Culture* 27:63–95.
———. 1995. "From One Computing Center to Another: Changing Agendas and the Rhetoric of Obviousness in the Transformation of Computing at MIT. 1939–1957." Unpublished manuscript.
Pannekoek, A. 1906–1907. "The Luminosity of Stars of Different types of Spectrum." *Koninklijke Akademie van wetenschappen te Amsterdam: Proceedings of the Royal Academy of Amsterdam* 9:134–302.
Paul, Erich Robert. 1993. *The Milky Way Galaxy and Statistical Cosmology 1890–1924*. Cambridge: Cambridge University Press.
Payne [-Gaposchkin], Cecilia H. 1924. "On ionization in the Atmospheres of the Hotter Stars." *Harvard College Observatory Circular*, no. 256.
———. 1925a. "Astrophysical Data Bearing on the Relative Abundance of the Elements." *PNAS 11*:192–198.
———. 1925b. *Stellar Atmospheres*. Cambridge: Harvard University Press.
Perry, Ralph Barton. 1944. *Puritanism and Democracy*. New York: Vanguard Press.
Peterson, H. C., and Gilbert C. Fite. 1957. *Opponents of War. 1917–1918*. Seattle: University of Washington Press reprint, 1971.
Phelps, Winthrop M. 1966. "The Cerebral Palsies." *Clinical Orthopaedics and Related Research* 44:83–88.
Philip, A. G. Davis, and David DeVorkin, eds. 1977. *In Memory of Henry Norris Russell*. Proceedings of IAU Symposium No. 80 (Dudley Observatory Report No. 13).
Pickering, Andrew. 1992. "From Science as Knowledge to Science as Practice." In *Science as Practice and Culture*, edited by A. Pickering. Chicago: University of Chicago Press.
Pickering, Edward C. 1886. "Photographic Study of Stellar Spectra." *Nature* 33:535.

———. 1891. "Preparation and Discussion of the Draper Catalogue." *Annals of the Astronomical Observatory of Harvard College* 26, pt. 1. Cambridge: Harvard College Observatory.
———. 1907. "The Scale of Observatory Work in the Future." *PopAst* 15:445–446.
———. 1909a. "The Light of the Stars." In *International University Lectures VI*, 67–82. New York: University Alliance.
———. 1909b. "The Future of Astronomy." *Popular Science Monthly* 75:105–116.
———. 1909c. "Distribution of the Stars." *Harvard College Observatory Circular* no. 147.
———. 1915. "Aid to Astronomical research." *Science* 41:82–85.
———. 1917. "Report of the Committee on Astronomy." *PNAS*, June 1917.
Pickering, E. C., and Antonia C. Maury, 1897. "Spectra of Bright Stars." *Annals of the Astronomical Observatory of Harvard College* 28, pt. 1. Cambridge: Harvard College Observatory.
Pickering, E. C., with Annie J. Cannon. 1901. "Spectra of Bright Southern Stars." *Annals of the Astronomical Observatory of Harvard College* 28, pt. 2. Cambridge: Harvard College Observatory.
Pierce, Newton Lacey. 1951. "Eclipsing Binaries." In Hynek, 1951, 479–494.
Pinch, Trevor J. 1986. *Confronting Nature: The Sociology of Solar-Neutrino Detection*. Boston: D. Reidel.
Plaskett, H. H. 1922. "The Spectra of Three O-Type Stars." *Publications of the Dominion Astrophysical Observatory* 1:325–84.
Plaskett, J. S. 1911. "Some Recent Interesting Developments in Asstronomy." *Annual Report of the Smithsonian Institution for 1911*, 255–270. Washington, D.C.: Smithsonian, 1912.
Plotkin, H. 1978a. "E. C. Pickering and the Endowment of Scientific Research in America. 1877–1918." *Isis* 69:44–57.
———. 1978b. "E. C. Pickering, the Henry Draper Memorial, and the Beginnings of Astrophysics in America." *Annals of Science* 35:365–377.
———. 1990. "Edward Charles Pickering." *JHA* 21:47–58.
Popper, Daniel M. 1967. "Determination of Masses of Eclipsing Binary Stars." *ARAA* 5:85–104.
Powell, Corey S., "J. Homer Lane and the Internal Structure of the Sun." *JHA* 19:183–199
Puiseux. 1916. "Review of Astronomy in the Year 1913." *Annual Report of the Smithsonian Institution for 1915*, 131–139. Washington, D.C.: Smithsonian.
Putnam, William Lowell, ed. 1994. *The Explorers of Mars Hill: A Centennial History of Lowell Observatory*. West Kennebunk, Me.: Phoenix Publishing Co.
Pyenson, Lewis. 1989. "Pure Learning and Political Economy: Science and European Expansion in the Age of Imperialism." In *New Trends in the History of Science*, edited by R.P.W. Visser et al., 209–278. Amsterdam: Rodopi.
Redman, R. O. 1927. "A Statistical Study of the Effect of the Mass-Luminosity Relation on the Hypothetical Parallaxes of Binary Stars." *MN* 88:33–51.
———. 1928. "A Determination from Binary Star Data of the Giant and Dwarf Branches of the Russell Diagram." *MN* 88:718–729.
———. 1930. "Y Cygni. Some Spectroscopic Results." *MN* 90:754–759.
———. 1938. "Line Profiles." *MN* 98:311–333.
Reingold, Nathan. 1968. "National Aspirations and Local Purposes." *Transactions of the Kansas Academy of Sciences* 71:235–246.
———. 1972. "Joseph Henry." *DSB* 6:277–281.

Reingold, Nathan. 1977. "The Case of the Disappearing Laboratory." *American Quarterly* 29:77–101. Reprinted in Reingold, 1991, 224–246.

———. 1979. "National Science Policy in a Private Foundation: The Carnegie Institution of Washington." in *The Organization of Knowledge in America 1860–1920*, edited by A. Olesson and J. Voss, 313–341. Baltimore: Johns Hopkins University Press.

———. 1991. "The Peculiarities of the American or Are There National Styles in the Sciences?" *Science in Context* 4:347–366.

———. 1991. *Science, American Style*. New Brunswick, N.J.: Rutgers University Press.

Reuben, Julie A. 1990. "In Search of Truth: Scientific Inquiry, Religion, and the Development of the American University. 1870–1920." Ph.D. diss., Stanford University UMI #9024358.

Roberts, Alexander. 1899. "Density of Close Double Stars." *ApJ* *10*:308–314.

Roberts, David Lindsay. 1998. *Mathematics and Pedagogy: Professional Mathematicians and American Education Reform, 1893–1923*. Ph.D. diss., Johns Hopkins University. University Microfilms International no. 9821188.

Robertson, H. P. 1937. "Dynamical Effects of Radiation in the Solar System." *MN* 97:423–438.

Robotti, Nadia. 1983. "The Spectrum of [Zeta] Puppis and the Historical Evolution of Empirical Data." *HSPS* *14*:123–145.

Roland, Alex. 1985. *Model Research: The National Advisory Committee for Aeronautics. 1915–1958 Volume 1*. Washington, D.C.: NASA SP-4103.

Rosenberg, Hans. 1911. "Über den Zusammenhang von Helligkeit und Spektraltypus in den Plejaden." *AN* *186*:71–78.

Rosseland, Svein. 1925a. "On the Distribution of Hydrogen in a Star." *MN* 85:541–546.

———. 1925b. "The Theory of the Stellar Absorption Coefficient." *ApJ* *61*:424–442.

Rossiter, Margaret W. 1971. "Benjamin Silliman and the Lowell Institute." *New England Quarterly* 44:602–626.

Rothenberg, Marc. 1981. "Organization and Control: Professionals and Amateurs in American Astronomy. 1899–1918." *Social Studies of Science* *11*:305–325.

Rubin, Jack. 1987. "Updating Historical $ and £ Data to Current Values." *Rittenhouse* *1*:82–84.

Russett, Cynthia Eagle. 1976. *Darwin in America: The Intellectual Response 1865–1912*. San Francisco: Freeman.

Sadler, Philip M. 1990. "William Pickering's Search for a Planet beyond Neptune." *JHA* *21*:59–64.

Sagan, Carl. 1974. "The Past and Future of American Astronomy." *PhysTod* 27:23–31.

Saha, M. N. 1920. "Ionisation in the Solar Chromosphere." *Philosophical Magazine* *40*:472–478.

Sánchez-Ron, José Manuel. 1994. *Miguel Catalán. Su Obra* Y *Su Mundo*. Madrid: Fundación Ramón Menéndez Pidal.

Schlesinger, Frank. 1899. "Suggestions for the Determination of Stellar Parallax by Means of Photography." *ApJ* *10*:242.

———, ed. 1913. *Dedication of the New Allegheny Observatory*. Pittsburgh: Misc. Scientific Papers of the Allegheny Observatory, n.s., 2, no. 2.

Schwarzschild, Karl. 1909. "Über das System der Fixsterne." *Himmel und Erde* *21*: 433–451.

Schwarzschild, Martin. 1958. *Structure and Evolution of the Stars*. Princeton: Princeton University Press.

Schweber, Silvan S. 1986. "The Empiricist Temper Regnant: Theoretical Physics in the United States. 1920–1950." *HSPS 17*:55–98.
———. 1990. "The Young John Clarke Slater and the Development of Quantum Chemistry." *HSPS 20*:339–406.
———. 1993. "Physics, Community and the Crisis in Physical Theory." *PhysTod 46*: 34–40.
Seares, Frederick H. 1922a. "J. C. Kapteyn." *PASP 34*:232–253.
———. 1922b. "The Masses and Densities of the Stars." *ApJ* 55:165–237.
———. 1939. "George Ellery Hale: The Scientist Afield." *Isis 30*:244.
Seeliger, Hugo v. 1900. "A Remark on the Articles on the density of the Algol stars in the Astrophysical Journal vol. 10, no. 5." *ApJ 11*:247–248.
Selden, William K. 1992. *Princeton Theological Seminary*. Princeton: Princeton University Press.
Servos, John W. 1990. *Physical Chemistry from Ostwald to Pauling*. Princeton: Princeton University Press.
Serwer, Daniel. 1977. "Unmechanischer zwang: Pauli, Heisenberg, and the Rejection of the Mechanical Atom. 1923–1925." *HSPS* 8:189–256.
Shapin, Steven. 1994. *A Social History of Truth*. Chicago: University of Chicago Press.
Shapley, Harlow. 1913a. "The Orbits of Forty-Four Eclipsing Binaries." *Science* 37: 29–30.
———. 1913b. "The Eclipsing Binary [epsilon] Aurigae." *Science* 37:646–647.
———. 1914. "On the Nature and Cause of Cepheid Variation." *ApJ 40*:448–460.
———. 1915. "Second Type Stars of Low Mean Density." *PNAS 1*:459–461.
———. 1927. "In Their Own Country. University's Emminent Astronomers, Pioneers and Leaders in All Phases of Work, Too Little Appreciated by Princetonians." *Princeton Alumni Weekly*, 8 April, 771–772.
———. 1957. "Dean of American Astronomers." *Sky and Telescope* 16:260–261.
———. 1958. "Henry Norris Russell." *BMNAS* 32:353–378.
———. 1969. *Through Rugged Ways to the Stars*. New York: Charles Scribner's Sons.
Shenstone, Allen G. 1961. "Princeton and Physics." *Princeton Alumni Weekly*, 24 February, 6–12; 20.
Shortland, Michael, and Richard Yeo, eds. 1996. *Telling Lives in Science*. Cambridge: Cambridge University Press.
Sitterly, Bancroft W. 1970. "Changing Interpretations of the Hertzsprung-Russell Diagram. 1910–1940." *Vistas in Astronomy* 12:357–366.
Sitterly, Charlotte E. Moore. *See also* Moore.
Sitterly, Charlotte E. Moore. 1977. "Collaboration with Henry Norris Russell over the Years." In Philip and DeVorkin, 1977, 27–41.
Slipher, V. M. 1927. "The Lowell Observatory." *PASP* 39:143–154.
———. 1933. "Spectrosocpic Studies of the Planets." *MN* 93:657–668.
Small, Henry. 1981. *Physics Citation Index 1920–1929*. Vol. 1. Philadelphia: Institute for Scientific Information.
Smart, William. 1931. *Spherical Astronomy*. 5th ed. 1962, Cambridge: Cambridge University Press.
———. 1945. "Arthur Robert Hinks." *Obs 66*:89–91.
Smith, Robert W. 1982. *The Expanding Universe, Astronomy's "Great Debate" 1900–1931*. Cambridge: Cambridge University Press.
———. 1991. "A National Observatory Transformed: Greenwich in the Nineteenth Century." *JHA* 22:5–20.

Smith, Robert W. 1994. "Red Shifts and Gold Medals 1901–1954." In Putnam, 1994, chap. 4.
Smyth, Piazzi. 1880. "Practical Spectroscopy in 1880." *Obs* 3:491–500; 523–529; 555–564.
Sokolovskaya, Z. K. 1971. "Otto Struve." *DSB 13*:115–121.
Söderqvist, Thomas. 1996. "Existential Projects and Existential Choice in Science: Science Biography as an Edifying Genre." In Shortland and Yeo, 1996, 45–84.
Sopka, Katherine Russell. 1988. *Quantum Physics in America*. New York: AIP.
Spence, Ian, and Robert F. Garrison. 1993. "A Remarkable Scatterplot." *American Statistician* 47:12–19.
Spitzer, Lyman, Jr. 1939a. "Spectra of M Supergiant Stars." *ApJ* 90:494–549.
———. 1939b. "The Dissipation of Planetary Filaments." *ApJ* 90:675–688.
———. 1977. "Russell and Theoretical Astrophysics." In Philip and DeVorkin, 1977, 3–8.
Spitzer, Lyman, Jr., and Jeremiah P. Ostriker, eds. 1997. *Dreams, Stars, and Electrons: Selected Writings of Lyman Spitzer, Jr.* Princeton: Princeton University Press.
St. John, C. E., C. E. Moore, L. M. Ware, E. F. Adams, and H. D. Babcock. 1928. *Revision of Rowland's Preliminary Table of Solar Spectrum Wave-lengths*. Washington, D.C.: Carnegie Institution Publication No. 396.
Stebbins, Joel. 1910. "The Measurement of the Light of Stars with a Selenium Photometer, with an Application to the Variations of Algol." *ApJ* 32:185–214.
Steins, K. 1936. "The Technic of Astronomical Computation according to an International Inquiry." *Acta Astronomica* 3, ser. A:76–134.
Sterne, T. E. 1939. "Apsidal Motion in Binary Stars." *MN* 99:451–462; "Apsidal Motion in Binary Stars (II) Distributions of Density." 99:662–669; "Apsidal Motion in Binary Stars (III) Limiting Ratios of Central to Mean Density." 99:670–672.
Stewart, John Q. 1922. "An Electrical Analogue of the Vocal Organs." *Nature 110*:311–312.
———. 1923. "The Opacity of an Ionized Gas." *Nature 111*:186–187.
———. 1924. "The Width of Absorption Lines in a Rarefied Gas." *ApJ* 59:30–36.
———. 1925a. "The Effect of Temperature Gradient on the Intensities of Fraunhofer Lines." *PopAst* 33:303–304. Reprinted from *Publications of the American Astronomical Society* 5 (1924):275–276.
———. 1925b. "The Width of Fraunhofer Lines." *MN* 85:732–738.
———. 1928. "A Theory of the Production of Dark Lines in Stellar Spectra." [Abstract.] *PopAst* 36:346.
———. 1941. "The 'Gravitation,' or Geographical Drawing Power, of a College." *Bulletin of the American Association of University Professors* 27, no. 1:1–6.
———. 1950. "The Development of Social Physics." *American Journal of Physics 18*:239–253.
———. 1958. "Henry Norris Russell." *MN 118*:311–312.
Stratton, F.J.M. 1924. *Records of the R.A.S. Club 1911–1924*. London: Taylor and Francis.
———. 1929. *Transactions of the International Astronomical Union* 3. Cambridge: Cambridge University Press.
———. 1949. "The History of the Cambridge Observatories." *Annals of the Solar Physics Observatory, Cambridge vol. 1*. Cambridge: Cambridge University Press.
———. 1957. "Henry Norris Russell." *Biographical Memoirs of Fellows of the Royal Society* 3:173–191.

Strauss, David. 1994. "Percival Lowell, W. H. Pickering and the Founding of the Lowell Observatory." *Annals of Science* 51:37–58.
———. Forthcoming. *Percival Lowell.*
Strömgren, Bengt. 1932. "The Opacity of Stellar Matter and the Hydrogen Content of the Stars." *ZAstrop* 4:118–152. Reprinted in *Publikationer og mindre Meddelelser fra Københavns Observatorium*, no. 83. Springer Verlag.
———. 1933. "On the Interpretation of the Hertzsprung-Russell-Diagram." *ZAstrop* 7:223–248.
———. 1938. "On the Helium and Hydrogen Content of the Interior of the Stars." *ApJ* 87:520–534.
———. 1940. "On the Chemical Composition of the Solar Atmosphere." In *Festscrift für Ellis Strömgren*, edited by K. Lundmark, 218–257. Copenhagen: E. Munksgaard.
———. 1951. "The Growth of Our Knowledge of the Physics of the Stars." In Hynek, 1951, 172–258.
———. 1972. "The Rise of Astrophysics." In Berendzen, 1972, 245–254.
———. 1983. "Scientists I Have Known and Some Astronomical Problems I Have Met." *ARAA* 21:1–11.
Struve, Otto. 1926. Review of C. H. Payne, *Stellar Atmospheres*. *ApJ* 64:204–208.
———. 1931. "Axial Rotation as a Major Factor in Stellar Spectroscopy." *Obs* 54:80–84.
———. 1935. "Some New Trends in Stellar Spectroscopy." *PopAst* 43:483–496; 559–568; 628–639.
———. 1943. "Fifty Years of Progress in Astronomy." *PopAst* 51:469–481.
———. 1945. Review of Martinoff, *Stellar Variability*. *ApJ* 101:380–381.
———. 1946. "The Copenhagen Conference of the International Astronomical Union." *PopAst* 54:327–339.
———. 1949. "The Requirements for a Graduate Student at an Observatory." *PopAst* 57:382–386.
Struve, Otto, and Velta Zebergs. 1962. *Astronomy of the Twentieth Century*. New York: Macmillan.
Sulloway, Frank. 1996. *Born to Rebel: Birth Order, Family Dynamics and Creative Lives*. New York: Pantheon.
Swings, Pol. 1968. "Remarks and Recollections of a Stellar Spectroscopist." *Bulletin of the Astronomical Institutes of Czechoslovakia* 19:257–259.
Szafraniec, Rose. 1970. "Henry Norris Russell's Contribution to the Study of Eclipsing Variables." *Vistas in Astronomy* 12:7–20.
Tayler, R. J. 1987. *History of the Royal Astronomical Society*, vol. 2. Oxford: Blackwell.
———. 1996. "E. A. Milne (1896–1950) and the Structure of Stellar Atmospheres and Stellar Interiors." *QJRAS* 37:355–363.
Thorp, Willard, Minor Myers, Jr., and Jeremiah Stanton Finch. 1978. *The Princeton Graduate School*. Princeton: Princeton University Press.
Todd, David P. 1897. *A New Astronomy*. New York: American Book Company.
Tscherny, S. 1927. "Störungen im System von [Beta] Lyrae." *AN* 230:157–159.
———. 1928. "Perturbations in the System of [Beta] Lyrae." *MN* 88:482–483.
———. 1929. "Die periplegmatische Bewegung im System von [Beta] Lyrae." *AN* 235:105–108.
Turner, H. H. 1912. *The Great Star Map*. London: John Murray.
———. 1913. "From an Oxford Note-book." *Obs* 36:382–386; 417.
———. 1922. "From an Oxford Note-book." *Obs* 45:271–272.

Unsöld, Albrecht. 1927. "Über die Struktur der Fraunhoferschen Linien und die Dynamik der Sonnenchromosphäre." *ZPhys* *44*:793–809.
———. 1928. "Über die Struktur der Fraunhoferschen Linien und die quantitative Spektralanalyse der Sonnenatmosphäre." *ZPhys* *46*:765–781.
———. 1930. "Über die Balmerserie des Wasserstoffs in Sonnen-spektrum." *ZPhys* 59:353–377.
Vakalopoulos, Apostolos. 1985. "Memories of Old Thessaloniki." Web essay adapted from *History of Thessaloniki*. Thessaloniki: Malliares Paideia. Posted in soc.culture.greek by S. Efremidis, arutha@athena.compulink.forthnet.gr.
Veysey, Laurence R. 1965. *The Emergence of the American University*. Chicago: University of Chicago Press.
Visher, Stephen Sargent. 1947. *Scientists Starred 1903–1943*. Baltimore: Johns Hopkins Press.
Vogel, H. C. 1890. "Spectrographischen Beobachtungen an Algol." *AN* *123*:289–292.
Wali, Kameshwar C. 1991. *Chandra*. Chicago: University of Chicago Press.
Walter, Kurt. 1933. "Die Bewegungsverhältnisse in sehr engen Doppelsternsystemen." *Königsberger Gelehrten Gesellschaft* 4:1–153.
Walter, M. L. 1990. *Science and Cultural Crisis: An Intellectual Biography of Percy Williams Bridgman 1882–1961*. Stanford: Stanford University Press.
Walworth, Arthur C. 1965. *Woodrow Wilson*. 2d ed. Boston: Houghton Mifflin.
Warner, Deborah Jean. 1986. "Rowland's Gratings: Contemporary Technology." *Vistas in Astronomy* 29:125–130.
———. 1994. "Tools, Toys, and Technomachismo: What Telescopes Reveal about American Culture in the Late Nineteenth Century." Unpublished manuscript, April 1994.
Waterman, Phoebe. 1913. "The Present Status of the Problem of Stellar Evolution." *PASP* 25:189–199.
Watson, James C. 1892. *Theoretical Astronomy Relating to the Motions of the Heavenly Bodies*. Philadelphia: Lippincott.
Weimer, Th. 1982. "Une instrument en voie de disparition: L'Équatorial Coudé." *JHA* *13*:110–118.
Weiner, Charles. 1969. "A New Site for the Seminar: The Refugees and American Physics in the Thirties." In *The Intellectual Migration: Europe and America. 1930–1960*, edited by Donald Fleming and Bernard Bailyn, 190–234. Cambridge: Harvard University Press.
Weinstein, Edwin A. 1981. *Woodrow Wilson: A Medical and Psychological Biography*. Princeton: Princeton University Press.
Wertenbaker, Thomas Jefferson. 1946. *Princeton 1746–1896*. Princeton: Princeton University Press.
White, A. D. 1896. *A History of the Warfare of Science with Theology in Christendom*. Reprinted, New York: Dover. 1960.
White, Benjamin V. 1984. *Stanley Cobb, A Builder of the Modern Neurosciences*. Boston: F. A. Countway Library of Medicine.
White, Harvey E. 1934. *Introduction to Atomic Spectra*. New York: McGraw-Hill.
Wigner, E. P. 1947. *Physical Science and Human Values*. Princeton: Princeton University Press.
Wildt, Rupert. 1939. "Electron Affinity in Astrophysics." *ApJ* *89*:295–301
Williams, John Rogers. 1905. *The Handbook of Princeton*. New York: Grafton.
Wilson, Robert E., and Edward J. Devinney. 1971. "Realization of Accurate Close-Binary Light Curves: Application to MR Cygni." *ApJ* *166*:605–619.

Withers, John J. 1929. *A Register of Admissions to King's College Cambridge*. London: John Murray.

Wood, Frank Bradshaw, Raymond Smith Dugan, and Newton Lacy Pierce. 1951. "Photometric Researches, Twenty-Four Eclipsing Variables." *Contributions of the Princeton University Observatory No. 25*. Princeton: Princeton University Observatory.

———. 1977. "The Study of Close Binary Stars." In Philip and DeVorkin, 1977, 47–49.

W[oolley], R.v.d.R. 1936. Review of *Theoretical Astrophysics*. *Obs* 59:318–319.

Woolley, R.v.d.R., and D.W.N. Stibbs. 1953. *The Outer Layers of a Star*. Oxford: Oxford University Press.

Wright, Frances Woodworth. 1987. "Constant Vigilance." Oral history by C. A. Whitney, privately printed. Harvard College Observatory Library.

Wright, Helen. 1966. *Explorer of the Universe*. New York: Dutton. Reprinted, New York: American Institute of Physics, 1994.

Wright, Helen, Joan Warnow, and Charles Weiner, eds. 1972. *The Legacy of George Ellery Hale*. MIT Press.

Wright, Monte. 1972. *Most Probable Position*. Lawrence: University Press of Kansas.

Yochelson, Ellis L. 1994. "Andrew Carnegie and Charles Doolittle Walcott: The Origin and Early Years of the Carnegie Institution of Washington." In *The Earth, the Heavens and the Carnegie Institution of Washington*, edited by G. Good, 1–19. History of Geophysics, vol. 5. Washington, D.C.: American Geophysical Union.

Young, C. A. 1884. "Pending Problems in Astronomy." *PAAAS* 33:1–27; *Science* 4:192–203.

———. 1885. "The Red Spot on Jupiter." *Sidereal Messenger* 4:119–120.

———. 1886. "Rotation Time of the Red-Spot on Jupiter." *Sidereal Messenger* 5:289–293.

———. 1888. A *Textbook of General Astronomy*. Boston: Ginn. Revised, 1895.

———. 1890. *The Elements of Astronomy*. Boston: Ginn.

———. 1891. "Address at the Dedication of the Kenwood Observatory." *Sidereal Messenger 10*:312–321.

———. 1892. "The New Spectroscope of the Halsted Observatory." *Astronomy and Astro-Physics 11*:292–296.

——— 1893. "Photography of Sun-spot Spectra." *Astronomy and Astrophysics* 12:649–650.

———. 1899. "Lane's Law of Increase of Temperature in a Gaseous Sphere Contracting from the Loss of Heat." *PopAst* 7:225–227.

———. 1900. "The Princeton Eclipse Expedition to Wadesboro, N.C. May 1900." *Princeton University Bulletin 11*:69–87.

———. 1902. *Manual of Astronomy*. Boston: Ginn and Company.

Zorn, Jens, ed. 1988. *On the History of Physics at Michigan*. Ann Arbor: University of Michigan Physics Department.

Zwiers, H. J. 1896. "Ueber eine neue Methode zur Bestimmung von Doppelsternbahnen." *AN 139*:369–380.

Selected Works by Russell

1898a. "A New Graphical Method of Determining the Elements of a Double-Star Orbit." *AJ 19*:9–10.

1898b. "Elements of the Planet DQ." *AJ 19*:147.

1899a. "Elements and Ephemeris of *Eros*." *AJ 20*:8; 31.

1899b. "Elements of *Eros*." *AJ* 20:134.
1899c. "Measures of the Diameter of Jupiter." *AJ* 20:77.
1899d. "The Densities of the Variable Stars of the Algol Type." *ApJ* 10:315–318.
1899e. "The Atmosphere of Venus." *ApJ* 9:284–299.
1900a. "The General Perturbations of the Major Axis of *Eros*, by the Action of *Mars*." *AJ* 21:25–28.
1900b. "The Great Inequality of *Eros* and the Earth." *AJ* 21:24.
1902. "An Improved Method of Calculating the Orbit of a Spectroscopic Binary." *ApJ* 15:252–260.
1905a. "The Parallax of Lalande 21185 and [Gamma] Virginis from Photographs Taken at the Cambridge Observatory." *MN* 65:787–800.
1905. With Arthur R. Hinks. "Determinations of Stellar Parallax from Photographs Made at the Cambridge Observatory. Introductory paper." *MN* 65:775–787.
1906a. "On the Light-Variations of Asteroids and Satellites." *ApJ* 24:1–18.
1906b. "Stellar Parallax Papers, no. 3: The Parallax of Eight Stars, from Photographs Taken at the Cambridge University Observatory by Arthur R. Hinks, M.A., and the Writer." *MN* 67:132–135.
1907a. "The Riddle of Mars." *SciAm* 97:25–26. [Anon.]
1907b. "Photographic Determinations of Stellar Parallaxes." [Abstract.] *Carnegie Institution of Washington Yearbook*, no. 5 . Washington, D.C., 1906.
1907c. With Zaccheus Daniel. "Venus as a Luminous Ring." *ApJ* 26:69–70.
1910a. "Determinations of Stellar Parallax." *AJ* 26:147–159.
1910b. "On the Determination of the Elements of Algol Variables." *Science* 32:882–883. Reprinted in *PAASA* 2 (1915): 32–33.
1910c. "On the Distances of Red Stars." [Abstract.] *Science* 31:878.
1910d. "On the Origin of Binary Stars." *ApJ* 31:185–207.
1910e. "Some Hints on the Order of Stellar Evolution." *Science* 32:883–884. Reprinted in *PAASA* 2 (1915): 33–34.
1911a. "A Study of Visual Binary Stars." *Science* 34:523–525.
1911b. *Determinations of Stellar Parallax*. Washington, D.C.: Carnegie Institution of Washington.
1912a. "Notes on the Determination of the Elements of Algol Variables." *Science* 35:708.
1912b. "On the Determination of the Orbital Elements of Eclipsing Variable Stars. I." *ApJ* 35:315–340.
1912c. "On the Determination of the Orbital Elements of Eclipsing Variable Stars. II." *ApJ* 36:54–74.
1912d. "Relations between the Spectra and other Characteristics of the Stars." *Proc. Am. Phil. Soc.* 51:569–579.
1912e. "The Eclipsing Variables W Crucis and W Ursae Majoris." *Science* 35:708.
1912f. "The Progress of Science in the University 1897–1912." In Davis et al., 1912, 164–168.
1913a. " 'Giant' and 'Dwarf' Stars." *Obs* 36:324–329.
1913b. "Heavens in July." *SciAm* 108:582; 588; 591.
1913c. "Notes on the Real Brightness of Variable Stars." *Science* 37:651–652.
1913d. "On the Luminous Efficiency and Color Index of a Black Body at Different Temperatures." *Science* 37:646.
1913e. "The International Union for Solar Research and the Astronomische Gesellschaft." *SciAm* 109:187; 194.
1914a. "On the Probable Order of Stellar Evolution." *Obs* 37:165–175.

1914b. "Relations between the Spectra and Other Characteristics of the Stars." *PopAst* 22:275–294; 331–351.
1914c. "The Solar Spectrum and the Earth's Crust." *Science* 39:791–794.
1916a. "Heavens in June." *SciAm 114*:588.
1916b. "Nebulae of Dimensions That Stagger the Imagination." *SciAm 115*:304.
1916c. "Percival Lowell and His Work." *Outlook 114*:781–783.
1916d. "Remarkable Surface Features of the Planet Mars." *SciAm 114*:358; 363–364.
1916e. "On the Albedo of the Planets and Their Satellites." *ApJ 43*:173.
1917. With Mary Fowler and Martha C. Borton. "Comparison of Visual and Photographic Observations of Eclipsing Variables." *ApJ 45*:306–347.
1918. "The Nebulae, Nature, Size, and Distance from Us." *SciAm 118*:192.
1919a. "Edward Charles Pickering." *Science* 49:151–155.
1919b. "On the Navigation of Airplanes." *PASP 31*:129–149.
1919c. "On the Sources of Stellar Energy." *PASP* 31:205–211.
1919d. "Variable Stars." *Science* 49:127–139.
1919/1920. "Some Problems in Sidereal Astronomy." *PNAS* 5:391–416. Reprinted 1920, *PopAst* 28:212–224; 264–275.
1920a. "The Determination of the Position of the Moon by Photography." *PASP* 32:105–111.
1920b. "The Probable Diameters of the Stars." *PASP* 32:307–317.
1921a. "A Superior Limit to the Age of the Earth's Crust." *Proceedings of the Royal Society [A]* 99:84–86.
1921b. "On the Accuracy with Which Mean Parallaxes Can Be Determined from Parallactic and Peculiar Velocities." *ApJ 54*:140–145.
1921c. ["Stellar Evolution."] *JBAA 31*:177–183.
1921d. "The Properties of Matter as Illustrated by the Stars." *PASP* 33:275–290.
1922a. "Notes on Ionization in the Solar Atmosphere." *ApJ* 55:354–59.
1922b. "On the Calculation of Masses from Spectroscopic Parallaxes." *ApJ* 55:238–241.
1922c. "Recent Advances in Stellar Astronomy. I. The Light of the Stars; II. The Sizes and Masses of Stars; III. The Constitution and Evolution of the Stars." *Rice Institute Pamphlet* 9. Houston: Rice University.
1922d. "The Theory of Ionization and the Sunspot Spectrum." *ApJ* 55:119–144.
[Brown et al.,1922]. With E. W. Brown, G. D. Birchoff, and A. O. Leuschner. "Celestial Mechanics." *Bulletin of the National Research Council* 4, pt. 1, no. 19, 1–22.
1922. With D. L. Webster. "Note on the Masses of the Stars." *MN* 82:181–182.
1923a. "Ionization and Pressure in the Reversing Layers of the Stars." *PopAst* 31:22–24.
1923b. "Some Results of a Study of Visual Double Stars." [Abstract.] *PopAst 31*:250–251.
1923. With W. S. Adams and A. H. Joy. "A Comparison of Spectroscopic and Dynamical Parallaxes." *PASP* 35:189–193.
1924a. "Series in the Arc and Spark Spectra of Titanium." [Abstract.] *PopAst* 32:230.
1924b. "Singlet Series in the Spark Spectrum of Aluminum." *Nature 113*:163.
1924c. "The Applications of Modern Physics to Astronomy." *JRASC 18*:137–164; 201–223; 233–263. Condensed review by H. F. Balmer.
1924d. "The Heavens in June, 1924, The Fall of the Nebular Hypothesis, and the Rise of a Plausible Substitute." *SciAm 130*:410.
1924e. "Ultimate and Penultimate Lines in Spectra." [Abstract.] *PopAst* 32:229–230.
1924. With K. T. Compton. "A Possible Explanation of the Behaviour of the Hydrogen Lines in Giant Stars." *Nature 114*:86–87.
1924. With J. Q. Stewart. "Pressure at the Sun's Surface." *ApJ* 59:197–209.

1925a. "A List of Ultimate and Penultimate Lines of Astrophysical Interest." *ApJ* *61*:223–283.
1925b. "Note on the Relations between the Mass, Temperature, and Luminosity of a Gaseous Star." *MN* *85*:935–939.
1925c. "The Intensities of Lines in Multiplets." *Nature* *115*:835–836.
1925d. "The Intensities of Lines in Multiplets." *PNAS* *11*:314–328.
1925e. "The Problem of Stellar Evolution." *Nature* *116*:209–212.
1925. With C. Moore. "On the Winged Lines in the Solar Spectrum." *ApJ* *63*:1–12.
1925. With F. A. Saunders. "New Regularities in the Spectra of the Alkaline Earths." *ApJ* *61*:38–69.
1925. With F. A. Saunders. "On the Spectrum of Ionized Calcium (Ca II)." *ApJ* 62:1–7.
1926/1927. With R. S. Dugan and J. Q. Stewart. *Astronomy*, vol. 1 (1926); vol. 2 (1927). Boston: Ginn and Company.
1927a. *Fate and Freedom*. New Haven: Yale University Press.
1927b. "On the Calculation of the Spectroscopic Terms Derived from Equivalent Electrons." *PhysRev* 29:782–789.
1927c. "On the Relations between Period, Luminosity, and Spectrum among Cepheids." *ApJ* *66*:122–134.
1927d. "Related Lines in the Spectra of the Elements of the Iron Group." *ApJ* *66*:184–216.
1927e. "The Arc and Spark Spectra of Titanium, pt. I; pt. II." *ApJ* 66:283–328; 347–438.
1927f. "The Relation of Physics to Astronomy." In *Physics and Its Relations*, 45–64. Poughkeepsie, N.Y.: Vassar College.
1927. With R. J. Lang. "On the Spectra of Doubly and Trebly Ionized Titanium (Ti III and Ti IV)." *ApJ* 66:13–42.
1927. With W. Meggers. "Analysis of Arc and Spark Spectra of Scandium (ScI and ScII)." *Scientific Papers of the Bureau of Standards*, no. 558. 22:330–373.
1928a. "On the Advance of Periastron in Eclipsing Binaries." *MN* 88:641–643.
1928b. "On the Determination of Dynamical Parallaxes." *AJ* 38:89–99.
1928c. "Sir Norman Lockyer's Work in the Light of Present Astrophysical Knowledge." In *Life and Work of Sir Norman Lockyer*, edited by T. M. Lockyer and W. L. Lockyer, 382–394. London: Macmillan.
1928. With W. S. Adams. "Preliminary Results of a New Method for the Analysis of Stellar Spectra." *ApJ* 68:9–36.
1928. With W. S. Adams and C. E. Moore. "A Calibration of Rowland's Scale of Intensities for Solar Lines." *ApJ* 68:1–8.
1929a. "Astronomical Books." *Saturday Review of Literature* 6:7.
1929b. "On the Composition of the Sun's Atmosphere." *ApJ* 70:11–82.
1929c. "Quantitative Analysis of the Sun." *SciAm* *140*:316–317.
1929d. "Stellar Evolution." *Encyclopaedia Britannica*, 14th ed., 21:375–379.
1929e. "The Highest Known Velocity." *SciAm* *140*:504–505.
1929f. "Where to Put It?" *SciAm* *140*:20–21.
1929. With Ira S. Bowen. "Is there Argon in the Corona?" *ApJ* 69:196–208.
1929. With A. G. Shenstone and L. A. Turner. "Report on Notation for Atomic Spectra." *PhysRev* 33:900–906.
1930a. "Fragmentary Molecules of the Sun." *SciAm* *142*:436–437.
1930b. "How Pluto's Orbit Was Figured Out." *SciAm* *143*:364–365.
1930c. "Hunting Pocket Planets." *SciAm* *142*:26–27.

1930d. "Life in Space and Time." In *Human Biology and Racial Welfare*, edited by E. V. Cowdry, 3–31. New York: Paul B. Hoeber.
1930e. "Revising Our Air and Our Water." *SciAm* 142:350–351.
1930f. "The Radio and the Spectroscope." *SciAm* 142:116–117.
1930g. "The Two April Eclipses." *SciAm* 142:274–275.
1930h. "What's in the Sun?" *SciAm* 142:212–213.
1930i: "Planet X." *SciAm* 143:20–22.
1930j. "The 'Green Flash' and Other Odd Phenomena." *SciAm* 143:114–116.
1930k. "Why Stars Twinkle." *SciAm* 143:180–181.
1930l. "Measuring the Distance to the Stars." *SciAm* 143:280–282.
1930m. "More about Pluto." *SciAm* 143:446–447.
1930. With R. S. Dugan. "Apsidal Motion in Y Cygni and other Stars." *MN* 91:212–215.
1931a. "Composite Polytropic Gas-Spheres." *MN* 91:739–751. Excerpted in Milne, 1931 154–155.
1931b. "Incandescent Refrigerators." *SciAm* 144:302–303.
1931c. "Notes on the Constitution of the Stars." *MN* 91:951–966. Supplement No. 9.
1931d. "Notes on the Constitution of the Stars: Addendum." *MN* 92:147.
1931e. "Refining Pluto's Orbit." *SciAm* 144:90–91.
1931f. "The Sun an Atom Builder—A New Theory." *SciAm* 145:232–233.
1931g. "Worlds from a Catastrophe." *SciAm* 145:92–93.
1931. With R. d'E. Atkinson. "Stellar Structure." *Nature* 127:661–662.
1932a. "The Composition of the Sun." *Annual Report of the Smithsonian Institution for 1931*, 199–214.
1932b. "The Master Key of Science: Revealing the Universe through the Spectroscope." *Technology Review* 34:279–281; 300–302.
1932c. "The Mystery of Unknown Lines." *SciAm* 147:14–15.
1933a. "A Rapid Method for Determining Visual Binary Orbits." *MN* 93:599–602.
1933b. "Opacity Formulae and Stellar Line Intensities." *ApJ* 78:239–297.
1933c. *The Composition of the Stars*. Oxford: Oxford University Press.
1933d. "The Constitution of the Stars." *Science* 77:65–79.
1933. With Charlotte Moore. "On the Masses of Giant Stars." [Abstract.] *PASP* 7:184–185.
1935a. "Abridged Record of Family Traits." CIW Eugenics Records Office and the Eugenics Society of the United States of America, attached to Paul Brockett to Members of the National Academy, 22 May 1935. Reel 11:9543–9550, PUL/HNR. [Abridged Record]
1935b. "Appendix I." In A. L. Lowell, *A Biography of Percival Lowell*, 203–205.
1935c. "Impossible Planets: The Investigation of Three Newly Discovered Stars." *SciAm* 153:18–19.
1935d. "The Analysis of Spectra and Its Applications in Astronomy." *MN* 95:610–636.
1935e. *The Solar System and Its Origin*. New York: Macmillan.
1935f. "World Astronomers Meet Again." *SciAm* 153:184–185.
1936a. "More about the New Lyttleton Theory of Planetary Origin." *SciAm* 155:266–267.
1936b. "New Light on the Origin of the Planets." *SciAm* 155:204–205.
1936c. "Orbit of the Binary Ref. Cat. 17^h 31." *AJ* 45:95–96.
1936d. "Polytropic Index and Photospheric Conditions." *MN* 97:127–132.
1936e. "Some Remarkable Double Stars." *SciAm* 154:314–315
1937a. "Model Stars." *Bulletin of the American Mathematical Society* 43:49–77.
1937b. "Review of *Theoretical Astrophysics*." *ApJ* 85:141–144.

1938a. "Address of Retiring President of American Astronomical Society, 30 December 1937: The Place of Approximate Methods in Astronomy." *PAAS* 9:108–114.
1938b. "Artificial Eclipses." *SciAm 159*:240–241.
1938. With Charlotte Moore. "A Comparison of Spectroscopic and Trigonometric Parallaxes." *ApJ* 87:389–423.
1939a. "Inside the Stars." *SciAm 160*:298–299; "What Keeps the Stars Shining?" *SciAm 160*:369–369; *161*:18–19.
1939b. "Notes on Ellipticity in Eclipsing Binaries." *ApJ* 90:641–674.
1939c. "Science and Freedom." *Princeton Alumni Weekly*, 10 March, 479–481.
1939d. "Stellar Energy." *Proc. Am. Phil. Soc. 81*:295–307.
1940. "A Famous Theory Weakens." *SciAm 162*:140–141.
1940a. With Charlotte Moore. *The Masses of the Stars with a General Catalogue of Dynamical Parallaxes*. Chicago: University of Chicago Press.
1940b. With Charlotte Moore. "The Systematic and Accidental Errors of Spectroscopic Parallaxes." *ApJ* 92:354–391.
1941a. "A Puzzle Solved?" *SciAm 165*:70–71.
1941b. "The Brightest Known Star." *SciAm 164*:22–23.
1941c. "The Distribution of Density within the Stars." *Annals of the New York Academy of Sciences 41*:5–12.
1942. "Present State of the Theory of Stellar Evolution." *Scientific Monthly* 55:233–238.
1943a. "Anthropocentrism's Demise." *SciAm 169*:18–19.
1943b. "Determinism and Responsibility." *Science* 97:249–253.
1943c. "Nature's Crossword Puzzles." *SciAm 168*:210–211.
1944. With Charlotte Moore and Dorothy W. Weeks. "The Arc Spectrum of Iron (*Fe I*): Part I, Analysis of the Spectrum; Part II, The Zeeman Effect." *Transactions of the American Philosophical Society* 34, pt. 2, 111–207.
1947a. "America's Role in the Development of Astronomy." *Proc. Am. Phil. Soc.* 91:10.
1947b. "The Ivory Tower and the Ivory Gate." In Wigner, 1947, 165–176.
1948a. "On the Distribution of Absolute Magnitude in Populations I and II." *PASP* 60:202–204.
1948b. "The Royal Road of Eclipses." In *Centennial Symposia* (*Harvard Observatory Monograph* 7), edited by C. Payne-Gaposchkin, 181–209. Cambridge: Harvard University Press.
1952. With J. E. Merrill. *The Determination of the Elements of Eclipsing Binaries. Contributions from the Princeton University Observatory* No. 26. Princeton: Princeton University Observatory.
1953. "Henry Norris Russell." In *Thirteen Americans: Their Spiritual Autobiographies*, edited by Louis Finkelstein, 31–45. Port Washington, NY: Kennikat Press. [Spiritual Autobiography]
1954. "Transcript of a Colloquium given at Princeton University Observatory." 27 April 1954. In Philip and DeVorkin, 1977, 97–113. The original transcript is at Princeton University Observatory, kindly provided by Lyman Spitzer.

• INDEX •